The Ancestry of Reason

How Consciousness Works

What It Achieves

How It Evolved

R. G. Naylor

The Ancestry of Reason
How consciousness works
What it achieves
How it evolved

By R. G. Naylor

Printed in the UK by Berforts Ltd.
17 Burgess Road, Ivyhouse Lane, Hastings,
East Sussex TN35 4NR

ISBN 9781910693841

© R. G. Naylor   2017
R. G. Naylor's right to be identified as
the author of this work has been asserted.

My thanks to all the friends who have contributed helpful comments.

Contents

1 Introduction 9
2 Reason and Consciousness 16
3 Clues to Where Consciousness is at Work 33
4 From Earliest Animals to *Homo Sapiens* - a Quick Sketch 50
5 An Introduction to Neurons and Nervous Systems 72
6 The Basic Mechanisms of Vision 90
7 Learning to See 107
8 Creating the Conscious Visual Moment 125
9 The Benefits of Conscious Sensory Experience 144
10 The Internal World 162
11 Subtler Forms of Reinforcement - Cognitive Emotions 176
12 Social Emotions 188
13 Social Lifestyles 205
14 Attention 222
15 The Neurology of Movement 235
16 The Neurology of Attention 251
17 The Neurology of Reward 270
18 Learning 290
19 Contextual and Episodic Memory 308
20 Sleep, Dreams and Memory 327
21 The Long-term Memory Store 344
22 Where the Thinking's Done 364
23 More About the Basal Ganglia 384
24 Some Guesswork About the Thalamus 400
25 Overview 412
26 Some Pivotal Innovations 424
27 A Tentative History of Consciousness 440
28 Language 458
29 The Evolution of Language 478
30 Endword 496
Bibliography 506
Glossary 564

Chapter 1

Introduction

Many years ago I lay on a Mediterranean beach, staring lazily at a perfect blue sky. Suddenly, far out over the horizon, a vapour trail materialised. Slowly, inexorably, the white ribbon unfurled, from an aeroplane too high to be seen, bisecting the blue. Then, just as the leading point of the trail was about to disappear behind the brim of my sunhat, there was a dazzling flash of light - a reflection of the sun off the invisible metal.

What a great image a novelist could make of that, I thought idly. Or better still, if a cunning film-maker could only capture it, in the background of a scene. It would stand, of course, for the traditional flash of inspiration or discovery, a subtler version of the light bulb in a balloon over a cartoon character's head. It could parallel the moment when the scientist stumbles on the insight that solves the crucial problem; or the moment when the hero realises that he's in love; or even the moment when the film's director begins to change the audience's perception of what is going on. The increasingly fuzzy vapour trail might make a good metaphor for an increasingly vague memory of the events that led to the flash of inspiration.

Then another vapour trail appeared, and then two or three more, crisscrossing each other. The sky began to look merely untidy. The transformation from perfect blue to messy chaos could also be used to metaphorical purpose, no doubt, but somehow the poetic potential didn't seem to be there any more.

I wondered why one incident should have a compelling quality, while a near-repetition of the same thing was merely tiresome. Obviously one can analyse the factors that gave the first occurrence its powerful charge - the appeal of snowy white on dazzling blue, the steady pace of the advance, the unexpectedness of the flash of light. But why precisely are these factors effective?

Plato taught that the world around us is a reflection of an ideal world. Freud advanced rather different ideas about the universality of certain symbols. And Jung effected a sort of uneasy marriage between the two with his theory of archetypes. But must there not be scope by now for a more thoroughly scientific explanation of the power of certain sorts of symbol?

I had been reading Robert Ardrey's and Desmond Morris's popularisations of the idea that the basis of our behaviour is shaped by our genes. Back then there were still thinkers who believed that the genetic influence was minimal, that the human mind was essentially a *tabula rasa*, shaped in all significant aspects by learning and cultural influences. But Ardrey's propositions had seemed essentially reasonable to me (and wouldn't have seemed strange to the nineteenth century author of *The Golden Bough,* Sir James Frazer). A few decades later it's taken for granted that appropriate behaviour - the sort that's likely to promote the survival of the genes, and usually also of the individual carrying them - is programmed into all animal species, including us, by the nature of muscles, nerves, sensory systems, and the brains that control them. Nowadays, indeed, a discipline known as evolutionary psychology pursues just this subject. It seems obvious that even humans, with our huge capacity for learning, are not wholly dependent on cultural knowledge handed down from generation to generation, let alone on rational analysis or random experiment, to find out what it takes to stay alive. Neither individuals nor species would survive on that basis.

Lying on the beach at Cassis I began to reflect that there must be a genetic basis not only for our behaviour but also for the way we think. The style and structure of our thoughts must be determined by the nature of the apparatus that produces them. Platonic ideals and Jungian archetypes might well be founded in the genetically determined structure of the brain. The way we do science and philosophy, or indeed anything, is also likely to be influenced. Natural selection has shaped our brains for promoting our survival, rather than for these more esoteric pursuits. There are likely to be inbuilt biases that are wholly appropriate to the essential functions of brains, but less so to the sort of challenges they face nowadays. If we could recognise how our neuronal hardware shapes the way we think we would have more chance of overcoming its limitations. Acknowledging where our perspectives are slanted by the apparatus we use to perceive the world, we might with luck achieve a more accurate view.

I would later discover that I was by no means the only person thinking along these lines. In 1970 Robert Sinsheimer wrote ….. *I think it is only logical to suppose that the construction of our brains places very real limitations upon the concepts that we can formulate.* In 1971 J. Z. Young considered the suggestion that *logical propositions owe their truth to our brain structure* - but decided that in the light of what was then known about the operations of the brain it was not a very profitable question. Long before that Charles Darwin had realised this implication of evolution. In one of his notebooks he wrote: *Plato says ….. that our 'imaginary ideas' arise from the pre-existence of the soul, are not derivable from experience - read monkeys for pre-existence.*

Even without probing into the workings of the brain it's possible to think of several examples of thought habits that have been shaped, in one way or another, by our genetic heritage. For instance, natural selection has endowed us with sensations and emotions that guide us towards actions likely to promote our survival and that of our genes. We are equipped to derive physical pleasure from appropriate activities, and pain or discomfort from what harms us. We also experience emotional pleasures and pains, which prompt us to behaviours such as taking care of our young, and living in more or less co-operative societies. These valuable emotions tend to focus our attention on things of positive or negative import, at the expense of the phenomena which have no effect on our welfare one way or the other. The behavioural choices to be made are sometimes difficult, however, and our emotions conflicted. This tends to lead to concepts of good and evil, and a constant struggle between them.

Another characteristic of the human mind that can be explained in evolutionary terms is the amount of attention we devote to sexual matters - the high proportion of stories and dramas that are concerned with relationships between the sexes, the gossip and the jokes. We are an unusual species in that the timing of our copulations is no longer pre-determined. Human sex hormones don't come into operation under the influence of shortening or lengthening days, nor do human females turn interesting colours, or give off powerful olfactory signals, irresistible to males, when their eggs are ready to be fertilised. Along with bonobos we are a species that can copulate at any time. Natural selection has favoured this behaviour pattern, it's believed, because it promotes bonding. The gratifications of copulation keep parents or social groups together, which means better care for offspring, enhancing their chances of survival. In humans pair-bonding must have become especially advantageous as populations grew in size and groups tended to fission more frequently, not least because it will have increased the chances of a reasonable balance of the sexes in both halves of a dividing group.

For copulation to acquire this bonding function males must have become sensitive to subtler signals, more frequently in evidence than the flamboyant periodical sort that characterise many primates; but at the same time they would have to be rather less powerfully affected by them. Female receptivity, meanwhile, ceased to be limited to the time of ovulation. Sexual congress became a matter for choice and decision, rather than a response to an appropriately timed stimulus. But if coition is not going to be automatically timed to coincide with ovulation then it has to happen often enough that there is a good chance of the coincidence occurring. Sex could only become a matter of choice if there was some means of ensuring that the possibility was never in danger of being forgotten. Moreover sex education becomes necessary, for the loss of the hardwired response means that young humans don't instinctively know what to do.

Bonobos, for whom social bonding rather than pair bonding is the advantage, have found one answer to the problem. Among them copulation and pseudo-copulation occur with extraordinary frequency, so an ovulating female is unlikely to miss out on the chance of fertilisation, while infants are educated about the process of sexual activity and conditioned to the habit at an early age. How far this behaviour is culturally transmitted, how far innate, is not clear.

For us humans culture and genes clearly interact. Our fascination with sexual matters must essentially be the effect of that genetically determined sensitivity to many small signals emitted by the opposite sex. But the stories of love and lust, the erotic art, the jokes, help to keep the reproductive act in mind, as one among the considerable range of behaviours now available to us, amongst which we make conscious choices. And they serve to educate the young. Did our ancestors perhaps once behave like the bonobos? It seems possible that they did, until language provided an alternative means of cultural transmission.

Turning to drier subjects, our relationship with numbers may also be attributable to evolutionary pressures. Glance at a disordered array of objects and you can take in immediately that there are six of them, or seven, or maybe you are capable of an immediate awareness of as many as eight. With slightly larger numbers it becomes a matter of *about eleven* or *around fifteen,* and with larger numbers still the estimates become ever more approximate. The limitations may be attributable to the limited potential offered by the architecture of the brain. But very probably we owe the instinctive grasp of number that we do possess to the eating habits of our distant ancestors. It's been shown that animals which feed on a succession of modest-sized items, such as fruit, are better at discriminating between rewards on the basis of number than carnivore species, which collect one large meal at a time. Number talent may have become additionally selectionworthy after hominids moved out of the forests, since it will have been seriously worthwhile to know how many lions or other pack predators were in the offing, so as to keep track of them all if possible.

The ease with which we aquire the idea of space as possessing three dimensions seems likely to owe a good deal to a brain which has evolved to suit the structure of the human body. That structure leads us to characterise the space about us as in front and behind, right and left, up and down, with the surface which supports us providing the anchor to which the dimensions can be related.

Imagine, by way of thought experiment, an animal which evolves in intergalactic space. It can move in any direction and in any orientation, and of course it may be approached by others from any direction. So this improbable species is spherical in shape, and its sensory organs are dotted about all over the surface of the body, since it's advantageous to be able to survey all directions at once. Such a creature would have no reason to classify space as three-dimensional. Perhaps it might be the way its eyes were distributed around its spherical body that influenced its conception of space. Points in space might be defined, say, by their relation to imaginary lines drawn through eyes on opposite sides of the body and connected to distant stars. The lines might be conceived of as stretchable, and a point might be defined by the degree of stretch required and the lines to which it was applied. And this procedure, which seems extremely demanding or perhaps downright impossible to our brains, might seem quite obvious to theirs.

Let's imagine this species happened to evolve a capacity for abstract thought. Einstein's idea of curved space-time might come quite naturally, and they might be much more comfortable than most humans with the concepts of multi-dimensional space that seem to be required by some branches of modern science. On the other hand a means of calculating the volume of a cube (supposing they had any reason to create a cube) might be an achievement that came very late in their cultural history.

These examples of the way our heredity shapes our thought seemed fairly obvious. There must be others, I felt, but to identify them one would clearly need to know something about how the brain works. So I set out to read up on the subject - without realising what I was letting myself in for.

It so happened, however, that it was a very interesting time at which to start on my education. Relevant disciplines growing at an ever-increasing rate, and buzzing with fascinating work. Neurologists were developing ever more delicate technology with which to pursue their investigations and record what goes on in nerve-cells. Psychologists were focussing on such matters as perception, cognition, and infant behaviour. Noam Chomsky was causing a stir with ideas about the innate basis of language. Ethologists were setting off to remote parts of the globe to record detailed observations of species in their natural habitats, providing information relevant to the evolution of human behaviour. In the years since then knowledge about the brain has grown exponentially. New discoveries have inspired more questions, and in every field the scale of investigations has moved into finer and finer detail. Moreover the various disciplines have drawn inspiration from each other's findings.

Particularly exciting have been the neurobiological insights provided by new technologies for probing the brain's workings non-invasively. Imaging techniques such as PET (Positron Emission Tomography) and fMRI (functional Magnetic Resonance Imaging), have made it possible to observe which areas of the brain are most active in the performance of a particular task. A technique called Transcranial Magnetic Stimulation provides a means of corroborating this information by seeing what sort of task is disrupted when a brain area is temporarily incapacitated. It has become possible to refine the knowledge derived from the ill-defined and often quite widespread effects caused by damage to the brain.

Huge potentials are being opened up, meanwhile, by the rapid advance of genetic knowledge and its application. Genetically engineered mice can be studied to discover what happens in the absence of one particular gene, or when the gene is inactivated only in one bit of the brain, perhaps at one particular stage of development. There's a means of revealing where a gene is being put to use. And sequencing the genomes of an ever-increasing number of species has provided scope for fruitful comparisons.

Another delicate technique applies a label to just one type of nerve-cell so that cells of this type fluoresce when they're busy. Or a single nerve-cell (a neuron) can be activated with a photon of light, so as to intervene in the activity of a neuronal network with great precision.

Another profitable approach has been the use of computers to model neuronal circuits. The models are necessarily simple, compared to what actually goes on in the brain. Nevertheless they have demonstrated that even a relatively modest network of interconnected units can imitate the brain's achievements in some degree - can learn, for instance, to recognise an object that does not always present a consistent two-dimensional image.

Knowledge is advancing at an enormous rate, and any account of the brain is certain to be rapidly outdated as gaps in the current understanding are filled in, or new refinements to existing knowledge bring radical shifts to old interpretations. As of now, the essential functions of some areas seem fairly clear, other areas remain enigmatic. Various major circuits have been identified, but there is an enormous amount still to be clarified. New details of how nerve-cells function at the molecular level are being uncovered almost daily. Nevertheless it's possible to draw a sketch that gives an idea of some of the brain's main features and their relationships.

Just as fascinating as the workings of the human brain is the question of how it evolved. How did we come to be capable of creative thought, as in the invention of tools and the practice of art, let alone of abstract thought? What evolutionary chances led us to the capacity for distilling out of our concrete experience notions like *happiness* or *weight*, and then more advanced concepts, like *energy* and *natural selection*; and not only arriving at these abstract ideas but also juggling with them, to create theories of cause and effect. How did we evolve into a species capable of conceiving that it evolved, and of studying the process?

As Theodor Dhobzhansky famously observed, nothing in biology makes sense except in the light of evolution. This is perhaps particularly true of brains. Evolution doesn't involve careful analysis of requirements or rational design. It can only fiddle about with what already exists, adding new components, modifying old ones so that they can perform an additional function, and just occasionally allowing some simplification - letting a component that's no longer useful fade away. The result, in the case of the vertebrate brain, is blindingly complex. But the accumulating evidence from comparative neurology, embryology and fossil skulls, combined with some logic about the order in which certain developments must have occurred, can point to some reasonable theories.

A century and a half after the publication of Darwin's *Origin of Species* many of the mechanisms of evolution have been sorted out. Most of the mutations which occur in genes probably have neutral or unprofitable or positively disastrous effects. But some give the organism in which they occur, and the descendants which inherit the slightly altered DNA string, an increased chance of flourishing. Over generations this means that the new version of the gene, if accident doesn't intervene, is likely to spread through the population. Other mutations may prove advantageous in the slightly altered context, and changes to the genetic recipe gradually accumulate. Occasionally an exceptional potential for change opens up when a gene or a whole stretch of genes, or sometimes the entire genome, is duplicated, and the 'spare' copies can be modified without interfering with the normal processes of the organism. Mutations may accumulate until their possessors are transmogrified into something so different from their ancestors that they must be regarded as a new species. Or a single species may split in two, exploiting different environmental niches.

Frequently the process of genetic change is driven by modifications to the habitat the species exploits. The climate changes, or the shape of the landscape alters, or new competitors or predators emerge. In the new context some versions of existing genes will be favoured more than others, and will become more common, and mutations that wouldn't previously have been favourable become so. The rate of evolution is accelerated.

It's not too hard to conceive how physical shape can be modified and elaborated by this process. But it's difficult to imagine our consciousness, our sensations, thoughts and emotions, our sense of self, as the product of an assemblage of nerve-cells, however complex, and to think that this too is the result of evolution. However, the evidence that can fuel the necessary imaginative leap is gradually emerging, and consciousness is becoming less mysterious. There's a pretty good explanation of how it's created, and there are some clear answers to the question of what functions it serves. It's even possible to make some intelligent guesses about what forms of consciousness are found in other species.

This book is an attempt to provide a rough sketch of the human brain, and to put together some of the evidence and ideas about how it evolved - along with a fair amount of guesswork. I hope it will be of interest to anyone who is curious about brains, but isn't lucky enough to have time to dabble at length in the scientific literature and wrestle with its obscurities. It's designed to be accessible to readers with little prior knowledge of neuroscience - though I should warn that it may be quite a hard read in places, for it's not possible to simplify a picture of the brain too far and still produce a meaningful account. Moreover studying the brain sometimes means taking new perspectives, looking at things from new angles, or questioning things that seem obvious. The first part of the book is principally concerned with defining just what it is that has to be elucidated, the second with some of the neuroscience that explains how consciousness – the essential foundation of reason - happens. My own experience is that studying the gradually accumulating evidence about the brain and its origins has been rather like working through a detective story. It's a challenge, but a very rewarding one, even though there is no one final answer to the mystery.

Chapter 2

Reason and Consciousness

Reason is a tricky word, with several different meanings - creating the apparent paradox that only an entity possessed of reason can be accused of being unreasonable. I define it as the ability to abstract general ideas from experience, to deduce relationships, and to create new ones. Reason sniffs out similarities between apparently dissimilar phenomena and reveals differences between apparently similar ones, all without reward other than the pleasure of discovery. It also identifies chains of causation. It can then use the knowledge creatively.

This definition is designed to include reasoning processes and results which are purely concrete and practical. The idea of using a stone to crack a nut, for instance, entails bringing together a knowledge of the nature of stones with a knowledge of the nature of nuts, and appreciating the possibilities of a sufficiently sudden change in their spatial relationship. Modifying an object to create a more effective tool is a more advanced example of the same thing. The chimpanzee who first peeled a twig and used it to fish in an ant or termite nest had realised both that the relationship of the twig to its bark was something that could easily be altered, and that the resulting probe would fit into a hole in a termite mound. Whether the first successful termite extraction was the result of conscious planning or serendipitous accident we can only speculate. But if it was an accident it encountered a mind that could appreciate the possibilities of the twig for extracting food from inaccessible places, and the concept was passed on to other minds that were equally capable of making use of it.

At the other end of the scale the most advanced form of reason deals in abstractions, in causes and effects, forces and functions and influences, in the consistencies underlying perceived events. It discovers relationships that are not directly observable by the senses, but must be deduced. This is essentially an operation of the same sort as the analysis and planning that goes into the manufacture of tools. It differs in working with generalised ideas rather than with a knowledge of concrete objects, and in being dependent on language, but it still involves bringing together disparate concepts. It's logical, therefore, to suppose that the capacity for abstract reasoning has evolved through an elaboration of the apparatus which is responsible for concrete reasoning.

A lot of the mental activity that goes into the production of a scientific theory, or a new tool, or a work of art, is certainly done at unconscious levels. However, the final products clearly reach the consciousness of their progenitors, and can be transferred to other conscious minds. So if creativity and rational thought are to be explained sensory consciousness must be accounted for first. Indeed explaining consciousness is the major part of the challenge.

For a long time consciousness was pretty much a taboo subject in scientific circles. There was no way of studying it directly and objectively in other people, or of measuring it, and introspection was entirely out of fashion. All that has changed in recent years, however. It's clear now that consciousness performs several useful functions, so the evolutionary advantages it provides can be identified. There's also a good deal of evidence as to how sensory consciousness is produced.

On the other hand it must be admitted that while in one sense we can understand the mechanics by which an assemblage of nerve-cells produces this effect we call consciousness we may never comprehend just why it should do so. It may be something we have to take as given, in the same way we accept the existence of gravity, or the fact that electrons have a negative charge. As the philosopher David Chalmers put it, this is the hard problem - *explaining how any physical system, no matter how complex and well-organised, [could] give rise to experience at all.* On current evidence I have to agree with another philosopher, Colin McGinn, who suggested that it's a problem we're not equipped to solve.

Consciousness is another ambiguous word, of course. It's something which can be suspended by sleep or anaesthesia or concussion, and there's the complication that many things are potentially available to consciousness at any moment but only a selection of them actually occupies it. Before consciousness can be lost, however, or used, it must exist. All the evidence suggests that it's something which has evolved, and evolved gradually. And we need to identify several levels.

The most obvious is sensory consciousness - awareness of the external world through vision, audition and olfaction, and of internal sensations of hunger and satiation, pain, pleasure, warmth, and so on. This sort of consciousness is pretty certainly shared with a good many other species (but by no means all) though the precise range of senses included and the degree of definition within each sense must vary considerably. Life for a creature with only this sort of consciousness can be conceived of as a sequence of sensations and emotions, a wholly egocentric sort of experience.

The next level entails an awareness that these sensations and emotions are changeable, and reflect current circumstances. It implies a new sort of memory, so that consciousness is no longer a matter of living exclusively in the moment. And it introduces the possibility of conceiving of other individuals of the same species as beings much like the self, experiencing similar sensations and activations. We might call it *consciousness of self and others.* It's something that, at the level of the individual, develops with time. It depends on maturation processes in the brain as well as on accumulating experience. As Ulric Neisser put it: *babies know what they want but they do not know who wants it*.

The major evidence for the existence of this sense of self in other species is that it makes social interactions subtler. For instance, it enables a mother to manage her infants with a certain amount of insight, recognising their limited knowledge and abilities. This sort of behaviour can most notably be observed among great apes, elephants and dolphins. As in humans, however, there is considerable variation in the degree of insight shown.

The ability to guess at the nature of another individual's perceptions and motivations also allows deceits to be perpetrated. Jane Goodall, who pioneered the study of wild chimpanzees, recounted how a bright young male in the troop she observed noticed a banana that was hidden from all the other chimpanzees present. Had he attempted to collect and eat it straight away he would pretty certainly have lost it to an older, dominant male. She received the impression that he carefully avoided looking any further in that direction until, the other bananas supplied to the troop being finished, he managed to initiate a general move away from the area. Then he sneaked back on his own and claimed the prize. Since then a good many clear examples of deceit have been recorded among great apes, along with quite a few suggestive observations in various monkey species. The subject is extensively explored in Byrne and Whiten's book *Machiavellian Intelligence*.

Deceit involves manipulating what another individual perceives, and rests on the ability to imagine another's viewpoint. The talent seems to coincide with the ability to imagine that objects might usefully be somewhat adapted, that current relationships might be altered, that roles other than the usual one might be played. In other words, it cohabits with the capacity for practical reasoning which can lead - in a species with appropriate manipulative ability - to the fashioning of simple tools. This sort of consciousness also begins to make language possible, since it is necessary to have a concept of another mind capable of receiving a message before there is any point in purposefully trying to frame one.

A third level of consciousness is consciousness of the self not just as a feeling, perceiving being, but as a thinking being. That means there is not only insight into the perceptions, emotions and motives of others, there is also the ability to take a step back from the action and reflect on these insights. Thoughts become detachable from the thinker, capable of taking the stage in their own right rather than just providing inspiration for action. The evolution of this mode of consciousness must have been closely interwoven with the development of language, which of course greatly facilitates such reflections. Furthermore, once thoughts can be contemplated as entities divorced from action they seem to cry out for expression and discussion. No doubt this was what motivated early humans to elaborate basic forms of language into structures governed by grammatical conventions.

Language in turn opened up the way for the evolution of a capacity for abstract thought, since abstract ideas need handles by which to grasp them. The thought must emerge, of course, before a word can be found for it, but it is difficult to get very far with an abstraction, let alone relate it to other abstractions, until there is a convenient means of storing it in memory and retrieving it again. Attaching it to the act of pronunciation and the sensory shape which is a word achieves this. Furthermore, a capacity for abstract thinking is generally only going to provide an evolutionary advantage if the results can be communicated to other members of the species. It's only likely to become established, therefore, in a species which already possesses a certain amount of language. The available evidence suggests that this sort of consciousness is pretty much confined to humans.

I shall argue that there is also a fourth, primitive sort of consciousness, merely a vague sense of current need and current purpose, which belongs at the beginning of this evolutionary progression - but that must wait for now.

What I have termed sensory consciousness appears to correspond to what Gerald Edelman and Giulio Tononi call primary consciousness and what Antonio Damasio calls core consciousness. Edelman and Tononi propose a 'higher order' consciousness which is *accompanied by a sense of self and an ability to explicitly construct past and future scenes.* Damasio writes of *extended consciousness* which has many levels and grades. It can create an elaborate sense of self, and provide awareness of a lived past and an anticipated future. The philosopher Daniel Dennett has similarly distinguished different levels of consciousness. All agree that language plays an important role in the most advanced form.

The second and third levels of consciousness must surely have evolved by elaborations of the mechanisms which produce conscious sensation. So before the question of abstract reason can be tackled it's necessary to consider what conscious sensation is, and how it might have come about. That's the main subject of this book.

We have this marvellous experience of a world out there, full of colours and shapes, textures and distances, rich in sounds, and enlivened with assorted odours. It's rather difficult to talk about, because we take it for granted that what we experience *is* the external world, in objective reality. Our sensory systems mesh together so well that the evidence offered by one is nearly always confirmed by the witness of another channel. Stretch your hand out towards what looks like a tangible object and the predicted tactile sensation occurs, nibble on something that looks and feels like an apple and it turns out to taste like one too, and provides the expected sense of nourishment. Our conscious sensory experience generally offers a very efficient guide to our environment, and enables us to fit our actions to it very effectively.

However, with increasing knowledge of nervous systems it has become apparent that the equation of sensory experience with reality isn't strictly correct. Our experience is not reality itself, but a model created in the brain out of the data which our sensory systems happen to be equipped to collect. (Other species collect somewhat different selections of data.) The greenness of leaves and grass is not integral to them, it's an effect created in the brain when light receptors in the eye register the wavelength of light that leaves and grass reflect. The sounds we hear exist only in brains. They are the transformation created by a marvellous neuronal mechanism when pressure waves travelling through the air wash up against the ear-drum. In short, conscious sensation models reality with what might be described as a set of coding systems. Similarly our pains and aches, feelings of comfort, sense of balance exist to provide a model of the current state of the self that will prompt appropriate behaviour. In 1973 Harry Jerison wrote: *A ..... modern way of thinking about the perceptual world of which either Man or other animals are conscious is as a construction of the nervous system designed to explain the sensory and motor information processed by the brain.*

Things that can be done without consciousness

All the evidence suggests that conscious experience is something that is created by a very substantial assemblage of nerve-cells (properly known as neurons). A great many species pretty certainly get along without it. In the simplest cases a sensory neuron is directly connected to the cell which controls a muscle, and activation of the sensory cell triggers a message which causes the muscle to contract. In the surf-clam, for instance, a light-sensitive neuron running around the edge of the body is illuminated when the shell is open and the animal is feeding. It connects to the muscle controlling the hinge of the shell, and if the amount of light reaching it is suddenly reduced - something which may indicate a predator passing overhead - the shell is quickly closed.

Behaviours of this sort are obviously fore-ordained by the architecture of the neuronal connections, and therefore determined by the genetic design of the species. We have a few such reactions ourselves. Sneezes and coughs are good examples. Another is the way one's leg jerks up in the air when the knee is tapped on a certain spot. That reflex is produced by a short connection from sensory cell to muscle which permits a rapid readjustment to muscle tensions when an unsteady footing threatens to throw us off balance. We notice these things happening, and sometimes we can make a conscious effort to suppress the action, but consciousness plays no part in its actual production. Moreover it's difficult to fake a reflex movement convincingly.

In many species the response to quite complex stimulus patterns, involving a combination of several features, may similarly be arranged through hardwired circuitry. It was the ethologist Niko Tinbergen who established the existence of these innate behaviours, experimenting with cardboard shapes and other models to discover just what it took to elicit a particular action. The evidence that the relationship between stimulus and response is genetically determined by the neuronal wiring is the fact that if the stimulus pattern contains the necessary elements it produces its effect, even though the normal context is missing. Infant herring gulls, for instance, gape hungrily when any sort of beak-shaped stimulus appears from above, and react most energetically to a shape that resembles an adult herring gull's beak - long, thin and yellow, with a red spot at the top. It doesn't matter if the rest of the gull is missing. All altricial birds (the sort that are helpless when they hatch and are fed by their parents in infancy) have some variation of this response, ensuring that they open their beaks to receive food when a parent turns up bearing it. No doubt they all react most strongly to the sort of beak proper to their species, but will also respond to a less than ideal stimulus, which means they don't starve if the parent's beak doesn't quite conform to the standard. (Once, hearing a squawk from under a bush while I was watering the garden, I found a newly fledged blackbird which, when I offered my little finger, gulped at that.)

In the breeding season a male stickleback builds a tube-shaped nest at the bottom of his patch of stream and tries to attract any pregnant female who passes to lay her eggs in it, where he will fertilise them, and then care for them. The sight of a rounded, silvery shape that approximates to the belly of a pregnant stickleback is sufficient to set off his courting display. He also defends his territory against rival males. Tinbergen showed that the male goes into its threat display - holding itself vertical with its tail uppermost and quivering violently - at the sight of any patch of red. In the streams which sticklebacks inhabit a patch of red is likely to be the breast of another male stickleback. In other surroundings strange things can happen. A fish that Tinbergen was keeping in an aquarium by a window went into a frenzied display at the sight of a passing Post Office van.

Often the stimuli which evoke these hardwired or reflex responses combine shape with movement. Day-old chickens crouch to the ground and freeze when a dark, roughly bird-like shape passes overhead if it has a short protrusion at the front and a longer one at the back. They don't do it if the long protrusion is at the front. It seems they are innately designed to try and hide from raptors, and not to waste time hiding from ducks and geese. Tinbergen identified many such hardwired responses, and many more have since been discovered by other ethologists.

Varied Types of Learning

In a great many species, however, many behaviours are obviously learnt. The detailed study of learning famously started with Pavlov, who was investigating digestive processes in dogs and measuring the rate at which they produced saliva when he noticed that his subjects salivated not only when their food actually appeared but also at the sight of the white-coated laboratory assistant who usually brought it. He then established that they reacted in this way to any sensory event which regularly occurred just before the appearance of food. Salivation is a genetically determined reaction to food, so he called it an unconditioned response, and the food an unconditioned stimulus. The bells, buzzers and other signals which the animals learnt to associate with food were conditioned stimuli, and the salivation which occurred in response to the conditioned stimulus was a conditioned response. To put it another way, a genetically programmed response comes to be evoked not only by the genetically ordained stimulus, but also by a stimulus pattern which has regularly been associated with it, and is likely to signal the imminence of the important event.

The ability to respond to a signal that indicates the probable nearness of something desirable is obviously useful. The ability to react to an advance warning of possible danger is even more valuable. This sort of learning is usually termed Pavlovian conditioning.

The discovery inspired a great flood of experiments along similar lines. In addition to those where the desirable or aversive event simply followed the cue there were many in which an action had to be performed to obtain the desirable. Numerous pigeons learnt to peck at a key to obtain a pellet of food or a drop of water, and rats learnt to press bars for similar wages. This is known as operant conditioning, and the reward which encourages the action is termed a reinforcement. Animals also learnt to perform an action in order to avoid an undesirable experience - usually a mild electric shock, which is a highly unnatural sensory event but has the double advantage of not doing the animal any perceptible damage, and of being easy to produce repeatedly at a precisely determined strength.

Learning to perform a piece of behaviour so as to avoid a noxious experience means learning to recognise a signal which indicates that it is imminent - learning, for instance, that when a light shows it's necessary to press a bar to avoid receiving a shock. Passive avoidance can work on the same principle - the bar can safely be pressed when the light isn't showing. In yet another sort of experiment the animal must learn that a reward for pressing the bar is available only when a light is on. In these cases there are in effect two stimuli to provoke the response - the bar which must be pressed and the light which signals the timing.

These experiments led on to more complicated ones. A pigeon might learn that there would be a food pellet if it pecked at the key when a certain shape appeared in a little window next to the key, but nothing if it pecked when another shape was on show. Or perhaps there would be a reward when a light to the right of the key was on, but not when a light showed to the left. Or there might be a punishment for pecking at the 'wrong' time as well as reinforcement for the 'right' response. This is discrimination learning. As well as its interest in the context of learning it's a protocol that has proved very useful for discovering just what sensory information different species are equipped to collect.

As the evidence accumulated it became apparent that much discrimination learning can be explained in terms of pre-programmed behaviour systems which are capable of a certain amount of modification. A certain sort of stimulus prompts a certain response, but the specification of the stimulus isn't very precise. The reinforcement serves to indicate more exactly the sort of stimulus that is worth a response, thus enabling the individual animal to adapt its inbuilt behaviour patterns to the potentials of the environment into which it happens to be born. Harry Harlow speculated along these lines in the fifties, suggesting that perhaps ....*conditioning does not produce new stimulus-response connections but .... operates instead to restrict, specify, and channelise stimulus-response potentialities already possessed by the organism.*

Newly hatched chicks, for example, peck at any very small, three-dimensional object below eye-level, a behaviour which points them towards picking up seed grains, but which may equally lead them to consume grains of sand. When an experimenter confronted some chicks with a pile of grain and a pile of sand they ate both in about equal amounts, and it took them some time to learn to distinguish food from non-food. After an hour with seeds and no sand, however, a distinct preference for seeds was established. But chicks left for an hour with sand and no seed didn't develop a taste for sand unless they were given an injection of high-calorie food direct to the stomach afterwards.

Under normal circumstances the chick's learning is guided by the mother hen, who may pick up seeds and drop them again in front of her brood, but the learning still takes time. Seeds come in many different shapes and colours, and there must be a good many different kinds which make acceptable food for chickens, so the advantages of the system are clear - the chick is born with a rough idea of what to peck at, but can adapt itself to whatever sorts of seed happen to be available in the locality.

The behaviour of pigeons in the standard experimental situation suggests an elaboration of a similar sort of basic pattern. The key in the experimental box is the only small round shape in evidence, and the nearest thing to a food-stimulus available. Indeed, it's the only conspicuous stimulus in sight, and of course the pigeon has been kept hungry in preparation for the experiment and is ready to respond to anything remotely appropriate. Actually acquiring food requires at least two pecks, one at the key and one at the food when it appears, but food and key are close together, so this is not a very great distortion of the normal way of things. And close observation showed that the pigeons treat the key as they treat the variety of reward for which they are working. When the bird is accustomed to receiving a food reinforcement it opens its bill just before contacting the key, and snaps it shut again as it raises its head. When it is used to working for a water reward its beak is almost closed as it hits the key, and remains there longer while the bird makes swallowing movements.

Moreover pigeons are very talented at learning to make all sorts of discriminations when pecking at a key to obtain food or drink, but they are pretty hopeless at learning to refrain from pecking at a key when they are only reinforced for abstention. In other words they find it very hard to suppress the normal food-getting action as a way of obtaining food.

In fact it has emerged that a species may be equipped with a large capacity for one sort of learning, and little or no talent for another sort. Thus pigeons readily learn to peck a key to obtain food or water, but not to avoid a shock - their inbuilt response to danger is to fly away. Similarly, a rat's natural reaction is to freeze when danger threatens, and rats are bad at learning to press a bar to avoid a shock (though they are slightly better than pigeons at adapting to such demands). Chaffinches will learn to press a key for food, or to perch in a particular place to hear some recorded chaffinch song, but it's very difficult to teach them to perch in a particular place for a food reinforcement, and impossible to persuade them to peck a key for song. A dog learns to take food when one auditory cue sounds, and to avoid it when it hears a different sort of sound, but doesn't, in this situation, make use of sound cues distinguished by coming from different directions. (Excellent guides to conditioned learning are to be found in the works of M. E. Bitterman, Robert Bolles, David Oakley and Paul Silverman.)

Obviously an essential requirement for operant conditioning and discrimination learning is a feedbeck mechanism which allows the results of an action to be related to the stimulus which evoked the action. These experiments in animal learning suggest that feedback channels tend to be specialised. A certain sort of stimulus pattern elicits a certain action, and feedback is provided only about the type of result that is likely to follow such an action. Similarly, only the sort of stimulus that is likely to be relevant will function as a cue.

Discrimination learning is thus often a matter of honing an innately determined (or unconditioned) stimulus to a sharper definition. Or perhaps the genetic recipe provides the specification for an ideal stimulus, but in the absence of the ideal an animal will respond to the best approximation available, and thus learn what makes an acceptable substitute and what doesn't.

Many species are not only pretty efficient at learning how to obtain rewards but also play the odds intelligently when rewards arrive only intermittently. They behave, in fact, with the logic that a human might show in similar circumstances. Those reinforced on every trial learn fast, but give up responding fairly rapidly if the reward ceases to arrive. Animals reinforced on a fixed ratio schedule (for every tenth action, say), or at fixed intervals, learn more slowly but go on trying longer after reinforcement ceases. Animals whose reward arrives on a totally irregular basis learn slowly but then keep working hard all the time, and persist for even longer with no reward.

Early experimenters tended to assume that the neocortex - that thick, rumpled blanket of grey matter which forms so conspicuous a component of the mammalian brain and is especially large in *Homo sapiens* - functioned as a store for all the learnt connections that constitute conditioned learning. But it turned out that Pavlovian conditioning is possible in some of the lowliest invertebrates. Furthermore rats whose neocortex was removed could still learn very effectively, and in some tasks progressed faster than normal animals. Such rats may also retain the ability to perform responses they learnt with a whole brain. This is not to say that the absence of the neocortex doesn't restrict what may be learnt; it does, very considerably. But the limitations seem to be attributable not so much to deficiencies in basic learning ability as to the fact that the most sophisticated parts of the sensory systems are lost, and lower parts of the brain have less capacity for analysing the sensory input so as to make the more difficult discriminations.

Beyond Behaviourism

The first scientists to conduct this sort of experiment were known as Behaviourists. Some of them seemed to think that all learning could be explained as conditioning produced by simple reinforcements (though the theory, paradoxically, doesn't allow for the existence of theories, and could logically only be tenable if they were persuaded to it by bribery). The improbable idea was laid to rest by Edward Tolman, using a variation on a standard experiment. Rats put in a maze soon learn to find their way to food or water, and the assumption was that it was the prospect of the reward which motivated the animal to learn the most efficient path through the maze. Tolman set one group of rats to learn a fairly complicated maze in the conventional way, while another group ran in the same maze without finding any food. After ten sessions a prize was provided for the previously unrewarded rats as well, and they immediately began to perform slightly better than those which had been rewarded all along. From the twelfth day onwards - as soon as they had had the chance to establish where the food was - they logged slightly fewer errors on the way to the goal. In a further experiment the most direct route to the food was blocked off, once it had been learnt; once again the animals which had initially explored the maze without reward found the best new route to the food more quickly than those which had always had food to aim for.

Thus it was demonstrated that rats study and learn practical geography for its own sake. Not only do they do it in the absence of reinforcement, but they do it more efficiently that way. It's as if the presence of food tends to distract attention from the business of exploration. (An introspective human may not find this surprising - we too tend to learn more of the geography of a strange city while wandering around it for pleasure than when hungrily searching for a restaurant.)

What a rat acquired when it learnt how to find its way through the maze, said Tolman, was a cognitive map - not just a memory of how to get from A to B but an idea of landmarks and spatial relationships. The hypothesis was tested with a simple, cross-shaped maze. A rat would be put in at one end of the cross, and food would consistently be provided at the end of the arm to the rat's right. When the rat had mastered this situation and consistently turned right the procedure was changed and it was put into the maze at the opposite end. For half the rats in the experiment the food was still in the same place, which meant they now had to turn left, and for the other half the reward was now in the opposite arm, so that they still had to turn right. If maze learning was a matter of conditioned responses the latter group should perform more efficiently in the new circumstances. If there was a cognitive map then the rats whose reward was still in the same place - identifiable by features of the laboratory, visible beyond the walls of the maze - should do better. It turned out that the group whose food was still in the same place got to it most rapidly, supporting the cognitive map theory.

The benefits to be gained from a cognitive map are obviously substantial, and the energy expended on exploration pays good dividends. The animal which has learnt its way around its environment knows the quickest way to the nearest shelter if danger threatens, where to find water or a safe place to sleep, the easiest route to a likely place for food. But what prompts it to explore in the first place, in the absence of any immediate concrete reward? Building an internal model of the environment must constitute a different sort of learning from the conditioned learning which merely serves to modify the application of instinctive behaviours.

This is not to say that conditioned learning and model-building are opposed. On the contrary, they work together. Successful models incorporate the fruits of conditioned learning. It may be conditioned learning that links food to the place where it's frequently found, and causes the site to be marked on the map.

Learning and sensory abilities

The evolution of learning abilities must have been closely intertwined with the evolution of sensory abilities. A species whose behaviour is entirely innate, composed exclusively of genetically determined responses to genetically determined stimuli, needs only such sensory apparatus as is required to register those stimuli. Any additional sensory equipment would be superfluous, indeed it might make the animal function less efficiently. But an organism capable of Pavlovian conditioning must of course be capable of registering the cue, or 'conditioned' stimulus as well as the unconditioned or innate one. Learning to make discriminations entails collecting enough sensory information to detect some difference between the relevant stimulus patterns. And where there is a capacity for building a cognitive map almost any new type of sensory receptor or new way of extracting information from sensory input may prove to be useful.

We humans have evolved the means of collecting enormous amounts of sensory data. How do we manage to make sense of it all, translating quite varied patterns of sensory input into consistent and meaningful percepts? For instance, we learn to recognise an animal as an ongoing entity, regardless of its current pose or the angle from which we happen to see it. We recognise a tune, regardless of whether it's played on the piccolo or the double bass, or performed by a double orchestra and massed choirs. We don't need to hear it in the familiar register; nor are we fazed by unusually fast or slow performances, unless the change is so extreme as to alter the whole mood of the piece.

All the evidence indicates that what we record is essentially a pattern of relationships. A tune is a pattern of tonal and rhythmic relationships between notes. A cow is a shape made up of components in spatial relationships some of which change, within certain parameters, and some of which don't. The distance between shoulder and knee or knee and hoof is always pretty much the same, but that between shoulder and hoof or nose and shoulder may alter, within certain limits. Thus we recognise a cow with equal ease whether it is standing up, walking, or lying down, and from almost any angle. We also recognise a distortion, exceeding the parameters of normal spatial relationships, which indicates that the animal is damaged, or dead. In addition, we come to recognise the way the spatial relationships alter as the cow moves. And we store the information that the cow is bigger than a rabbit, smaller than an elephant.

The bundle of sensory information that enables us to recognise the same essential pattern of relationships in many different manifestations can be termed a *schema* (borrowing a usage coined by Frederick Bartlett and adopted by Jean Piaget). The schemata created in different sensory channels become organised, of course, into multisensory schemata. We remember that cows moo, and give milk, that their short hair has a certain sort of feel, and we learn that it's important to distinguish a cow from a bull. Recording all this information must require a good deal of space, and we can deduce that this is what the complex circuitry of the neocortex is needed for.

Relationships, spatial ones, are of course the very essence of maps. Making a cognitive map means learning to recognise those features of the landscape which are conspicuous enough to function as landmarks, and establishing how they relate to each other. The positions of less conspicuous but more vital things - food sources, water, shelter - can then be related to the pattern of landmarks.

Three things make the creation of the map a particularly challenging business. One is that the landmarks currently visible must be related to those no longer within sight. The second is that the map must be capable of being rotated - the landmarks and the relationships must be recognisable from any angle, and be usable as guides to various goals. The third is that an animal needs to keep track of its own frequently changing position on the map.

The benefits of a cognitive map are huge. Its possessor is not limited to reacting to whatever is within reach of its sensory systems, or else wandering about at random until it encounters something that prompts a response. It can set off for some distant place where particularly tasty food is often to be found, or for a remembered spot of comfortable shade, or a secure place to sleep. In other words it can apply the knowledge gained through past experience to take itself to areas beyond the current range of its senses.

There are other means of directing travel to useful goals that are out of sensory reach. Several species of insect, such as bees and ants, have been shown to be pretty good at getting around in an organised fashion. But the general consensus is that they do it by a means more like conditioned learning. They learn to head towards a landmark, and then to change course towards the next landmark, and they can record how far they have travelled from home. But they don't appear to recognise an asymmetrical landmark from different sides, or to have an overall picture of the landscape which would allow them to take shortcuts.

This navigational system is more extravagant in terms of the energy spent on locomotion, but must require much less brain space than a cognitive map, and much less time spent on learning. In the species in which it's been studied it's augmented by the ability to register the polarisation of light, so that even on a cloudy day the sun provides an aid to orientation. Ants also leave a trail of chemical marker which they can follow to return to the nest, and to guide future excursions. The zigzag flight of bees and wasps, meanwhile, probably gives them a wider view of a landmark than a more direct approach would, as well as providing motion parallax, so that the landmark can stand out from its background.

A hypothesis

Conditioned learning, as defined above, requires a relatively simple form of memory. When innate responses are modified by the effect of reinforcements all that's needed is to be able to suppress the response to the unprofitable variants. For building a cognitive map a good deal more is necessary. Features currently in view must be related to features no longer present, which means remembering where one has just come from, and what lies behind one's back. There must also be some record of how much time and effort has been expended on getting from A to B to C. These are the threads that connect the place where one currently is to the place one has visited before and to which one now thinks of going. And most importantly, to take advantage of a cognitive map it's necessary to be able to exercise recall - to refer to areas on the map other than those within sensory reach. Memory must become a store out of which records can be retrieved, not just matched up against current input.

Something else is implied too. An animal whose lifestyle is dictated by hardwired stimulus-response reflexes doesn't get to make choices. Its repertory of responses is accompanied by a hardwired means of deciding which takes priority when more than one potential stimulus presents itself. When two stimuli coincide the response will be made to the more urgent one (escaping being eaten, for instance), or if that criterion is not applicable, to the one that is most urgent in the context of the animal's current needs. If the animal is confronted with two stimuli of the same sort it will respond to whichever is exciting its sensory receptors most strongly. Sometimes this will be the nearest one, and if it's a food stimulus the response will result in its disappearance, and the animal will be free to go and deal with the next nearest. Where innately determined stimulus-response patterns can be modified by conditioned learning the methods of deciding between stimulus patterns must be much the same, except that the weighting given to the various possibilities is influenced by past experience.

For the species with a cognitive map the same priorities must be applied, but new possibilities open up. Using its capacity for recall an animal can make a prediction as to where it will find a satisfactory supply of food, and on that foundation make a plan. If memory provides the prospect of really tasty food on the other side of the hill it may be worth passing a less exciting meal on the way, suppressing the response to a present stimulus in favour of saving up the appetite for something better. The means of making predictions and plans leads on to the possibility of making choices. As more and more sensory information was collected, the power to make choices will have become increasingly significant.

Our hardwired responses are produced without conscious choice, and in many species a good deal of behaviour clearly depends on genetically determined neuronal wiring and can be assumed not to involve consciousness. Pavlovian conditioning does not necessarily require consciousness either, for humans can be conditioned to produce certain responses without ever noticing the conditioned signal, the unconditioned stimulus or the response. For many species this is probably also true of operant conditioning. Discrimination learning poses a more difficult question, but the best guess is that it depends on just how complex the stimulus patterns to be distinguished are.

However, there are several reasons for supposing that some form of consciousness is necessary for making and using a cognitive map. Firstly, how can the current location be related to a place that is now out of sight? The answer I find easiest to conceive is that the scene currently occupied can be related to those recently passed through because one is a conscious perception and the others are conscious recollections. It's difficult to imagine unconscious recollection, and equally difficult to imagine totally unconscious planning. An animal which chooses to head for somewhere beyond the reach of its senses surely has some sort of conscious idea - even if it's only the haziest possible awareness of the goal, and of the need which it's expected to answer. This, it seems reasonable to suppose, is what keeps the animal on its path, carrying out the plan, and resisting the distractions that crop up along the way. There's no way of proving this, of course. But I think we can be sure that when a vertebrate behaves in this purposeful manner it must at least be using a more primitive version of neuronal mechanisms that in many species have been expanded to produce sensory consciousness.

How might the evolution of the capacity for firstly conditioned learning and then schema-building and consciousness have come about? Paul Rozin suggested that new adaptive specialisations appear at first in particular contexts. Each innovation evolves as an answer to a particular problem, as it were, and *tends to manifest itself only in the narrow set of circumstances in which that problem arises*. This must certainly be true of the capacity for conditioned learning, which clearly can only happen where there is both a sensory mechanism to register the consequence of some particular sort of response and a channel to transmit a message about that consequence to the site where responses are triggered. The neuronal connections which convey different sorts of feedback must have come into being one at a time, resulting in the very specific learning abilities of different species.

It's reasonable to guess that schema-building and consciousness evolved in the same way, functioning only in one context to begin with, and then in some lines of descent being extended to guide other sorts of behaviour. Each extension to consciousness is likely to have required the addition of some new wiring in the brain. If so, then the nature of the activities in which a species can make use of consciousness must be determined by the particular range of connections its brain has acquired. Thus there may be species which, say, use consciousness only for building a cognitive map, can apply some discrimination learning to their eating habits but only Pavlovian conditioning to the business of escaping from danger, and which govern their relations with conspecifics entirely by means of genetically determined responses.

This list of possibilities is not random. A conspecific is very much a genetically determined stimulus pattern. Its appearance, vocalisations and odours are dictated by the same set of genes which shapes the sensory systems that must register them. Consequently appearance, sounds and smell can be entirely predictable. A purely hardwired system must be very adequate in species which don't distinguish among individual conspecifics.

For avoiding danger a hardwired system can be similarly effective. In fact it's likely to be the best bet, since the hardwired response must be pretty efficient if generations have survived to pass it on. Experimenting with new ways of dealing with threats is likely to be risky. On the other hand, learning to react to the stimulus which signals the possible approach of danger, so that evasive action can be taken sooner, is clearly a very useful talent. There are obvious advantages, too, in being able to improve on the hardwired indication of what's edible.

But geography, at the local level, is unpredictable. The only sort of rules about pathfinding that can be written into the genes are relatively generalised ones. Quite a few of the more elementary aquatic species work on the principle of heading up towards light when food is needed and staying in the dark at other times, or heading one way in the morning, the other in the evening. Some rules that work on a global scale guide the migrations of birds, turtles, and other species equipped with a magnetic sense, or with the ability to register the changing pattern of the stars - rules like head north in the spring and south when the days grow shorter. But there is no way the genes can encode the particular pattern of landscape into which an animal will be born. So a capacity for the most complex form of learning would be most valuable in the context of getting about.

The guess, then, is that in the evolutionary history that led to *Homo sapiens* the capacity for conditioned learning probably evolved first in relation to the form of behaviour where it was most useful, and gradually spread to other areas of life. And the capacity for creating a conscious sensory model of the world grew in the same way. Each extension enabled the individual to adapt to the environment in a more thoroughly flexible way. This, said Kenneth Craik, an influential psychologist and philosopher, *may be one of the functions of consciousness - to permit a greater 'elasticity' and flexibility and unity of response*.

However, the adaptability of the individual mustn't be confused with the adaptability of the species. Bacteria may have a very limited and fixed repertoire of behaviour, but they reproduce at furious speed when circumstances are favourable, and have plenty of opportunity to produce genetic adaptations to changing circumstances. The capacity for conditioned learning is only profitable, and therefore only exists, in species which live long enough to derive some benefit from having learnt something. And the construction of a cognitive map requires an investment of time and energy that is only worthwhile if the individual's potential lifespan allows the rewards to be reaped. The prospects for the survival of a species' genes may not be so very different, therefore, whether the strategy is one of short existences, huge numbers, high wastage-rates and frequent genetic mutations, or whether it is one of long lives, more modest numbers of individuals, and an investment in learning which increases the chances of individual survival.

We should note, too, that while the capacity for constructing really sophisticated schemata, in other words the capacity for advanced reasoning, increases adaptability yet further, it's a high-risk strategy for survival. The capacity to reason is also the ability to produce ideas that are bad for the prospects of the individual's genes, and in some cases may threaten the whole species - to decide that copulation is wicked, for instance, that suffering is inherently better than pleasure, that there is a god who is pleased when we murder fellow humans, or, as some misguided and rapidly deceased individual apparently once did, that it would be a good idea to eat nothing but carrots. Reason can undermine the guidance provided by our genes. The bacterial strategy for passing genes on to many generations of descendants may well be sounder than ours.

Recapitulation

Reason can be defined as an ability to perceive relationships of all sorts, and to juggle with them and create new relationships - whether by making a simple modification to a twig to turn it into a tool, or by constructing an abstract idea.

Rational thought requires consciousness, and consciousness will be the subject of much of this book. It can be subdivided into at least three levels.

The most obvious is sensory consciousness, out of which the more complex levels must have evolved. The last to evolve is closely intertwined with the invention of language.

Even in humans not all behaviours require consciousness, and some learned responses can be performed without it.

Learning has been studied extensively, in various species. Some forms, which answer nicely to the term conditioning, concern when to produce a certain response.

Much conditioning seems to entail refinement of the stimulus pattern to which a hardwired response is produced.

Learning the geography of the local environment is a much more complicated business, but extremely valuable. I hypothesise that it was for this application that some of the machinery of consciousness first evolved.

A map is a pattern of essential relationships, and can be regarded as a form of schema, to adopt a term used by Piaget and Bartlett.

A capacity for building schemata was gradually extended to other subjects, in addition to the layout of the environment.

The essence of a sensory schema is that it records the essential features of stimulus patterns and their relationships, allowing unimportant variations to be discounted.

Chapter 3

Clues to Where Consciousness is at Work

Geographical learning, or cognitive map-making, is the most widespread behaviour which seems to demand some degree of sensory consciousness. It's hard to see how the relationship between current surroundings and places now out of sight could be deduced without consciousness, or how a goal could be chosen, and remembered during the journey. Furthermore, mapmaking is a particularly valuable talent, so it seems reasonable to suppose that this is the context in which consciousness first appeared.

If the hypothesis is correct there should be a large number of species which are capable of making cognitive maps, but don't apply sensory consciousness to other purposes. They can be supposed to have some sort of awareness of where they are and where they're going; but they don't have to think, as it were, about eating, or their interactions with other animals. They perform those behaviours as automatically as we withdraw a hand from a hot surface, without needing to consider alternatives or make a decision. To put it another way, an animal whose consciousness is only concerned with a cognitive map thinks about where it goes to eat, but not about what it eats. Its plan to make for a place where food is likely to be found may involve a conscious sense of hunger, and perhaps a memory of past satiety in that place, but no conscious idea of a particular sort of food.

There should also be species which apply sensory consciousness in a wider range of contexts, but the range of behaviours for which they can exercise prediction, planning and choice may still be narrower than is the case for us (even after allowing for their narrower range of potential behaviours). The logic of evolution indicates that our closer relatives should certainly use sensory consciousness quite extensively. And there's nothing to say that something broadly similar hasn't evolved in other, distant branches of the animal kingdom.

It's often very hard to imagine that other mammals could be operating without conscious sensation and conscious intentions. The dog trotting down the road with a purposeful air so obviously seems to know where it's going. Monkeys and squirrels, leaping from branch to branch, can hardly be simply responding to stimuli in a pre-programmed manner, and if they had to perfect an inborn behaviour pattern by a conditioning process they would be very lucky to survive the education. A young cat eyes the top of a fence, crouches a little deeper as if about to take off, pauses to study the distance again - no question but that it's contemplating whether it can produce a powerful enough leap.

But these subjective impressions carry little weight in science. How can the possibility of other species' consciousness be more objectively investigated? The question is less impenetrable than it might seem. It's certainly not difficult to deduce that a good many species have a very efficient grasp of the geography of their home territories.

The idea of the cognitive map is founded on the fact that rats go in for exploring just for the sake of it, and learn more geography that way than when they're distracted by the presence of food. Species which waste energy are unlikely to flourish, so it's safe to assume that if an animal spends a lot of time on locomotion and on investigatory behaviours without showing any interest in obtaining immediate rewards, then it's building or updating its cognitive map. Even if many species cannot be as conveniently observed as the laboratory rat, it's obvious that many of them go in for exploration. Domestic dogs and cats provide convenient examples. They check out their neighbourhood regularly, obviously know their way around, and since they are generally more than adequately fed at home they can't be prompted by a serious need for food or drink.

It also seems appropriate to attribute mapping ability to a species which defends a territory of any size - a territory, that is, which significantly exceeds the range of the species' sensory systems. Territorial animals make their way around their fiefdoms in a confident, purposeful and efficient manner, and they clearly remember where the boundaries are, the borders within which a neighbour becomes an intruder. In some cases the boundaries are also marked out on the ground, by labelling convenient features with urine or faeces, or with scent from a special gland. Such odour-marking may perhaps serve as a useful aid to the animal's own memory, as well as a notice to potential trespassers. But the animal needs to have a pretty good idea of how to find its way around its boundary markers if they are to fulfil their function properly.

There are also other lines of evidence. In a good many mammalian species mothers leave the infant young behind in some sort of shelter while they go off looking for food, roaming far afield in the process. This would be an impracticable proceeding without an efficient and reliable means of finding the way back. Moreover the parent often travels a fair distance in a meandering fashion, zigzagging here and there, and then makes her way home by the most direct route available. Such skilful navigation certainly indicates the possession of an efficient cognitive map.

The same argument shows that birds must be equally well equipped. They too have to look after their offspring, and are absolutely obliged to deposit their eggs somewhere. In some species parents can identify their own nest in a crowded colony of thousands, but don't recognise their own chick if it strays outside the nest, which suggests that they are much better at learning detailed topography than they are at learning to distinguish individual chicks.

More amenable to experimental investigation is the fact that some bird species go in for hoarding food against hard times, and are pretty good at recalling where they hid it. Research with a seed-hoarding corvid called Clark's nutcracker indicates that it uses a spatial pattern of landmarks to remember where its stores are - thus satisfying the definition of a true cognitive map. It's arguable, anyway, that a fair degree of mapping talent must have developed by the time avian ancestors took to the air, for flying seems an impractical proposition without an efficient means of aiming towards suitable places for coming down again.

But there are grounds for tracing cognitive mapping ability much further back than this. Several species of reptile demonstrably share the gift. Crocodiles remember where they laid their eggs and return to care for the hatchlings, and some species use the same nesting site every year. Some species of snake regularly find their way back to communal hibernating sites. Turtles, confronted with the cross-shaped maze experiment, use surrounding landmarks to find their way to the food just as rats do. A similar ability has been found in goldfish.

It seems likely that the fish which first ventured out of water and gave rise to tetrapods were endowed with at least some mapping ability before they went any distance on land, since they must have needed, like many modern amphibians, to be able to find their way back quite regularly. And a knowledge of the terrain would certainly be a valuable asset for any fish frequenting the shallows at the edge of the ocean.

It's tempting, in fact, to suggest that mapping ability began to develop quite early in vertebrate evolution, and may have been one of the factors that contributed to the success of one particular ancestor rather than another. The useful talent seems to have evolved more than once, for it's found in at least one other phylum. Some species of octopus establish temporary dens and return directly to them after extensive foraging expeditions.

For a terrestial species which moves about with any degree of speed, and especially one with a raised centre of gravity, knowledge of geography is not in itself sufficient. It's also important to know something about the nature of the ground underfoot and the solidness or otherwise of the objects that make up the environment. What sort of ground provides a firm footing, and what sort of visual input indicates stickiness that will slow down a running animal, or crumbliness that will slide away under impact? What sort of vegetation will bend and allow an animal to push a way through, what is an impenetrable and perhaps prickly barrier? What presages a surface that will hurt if you come up against it too hard? These are things it's valuable to know when a route is being chosen, or a journey planned. They are particularly necessary both for prey animals fleeing a predator, and for the pursuing predator.

In fact it's probably only worth evolving the means of rapid locomotion, or growing large, if there's already a capacity for learning about the nature of the local terrain, so that reasonably reliable decisions can be made about where to put the feet. In many mammalian species close investigation of the immediate surroundings begins at an early age, while the animal is still under mother's protection. Careful prodding, with nose, whiskers and feet, educates the youngster about tactile values. When it grows old enough to leave mother, to explore more widely and form its own cognitive map, it will have some idea about what bits of terrain to mark on its map as best avoided.

For primates which climb high into the trees and are heavy enough to suffer severe damage if they fall to the ground this sort of knowledge is particularly vital. They must have a sound knowledge of what sort of branch or creeper will be strong enough to support them, and what's likely to give way under their weight. Young primates naturally show a great deal of curiosity about the world around them, and spend a lot of time investigating its properties.

Whether young birds need to explore the nature of their surroundings too, or whether they can afford to learn the hard way, is a moot point. I've only once seen a bird try to land on a twig that was inadequate for its weight, and that was a thrush which had been distracted by hearing another thrush singing on his territory. When the twig swung down to the vertical and tipped him off he easily rescued himself with a quick flap of his wings. For many birds the occasional error of this sort perhaps isn't too expensive.

In some species, then, exploratory behaviour covers not only the geographical lay-out of the environment but also some careful investigation of the tactile nature of its components. The knowledge thus gained underlies the predictions that allow a heavy animal to race across the ground at speed, or to leap through the trees. The guess is that conscious vision combines with conscious tactile feedback to create schemata which relate visual stimulus patterns to tactile quality.

To recapitulate: where behaviour is determined wholly by the nature of the wiring from sensory receptors to motor neurons there's nothing to learn. Where genetically preprogrammed responses can be modified by conditioned learning animals learn by chance, in the process of going about their business, but have nothing to gain by taking the risks involved in seeking for learning. But sophisticated sorts of learning, such as the construction of a cognitive map, must involve a substantial investment in neuronal hardware, and species which have made this investment need to put a further investment of time and energy into acquiring the knowledge that will make it useful. Consequently the talent must come accompanied by a drive to use it, which ensures that it earns its keep. Without that natural selection would not have permitted the costly machinery to evolve. Thus species which explore put a lot of effort into it. They investigate their territory thoroughly to begin with and check it over regularly, keeping up with any changes that result from storm or flood, other animals' activities, or the passing seasons. In some species youngsters also invest time in tactile experimentation.

This gives us a means of deducing in what other areas of activity a species might use knowledge gained through conscious sensation. We look for a parallel to exploratory behaviour - for activity apparently undertaken for its own sake, energy expended for no immediate reward, learning that's achieved in the absence of detectable reinforcement. That's a pretty good description of the sort of behaviour we call play. The type of play a species indulges in should give us a pretty good idea of what we want to know.

Play

Play occurs only in species with brains of a certain size (after adjusting for the fact that larger bodies need larger brains) and it seems to be especially provided for by a period of juvenility which intervenes between helpless infancy and competent adulthood. The length of this juvenile period correlates closely with brain size, reaching its peak in the great apes and of course in humans.

The implications of this were perceived over a century ago by Karl Groos, a psychologist: *Now we see that youth probably exists for the sake of play. Animals cannot be said to play because they are young and frolicsome, but rather they have a period of youth in order to play; for only by so doing can they supplement the insufficient hereditary endowment with individual experience.* For a large chunk of the twentieth century, though, play was carefully ignored. Difficult to define or to measure, it was classed as a hopelessly unscientific subject. Gradually, however, its importance came to be recognised. It was realised that the more a species plays in youth, the more adaptive and creative the intelligence it shows in adulthood. Moreover deprivation experiments demonstrated that when individuals of species that normally play are denied the chance to do so, they grow up to be abnormal and malfunctioning adults. By 1988 Arne Friemuth Petersen could report: *comparative psychologists now hold that a capacity for imaginative problem solving, as well as inventiveness, can only evolve in animal species whose young remain immature for a sufficiently long period.*

The most obvious sort of play is sheer physical activity, or motor play, involving rushing about, rough-and-tumble with playmates, and perhaps the use of twigs or other suitable objects as playthings. The knowledge a young animal acquires from such activity is vital to the business of planning and choosing actions - a knowledge of its own capabilities. The primate swinging through the tree-tops must not only know which branches will bear its weight, it also needs to know how far it can jump. The predator needs to know when to spring so as to have a good chance of bringing down a moving target. The fleeing animal has a better chance of survival if it recognises which obstacles it can jump or scramble over and which to avoid. Just as we humans are well-advised to know how agile we are before trying to jump across a stream or leap on to a moving bus, so the high-living primate or the fleeing prey must avoid trying something beyond its capacity. Motor play is the means by which a youngster discovers just what it can do, and just what choice of actions is open to it.

The learning is a little complicated since the animal begins its experimentation before it's fully grown. As it gets bigger and stronger its capabilities increase and it can achieve more ambitious feats, so its ideas of its own physical potential must gradually be adapted. The youngster needs to play pretty regularly simply to keep up with its expanding size and power. Once it's fully grown, however, it can afford to give up motor play, and most species do.

Primates have extended childhoods - several years in the case of the great apes - and devote a good deal of time to play. A period of juvenility with time for a significant amount of play is also common among mammalian predators. The knowledge thus gained is part of the essential underpinning for the tactics and strategy that will be demonstrated by the adult hunters. And if carnivores spend quite a bit of their youth trying out the skills they will need in hunting, young herbivores likewise spend time (though not so much) practising the skills they will need to escape, and acquiring the knowledge that may enable them to take the best route when chased. In some species, meanwhile, not only running and chasing but all the movements in the repertoire may be tried out in play, including those that will later be employed in mating and in male competition for females.

Play has been observed in numerous mammalian species, including some that might seem unlikely gambollers, such as hippopotami and tapirs. The large-brained cetaceans go in for a great deal of it, and (like apes) sometimes continue to play in adulthood. It seems to be less common among birds, but it's found in some of the larger, longer-lived species, such as crows, parrots and gulls, and something that seems to qualify as play has been observed in one or two reptiles.

We humans are obviously prime examples of adaptable motor ability that is perfected through experimentation. Introspection indicates that the knowledge is phrased in terms of a memory of physical sensation, which serves to measure the amount of effort used in a task, and allows us to estimate the amount which will be required for a similar task. We develop what might be called motor schemata, derived from conscious sensations of muscle contractions and bending joints. Other species which indulge in motor play can be supposed to hold the knowledge which allows them to plan and choose their actions in the same form. An animal contemplating a jump surely thinks about the effort required, as we do, in terms of muscular sensation. Where a species indulges in motor play it's fair to deduce that this sort of information is available to consciousness.

The other vital thing that infants learn when they play is how to use feedback from other senses. For us the most important form of feedback about the results of our actions is visual. Our conscious visual perception of an object allows us to choose to reach for it. If we fail to make contact it's the conscious visual experience that reveals by how much we have failed, and how to adjust the movement when we reach again. Without some such feedback there could be no learning from failure - which means no learning.

The hypothesis, then, is that visual consciousness was used first only for mapbuilding. I suspect it was linked to conscious feedback about some longlasting effect such as the state of fullness in the stomach, or perhaps internal temperature, something which could readily be correlated with the ongoing sensory input from the environment. It found new applications when additional forms of sensory information were linked to it. A conscious sense of touch meant that it was possible to learn about the tactile properties of different parts of the visual environment, and use the knowledge for planning. When information about the position of the body and the deployment of the muscles was promoted into consciousness it became possible to evolve a more flexible musculature and learn how to use it, instead of depending on a wholly preprogrammed repertoire of movements. A significant aspect of that development was that vision became a feedback sense, like the internal and contact senses.

The more a mammalian species gambols in youth, the clumsier the infant's initial movements. In some species only the essential business of finding the nipple and sucking on it is done efficiently, and even that improves with practice. Some of the earliest efforts at limb deployment may be so unco-ordinated, so experimental, that they can't be said to qualify as play, though they obviously constitute the first part of the learning process, the earliest explorations of motor possibilities. The need to practice and develop motor skills correlates with the possession of complicated musculatures, which require complex controls, and can be adapted to cope with varying circumstances. Species equipped with simpler muscular arrangements and purely hardwired control systems can be capable of impressive feats - catching fast-moving insects, for example. But these innate behaviours are stereotyped. In the species which play, individual animals tend to develop their own distinctive style of movement.

Object Play

For species with some manipulative ability suitable objects often become playthings. Young carnivores jump at falling leaves, and roll small stones about with their paws. Dolphins dart after strands of seaweed, and try tennis strokes with their beaks. Some birds play games with falling feathers. But it's among primate species that the scope for object play is greatest. It's doubtless from such play that the knowledge is gained which inspires the use of objects as tools, and among the larger-brained species makes possible the production of tools.

No-one has yet had the luck to observe the invention of a new tool by a primate in the wild. Quite a few suggestive experiments have been done, however, with apes and monkeys kept in varying degrees of captivity. Back in the thirties Wolfgang Kohler spent many years studying a group of chimpanzees kept in a large enclosure, with individual cages for sleeping. In one experiment he left bananas outside the cages, out of reach, and gave the animals two sticks which could be fitted together and turned into a tool long enough to pull the bananas within reach. The result was not impressive. Only the dominant male of the group worked out the potential of the sticks for obtaining the fruit

Some years later another student of animal behaviour, Paul Schiller, extended this experiment to see what happened if the animals were allowed to play with the two-part stick before confronting the problem of using it to rake in food. In the first part of the experiment nineteen of the twenty adult chimpanzees who were given the sticks to investigate discovered how to fit them together within fifteen minutes, and the majority of them did it almost straightaway. When there was a banana to be hooked into the cage, however, only some of those chimps who had so readily fitted the sticks together earlier thought of doing it again. Those that did, moreover, took far longer to get around to joining the sticks together in the presence of the banana than they had when there was no reward in sight. The animals experimented more creatively, Schiller concluded, when their attention was fully concentrated on the potential tool components, and there was no material reinforcement to distract them.

This parallels the experiment in which rats explored a foodless maze more effectively than one in which a material reward was provided. Once again, the conclusion seems to be that this more sophisticated sort of learning is done for its own sake. It doesn't require material incentives, so it must be rewarding in itself. But a concept must be firmly established by much practice if it's to come readily to mind when it's needed, at a time when attention is centred elsewhere.

Repetitive manipulative play in youth provides the perfect opportunity for realising the possibilities of objects as tools. Play is an activity only undertaken when an individual is well fed, and not particularly interested in more food, so a piece of food can be a plaything rather than a distraction. These are the sort of circumstances under which technological innovations for food-getting are most likely to arise. The innovators, therefore, tend to be older juveniles, who have developed good muscular control but are still free of adult responsibilities and thus have leisure and energy to spare.

This certainly seems to have been the case for one famous bit of innovation, among a colony of Japanese macaques on the island of Koshima. This group has been under continuous study for over fifty years now. Originally the monkeys inhabited a heavily wooded and mountainous terrain, so in order to tempt the monkeys to an area where they could more easily be observed the primatologists took to scattering wheat in a sandy clearing by the sea. It was a juvenile who discovered, almost certainly in the course of play, that if she threw the wheat in the water it would float, while the sand was washed off and sank. The habit of washing the wheat in this way was copied by her contemporaries, and by younger monkeys, and became established in the culture of the group.

Later on the scientists provided sweet potatoes too, and the same monkey initiated the practice of washing the dirt off them in the brook, or in the sea. The practice has continued, even when the sweet potatoes have already been washed. Since the washing is now always done in the sea the experimenters speculate that perhaps the monkeys appreciate a salty flavour. The habit of venturing into the sea is also something that was initiated by juveniles - the original adults refused even to get their feet wet. But as the colony grew, and artificial provisioning was reduced to avoid the danger of the island becoming seriously overcrowded, the monkeys took to catching fish for food. (The habit largely died out again a few years later, when the population pressure reduced).

In the Jigokendain Monkey Park, elsewhere in Japan, macaques also live in fairly natural conditions. Here primatologists have been able to study how good they are at developing tool use when given a little prompting. They placed a piece of apple in a transparent tube, and left a hooked stick touching the apple. When half a dozen monkeys had learnt to pull the apple out they left the stick further from the apple, then outside the tube, then yet further away. Next they tried leaving assorted sticks around, so that the monkey had to choose an appropriate one, and finally the monkeys were left to find natural sticks. Only a couple of individuals mastered the most demanding of these tasks, but one of them, called Tokei, added her own variations. On five occasions she was observed bringing a long stick to the pipe and biting it to a suitable length. And once she pulled up a shrub, plucked off the leaves, bit off the root, and used that.

Then, when no stick was available, Tokei began throwing a stone so that it pushed the apple to the other end of the pipe, carefully selecting an appropriate size. After a long interval two other monkeys copied this technique, and later yet a third. Other individuals didn't attempt it, but sometimes stole the fruit before the stone-thrower could get it. Tokei, who held a high rank in the hierarchy, threw with less momentum when other monkeys were around, and could usually chase a would-be thief off before they could reach into the tube to gain the prize. One lower-ranking operator learnt not to try for the fruit in the presence of others, and the other often lost what they had worked for.

When the stone-throwers became mothers an easier option emerged. The infants could crawl into the pipe, and if they brought out the piece of apple the mother might take it off them. Only Tokei, however, actively pushed her infant into the pipe.

Clearly it's significant that both these populations of macaques live in a more or less natural environment, where juveniles have plenty of opportunity for object play. The Jigokendian monkeys nevertheless took quite a while to work out how to get the food out of the transparent tube, and the Koshima troop had been eating grain for some time before the washing trick was thought of.

Since these experiments began it's been discovered that some monkey species use tools in the wild, mostly for obtaining valuable, protein-rich foods that would otherwise be inaccessible. Long-tailed macaques use stones or pieces of wood to crack nuts, snails and crustaceans, and some that live by the sea use stone hammers to chip oysters off the rocks at low tide. They mostly use chunky stones for cracking jobs, thinner ones for dealing with oysters. The more dextrous individuals use carefully angled blows.

Long-tailed macaques which frequent the Monkey Temple in Phra Prang Sam Yot are liberally fed by human visitors, and close encounters aren't discouraged. They've developed such tricks as stealing human hairs to floss their teeth with.

Bearded capuchins are also notable tool-users. In some groups they crack nuts by placing them on a rock worn slightly hollow by much use, and raise a heavy stone above the head before bringing it down on the nut, so that muscular force is greatly amplified by gravity. Like macaques they choose their tools quite carefully. Given a choice of two similar-looking stones of different weight, they selected the heavier one. In other capuchin societies the nut may be banged on wood, or fragments of bone are used to get at the kernel. One individual was observed employing a stone to shape a bit of bone to greater efficiency.

The enthusiasm for object manipulation in capuchins was underlined when some captive ones were given lumps of clay to play with, along with paint, feathers and leaves. Some of them spent up to half an hour moulding the clay, and decorating the results.

The great apes employ tools not only in pursuit of otherwise inaccessible food but for other purposes too. Orangutans, for instance, have been observed using leaves for cleaning the self, for protecting the hands when picking up spiny fruit, and for making interesting noises. Branches may be used for swatting bees while the nest is raided for honey. Chimpanzees make sponges out of chewed-up leaves to sop up water from holes in tree trunks. Gorillas have less use for tools, but prepare some of their food (such as nettles) in ways that require a similar sort of dexterity and intelligence.

Tool-using techniques, in both apes and monkeys, vary between different groups, so they are clearly acquired by imitation, not genetically determined. Individuals brought up in close association with humans spontaneously copy quite a few of our tool-using activities. A few apes, given paints to play with, have taken to creating pictures, and some have even given titles to the results - though to the human observer they generally seem to qualify as abstract rather than representational works.

(Photographs of monkeys using elementary tools can be found in the central colour section.)

Social Play

In social species play tends to be a social activity, when playmates are available. Human children often invent an imaginary companion if no real ones are around. In other social mammals play is most commonly seen when two or more juveniles get together. It's notable, moreover, that the longest childhoods are found in those species which live in groups. It can be deduced that play is important in developing social as well as physical and cognitive skills.

This conclusion was substantiated by Harry Harlow, who arranged for captive rhesus monkeys to grow up under various circumstances of social deprivation. He found that a monkey brought up with only its mother for company, and allowed no opportunity to interact with other young monkeys, fared very badly when, as an adult, it was finally introduced into a group. Such socially deprived individuals had no idea how to cope with their companions, looked miserable, and were sexual failures. Youngsters separated from their mothers but allowed regular sessions in the company of other young monkeys also showed serious handicaps as a result of the deprivation, but their social life was not so severely affected as that of the play-deprived group.

For social mammals a fellow member of the same group is probably, as Nick Humphrey emphasised, the most demanding and challenging stimulus there is. It's an inconsistent stimulus pattern, which tends to react differently in different moods and circumstances. Moreover, a social lifestyle requires a delicate balancing act between protecting one's own interests and maintaining the viability of the group. Social life, mammalian style, is probably only possible because a good deal of youth is spent studying for it. This is especially true of the complex social life practised by apes and monkeys.

Play also functions, it's been suggested, to promote social bonding. The group that plays together stays together, as Robert Fagan put it. No doubt play can serve this function because it's a pleasurable activity, and so inspires pleasant feelings (mostly) towards the playfellows. It helps to ensure that by the time the bonds that bind a youngster to its mother are loosened, the more flexible bonds that tie the individual to the whole group will have been established.

It must be significant that peccaries, which live in groups and show an exceptional amount of co-operative behaviour over food-sharing, defence and nursing, are enthusiastic about play. This is one of the few species where the adults play too. Each herd has a playground, an area about two metres in diameter where the regular activity wears away all the vegetation. The main bouts of play take place here, but brief bursts of a minute or two can happen almost anywhere in the course of the group's excursions.

The bonding process is conventionally regarded as a matter of establishing emotional ties, but it must also involve perceptual learning. A basic requirement for the success of a social unit is that the members should be able to recognise each other. This is a challenging task in many ways. The differences which distinguish individuals of the same species are generally quite modest. In addition, visual, aural and olfactory characteristics can all undergo gradual modification as the individual grows and ages. The urge to play prompts youngsters to pay close attention to their fellows, and probably helps to establish foundations for this particularly difficult perceptual task.

In some social mammals it's been demonstrated that group members can recognise each other not just as co-members but as individual characters with distinctive behaviour patterns. Play clearly provides much of the knowledge of character on which chimpanzee politics is founded. The playful interactions of juvenile social carnivores, meanwhile, must greatly facilitate co-operative hunting later. The time a species spends in play correlates not only with the sophistication of its motor activity but also with the degree of subtlety in its social arrangements.

A feature of social play is that there is a good deal of unforced role-swapping. One individual chases another, and then, without any ado, the chaser willingly becomes the chased. This experience of role-playing must be a good grounding for the changing nature of social life. A lively juvenile playmate may grow into a powerful male, work his way up to be leader of the pack, then be replaced and become an unimportant has-been. Some individuals may be friends most of the time but competitors occasionally. The role-swapping of play prepares the juvenile mammal for the fact that social relationships are not inherently fixed.

The more complex role-playing of human childhood - mummies and daddies, doctors and nurses, cowboys and Indians - provide a means of beginning to understand the elaborate roles of adulthood. Perhaps there is something of this in the play of other species too. Maybe young ungulates benefit from learning something of what it feels like to be chased, under circumstances where the physical sensation is not overlaid and dwarfed by fear.

It may not be too fanciful to see, in the changing roles that a social animal plays and in the role-swapping of juvenile play, a parallel with the business of technological innovation. Social animals must become accustomed to the idea that individuals play different social roles at different times, and sometimes switch roles quite suddenly. A mind that can cope with these re-assignments of status is perhaps prepared for the idea that a stone or a twig might also change its role.

Missing out on social play doesn't lead only to social handicap. Dorothy Einon found that rats which lack companions to play with when they are young show below-average intelligence as adults. They're extremely active, but excessively wary of novelties, and their performance on standard memory tests is below average. They're also very inflexible, taking much longer than a normal rat to discontinue or reverse a learned response. Einon concluded that play promotes versatility in adult behaviour, and improves an animal's ability to learn.

The definitive features of play were noted by Robert Fagan. It consists of behaviour which is undertaken out of its normal context, and which doesn't lead to its natural consummatory conclusion. Young animals chase each other without any purpose other than chasing, fight without trying to do any damage, mount without copulation; and dolls' tea-parties don't usually involve the actual consumption of food or drink. Play actions are often exaggerated in form, and undertaken with excessive energy. And the sequence may differ from the normal, functional performance, or include acts from different contexts.

Even without these guidelines we find it easy to recognise play in other species. It's made easy by the fact that the participants need to make their intentions plain. It's especially important that an invitation to a play fight be clearly distinguishable from a challenge to a serious one. As the capacity for play has evolved, therefore, there has also evolved a signalling system which indicates *this is play*. And in the species which sometimes play as adults the signals are well developed. A play-face, with mouth dropping open in something like a grin, has been well documented among various primates and carnivores. There's also, in quite a wide range of species, a jump-and-crouch-and-pause-with-head-on-one-side which constitutes an invitation to play. We instinctively use, and recognise, fairly similar means of indicating playful intentions, so it's not too hard to read the messages.

Play and Adaptability

Play, then, goes with modifiable motor patterns, practical ingenuity, and demanding social lifestyles. In species which possess one or more of these characteristics there's a period of juvenility during which the necessary practice can be carried out, the necessary ideas developed. In line with this, Konrad Lorenz noted that the species which go in for a lot of play are those which occupy varied types of habitat, and those which are not specialised in their diet but eat many types of food. Such species have larger brains than those which are more set in their ways and depend more heavily on innate behaviours.

It's now widely accepted that the correlation that applies to species holds good for individuals too. Both among humans and apes, the most creative individuals tend to be found among those who remain most playful in adulthood, and playful females seem to make the most successful mothers, educating their young most effectively. The urge to explore and to play is also the urge to investigate, to mess around, to see what can be done with something, and this curiosity is what drives art, science and technology. (The connection between play and art was noted by Herbert Spencer, in his *Principles of Psychology*, published in 1855.)

The link between the evolution of larger brains and the evolution of play behaviour is a two-way one. Play is something that happens when nothing more important is going on. Youngsters don't play when they are hungry, thirsty, tired or frightened. They play when all wants have been satisfied, and when mother is at hand to ensure that they are safe, or when they are within easy reach of the security of the den. The immediate requirements of survival always come before the investment in education. When times are hard - when food is scarce and all the energy has to be devoted to finding it, or when there are many dangers about - then the time given to play is reduced, or there is no play at all. At such times the youngsters lack either one or both of the two essential ingredients, a surplus of energy and a sense of confidence.

However, it has been observed in both humans and rats that if the young are deprived for a relatively short time of the usual chances to play, and are then provided with a suitable opportunity to make up for it, they play more exuberantly and for rather longer than usual. This suggests that there is a natural ration of play for each species, one that is perhaps dictated by the amount of energy that can be spared from the more basic requirements of surviving and growing.

If the bigger-brained species are the ones which need to do the greatest amount of this sort of learning they are also the ones which, thanks to that learning, are likely to get through the important business of the day, the consumption of food, with time to spare - time in which the adults can laze around and the youngsters can play. Moreover a sufficient surplus of calories is required to allow for the exuberant expenditure of energy which play represents. To obtain adequate nourishment herbivores have to spend a great deal of time chomping, so play for them is largely confined to the period when a high input of calories is being obtained quickly and easily, by suckling.

The same consideration explains why primates in captivity, who don't have to search for food, tend to be more inventive than those in the wild. They have almost unlimited time in which to investigate and play - if their environment provides adequate space, things to play with, and interaction with conspecifics.

The argument, then, is that play is the means by which an animal rehearses the behaviours for which it is equipped to be flexible and adaptable, those about which it learns to make predictions on which to base choices and plans. And conscious sensation is needed for such activities.

A species which explores its surroundings, we can infer, enjoys some degree of consciousness in at least one distance sense - probably vision or olfaction, or both - for finding its way about. It may also be supposed to have some degree of consciousness in at least one feedback sense, which provides the means of marking useful places on the map. A species which investigates the tactile properties of its surroundings can be deduced to have conscious tactile sensation too, and the means of linking it to conscious visual sensation. A species which indulges in motor play must be conscious of the information provided by muscle-stretch detectors and joint-position detectors. And in a species which lives in complex social groups the capacity for managing social relationships is developed through play, and interactions among group members involve conscious emotion.

This conclusion doesn't seem outrageous. It's certainly very hard to conceive of an animal without any conscious sensation indulging in play. A further argument is that play is generally neither a response to an environmental cue, nor an action set in train by an internal need such as hunger or thirst. The impetus for initiating any particular bit of play has to be an inspiration conjured out of the animal's brain. It implies the availability of some of the mechanisms that are necessary for making a plan.

Jerome Bruner, Alison Jolly and Kathy Sylva, in their preface to an excellent collection of papers on the subject of play, came to the same conclusion that Karl Groos had reached eighty years earlier: *play is the principal business of childhood.* It was the increasingly detailed observations of wild apes and monkeys, they suggested, which led to a realisation that the role of play during immaturity *seems to be more and more central as one moves up the living primate series from Old World monkeys through the Great Apes to Man - suggesting that in the evolution of primates, marked by an increase in the number of years of immaturity, the selection of a capacity for play during those years may have been crucial.* In other words the evolution of increasingly large brains among primates was accompanied not only by an increase in the number of years of immaturity, but also by natural selection operating on the capacity for play.

Conditioned learning links a certain sort of feedback to the action evoked by a certain sort of sensory stimulus pattern, via channels which are genetically ordained. Modern neuroscientists largely agree that sensory consciousness functions as a means of bringing a wider variety of information together. Discovering the correlations between more varied inputs involves experimentation, while establishing the reliability of the correlations and fixing a record of them requires repetition of the experimental behaviour. This is what play achieves.

Exploration and play offer a useful, though not necessarily comprehensive guide to where sensory consciousness is at work. Can we also guess where it is absent? I think there is one indicator available, at least. Genetically determined responses to genetically determined stimulus patterns are often insensitive to the size of the stimulus - as they can afford to be if, in the habitat which the species normally occupies, there is no scope for ambiguity and confusion. In the streams in which British sticklebacks evolved, for instance, any moving patch of red was likely to be another male stickleback, and there was no need to allow for the possibility of a passing Post Office van. Similarly, the mating system of some species of frog involves all the males gathering round a pond at mating time and croaking their loudest. When a female arrives she is attracted to the loudest vocalisation, which means that she is likely to choose a strong and fit partner. If there is nothing else around that sounds quite like the mating call the system by which the female registers the signal does not need to incorporate any control for scale.

A system like this does, however, leave scope for misapplication. Cuckoos can get other species to bring up their young because the small birds in whose nests they deposit an egg have hardwired responses to eggs and nestlings which include no firm specifications as to size, so they continue to feed the alien even when it grows bigger than them. Indeed, with genetically determined systems an oversized stimulus may evoke a particularly strong response. Once, on a beach in Goa, my companion playfully started chasing a tiny crab, about the size of my fingernail. It retreated at first, but then started to defend its territory and became extremely belligerent, waving its fighting claw in the air, making rushes towards our feet, and then standing on tip-toe for greater impressiveness and waving again, until we retreated. This potentially rather dangerous lack of a sense of proportion can safely be assumed, I think, to go with a lack of sensory consciousness in respect of the relevant behaviour.

Recapitulation

The ability to build a cognitive map is extremely valuable, and seems to be pretty widespread among vertebrates, and it's hard to imagine how mapping could be done without some degree of sensory consciousness.

We can guess that animals which put energy into exploring for its own sake have at least some degree of sensory consciousness.

Some species explore tactile aspects of their environment, as well as its layout, which implies a conscious tactile sense that can be linked to conscious visual information.

Many larger-brained vertebrates grow slowly into adulthood, and pass through a period of juvenility during which they devote time to play.

Motor play is common, providing a means by which an animal finds out what it can do, thereby putting itself in a position to choose and plan its actions.

It's also the means by which a repertoire of quite rough-hewn movements is transformed into an extensive range of finely controlled ones.

Motor play must entail a conscious sense of the feel of movement, and a conscious feedback sense to indicate when an action has fulfilled the plan, or, most importantly, how it failed.

It's noteworthy that the feedback here can be provided by vision or hearing, both distance senses.

Suitably endowed species also indulge in play with objects.

In social species the young play together and learn how to live together.

The length of the period of juvenility, the amount of time devoted to play, the size of the brain and the adaptability of the species are all correlated.

It's reasonable to deduce that the sort of play that the young of a species goes in for provides a strong clue as to the areas in which sensory consciousness must be present in that species.

On the other hand a tendency to respond to a drastically oversized stimulus suggests that sensory consciousness is not involved in guiding that type of action.

Chapter 4

From Earliest Animals to *Homo Sapiens* - a Quick Sketch

How could such complexity as the human brain be the outcome of an evolutionary process - merely the result of umpteen generations of sometimes sloppily reproduced and from time to time expanded strings of DNA, with the results ruthlessly pruned by natural selection? Some mutations are disastrous, or at least unhelpful, some neutral. Only a few, it's thought, are profitable enough to improve their possessors' chances of giving rise to many descendants, so that over generations these genes gradually spread through the population. And chance must play a big role in whether that actually happens. Could our consciousness, our capacity for thought and imagination, really be the consequence of this haphazard process? As more and more discoveries are made about genes, brains, and related matters the idea is beginning to look less implausible.

Several of the means by which genetic recipes expand have been identified. Most notably, a gene or a larger stretch of DNA may be duplicated, and in rare events even a complete genome. The 'spare copies' have then been able to mutate into new forms, and do it without upsetting existing mechanisms. (A couple of duplications in hominid ancestors seem to have played a part in increasing the size of the human brain.)

Quite a few of the proteins which form modern animals must date back to early stages of evolution, for the genes which carry instructions for them exist in very similar form in many widely separated species. Some, indeed, vary only moderately between mammals, fruitflies and yeast. The main transmitters by which nerve-cells communicate with each other are also very ancient. Major variations in one of the basic components of animal life must be more likely to be disastrous than otherwise, so it's not surprising if many of the recipes have undergone only minor changes. But once a protein has become established in one role it has often found new, additional uses elsewhere. The same gene may be activated at different times in different parts of a developing animal.

Older genes include those responsible for the signals that determine how a body is shaped. Those that pattern the front-to-rear axis of bilaterally symmetrical species seem to date back at least to the origin of such species, as do those which differentiate the upper and lower surfaces of the body, suggesting that all extant bilateral species arose from a common ancestor.

The greatest scope for variation, meanwhile, is obviously to be found among those genes that govern just how long the protein-recipe genes are active, determining how big an animal grows, how its components are proportioned, and how much is produced of the various substances that provide communication between cells. At least this is true in larger species. The bigger an animal is, the greater the chance that a mutation leading to a little further expansion of this or that part will not reduce viability. The tiny nematode worm, *Caenorhabditis elegans*, around 1mm. long, is composed, in its commoner form, of 959 cells, of which just over 300 are nerve-cells. In something of this size the addition of just one extra cell seems as likely to mess up the organisation as to prove an advantage. In an animal made up of millions or billions of cells, on the other hand, it should be possible to fit in quite a few extra without causing disruption. If legs grow a fraction longer than the standard, for instance, there may or may not be advantages in standing a little taller or being able to run a little faster; only when food is hard to come by will there be a significant disadvantage in having a little more flesh to feed.

What's true of whole organisms must also be true of brains. Once they get to a certain size the chances that a bit of expansion here or there will prove advantageous rather than otherwise begin to improve. Usually, though, it's the components that form later that do the most flamboyant expanding. The most essential areas, the ones crucial to the basic functions of an embryonic organism, necessarily develop first, and there is limited scope for variation there, except insofar as the size of the brain must be fitted to the size of the animal. With the essential systems in place there is room for evolution to play about with the later-developing parts of the brain. These tend to be the areas which are responsible for a species' sensory and behavioural specialities.

The possibilities are especially well demonstrated by fish brains. They are all built pretty much according to the same general vertebrate plan, but the relative size of the component structures can vary enormously, and it's clear that the expansion of a particular structure is what has made specialisations possible. The large-finned shark (right) devotes a great deal of space to processing olfactory input, while in that adept catcher of flies the trout (below) the visual part of the brain is very large. The olfactory area is dotted in these sketches, small circles indicate the visual area, and the non-olfactory part of the forebrain is hatched. Fish which scoop up gravel in their mouths and extract the particles of food hidden amongst it have a greatly expanded version of a hindbrain nucleus which correlates gustatory and tactile input. Other areas are expanded in other specialists.

The distinctive feature of mammalian brains is the conversion of part of the forebrain into a six-layered quilt, the neocortex. This is modestly sized in some species, much bigger in others, especially primates. In humans it spreads out over the rest of the brain, covering it almost entirely, with only the wrinkly structure known as the cerebellum peeping out from underneath. (*See the colour section.*)

In the developing embryo the main neurons or nerve-cells of the neocortex are formed along the walls of cavities filled with spinal fluid, known as the frontal ventricles. In the first stage the cells keep dividing and dividing, resulting in repeated doublings of the number. Then a second stage begins, when one of the new cells produced by each division sets off on a migration to its neocortical destination, while the other remains behind to divide again. The neocortical nerve-cells are organised in columns, and Pasko Rakic adduced evidence to suggest that the duration of the first set of divisions determines how many columns there will be, and the second how large they will grow. Even a brief postponement of the termination of either stage can obviously have a substantial effect.

An investigation of gene expression in humans, chimpanzees, orangutans and macaques in liver, blood cells and brain showed that in all three areas the pattern of expression varied according to the species. But the most prominent difference was found in the human brain. And tellingly, it's a gene thought to influence the migration of neurons in the developing cortex which has changed fastest since hominid ancestry diverged from that of chimpanzees.

A factor which must surely have made a significant contribution to the evolution of size and variability in the mammalian brain is that neocortical neurons are produced in notably larger quantities than are actually required. Those that successfully form connections and get integrated into circuits survive. Those that don't find useful employment fade away. Thus unused neurons don't become a burden, requiring nourishment while contributing nothing in return. This implies that there is no great pressure to suppress any tendency to produce increasing numbers of neurons. The danger that a large neocortex might prove unprofitably expensive is avoided. (It must be significant that the major part of the shaping process is carried out while the growing infant is being sustained by mother's milk, and is under her protection, a period in which an immature brain can be modified in relative safety.)

Furthermore, although the essential layout of the neocortex is genetically determined it is nevertheless adaptable. In congenitally blind or deaf people, for instance, the space which lacks its normal sensory input is taken over by another sense. It thus seems likely that when mammalian sensory equipment was expanded, or body-plans modified, there was a good chance that neocortex could cope to some extent with the additional data. The new input could compete for space with the long-established sorts, and perhaps make use of some of what would previously have been excess neurons, reducing the wastage rate. Certainly things work the other way around when a species adopts a lifestyle that makes a type of sensory equipment unnecessary. Blind mole-rats, which live in lightless underground tunnels, are born with vestigial eyes, and pathways to what should be the visual part of the cortex; but in the absence of input the visual pathways are lost, and axons from elsewhere take over the vacated area.

The neocortex doesn't grow in isolation. A good many other structures expand in proportion with it, at least to some degree, and especially those which are most closely linked with it. Minor variations in the relative sizes of some of these structures reflect specialised behaviours. An area particularly implicated in learning how to find a way around tends to be a little larger, for instance, in species which cache stores of food, and need to remember where they are. But mammals don't exhibit such flamboyant variations in subcortical brain patterning as are found among fishes. This might be because mammals haven't been around nearly as long as fishes - of which there are far more species. But it's perhaps more probable that significant alterations to the mammalian subcortical structures became superfluous once the superbly adaptable neocortex came into being.

In mammals sensory specialities are reflected in the amount of neocortex devoted to them. Primates have an extra large ration of visual cortex, dogs an extended olfactory cortex, and in rats and mice a large proportion of the whole deals with input from the whiskers. In the more elaborate brains an increasing amount of cortex combines information from more than one sense. A further trend is that as the neocortical area devoted to a sensory channel enlarges it becomes increasingly subdivided into specialised subsectors. Humans have more visual areas than monkeys, for instance. It's thought that there's a limit on how big one 'processing unit' can grow, beyond which the number of local interconnections becomes impossibly dense. Once that's reached it's more practical to convey the results to another subsector for the next round of processing. The same factor is probably responsible for the way the two cerebral hemispheres become increasingly specialised as brains grow larger.

Brain evolution isn't just a matter of varying the relative sizes of the components, of course. Many other factors are at work - modifications to the interconnections, to the distribution of the substances by which nerve-cells communicate, and subtle changes to the way one or other type of nerve-cell works. This sort of thing occurs more frequently than the more obvious changes to brain structure, and quite subtle variations can result in significant changes to behaviour.

As brains grow larger and more subtle, meanwhile, they begin to influence the selective pressures that will shape further evolution. A capacity for learning through classical reinforcements can do something towards this. For instance, an animal which has learnt to recognise a certain stimulus pattern as an indicator of a satisfactory form of food is likely to frequent the places where such rewards are found. And sometimes it must happen that this form of food is found particularly rewarding less because of its intrinsic qualities than because the individual is better adapted to digesting it than many of its conspecifics. Or it may be better adapted to obtaining it, and therefore doesn't have to work so hard or compete very heavily to feed well. Hanging out where it can best exercise its talent, the animal is likely to encounter breeding partners with a similar skill, which have learnt to pursue the same rewards. That makes a new adaptation of this type more likely to be passed on to future generations.

A cognitive map further improves an animal's ability to concentrate its activities in the habitat to which it is best suited. Again, this increases the odds that such adaptations will become established in future generations. Moreover if any additional modifications occur which suit their possessor further to that sort of environment they too are likely to be preserved. And as the individuals of a species begin to frequent two rather different habitats, each individual opting for the one to which it is better suited, and probably finding a mate with similar talents, the species may eventually split into two.

Sometimes this seems to happen quite rapidly. A species of lizard introduced to an island was found to have divided into two different lines within thirty six years. They differ in head shape, bite strength and digestive tract structure, and pursue different diets. Almost certainly there was already a fair amount of variation in the genes governing these traits, which could be channelled into two separate populations as individuals investigated the new terrain, learnt to exploit the potentials to which they were better suited, and tended to meet mates with similar preferences. Further adaptations would then become selectionworthy. But the same developments might equally be initiated by a genetic mutation that made a shift to a neighbouring terrain advantageous for those individuals in which it occurred.

The evolution of new species must also have been assisted a little where infants had a capacity for learning about the appearance or sound of the parent that looked after them, and for deriving reinforcement from those stimulus patterns. That would produce a situation where the deviation from the standard pattern which defined a potential mating partner would tend to be attractive to the offspring of the original deviant, rather than offputting, and if the characteristic survived into subsequent generations there would be a chance of its becoming more emphasised. Such variations in appearance or sound or odour might sometimes lead to new species.

Finally, the development of social lifestyles produced, for the mammals concerned, an additional selective pressure. Social living provides many models for imitation, and imitation begets cultures. An individual which is not very good at doing the things its fellows do is liable to be disadvantaged. A hamfisted chimpanzee which is less successful at fishing for ants than its colleagues may be a little less well nourished, and consequently a little more vulnerable to infections, a little less likely to breed successfully.

This sort of pressure will have become stronger as culture played an increasing role in influencing behaviour. It will have become particularly powerful as first hominids and then *homo* evolved. Being bad at maintaining an upright stance will have become a disadvantage when many of the other members of a group were walking upright, and thus able to get an earlier warning of danger. Lacking a talent for language would be a disadvantage if other group members were communicating more successfully. Where there was culture adaptations that fitted the individual to the culture will have become important. A material culture would have to be reasonably advanced before a group became capable of supporting a range of thoroughly varied talents.

The influence of behaviour on evolution was first remarked over a century ago, and named the Baldwin effect, after one of the first people to propose it. Discounted for quite a long time the idea has now come back into favour. It's arguable that the Baldwin effect may have made a significant contribution to some of the most notable evolutionary events. Every now and then a line of animals has gradually accumulated modifications to physical shape or functioning which produced the potential for a drastic change in behaviour - made it possible to spend longer and longer periods out of water, for instance, or to jump from a small height without suffering damage. The modifications which turned an aquatic animal into a terrestial one, or a reptile into a bird, will have been spread out over many generations. But the selection process will only have favoured them if the animals were making a habit of emerging from the water on to the land fairly regularly, or risking falling off high places. Only in animals equipped to learn that there were benefits to be gained from such ventures, so that they undertook them pretty regularly, will natural selection have been able to work.

It seems possible that one notable innovation of more recent times owes something to the Baldwin effect, and even to culture. The evolution of hands, with fingers and an opposable thumb, may well have been facilitated by a growing capacity for motor learning in primates. That should have meant that the possessor of digits slightly longer than those of its parents could work out what it could do with them. The primate's capacity for imitation, meanwhile, would have meant that once one individual in a group had discovered a trick that could be performed with fingers other individuals could benefit.

In short, bigger brains and longer infancies mean that the opportunity for combining new genetic mutations occurs less frequently, but the process becomes less random.

Evidence from inter-species comparisons

As the genomes of more and more species are sequenced analysis of the similarities and differences should make it possible to work out a pretty good idea of just how vertebrate brains expanded and elaborated. Several interesting discoveries have already been made, such as the fact that quite complex means of transferring messages between nerve-cells had been established before brains underwent any major expansion. Among the proteins involved in this job those that evolved more recently have a more variable pattern of distribution in the brain than the older ones. They include some that are particularly associated with learning. Comparisons of gene expression in humans and chimpanzees is proving particularly enlightening.

Genomic and proteomic studies of this sort are in their infancy, but much is being learnt from the study of developing embryos. This can be put together with paleontological evidence, though that is necessarily rather limited. Fossilised skulls say something about the brains that occupied them, but usually not a great deal, since early brains generally didn't fill the skull tightly enough to leave telltale impressions on its interior. Sensory organs, luckily, have often been preserved in recognisable form, and sensory organs carry some significant implications about the brains to which they report. Further deductions can be made from the appendages which were gradually added to the original, basic vertebrate bodyplan, appendages which would have required direction from the brain.

A great deal, however, must be inferred from comparisons among extant species. Parallel or convergent evolution does occur from time to time, but in general the arrangements we share with fish are a great deal older than those we share only with other tetrapods, and those in turn are older than those we share only with other mammals. Inconveniently, however, there are no extant species which might provide clues as to how some of the most interesting developments proceeded. When a revolutionary genetic innovation appears, bringing great success to its possessors and creating a new niche for them, the result is usually a succession of increasingly well-adapted variations, so that many assorted species evolve but none of the intermediate ones survive.

Vertebrates form a subsection of the chordate phylum. Chordates are defined as bilaterally symmetrical animals distinguished by, among other things, a stiffener along the back made of collagen and called a notochord, and a single nerve-cord running underneath it in a fluid-filled canal. Unfortunately, from the point of view of evolutionary studies, the vertebrate elaborations on the basic chordate plan have been so successful that very few non-vertebrate chordates remain in existence, and they are very elementary creatures with very basic nervous systems.

The surviving non-vertebrate chordates which probably give the best idea of what an ancestor of the vertebrates might have been like are the various lancelets, also known as *Amphioxus* or *Brancheostoma*. These slender little creatures, about five centimetres long, are constructed, like vertebrates, in a segmented fashion. They can swim reasonably well, but the adult leads a mostly sedentary life. It makes a burrow in sand, into which it retreats, backwards, if danger threatens. At other times it stands swaying in the current, its tail anchored in the burow.

The larvae are more active, living in open water, where they move to higher levels during the day and retreat downwards at night.

Amphioxus are filter-feeders. There is a large pharynx with slits along the side, lined with little hair-like projections called cilia, which wave water through the cavity. Food particles stick to the mucus-covered cilia and are transferred to the digestive tract, while the water flows on into another cavity and exits at the rear. Blood is pumped around the animal by a very simple network of blood vessels, and oxygenated as it passes through the tissues between the pharyngeal slits.

There is nothing that can be defined as a head (amphioxus means 'pointed at both ends'), and what might be called the brain is no thicker than the nerve-cord behind it, with no bumps or swellings. However, genes which in vertebrates are associated with the basic three divisions of the brain - forebrain, midbrain and hindbrain - are expressed in the appropriate regions, though the genes associated with the front of the forebrain are missing.

Sensory arrangements include a fibrous growth around the mouth, which keeps out large particles. Tactile and chemosensory receptors on it presumably serve to prompt the closure of the mouth when something noxious comes along. There are two eyes, one behind the other, pretty certainly one for a modest form of vision and the other, somewhat differently organised, for regulating an internal clock. A gravity-detecting arrangement has been identified. An indentation at the front of the head might be a nasal cavity, except that there is no evidence of any processing of olfactory input. And there are cells resembling those which in more complicated brains are involved in the release of hormones and the control of internal conditions. The animal is sensitive enough to touch, meanwhile, to indicate that the body surface is fairly well supplied with tactile receptors.

We cannot know how much evolutionary change amphioxus has undergone during the aeons in which the vertebrate line was evolving. It may have altered very little, having established a satisfactory niche at an early stage, or it may have lost some of the features possessed by its forebears in adapting to its modest way of life. But the fossil evidence suggests that it has been around in much the same form for a considerable amount of time. In any case it's pretty much - as Chico Marx is reputed to have said, when told the poker game he was planning to join was crooked - the only game in town. The only other contenders are the tunicates, or sea-squirts. Genetic evidence suggests they might have been closer to the ancestors of vertebrates than *Amphioxus*, but they are even simpler animals. The larvae are active, but in adulthood they absorb their notochord and settle down to a completely sessile life on the sea-bottom.

The next extant species up on the scale of complexity are the lamprey and the hagfish, or slime-eel, whose brains are tiny (a fraction of a percent of total body weight) but built pretty much on the usual vertebrate plan, apart from a few distinctive variations in the hagfish. So the chain of clues to brain evolution that can be obtained from comparative neuroanatomy is interrupted by a great chasm. And there is only limited fossil evidence to bridge the gap.

The fossil record

The first chordates probably appeared back in the Precambrian epoch, a time when all sorts of new and wonderful creatures were coming into existence, many of which didn't survive into the Cambrian. The Cambrian left us a few fossils of non-vertebrate chordate species, notably *Pikaiea gracilens*, a little thing that may have looked rather like *Amphioxus*, but is thought to have had a rather more mobile lifestyle.

A lot of the changes that would be required to turn an animal modelled on *Amphioxus* into something more like a vertebrate are immediately obvious. In addition to a spine - cartilaginous in some cases, bony in most - a vertebrate has a well-developed olfactory apparatus, and a pair of well-defined eyes. Vertebrates also tend to go in for more energetic methods of acquiring food, looking for it rather than waiting for it to arrive, grabbing it rather than just filtering it out of the water. Important talents that go hand-in-hand with these are an improved method of extracting the nutrients from food and pumping them around the body, and separate arrangements for acquiring oxygen.

Some of these advances may have been in place by around 530 million years ago, on the evidence of a small fossil animal from China, which was probably a chordate. *Haikouella* had both stomach and gills, a heart and a circulatory system, and a well developed musculature. It also had a swelling at the top of its central nerve-cord big enough to be termed a brain, and paired eyes, though without iris, lens or cornea. Tentacles around the mouth suggest it was probably still a filter feeder, but the flow of water through the body was powered by muscles rather than cilia. Genetic evidence suggests that some particularly interesting developments were occuring around the time of *Haikouella's* existence.

Various other fossils from the Cambrian period offer hints as to possible chordate evolution and vertebrate ancestry, but the next solidly informative evidence comes from the Ordovician. By 470 or 460 million years ago there were unambiguous nasal openings, and paired lateral fins. Dorsal and ventral fins came earlier, but it was adding such appendages on either side which will have made the big contribution to steering ability, and will also have required a considerable advance in the neuronal control system. Moreover by this time there were jaws, so that food could be firmly grabbed.

These innovations - paired fins, paired eyes, a developing olfactory sense, and jaws - are clearly closely linked. *Amphioxus* swims rather clumsily, its body forming deep curves and its head waving from side to side, so its single eye must cover quite a wide field of view as it moves. Presumably the larvae can steer towards the light quite effectively by ensuring that the illumination fades by about the same amount at the end of rightward and leftward sweeps. A certain degree of darkness may prompt a turn away - an arrangement which should also help the larva to avoid collisions.

Later chordates evolved narrower blocks of muscle, more elaborately shaped, which enabled them to swim more smoothly. Carving a straighter path through the water would make it worth having the wider view provided by two eyes. In addition, better distance senses for indicating where it was worth going will have become valuable as locomotion became more efficient, and as distance senses improved there were greater advantages to be reaped from better locomotion. Potential food, or the cues that suggested it, could be registered at greater distance, and approached rather faster.

With sensory information good enough to guide locomotion, the improved steering provided by lateral fins became a good investment. Jaws, meanwhile, made it possible to tear at vegetation, and seize chunks of nourishment instead of just sucking in tiny fragments. Thus sufficient energy could be obtained to power the greater amount of locomotion necessary to find larger supplies of food.

Stationary food can be sniffed, tasted and touched before it's seized. Grabbing something moving and clamping it in the mouth is only a profitable tactic if there's a sensory system which can establish in advance that it's pretty definitely food. It's a behaviour that's likely to have developed only after the evolution of a capacity for registering visual objects, or at least moving visual objects (as opposed to the general distribution of light and shade). Patches of photoreceptors must by then have evolved into proper eyes, recessed into indentations and provided with a light-focussing protective cover, perhaps even an adjustable lens. With contrast vision operating in conjunction with efficient locomotion, good steering, and jaws, the potential for active predation on moving prey opened up.

Jaws come close to qualifying as one of the defining characteristics of vertebrates, since only two groups of living vertebrates lack them, the hagfishes and the lampreys. Neither has lateral fins, supporting the idea that there was a mutual influence in the evolution of these features. Hagfishes, or slime-eels, are thought to be the descendants of a line which split off from other ancestral vertebrates very early on, since they differ from all other extant vertebrates in several ways. They live in deep ocean, making their living largely by rasping the flesh from dead animals, and are heavily dependent on olfaction for finding their food, although (like lampreys) they have only a single nasal opening.

Lampreys, much more visual animals, are parasites, using their circular mouths to suck blood from living fishes, though they spend much of their lives as filter-feeding larvae. Apart from the absence of jaws and lateral fins, and the single nasal opening, lampreys are built pretty much according to the same plan as other vertebrates, and so are their brains. The introduction of new evidence regularly sways opinion as to whether lampreys are more closely related to hagfishes or to the main vertebrate line, but they are obviously positioned somewhere between the two. (*See page ii of the colour insert for pictures.*)

A duplicated nasal opening must have evolved after lampreys split off from the main line of descent, for they have only one. Paired eyes had certainly been established by the Ordovician period, along with an effective olfactory system. Much of the evidence comes from a large group of fossil species known as ostracoderms (shell-skins). They derive their name from the fact that, having acquired the ability to incorporate minerals into living tissue, they wore the result on their backs, in plates, rather than using it for internal support. The ostracoderms thrived for about 100 million years, and multiplied into nearly 200 species. It had been thought that perhaps one of them converted exterior shell into interior bone, and was the ancestor of nearly all vertebrates. The theory was crushed, however, by the discovery of a fossilised fish with a well-developed internal skeleton which has been dated to 419 million years ago.

The traditional classification system puts a great divide between the jawless species and all the rest. The next big division is between bony fish and cartilaginous fish, such as sharks and rays. Fossil evidence which might shed light on when and how the split occurred is unfortunately minimal. One hypothesis is that the cartilaginous ancestors once had bone, but renounced it in favour of cartilage, which made for faster swimming. Intriguingly, some of the cartilaginous fish have remarkably large brains. In terms of brain-to-body ratio some are in the same ball-park as the smaller-brained birds and even mammals. Some of the largest brains belong to predatory sharks, which devote a large proportion of the space to analysing olfactory input, and can track their prey by smell from great distances.

Bony fish, in turn, have been subdivided into two main groups, of which by far the larger is the ray-finned fishes, or teleosts - some of which also have quite large brains. The smaller one, which in the modern world embraces lungfish and coelocanths, is defined as the lobe-finned fishes. It's to this group that the ancestor of the tetrapods belonged. Paired, fleshy, bilateral fins gradually grew longer and were re-morphed into limbs, developing joints and digits, while the bone structures which supported them and attached them to the spine grew stronger, as revealed by a series of fossils. The pectoral fins were transformed somewhat in advance of the pelvic ones. It's thought that the initial benefit of the modification lay in allowing the fish to move around on the bottom in very shallow waters, feeding on the vegetation or invertebrates found in this well-lit situation, with some species perhaps using their front fin-legs to uncover buried food.

No doubt some of them sometimes stuck their heads out of water to grab food from just beyond the water-edge, as some modern fish do. And no doubt the dashes on to dry land eventually grew longer, creating selective pressure for an ability to extract oxygen from air as well as water. The evidence so far available suggests that legs were quite well developed before there were any lengthy excursions. It was around 380 million years ago that vertebrates first became properly established on dry land. The illustration shows, as fossil skeletons with bodies reconstructed around them, some of the stages of the transformation from fish to land animal. It's taken from *Gaining Ground: the Origin and Evolution of Tetrapods* by Jennifer Clack, published by Indiana University Press.

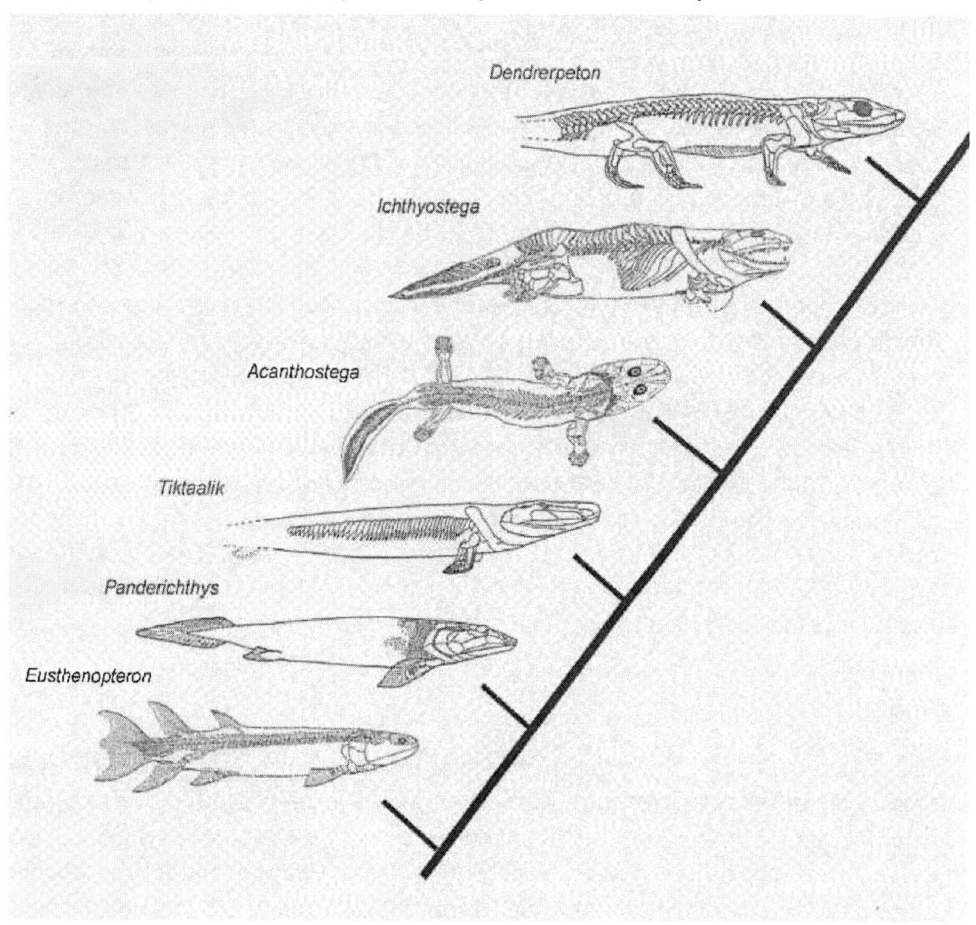

Various modifications to sensory systems became necessary in the new environment. Olfaction had to be adapted to deal with molecules carried by air rather than water, and the auditory system to accomodate the fact that soundwaves are less effectively conveyed by air. Moving on to land must also have exerted some pressure on the neuronal system governing navigation, since the relationship with gravity becomes much more forceful out of water - although large feet, a low centre of gravity and a slowish rate of locomotion no doubt limited the dangers of stepping on inadequate support.

Another consequence must have been that it was no longer necessary to be constantly counteracting the effects of waves and currents. It became possible to relax completely, without needing to be glued to an immobile surface or jammed into a small space. Motor systems could be turned off more completely between bouts of activity, creating an increased potential for sleep.

Early tetrapods were doubtless still very dependent on their ancient aquatic environment, needing to revisit it regularly to keep their skins moist and to maintain their internal fluid balance at the correct level. And like most modern amphibians they must have needed to lay their eggs in water. The evolutionary innovation which defined the next big step, the appearance of reptiles, was the development of a covering for the egg which enabled its water content to be maintained in air. For reptiles and birds this meant a hard shell, lined with a membranous sac, termed the amnion. In mammals the shell became unnecessary and the amnion was elaborated. Reptiles, birds and mammals are thus classified as amniotes, amphibians and fish as anamniotes. Internal fertilisation of eggs and even internal maturation of embryos seem to have evolved several times, in various fish and in assorted amphibians, but the amniotic sac was new.

Not only the eggs but also the animals themselves acquired a new, tougher, watertight covering, which meant they were not obliged to stay close to water, or at least to damp places. The freedom to roam made reptiles extremely successful, as witness the numerous species of dinosaur, and their long survival. Some reptiles grew very large, and brains expanded too in many cases. The biggest brains are found among the crocodilians. The ancestry of both crocodilians and birds can be traced back to the dinosaurs.

Mammals

Mammals have been thought to have evolved from a branch of the reptile family that otherwise died out. The discovery of a fossil skull that looks mammalian, and dates to about ten million years before any distinctively reptilian skulls, cast some doubt on this assumption, hinting that mammals might have evolved more directly from an amphibian ancestor. But extant amphibians offer little indication of what such an ancestor might have been like.

A crucial development in the evolution of mammals was the development of a new sort of body-covering, hair. The hairs may not have been very thickly distributed in the first place, and a likely theory is that their first function lay in extending the sense of touch. In modern mammals each hair has a nerve-cell wound around its base, which is activated when the hair is deflected, so that actual contact with the skin is not needed to reveal the presence of something close by. This extension to the tactile system may have been one of the factors that gave early mammals the confidence to remain active longer as twilight fell.

As the hair grew denser it provided useful insulation, which made it practical to stay active yet later, even in cooler climates. It also opened the way to another development. There may already have been a certain degree of warm-bloodedness (as in some reptiles), and with insulation it became practicable to improve on the system. Ultimately the means of generating heat internally became efficient enough to maintain a constant temperature. With insulation and improved heat production the hairy animals could establish new niches, temporal or spatial, where there was as yet little competition.

The evolution of a new, extra-sensitive type of photoreceptor in the eye, responsive to a wide range of wavelengths of light, contributed to the ability to exploit these new niches. Mammals also tend to have rather more forward-facing eyes than most reptiles do, so that the overlap between the two visual fields is larger. Within that overlap the chance of a photon encountering a photoreceptor is doubled, and it's thought that the arrangement evolved as an adaptation to operating in low light levels. It was probably for the same reason that mammals ceased to use their eyes independently, as birds and reptiles do, but are obliged to focus both in more or less the same direction.

Even when modified for low light levels, however, vision still can't provide the sort of guidance it does in daylight, and is useless in complete darkness. The extended tactile system, meanwhile, is only helpful at very close quarters. Part of the deficit was made up by new olfactory receptors, and a substantial expansion of the olfactory processing system. There were also refinements to the structure of the ear, leading to increased amplification of sounds and a greater sensitivity to high frequencies.

The early mammals adopted a variety of lifestyles. Some became arboreal, while one of the bigger species seems to have preyed on small juvenile dinosaurs. A large proportion of these early mammalian species failed to survive, however. It wasn't until after the extinction of the dinosaurs that mammals took off in a big way, and a wide range of sizes and specialisations developed. Even today, however, over half of all mammalian species are largely nocturnal, and a considerable majority of the smaller ones.

Maintaining a steady or fairly steady temperature made it possible to look for food when other animals couldn't. But it also made a great deal more food necessary, in order to fuel the central heating system. Reptiles need many fewer calories than mammals, and some species can make one good meal last for days, or even months. Mammals need to eat more regularly (apart from special adaptations for hibernation). Eating enough for heating purposes involved enlarging the gut, while the expanding brain required extra energy too. Metabolic systems became efficient enough to allow mammals to acquire larger muscles as well, which must have helped in the acquisition of all the extra fuel, but in also used up some of it. An increase in atmospheric oxygen levels played an important part in making such metabolisms possible in the first place, and supported larger body sizes.

Thanks to their efficiency in the acquisition and processing of calories mammals could afford to develop their defining characteristic, feeding the young with a special, very rich form of nourishment provided by the mother. The species in which milk production materialised must have been one that was already giving its infants a certain amount of care, and maintaining contact with them. But the innovation created the potential for young to be born without the capacity to do anything more than latch on to a nipple, and thus opened the way for brains that did some of their developing under the influence of the environment, without having to engage in the distraction of food-finding.

Early mammals pretty certainly laid eggs, like modern monotremes such as the duck-billed platypus. It's thought that milk first functioned as a means of keeping the eggs damp, oozing from an abdominal patch as it still does in monotremes. Gradually the supply must have come to last beyond the hatching, and to provide an increasing amount of nourishment. This seems to have allowed eggs to decrease in size, provisioning the developing embryo less thoroughly than do bird and reptile eggs, since the deficiency could be made up rapidly after hatching. The next step must have been the replacement of the egg-yolk by a primitive placenta, so that the development of the embryo could be supported internally for a short while, as in modern marsupials, such as kangaroos. From this evolved a fullscale placenta, providing the sort of sustenance that allowed for more extended development within the womb, as in our sort of mammal, the most numerous kind, termed eutherian.

In birds and reptiles three genes govern the transfer of essential nutrients from the maternal liver to the developing egg-yolk. They have all been lost in mammals, apart from one which remains functional in monotremes. A genetic analysis suggests that in the line leading to eutherian mammals one disappeared around 170 million years ago, a second around 140 million years ago and the third 70-90 million years ago.   The successive innovations in maternal nourishment of developing young must obviously have occurred before these dates.

Neither fur nor nipples are much preserved in the fossil record, but the bone structures of the ear are helpful in distinguishing mammals from non-mammals. Teeth also offer helpful evidence. Reptiles are born with teeth, and they are replaced regularly throughout life. Mammals are born without teeth, usually acquire a first temporary set, and these are replaced over a fairly short period with a permanent set. Growth is very rapid in the early, toothless period, when the nourishment is being supplied by mother. So it's often possible to identify a skull or fragments of a skull as having belonged to a mammal, even if the rest of the skeleton is missing. When remnants of both skull and postcranial skeleton are found the ratio of brain size to body size is taken as a defining characteristic. Fossils taken to be mammalian have been dated to more than 210 million years ago.

It seems likely that maternal provision of milk to infants could only have evolved successfully in an animal that had already acquired a certain degree of brainpower and cognitive development. If the genes for milk production are to be passed on the benefit has to go to the infants which inherit them, so mothers have to be equipped with sensory systems adequate for identifying the correct infants. Then a further increase in cognitive powers is worthwhile. An infant that can recognise its mother can keep company with her after it has become mobile, and benefit from imitating her actions and reactions. Once mothers and infants are tied together by this new means of nurture there are all sorts of potential uses for larger brains.

Nevertheless, although the largest vertebrate brains (as measured in proportion to body weight) are found among mammals, at the lower end of the scale there is a considerable overlap with cartilaginous fishes, birds and reptiles, and some with the ray-finned fishes.

The immediate forebears of *Homo sapiens*

Primates were in existence in the Palaeocene epoch, and becoming widespread, and by about 55 million years ago they were looking something like modern primates. Fossils with ape-like features are found at sites dated to over 35 million years ago. Genetic and fossil evidence suggests that the last ancestor shared by humans, chimpanzees and bonobos probably lived somewhere between six and ten million years ago. But interpreting the fossil record to uncover how hominids descended from an ape-like ancestor is a matter of ongoing guesswork. New fossils are discovered regularly, frequently shaking up established ideas, and it's generally agreed that yet more are needed to resolve the controversies. Probably, too, the numerous species names which have been awarded to the fossils so far unearthed will need to be reduced and rationalised.

One thing which seems clear is that the ape species which represents the first ancestor we don't share with modern apes gave rise to a whole clutch of species, all but one of which eventually went extinct. Some of the fossils that date back seven, six or five million years ago offer clues which suggest some of these species could walk upright when it suited them, though still well adapted for climbing trees. This fits well with the partially wooded landscapes in which these species seem to have lived.

The Australopithecines, of which there were several species, had begun to emerge by five million years ago, and were around for at least three million years. They were generally between about one and one and a half metres high, and their cranial capacity ranged from about 400 to 550 cm.[3] Some at least of the Australopithecines, on the evidence of leg bones and joints, stood upright rather more than apes usually do, but retained long arms, curved fingers, and other adaptations for tree climbing. Relevant foot-bones, unfortunately, are often missing. But footprints preserved from around 3.6 million years ago at Laetoli in Tanzania, indicate that at least one of these hominids was accustomed to walk upright. (*See page ii of the colour insert.*)

Another group of hominid species is in evidence in the latter part of this period, usually classified nowadays as *Paranthropus*. They had robust bones, powerful jaws for crushing tough food, and ridges along the central line of the skull to which the jaw muscles were attached.

Species that are classified as *Homo* emerged around two and a half million years ago, in East Africa, according to the available evidence, and spread southwards from there during a warm period. They developed fully upright locomotion, and the long arms which were so useful for climbing trees grew shorter. Both these developments must relate to the fact that, the forests having shrunk, they were colonising open grassland. Here the longer view provided by the upright stance would have been reassuring. It's been suggested that another factor promoting verticality might have been that it enabled them to stay active longer in the heat of the day, with only the top of the head fully exposed to the sun (but protected by hair), while the erect body caught any breeze that was going. This might have allowed them sometimes to scavenge meat from the kills of carnivores, left unguarded during the siesta hours.

The really distinctive feature of *Homo*, of course, is that foreheads grew higher and wider, and the brain inside the expanded skull grew larger. Perhaps this correlates with the extended uses that could be found for hands when they no longer had to double as feet. Tools of some sort were already in use, as among modern apes and monkeys. Animal bones bearing marks which show that the flesh had been scraped off have been dated to nearly 3.4 million years ago, and some basic stone tools were discovered at an archaeological site dating to 3.3 million years ago. From about 2.6 million years ago a variety of tools are found in association with the fossils of both early *Homo* and late Australopithecines, tools which, though simple, are more advanced than anything an ape has so far been found using, implying somewhat greater manual dexterity and perhaps a more creative imagination.

An early form of *Homo* was named *habilis*, on the assumption that making tools was a definitive trait, a later one *erectus,* on the assumption that walking upright was. The range of tools produced expanded to include hammers, anvils, scrapers, blades and choppers, some of which show signs of the varied sorts of wear that would have been caused by use for woodworking, butchery, and cutting plants. (Analysing the different forms of microwear resulting from different types of use is one of the techniques of modern archaeology.) Some of these Early Stone Age products will no doubt have been used to make other tools in less durable materials, such as wood, bone and horn, and it's likely that these included digging sticks for getting at edible roots.

The evidence of food processing is particularly significant. Root vegetables and raw meat need a lot of chewing, as Zink and Lieberman have pointed out. Pounding or slicing reduces the effort required considerably, which will have enabled our ancestors to eat more of these high quality foods while spending less energy on chewing and on digestion. Thus they will have been able to fuel their brains more efficiently, so that larger brains became practical. Significantly, *Homo erectus* is characterised by teeth which are smaller than those of other hominids, as well as by a larger brain.

The invention of cooking contributed to these trends. By 1.6 million years ago *Homo erectus* was taking meat back to the campsite for butchering, and teeth, jaws and gut had all shrunk, while the brain had expanded notably. Unfortunately it's difficult to be sure that occasional traces of ash and charred bones represent evidence of cooking fires, rather than natural events, but it seems more likely than not. Certainly ash and burnt bone fragments have been found in million-year-old sediments.

Homo erectus seems to have been a fairly successful hunter well before this, aided by some significant modifications to the shoulder that enabled more powerful throwing. The hand, too, had been gradually modified to provide a more flexible grip, and by around 1.76 million years ago the stone tools were becoming more elegant, among them large bifacial handaxes and cleavers. This new style is known as the Acheulian, after St. Acheul in France where such tools were first discovered, but similar artefacts are found throughout Africa and in much of Asia and Europe. A notable advance in fine manual control is certainly revealed, and perhaps we may deduce a dawning aesthetic sensibility. Cranial capacity was still increasing, and ranged from 800 to 1400cm.^{3} over some million and a half years.

The larger brains must have helped their owners to migrate into new environments, by way of coping with the major climatic changes that occurred over this period. Several species of *Australopithecus* and *Homo* were successful enough to expand into Europe and Asia. The most widely accepted theory at the time of writing is that *Homo sapiens* also first came into being in Africa, before similarly moving outwards. And new, improved variations twice appeared in Africa, and expanded in the same way, absorbing or replacing previous populations. The current version is termed modern *Homo sapiens*, or *Homo sapiens sapiens*. (Cousin species also emerged – Neanderthals, Denisovans and *Homo Floriensis*.)

Around 300,000 years ago evidence of major advances begins to emerge, defining the beginning of the Middle Palaeolithic period, or middle Stone Age. There are indications of more sophisticated methods of splitting stones, more small tools, and the occasional point still bearing traces of mastic, suggesting that composite tools were being made. There's the odd hint of activity with aesthetic rather than practical purpose, such as the remains of blocks of pigment found in a cave in South Africa. There's also definite evidence of the controlled use of fire. These developments are associated with archaic forms of *Homo sapiens*, which sported modern-sized brains with bodies rather more robust than today's.

Larger brains capable of obtaining food more efficiently, and processing it with simple tools, must at some point have worked out the advantages of saving any spare for consumption in leaner times, making for a more consistent supply of nourishment. The cognitive developments that made co-operative hunting and gathering possible as a standard practice are also likely to have been accompanied by food-sharing, to the benefit of pregnant and lactating females and dependent offspring. The more reliable supply of food, and especially the provision to growing brains, made it practicable for brains to expand further.

The Late Palaeolithic is defined as 50,000 to 10,000 years ago, and was inhabited by anatomically modern humans. In sites from this period bows and arrows have been found, spear throwers and harpoons, and the earliest evidence of building, in the shape of post holes, with hearths and tool remains sometimes found within the outline. There was painting, carving, engraving, and personal adornment, using all sorts of materials. Some of the carvings, along with needles and a faint trace of woven textiles imprinted on fired clay, indicate that clothes were being worn. Burials were becoming more common, sometimes with tools or jewellery in the grave. The number of sites found suggests that populations were growing denser. This latest form of *Homo* spread not only through Europe and Asia and as far as Australia but also into the Americas.

Language

Since modern apes can cope with words the hominin species which had slightly larger brains and made somewhat more complex tools could certainly have done so too, if the inspiration occurred. And it seems likely that words were first invented somewhere around the time of, or even before, the first stone tools that show evidence of serious modification of the raw material - the mental tool perhaps preparing the way for the material one. Though words would have been put together in only the most elementary way for a long time, they can nevertheless be supposed to have done a great deal to drive the expansion of the hominid brain. They could only have been created where there was already an appreciation of the different perspectives of other people, but the new means of communication would help to draw attention to these differences. Tuition of the young probably came to be more widely practised as a result. Words would have made it possible to indicate important but not necessarily obvious factors in tasks such as toolmaking and hunting, and adding words to a practical demonstration would tend to make the important points more memorable. Being able to pass on practical ideas more easily would have made it worthwhile to evolve larger brains for conceiving new ones. Equally important, proto-language will have smoothed social relations, and made it easier for groups which possessed it to unite against those that didn't. If they multiplied more successfully than their neighbours the talking animals must also have found it rather easier to spread into new habitats, and adapt their ways to new conditions.

The syntax of language seems likely to have evolved in line with the syntax of action - especially as both seem to depend on much the same bit of brain. As the means of putting together longer sequences of action for the manufacture of more complicated tools evolved, so longer sequences of words could no doubt similarly be constructed. Where there were manufacturing tasks with subroutines there could have been the occasional sentence with a subsidiary clause. By the time huts were being built, and especially huts large enough to require the co-operation of several individuals in performing an ordered sequence of operations, words may have been strung together in a similarly ordered fashion.

The driving force behind the brain expansion that characterises the evolutionary development from an Australopithecine ancestor to modern *Homo sapiens* must have been an increasing urge to use the brain, leading to more exploration, more experimentation, more play. The willingness to indulge even in adulthood in playful behaviour which brought no obvious profit is likely to have been there already in the last common ancestor of humans and chimpanzees, judging by the behaviour of happy, well-fed apes in captivity. In the human line it brought discoveries and inventions that led to more efficient behaviours and reduced the amount of time and energy needed to find nourishment. This created scope for dreaming up further technological innovations, and for activities which improved group cohesion. Time and energy could be used for shaping more aesthetically satisfying tools, for educating the young, for gossip, which gradually expanded into group history and myth, and for thinking up yet more interesting ideas, such as carving little figures, or painting on cave walls, or decorating the self. The more efficient *Homo* became at obtaining food, the more time there was for entertainment of various sorts. One form of entertainment was thinking of ways to improve efficiency further.

Technological advances - especially the invention of agriculture - combined with the potentials that grew from language to create scope for ever larger groupings. Larger groupings provided increasing possibilities for ideas to encounter other ideas and breed with them. These processes obviously developed exponentially, very slowly at first, but ultimately leading to the manic rates of technological innovation which rule today in the more advanced societies.

Recapitulation

Some of the means by which genetic recipes are expanded and adapted are becoming clear.

As animals increased in size and complexity there was scope for new sorts of evolutionary change.

Increasing learning ability probably did much to support natural selection, thus compensating for the concomitant longer lifetimes that reduced the opportunities for genetic variation.

We can learn something from fossil skulls, and make some deductions from the sort of sensory apparatus and locomotory appendages that have been preserved in association with them.

Much can also be learnt from comparing the brains of extant vertebrates, though unfortunately there is no evidence covering some of the most interesting bits of the evolutionary process, such as the transition from simple chordates to vertebrates.

Genes and fossils combine to tell something about how and when mammals evolved.

Similar evidence as to the evolution of primates and especially *Homo* is building up at such a rate that conclusions are frequently being revised.

The chronology revisited

More than 540 million years ago (Precambrian): first chordates
~ 530 mya(Cambrian): *Haikouella* – with stomach, gills, heart, muscles, paired eyes
~450 mya (Ordovician): early land plants
380 - 370 mya (Devonian): early amphibians
?340 mya (Carboniferous): early reptiles
By 240 mya (Triassic): mammals and birds emerging
Before 170 mya: mammals began producing milk for their young
Before 70 – 90 mya: eutherian mammals emerged
Before 55 mya: first primates
Before 35 mya: first apes
6-10 mya: the last common ancestor of humans, chimpanzees and bonobos
4 – 2 mya: *Australopithecus*
3.39 mya. simple tools probably being used
3.6 mya: footprints of upright-walking *hominid* at Laetoli
2.6 mya: a fair variety of tools coming into use
2 mya: *Homo erectus* had evolved shoulder modifications that made for more powerful throwing, and was probably an effective hunter.
1.76 mya: The hand had been modified for a more flexible grip and more stylish stone tools were being made.
By 1.6 mya there's evidence of meat being taken back to a campsite for butchering.
~ 300,000 years ago: the middle Palaoelithic period begins, with early forms of *Homo sapiens* making more sophisticated tools, and showing signs of interest in aesthetic activities.

Chapter 5

An Introduction to Neurons and Nervous Systems

In Elizabethan times the procession of the stars and planets around the heavens was conceived of as a stately dance, illustrating divine order. In a parallel metaphor I like to see the succession of brief molecular gyrations which constitute the processes of life as a cross between square dance and hot jive, performed on a minute scale and at unimaginable speed. Thanks to the tiny electrical charges on their surfaces one molecule can grab another, contort itself to twist around with it, and let go, perhaps throwing it to another partner. Some of the steps are repeated thousands of times per second.

Interactions like this are the means by which a cell imports what it needs to fuel itself and to function, and throws out its rubbish. The cell is walled by a membrane composed of molecules of a fatty material called lipid, and the various pumps which shift stuff in and out of the cell are dotted about among them. Commonly there is a sort of fishing hook protruding on one side of the membrane, and when this comes into contact with the type of molecule in which the pump specialises a tiny electrical bond is formed. The act of forming the bond makes the protein pump change its shape, so that the captured prize is drawn into the channel. This movement in turn breaks the bonds that caused it, and the captive is thrown out on the other side of the membrane, while the channel and the fishing line resume their original form, ready to repeat the action.

It happened that the particular selection of molecular materials involved in building a cell gave the very earliest (single-celled) organisms a negative charge in relation to the seawater in which they evolved. This situation created the potential for signalling systems that allowed them to register certain sorts of encounter with the outside world, and to react. Key players in the signalling system are ions - atoms which have one or two extra electrons in their outer shell, giving them a negative charge, or lack one or two, giving them a positive charge. Ions play many roles in the management of cells, and the cell's pumps operate to maintain the various sorts at consistent levels, higher than the concentration found in the outside medium in some cases, lower in others.

In addition to the pumps, cell membranes are furnished with channels that function as gates. Some types can be opened by mechanical force, others by an encounter with a particular type of molecule that operates as a key. The brief opening allows in a small flood of ions, of one of the varieties found in higher numbers outside the cell. The width of the channel and the electrical charges along its lining determine which sort. The ions have a brief and focussed effect before they are pumped out again, an effect which has been harnessed to produce many varied consequences.

A single-celled inhabitant of fresh water, the paramecium, provides an example of how such signals work. It's covered all over with tiny hair-like structures called cilia, which it waves in unison to row itself about. If the paramecium bumps into something the impact on its front end causes gates in its membrane to deform, letting in calcium ions, which have lost two electrons from their outer shell and thus have a strong positive charge. This reverses the usual negative charge on the inside of the membrane, causing the microscopic oars to go into reverse, so that the creature backs away from whatever it has bumped into. Conversely, an impact at the rear opens gates which let potassium ions out. Their departure increases the negative charge along the inside of the membrane, which makes the oars beat more strongly, and the paramecium accelerates away.

While this sort of gate is opened by brute force others are persuaded by subtler means. A chemical key engages with a protrusion on one side of the channel and causes a change in its conformation, allowing ions to pass through. Chemical gate-openers are good for identifying food, and in a single-celled creature for admitting it. Other sorts may respond to substances emitted by conspecifics, which thus act as signals. Many species of bacteria, for instance, communicate well enough to organise themselves in sheets spread out over a substrate, allowing each individual decent access to whatever nourishment is going. In some species a substance released by individuals which have encountered food functions as a signal that arouses nearby dormant individuals.

A signalling substance and a receiving mechanism that reacts to it are unlikely to have evolved simultaneously, so it seems a good bet that the first substances to provide a means of communication between individual unicellular organisms started out as something else - probably as waste products. Some such excretion must have been transformed into a signal by the evolution of a receptor. The multiplication of receptors and signals was obviously a crucial contribution to the evolution of more complex life-forms. Communication between individual members of a single-celled species living in co-operative assemblies must have been well established before true multicellular organisms could come into being - organisms, that is, where component cells permanently suppress a part of the genetic instructions they carry, and function as specialists. Each cell, in such an organism, must know what job it's supposed to do, and each specialist must be sited in the right place. A number of the molecules which function as signals governing the development of multi-celled animals are also found in modern, single-celled creatures, helping to confirm that it was the evolution of new receptors for existing substances that made multicellularity possible.

Organising the functioning of multicellular organisms requires further intercellular communication. One method is used by plants and fungi as well as animals, and must be the most ancient. Hormones are circulated around the whole organism, but the message is registered only by cells which are equipped with the appropriate receptor. The animal kingdom evolved another, faster method of conveying signals, in the shape of cells especially dedicated to delivering messages to specific targets - neurons, the cells of which nervous systems are composed.

The basis of neuronal function lies in the fact that these cells exaggerate the negative electrical charge which is common to all. They do this by means of pumps in the membrane which provide tight control over four types of ion, three of them positively charged - calcium, sodium and potassium - and one negative, chloride. They pump in potassium, keeping it at a higher level than in the intercellular medium, and pump out sodium and calcium, which are consequently more densely distributed outside. One pump swaps sodium ions for potassium, ejecting three of the former for every two of the slightly larger potassium ions which are imported, a deal which, despite the fact that both these ions are positive, helps to swing the overall balance in the negative direction. The result is that neurons in a 'resting' state carry a charge of between -75 and -40 millivolts in relation to the slightly positively charged medium that fills the space between cells. Since the negatively charged molecules inside the cell are attracted towards the positive charge outside they tend to end up clinging to the membrane, and the negative charge is most concentrated there.

A typical form of neuron or nerve-cell is shown on the left. A whole bunch of delicate fibres known as dendrites branches off from the cell body (where the nucleus containing the genetic material is situated) and the arrangements for receiving signals are found on these. A single fibre called the axon, branching towards the end, passes on messages to other neurons.

The connections through which signals are received and despatched are called synapses. At the synapse the two cells are separated by a minute gap. On the axon side a chemical known as a neurotransmitter is held in little vesicles. When a sufficiently powerful signal reaches this point some of the vesicles fuse with the membrane and release transmitter into the gap. On the receiving side there is a cluster of gates in the membrane which are opened by the transmitter, gates which admit sodium ions and in some cases calcium ions as well. These positively charged ions sweep up the inner side of the membrane, attracted to the negatively charged molecules which are nestling there and cancelling out their effect. If the sodium current is strong enough it reaches a place where there are more channels through the membrane, of a different sort. These are opened not by a chemical transmitter but by the depolarisation of the membrane (the switch from negative to positive charge). They let in more ions, so that the current travels on further. A diagram of a synapse and a micrograph are on page iv of the colour section.

The more complex the neuron's web of dendrites the more synaptic inputs are likely to be necessary to produce a current which reaches as far as the cell body. There the membrane is particularly well supplied with voltage-gated channels, as is that at the top of the axon. If the wave of depolarisation gets that far it can continue down to the output synapses.

There the depolarisation causes calcium gates to open, and if the influx of calcium ions is strong enough a release of transmitter is triggered, sending a signal on to the next neurons in line. The wave of depolarisation which sweeps down the axon and may result in the release of transmitter is known as an action potential. In some of the more complex neurons a wave of depolarisation strong enough to reach the axon-end and qualify as an action potential can also be initiated at certain points along the major dendritic branches. The initial dendritic excitations which may or may not prove to add up to a current which reaches the axon-end are termed excitatory post-synaptic potentials.

During the brief course of the positive current through the neuron, voltage-triggered potassium gates have opened and allowed potassium ions to rush out. But these channels close again very rapidly, and the sodium/potassium exchange pumps work hard to re-establish the previous balance of ions in the cell. If the negative charge on the interior of the membrane is inadequate it must be restored before another impulse can occur. Nevertheless, some nerve-cells are capable of producing hundreds of impulses per second.

Both membrane-pumps and membrane-gates come in considerable variety. Voltage-gated channels, for instance, have varying degrees of sensitivity, and stay open for different lengths of time. Different neuronal types employ different selections. As a result the rate at which impulses can be produced varies over a wide range. Moreover many neurons are a little bit leaky, and consequently fire at a low rate even without synaptic input, so that it's a change of rate that carries significance. (In many cases the 'resting potential' is more of a theoretical construct than a reality.) Some cells pulse away pretty steadily, functioning as rhythm generators and contributing to the control of such vital actions as breathing and the heartbeat.

Many of the neurons involved in cognition tend to fire bursts of impulses, rather than single efforts, and many can switch between the two styles. Some can only be activated by an appropriate number of dendritic inputs arriving within a very short time, but others are able to sum up inputs over a longer period (though it's still measured in milliseconds). And given prolonged input some neurons gradually reduce their response while others grow ever easier to excite.

Some neurons get their input from sensory receptors. Others synapse on to muscles, prompting them to contract. In all but the least sophisticated of species sensory neurons and motor neurons are for the most part separated by a chain of contacts that can include few or many intermediary neurons. The more of them there are, the more processing the sensory input gets and the more thoroughly the evidence can be weighed before a decision to act is taken.

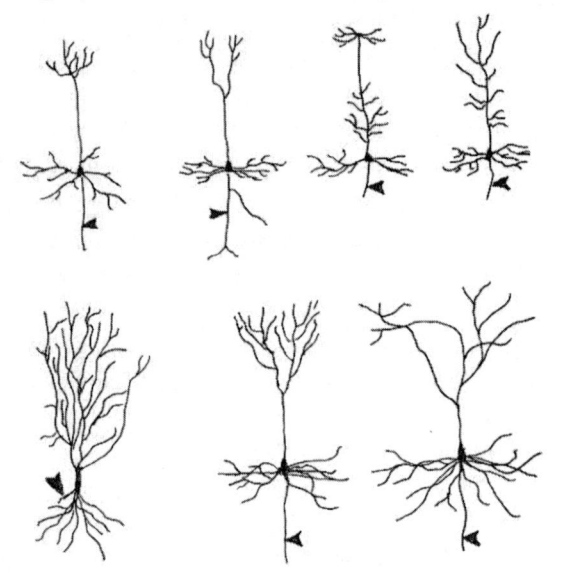

Neurons come in many shapes and sizes. The illustration shows a small selection of the types found in the brain. The arrowheads indicate the axons.

The synapses which can lead to a signal being passed on are called excitatory. The transmitter molecules emitted at other synapses produce the opposite effect. When they hit the receptors on the post-synaptic neuron they open channels which let in chloride ions. The resulting increase in the negative charge of the receiving cell makes it less likely to fire in response to excitatory input. These synapses are called inhibitory, as are the neurons which send such messages.

In simple animals inhibitory connections play a straightforward role in the control of movement, which was obviously their earliest function. The inhibitory message suppresses a previously ongoing activity when sensory input suggests it's not currently appropriate. In the surf-clam the shadow overhead which might indicate a passing predator prompts not only an excitatory message that brings about the closure of the clam's shell but also an inhibitory one to the neurons organising feeding movements. Inhibition can also ensure that two incompatible actions are not attempted simultaneously.

In complex neurons it does a subtler job, adding enormously to the possibilities for information processing. Whether a neuron passes on a message to other neurons depends on the balance between excitatory and inhibitory inputs. In the large neurons which form the major processing units of the brain inhibitory synapses are widely distributed, and their effect depends on where they are sited. Those that synapse on dendrites influence only the consequences of excitatory input in the same locality. Inhibition delivered close to or actually on the cell body is more powerful, as is that which arrives either on the initial segment of the axon or at its output end. With inhibitory inputs at all these levels, as well as thousands of excitatory inputs, a single neuron can perform quite elaborate calculations. The result is passed on to many more such neurons, which in turn combine it with a great many other inputs.

In vertebrate brains the major excitatory transmitter is glutamate, and the major inhibitory transmitter is GABA. (GABA is short for gamma-amino-butyric acid; after this example I shall provide such information only in the glossary, on the assumption that most readers either already know what the abbreviations stand for or won't really care.)   Glutamate and GABA, however, are but leading players in a large cast.

The substances that participate in the communication between neurons don't all serve to trigger the admission of ions and contribute directly to excitation or inhibition. Sometimes their impact on a receptor sets off a chain of reactions within the cell and modifies its state in one way or another, often by slightly altering the way one of the pumps or gates works. This may make the cell more susceptible to excitation, or less; it may alter the speed with which the cell recovers after each impulse, or the degree to which it does so. The consequence is often an alteration to the style in which the cell fires, switching it, for example, from single impulses to bursting mode, or *vice versa*.

The receptors which modulate the metabolism of the cell in this way are called metabotropic, while the sort that let ions in are called ionotropic. The series of reactions produced by a metabotropic input takes more time than the effects of an inflow of ions, and the results can be quite longlasting. The complication is that some transmitters can operate on both ionotropic and metabotropic receptors, so their effect depends on which type is activated. The receptor type may also determine whether the chances of the receiving cell producing an action potential are improved or reduced.

Over a hundred substances have been identified which serve to activate either ionotropic or metabotropic receptors, or both. Some of them seem to work in tandem, both released into the synapse at the same time. Many of them are molecules which serve other functions elsewhere. And many are also found in invertebrate species, though not always playing the same role. In vertebrates, for instance, the transmitter that conveys an activating signal from neuron to muscle is acetylcholine, whereas in insects it's glutamate.

Only a limited number of transmitters function ionotropically, but the receptors for them come in great diversity, each general type being composed of subunits that can be assembled in different combinations. However, there seem to be only three basic families of ionotropic channels, so at least there is a clue as to how all this diversity began.

Some neurons communicate only with others in the immediate vicinity. Others have long axons which carry their messages to distant destinations. The longest belong to the cells which carry instructions from brain to muscle, and to the neurons which carry sensory information from the periphery in the opposite direction. In a large species, such as whale or giraffe, these nerve fibres are measured in metres. Axons which cover long distances join up in neat parallel formation with their fellows and form the ropelike structures which are nerves.

The speed with which impulses travel along axons varies considerably. It can be anything from one to a hundred metres per second. One factor is the width of the fibre - large diameters mean faster currents. In some invertebrate species vital messages are transmitted by extra-wide fibres. (It was these large cells that provided the first insights into neuronal communication). Speed can also be achieved by insulating the axons, which additionally prevents cross-talk between adjacent axons. Cells called glia wrap around them, except where the voltage-operated gates are, and prevent leakage. (The neuron depicted earlier in this chapter is shown with this axonal insulation.) Urgent messages, such as the fact that you have just trodden on something sharp, travel in insulated fibres. Messages that don't require such instant action, such as stomach ache, travel second class in naked axons. The white matter of the brain is composed of the numerous insulated axons making their way between different areas, while Poirot's "little grey cells" are actually the cell bodies and their dendrites, along with axons that provide local communications.

Just to add to the complexity, meanwhile, there are some connections between neurons that don't involve chemical transmitters. Instead a tiny tunnel links the two cells and allows the ionic current to flow straight through. This is known as a gap junction. Here there's no scope for modifying the message that's passed on, but without the tiny pause for releasing transmitter and opening channels the transmission is faster. Gap junctions are found where extreme speed is valuable - triggering escape responses in some insects, for instance.

The Earliest Nervous Systems

Modern animals offer some clues which allow us to make an intelligent guess at how nervous systems must have begun. The simplest multicellular animals around today are the sponges - so plant-like in form and immobility that it was a long time before they were recognised as animals. Sponges live by extracting particles of food from the water which they wave through their many tiny holes with flagella - tiny whiplike protrusions - extruding the rubbish through a larger hole. They also have contractable cells which can reduce the size of the holes, by way of defensive action, in response to chemical or other stimuli impacting on them. These cells thus combine the functions of neuron and muscle, something which seems to be unique as far as modern animals are concerned. Interestingly, both GABA receptors and metabotropic but not ionotropic glutamate receptors have been identified in a sponge. Almost as elementary as the sponges are the placozoa, tiny disc-like animals which creep across the sea floor.

Not surprisingly there is no fossil evidence to indicate how multicellular animals first emerged, although there are detectable remains of single-celled organisms, bacteria-like forms that spread across the ocean floor in mats, the earliest of which have been dated to three and a half billion years ago. These were cells of the simpler, relatively unstructured kind, termed prokaryotic; multicellularity would arise among those which evolved a more complex organisation, with their genetic material enclosed in a nucleus – eukaryotic cells.

It's a reasonable guess that multicellularity began with increasingly close co-operation among eukaryotic cells which spread across the sea floor in a similar fashion to the microbial mats.   A more organised form of togetherness seems most likely to have evolved among neighbours which were already rubbing up against each other.   There would be patches of closely related individuals, descendants of a single ancestral cell and thus genetically identical or very similar. The means of conveying the signals that would promote co-operation would evolve more easily in these circumstances.

For such rock-clingers surfaces would become crowded, and there would be competition for space, and for nutrients, so there could be a great advantage in co-operating to create a means of rising above the surface.   Fossil evidence indicates that some very early multicelled animals were pancake shaped, and presumably grazed on the microbial mats.   Other organisms elevated themselves more thoroughly by forming streamers, attached to the rock by a holdfast – though it's not entirely clear whether they qualify as animals.   A reasonable hypothesis is that the ancestors of many extant multicellular animals took a different route, and began by co-ordinating themselves into baglike forms, with openings through which the nutrient-bearing water could be wafted by waving flagella or cilia (the same devices which various single-celled creatures use as a means of propelling themselves through the water).   This would give them access to particles of nourishment that passed by out of reach of the flat sheet of non-co-operating organisms.   By way of further benefit, perhaps the food-absorbing channels were concentrated on the inside of the bag, while the cell-walls that formed the outside became a little tougher and more protective.

The sponges, it may be supposed, found one way of exploiting the possibilities thus created - and they were successsful enough to give rise to several thousand species in today's world.   But there was another way of doing it which had greater evolutionary potential.   Instead of many holes the bag had a single opening to wave the water through.   And at some point the bag wall became a double layer of cells, and the design was improved by a division of labour.   The receptors which registered data from the outside world were concentrated in the outer layer, while the cells of the inner layer became more contractable, capable of re-shaping the whole bag in response to messages from the outer layer.   This more sophisticated construction brought the advantage that one sensory cell could contact several movement-producing cells, and similarly one movement-producing cell could receive input from several sensory cells, an arrangement which allows great improvements in sensitivity, strength of response and variety of possible responses. Thus a distinction between neurons and muscle-cells was initiated.

Even the earliest single-celled creatures will have had some shape-changing potential, reproducing just as modern cells do by dividing themselves in two after their DNA had been duplicated. Probably the mechanism was expanded in the bag-animals, allowing them to shrink away from unwelcome sensory inputs, and perhaps to pulse gently, increasing the flow of water past the food-absorbing channels. As the potential for movement was strengthened perhaps it became possible to roll around if the current changed, so that the bag's opening faced towards the source of nourishment again. And as bags proliferated perhaps some of them acquired the means of rolling away from the neighbours, to a new, less crowded space. Continuing with the guesswork, it's reasonable to suppose that this fairly organised conglomeration of cells acquired the means of monitoring the density of food particles in the inflow of water. A diminishment in the food supply could have been something else that would trigger the roll to a new, less crowded environment. Possibly the rolling would continue until the number of food particles hitting the receptors reached a decent level again.

The first recorded traces of multicellular animal life are tracks (amazingly preserved) across the microbial mats. They imply a slow, laborious progress across the food supply. Later tracks indicate a more efficient wriggling across the sea-floor, and there is evidence of some early burrowers.

The sort of species indicated by these clues, and by some of the most ancient fossils, can be imagined as elaborations of the hypothetical bag-animals, often elongated versions. Some would acquire legs, which proved a particularly successful idea, leading eventually to the arthropods, a phylum which contains far more species than any other. But some must surely have been knocked off their supporting surface to drift freely in the water. Those that could still, under these new circumstances, turn their opening towards the current bringing food would have the best chance of survival. Then, perhaps, the simple neuromuscular arrangements which could reorient the animal or produce a withdrawal from potential damage came to be co-ordinated into an elementary sort of locomotion, making successful survival yet more likely.

(An important feature of this hypothesis is that some control of orientation was in place before the means of locomotion evolved. This may not have been true of whatever made those tracks across the microbial mats, which look somewhat random – but with food all around a lack of steering ability wouldn't matter.)

Making use of all three of the sea's dimensions should have meant less competition, for the time being, and therefore less pressure for genetic innovations than life on the sea floor. But freely drifting animals with no hard parts are particularly unlikely to be preserved as fossils, so there is little chance of finding solid evidence for the hypothesis. However, there are still animals around which seem like elaborated versions of a basic bag design. They are found in two phyla, the comb jellies, or Ctenophora, and the Cnidaria, among which are hydra, sea-anemones, and jellyfish.

In one of the more sophisticated Cnidaria, the *Scyphozoa* jellyfish, there is a network of motor neurons linked to a number of centres which contain pacemaker neurons as well as light-sensitive spots and gravity receptors. The synapses of the network are bidirectional, and the connections are so numerous that many fibres can be cut without affecting its functioning. The steady beat of neuronal impulses around it, transmitted to the muscles, is what keeps the jellyfish expanding and contracting its bell. Tactile and chemical receptors are found on another diffuse nerve net. Their inputs are reported to the nearest centre, and an ominous message causes an increase in the rate of firing in the nearest part of the neuronal network, which spreads around the bell. The result is that the jellyfish tips over a little, and then as the message spreads around the network accelerates away from the threat, with its stinging tentacles trailing behind it.

The idea of the ancestor of most if not all multicellular animals as a double-walled bag is derived from the fact that this is the structure which emerges at a very early stage of embryonic development, common to all the species in which it has been studied. The fertilised egg develops first into a ball of cells, then into a hollow sphere which is composed of two layers, the outer one known as ectoderm, the inner as endoderm. Skin and most of the nervous system develop from the ectoderm, and in species like the jellyfish muscles develop from the endoderm. In bilateral species, with their much more complicated structures involving many additional forms of tissue, there's also a middle layer, and it's from this mesoderm that muscle is derived.

Just how sponges, ctenophores, cnidarians and bilateral animals are related is a matter of much discussion. It's currently being fuelled by an increasing amount of genetic evidence. Some of this hints that a proportion of the genes which underpinned the evolution of elaborate bilateral species may already have been present in their predecessors - a not unreasonable idea, but not one that answers the question.

In more complex animals two types of nervous system and two sorts of muscle deployment become increasingly distinguishable. There are regular contractions that do such things as propelling food through the digestive system, providing for the intake of oxygen, and distributing essentials around the body by means of a circulatory system. And there are irregular movements that occur in response to sensory input. It's the elaboration of the latter that's more relevant to the evolution of brains and of reason. Nevertheless, it's worth noting that brains, when nothing much is going on, still pulse away quietly, much in the manner of hearts and lungs - and jellyfish. Indeed so do many individual neurons if they're not inhibited.

Sensory Receptors

A multitude of sensory receptors exists, but they can be divided into just a few groups. Probably the most versatile class is comprised of those which are stimulated by mechanical pressure. These of course include tactile receptors, which in the more sophisticated species come in several varieties, including some which have thresholds high enough to respond only to stimuli likely to cause harm. Activity in these latter tends to correlate, in us, with the experience of pain, but officially they're known as nociceptors, responding to noxious inputs. There are also several sorts of internal mechanoreceptor, monitoring the positions of limbs and joints, the pressure in the alimentary canal, and other vital matters.

In mammals, as already noted, hair adds an extra, very delicate dimension to external touch. Receptors wound around the roots of a hair send off signals when the hair is deflected, even though the skin itself is not touched. Many nocturnal and burrowing animals get a major part of their information about their surroundings by such means, and especially through that extra-size version of this tactile device, whiskers. Feathers operate in the same way for birds, enabling them to gauge the strength and direction of air currents, and flying insects have similar mechanisms.

Fish are equipped with something comparable, which enables them to sense and adapt to water currents, and to detect disturbances in the water which may presage danger. The lateral line system consists of a line of pores which most commonly runs along the sides of the body and across the head, forming a tiny canal in the skin. The canal is lined with cells which are covered in those tiny, hairlike protrusions called cilia. When these are bent messages are despatched to the brain indicating the direction of the water pressure.

Hearing may have evolved as a development of the lateral line system, enabling fish to register more delicate disturbances of the water, and those of more distant origin. When vertebrates moved on to land the mechanism was adapted to the even more delicate business of detecting waves of disturbance travelling through air. When a wave of intensified air pressure hits the ear-drum the effect is amplified by a cunning arrangement of little bones, so that a stronger pressure wave is transmitted to the liquid contained in the ear canal, a long, thin, curled tube which grows gradually narrower. One wall of the tube is formed by the basilar membrane, stiff at the beginning and increasingly flexible further up. Pressure waves which follow each other at high frequency in this enclosed space are powerful enough to bend the wide and stiff end, and thus exhaust themselves. Those that arrive at lower frequencies make their indentations increasingly further up. In other words the first part of the basilar membrane is tuned to a high tone and later parts to increasingly low tones.

The other side of the membrane is lined with hairs, or stereocilia, which engage with more cilia on an opposing membrane. The base of each of the latter is swathed in neuronal dendrites, and the message to the brain when a set of hair-cells is deflected indicates the pitch of the sound that has caused the movement.

Above the cochlea is a trio of semi-circular canals which provide the senses of gravity and acceleration. Here the weight of liquid dragging on the hair-cells indicates which way is down. And when the head turns, or the whole animal moves through space, the liquid moves less slowly than its container, slurping just as liquid in a bottle does, thus providing a report on the movement. In a whole variety of species, ranging from minute creatures living in mud at the bottom of tropical oceans to many species of birds, a similar mechanism provides a sense of the earth's magnetic field. Tiny fragments of a magnetic mineral are contained within a hair-lined space, where they function like a compass needle and their movement is registered by the hair-cells.

A second major class of receptors is concerned with detecting chemicals. Each type is shaped to fit around a distinctive part of the sort of molecule in which it specialises. When a receptor engages with such a molecule the result in some cases is the opening of a channel that carries an ionic current, in others a series of reactions within the cell, as at a metabotropic synapse. One taste receptor responds to a substance that also functions as a neurotransmitter, the amino acid glutamate, but much larger amounts of it are required.

Taste and smell are the obvious chemical senses. In addition there's something known simply as the chemical sense. Many aquatic animals have receptors for this distributed all over their bodies, enabling them to make a rapid retreat from noxious substances. We land animals have less need of an all-over early warning system, being protected by the wrapping that enables us to maintain our water content when surrounded by air. However, the defensive chemical receptors are still readily accessible in the tender linings of our various orifices. When stimulated they trigger the production of mucus to wash the offending substance away. They're also responsible for the stinging sensation which may prompt additional, voluntary responses.

Various other chemical receptors, meanwhile, function to monitor internal chemical conditions. The information they provide is used to regulate important housekeeping matters such as the retention or excretion of liquid, the need for some particular type of food input, or for a better supply of oxygen.

Since both taste and olfaction evolved in an aquatic environment, where all signals are waterborne, the distinction between the two may seem a little hazy. They're distinguished by reporting to different brain areas, reflecting the fact that they guide different actions. Taste indicates what to eat, olfaction indicates what to approach, what to avoid. One reveals the definite existence of food, the other registers molecules that have escaped rather than the thing itself, offering clues about desirables, or threats, which are still distant.

Olfactory receptors come in great variety, though the number varies considerably between species. We humans probably have around five hundred types. Rodents, which depend so much more heavily on this sense than we do, have more than twice as many. Dogs are still better equipped.

The receptors sit in the olfactory epithelium, and report to the olfactory bulb just above it, at the very front of the brain. Most mammals, and some other land-dwelling vertebrates, also have a secondary odour-sensing system. Its receptors occupy something called the vomeronasal organ, and report to the accessory olfactory bulb. They're distinguished by being shaped rather differently, and while olfactory receptors respond best to lighter molecules that can drift freely the vomeronasal receptors are good for heavier ones, found near the ground or actually deposited there. Those snakes that hunt their prey with a forked tongue, waving it as if tasting the air, are using their vomeronasal organ.

It was thought at first that the mammalian vomeronasal organ's function was confined to registering the pheromones emitted by conspecifics, odours which often affect the receiving animal's hormonal activity as well as its behaviour. But as more is discovered the picture is becoming increasingly complicated. Some pheromones turn out to be registered by olfactory receptors, while some carnivores have vomeronasal receptors that are sensitive to the smell of prey. And some fish have the vomeronasal type of receptor dotted about among their olfactory ones, rather than in separate organs. It seems possible the crucial distinction between the two sorts of receptor lies in the structure and the volatility or otherwise of the odour molecules with which they engage, rather than in the function they serve for their possessor. (We humans still have a vomeronasal organ, but whether it's functional is a matter of controversy.)

Taste receptors have multiplied much less prolifically than those for odours. We humans possess only five types. These register the substances we recognise as sweet, salty, sour and bitter, along with glutamate. Natural plant poisons taste bitter - though not everything that tastes bitter is a poison - so the significance of that receptor is obvious. Glutamate represents protein. It can be no coincidence that the three other tastes indicate substances that are good in appropriate quantities but harmful if taken in excess. And all five tastes signal substances for which the digestive system can usefully prepare itself, increasing the production of saliva, stomach acid, or insulin to levels not necessary for blander food.

Our particular range of taste receptors is probably fairly typical for an omnivore. Several species of carnivore have been shown to possess a range of amino acid receptors, but none for sugars. Herbivores, on the other hand, are likely to lack amino acid receptors, and those fructivores which can digest only ripe fruit may be particularly well endowed with sugar sensitivity.

The extensive inventory of flavours we experience depends on combining the limited gustatory data with olfactory information - which is why a cold in the head can ruin one's appreciation of food. Receptors for the defensive chemical sense make a contribution too, responding to things like alcohol, and capsaicin, an ingredient in chillies. Some types of heat-recording receptor can also be activated by capsaicin, which is why chillies taste hot, while one type of coolness receptor is activated by an ingredient in mint. The total taste experience is also influenced by input from tactile receptors in the mouth.

Temperature sensors come in several varieties, sensitive to different degrees of warmth or coolth. In two families of snakes, the boas and the pit vipers, heat detection has been developed to a very sophisticated level, so that patterns of hot and cold can be registered in the way that other species register light and dark. For these snakes the infra-red 'eye' in the middle of the forehead (the pit that gives the pit vipers their name) plays a bigger part in locating and catching prey than do the two conventional eyes.

Light receptors contain a supply of molecules which change shape when hit by a photon. That triggers a string of chemical reactions which multiply the effect. The result, in most cases, is to inhibit the cell. Photoreceptors don't fire action potentials, as most neurons do, but simply emit less or more transmitter according to how much light is hitting them. This means that the message they pass on can be a very sensitive indication of the current input.

Photoreceptors come in many varieties, most of them maximally sensitive to one particular wavelength of light, and much less affected by longer or shorter wavelengths. The optimal wavelength can be shifted by quite a modest change to the genes which code for the light-sensitive pigment, it seems, for there are many slight variations between species. We humans, like some of the other primates, have three types of wavelength-sensitive receptor, a short-wave one (which is activated by blue light), a medium-wave one (green) and a long-wave one (red). (See the chart on page iv of the colour insert.) Many species are much better endowed. Fish, amphibians and reptiles may have up to five different types, including one in the ultra-violet range. Some invertebrates are reported to have even more. Mammals appear to have lost all but the short and medium wave photoreceptors towards the beginning of their history, when they carved out a new niche by becoming nocturnal.

The development of this new lifestyle must have been greatly aided by a different type of photoreceptor, one that was activated by a much wider range of wavelengths and was a great deal more sensitive, capable of responding, on occasion, to a single photon of light. Our long-wave receptor, meanwhile, is thought to have evolved by means of a duplication of the gene for the medium-wave one, followed by some modification. This development must have been relatively recent, since not all New World monkeys possess the means of registering red light. It may have happened more than once, since two sorts of red receptor are found in humans, with slightly different peak sensitivities.

Gathering information about objects and environments is the obvious use for photoreceptors, but not the only one. Most or perhaps all living creatures are equipped with an internal clock, which is kept in alignment with the light/dark cycle by a light-sensitive mechanism. This uses yet another type of photoreceptor, with a different light-absorbing pigment and a different sensitivity, and in many species it occupies a separate, independent eye, or in a few cases a pair of them. The complex eye of the vertebrate seems to have evolved by means of a fusion between what were originally two eyes, one for clock-setting and one for vision, as we shall see in the next chapter.

Brains

The biggest range of sensory receptors and the most complex nervous systems evolved in bilateral animals, possessed of a front and a back as well as a top and a bottom. The front - defined by the direction in which the animal usually travels - was the obvious place for the receptors for distance senses to be situated. How much energy is spent on locomotion must depend on factors such as whether food is needed, and whether it's the season for searching out mating opportunities, so the internal monitors and the hormone control system were located for'ard too. And while escape responses might be organised locally it was helpful to report them to the same region, so as to co-ordinate them with other muscle deployments. Bilateral creatures thus acquired a conglomeration of cell-bodies at the front for dealing with all these matters. In some species the conglomeration grew big enough to be called a brain.

In the vertebrate the nervous system begins to take shape when the embryo has developed into a hollow sphere of cells - three layers of them, as we saw earlier. A groove forms along the top of the sphere, and its edges fold over to form a tube, lined with cells from the outer layer, or ectoderm. They give rise to the cells which become the brain and spinal cord, while the central space becomes the fluid-filled centre of the cord and the ventricles of the brain. A ridge forms along the join at the top of the tube, meanwhile, which provides the cells that become the peripheral nervous system. As the process of cell division advances the top end of the tube swells more than the rest, and crossways grooves appear, creating three bulges. These bulges become the major divisions of the brain - forebrain, midbrain and hindbrain.

Two more crossways dents appear, subdividing hindbrain and forebrain. The rearmost bump will be concerned with overseeing the most fundamental activities of vertebrate life, such as breathing and the heartbeat. Bulge number two is where neurons carrying messages from the various mechanical senses will make their reports, and also those bearing information from the viscera and those concerned with taste.

The midbrain receives some direct input from the eyes and some relayed information from senses that have reported initially to the hindbrain. It also contains structures that play a role in managing movement. As the hindbrain and midbrain remain relatively slender in mammals while the rest of the brain expands luxuriantly to envelope them they are sometimes referred to as the brainstem.

The second bulge from the top becomes the diencephalon, or 'betweenbrain'. One of its most conspicuous components is the thalamus. In the mammalian brain this is a complex structure. One small part of it processes input from the retina before relaying the result to the neocortex. Auditory and somatosensory data that has already received some processing in lower brain areas goes to other sectors, and is similarly passed on. Most of the thalamic nuclei, however, perform more obscure functions, as yet imperfectly understood.

Below the thalamus is the hypothalamus, which monitors internal conditions such as temperature and the current degree of need for food or liquid. It also adapts the organism's hormonal tuning to prepare for things like copulation, pregnancy and parturition, and for fights and other stressful situations. This is also where the central circadian clock is situated.

The vertebrate brain, like the body, is a mirror-image structure, and there are two of most of its components – two thalami, two hypothalami, and so on. An exception is the cerebellum, symmetrical but singular, which grows out behind the hindbrain. The sketch shows hindbrain, midbrain and diencephalon in the mature human – where they are almost completely covered by the telencephalon, or end-brain, the structure that grows from the topmost bulge.

Here there is only one direct sensory input, the olfactory, which goes first to the olfactory bulbs. There are also several other significant structures, but in mammals the most conspicuous feature is the thick, multi-layered quilt of cells that covers it. This is known as the cerebral cortex, or - because its absence in non-mammals indicates that it's a comparatively recent development - as the neocortex.

The neocortex is large in those mammals we think of as particularly intelligent, especially in primates and cetaceans, where it spreads out over the rest of the brain. In *Homo sapiens* it's so large that only the cerebellum peeks out from underneath. In many of the larger brains the sheet of cells has had to be crumpled up very considerably to fit inside the skull. (The arrangement is not inconvenient, since it allows strongly connected areas to approach each other more closely than they otherwise would.) The resulting bulges are called gyri, and the crevices between them sulci. (Pictures of a real brain are to be found on page v of the central colour insert.)

The neocortex comprises two sheets of cells, one for each side of the brain, forming twin hemispheres, connected by a substantial exchange of axons through the corpus callosum. The biggest fissures caused by the crumpling divide each of these two cerebral hemispheres into four lobes. From front to back these are the frontal, temporal, parietal, and occipital. The two hemispheres fold inwards at their edges, and the lower sketch shows the inner side. The dotted area represents the corpus callosum.

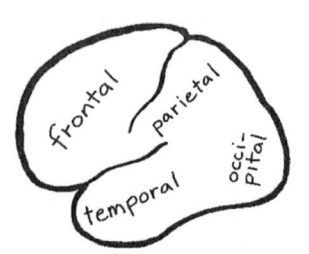

The general principles of organisation are pretty consistent in all mammalian species. Visual information goes initially to the occipital lobe, auditory information to the temporal lobe, somatosensory information to the parietal. The frontal lobe evaluates the significance of the information, whatever its nature, decides when action is needed, and organises movements.

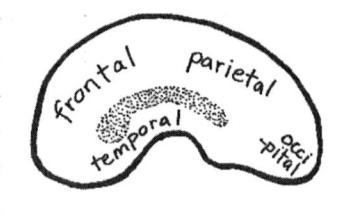

In smaller mammalian brains the primary sensory areas take up a good deal of the space. As the brains get larger the output of the primary sensory areas is passed on for more and more additional processing. This entails a corresponding increase in the number of areas receiving more than one sort of sensory information, regions which in the early days of brain exploration were termed 'association areas'. As the association areas grew, so did the frontal lobe, and most notably the forward part of it.

That's a very rough sketch of the ground plan of mammalian brains. The reality is a complex three-dimensional jigsaw to which two-dimensional diagrams can't begin to do justice. A further complication is that the anatomists who first explored the brain had only a few clues as to what many of its various structures did, so many of them were named for their appearance - which is helpful enough for subsequent dissectors, but less so for those of us who merely read about the brain. A few structures are distinguished by colour, so there's a red nucleus, a locus coeruleus (blue), a substantia nigra and a periaqueductal grey. Others were christened for what their shape resembles (at least in the human brain) - the amygdala (almond), the hippocampus (sea horse), the pons (bridge) and the colliculi (little hills).   And sometimes imagination failed, so there is a zona incerta, and - my favourite paradox - the substantia innominata.

After that brief introduction to the brain the next step is to take a closer look at just one set of the brain's pathways, so as to give some idea of how its neurons process sensory input.

Recapitulation

The cells which constitute most life-forms are bounded by a fatty membrane, in which are protein pumps which import useful molecules or export unwanted ones.

There are also gates, opened by mechanical force or a chemical key, which let in positively or negatively charged ions to effect brief changes in the way the cell works.

All cells are somewhat negatively charged in relation to the intercellular medium, but neurons, the cells which transfer messages around the body, have exaggerated this charge.

When an excitatory transmitter opens the gates at the synapse connecting one neuron to another a wave of positive charge sweeps along the inside of the dendritic membrane.

If excitation occurs at a sufficient number of synapses the wave of depolarisation may reach the cell body and continue along the axon leading from it. This may lead to the release of transmitter at the far end of the axon, starting the whole process over again in the neurons on which it synapses.

At any stage, however, the excitatory message may be counteracted by input from inhibitory neurons. Whether a message reaches the end of the axon or not depends on the balance between excitatory and inhibitory inputs.

The developing embryo initially takes the form of a hollow ball, three layers thick in bilateral animals, two in simpler ones. The nervous system, along with skin, will develop mainly from the outer layer in both cases.

Gradually an increasingly varied range of sensory receptors evolved. Those that respond to mechanical pressure provide us with hearing, a sense of gravity and of acceleration and information about the disposition of the body, as well as a tactile sense.

Chemical senses are olfaction and taste, the general chemical sense which responds to potentially harmful substances, and various systems which monitor internal conditions.

Light receptors regulate an internal clock as well as contributing data about the external world.

Chapter 6

The Basic Mechanisms of Vision

The mammalian neuronal system that's perhaps best understood is the one involved in vision. That's far from saying that it's thoroughly understood - new subtleties are being discovered all the time. But enough is known to give an idea of some general principles of brain function.

The Retina

At the back of the eye the rays of light reflected from the outside world, refracted by the protective cornea and more finely focussed by the lens, cast an upsidedown image on to a sheet of light-sensitive cells. In one human eye there are around 125 to 130 million of these photoreceptors. The majority are rod-shaped, containing a pigment which reacts to light of a wide range of wavelengths. These are highly sensitive, especially in poor light, and the chances of a response to very weak input are further increased by the fact that several photons arriving at a rod within the space of about 100 milliseconds can combine their effect. The rods are entirely responsible for what we see under minimal illumination, when the world becomes devoid of colour.

The photoreceptors which operate in bright light are cone-shaped, and there are only about seven million of them. They need about a thousand times more light than a rod to activate them, and they are choosy about the wavelength of light to which they respond. We have three varieties, each producing its maximum response to a different wavelength. Cones respond at a very fast rate, registering pulses of light at up to 55 per second (55 Hertz),

This micrograph, reproduced from a paper by B. Boycott and J. E. Dowling in the Philosophical Transactions of the Royal Society of London, Series B, shows a section through a human retina, with receptors at the top. The scale bar is 50 micrometres.

Receptor cells are packed very densely in the centre of the primate retina, and especially in the centremost area, about the size of a pinhead, called the fovea. Towards the edge the density decreases noticeably, which is one of the reasons we can only discriminate fine detail in the centre of the visual field. The central clustering is mainly of colour-sensitive receptors, especially those for the two longer wavelengths, which are what the lens focusses most effectively.

The photoreceptors synapse on to long, thin neurons known as bipolars, which come in several different types. Some are activated when the little spot of retina about which they report is illuminated by the type of light in which the contributing photoreceptors specialise, and are known as ON bipolars; slightly fewer are activated by an absence of that sort of light, and are termed OFF bipolars. They get their input from a varying number of receptors - in the case of cells in the fovea only one. Running perpendicularly away from the photoreceptors, the bipolars connect in turn to large neurons called ganglion cells, each bipolar cell synapsing on to several ganglion cells

Influencing the communications at either end is a fine network of cells. The ones weaving among the receptor-bipolar connections are known as horizontal cells, those among the bipolar-ganglion connections as amacrine cells. Both come in several varieties, and their functions are still far from fully worked out. One sort of amacrine cell serves to transfer the signal from rod bipolars to ganglions, summing the inputs from many rod bipolars before they are added to the input from cones, an arrangement which no doubt makes it easier to balance the very different dynamics of the two receptor types. Other amacrine cells adapt the system to suit the current level of illumination by modifying the way information is transferred to ganglion cells. A coloured diagram of the retina is on page vi of the central insert.

Some intimation of the scale of neurons can be derived from the fact that all these cells - ganglions, bipolars, amacrines and horizontals - are situated in front of the photoreceptors. There's a network of tiny blood-vessels too, as you can sometimes see for yourself when the optometrist's torch shines into your eye. The whole collection is thin and transparent enough for light to pass through it.

There are around a million and a half ganglion cells in one human eye, collecting the signals from those 130 million photoreceptors. They come in several varieties - at least seventeen in primates. The basic classes are distinguishable by the size of the cell body, the shape and extent of the dendritic bush, and the speed with which their axons conduct impulses. Not all are exclusively concerned with vision. Some also provide signals about the overall level of illumination, used to control the size of the pupils and to keep the brain's biological clock synchronised to the diurnal alternation of light and dark.

The most conspicuous types of ganglion cell function to filter contrasts out of the data provided by the rods and cones. One sort, known as a parasol cell, has a wide spread of dendrites. The patch of photoreceptors which feeds into it is organised into a centre and a surround. In some cells the data from the ON bipolars and the OFF bipolars is organised in such a way that the ganglion cell is excited when the centre of the receptive field is illuminated, but the rate of firing is reduced if the light overflows on to part of the surrounding area, and if the whole receptive field is illuminated the cell fires only feebly, much as it does when there is no input. Other parasol cells are excited when the centre of the receptive field records darkness and the border is illuminated, but the effect of darkness in the central area is much reduced by darkness in the surround. The former sort are designated ON-centre cells, and the latter are known as OFF-centre.

Their passion for contrasts reflects the pattern of excitatory and inhibitory inputs from the bipolars. These are balanced in such a way that the ganglion cell can be excited not only by a spot of light (or darkness) but also when an edge between light and darkness is appropriately positioned on its receptive field, or when a light or dark line crosses the centre. The input relayed to both centre and surround comes from both red and green cones and from rods, so the light which excites an ON-centre cell can be of any wavelength, and an OFF-centre cell is similarly sensitive to a contrast defined purely in terms of luminance.

Another type of ganglion with a smaller spread of dendrites works on a similar principle but receives input only from cones. One variety is excited by red light in the centre of its field, but the excitement is counteracted by inhibition if the red also extends across the surround. The cell's response is enhanced, on the other hand, if more green than red is reported from the surround. These cells therefore respond to borders marked by different amounts of longwave and mediumwave light, rather than to a simple variation in luminance. Other ganglions have GREEN-ON centres, and combine them with surrounds that prefer red wavelengths. The receptive fields of all these contrast-sensitive cells are heavily overlapped, so the same receptor can contribute to many different ganglions.

In short, comparisons between the inputs from groups of adjacent receptors are used to reveal contrasts, and thus borders. For blue light things are a little different. Ganglions which receive the output of the short wavelength receptors don't have a centre/surround organisation. And here the inhibitory input comes from both red and green receptors. This somewhat lopsided arrangement no doubt reflects the comparatively recent evolution of our red receptors, which were cobbled into the less colour-sensitive system of our nocturnal ancestors.

The BLUE-ON ganglion also has a partner, the BLUE-OFF ganglion, which has the same input pattern but in reverse - excitation from both red and green receptors and inhibition from blue receptors. It's the activation of BLUE-OFF cells which leads to the sensation of yellow. There's also a type of ganglion which simply responds to blue light, without any opposed influence.

Both types of contrast-sensitive cell have small receptive fields in the centre of the retina and larger ones towards the periphery. A ganglion cell in the fovea may get the input to its central field from just one receptor. This is another of the factors that makes the fovea good for seeing detail, and leaves the periphery ill-equipped for it.

Parasol cells are distinguished not only by receiving input about all detectable wavelengths of light and by having a spread of dendrites that gives them receptive fields of a medium size. They also respond to input very rapidly, but the response is not maintained, so they are only maximally excited by novel input. A subset is wired so as to respond briefly to the advent of either light or darkness, which means that a stimulus passing across the receptive field produces a double burst of excitation. This makes them particularly sensitive to movement. Novelty and movement are urgent matters, so these cells have sturdy axons which conduct impulses fast.

The smaller ganglion cells which identify red/green contrasts and have smaller receptive fields convey their messages in medium-sized axons. The blue and yellow specialists have still smaller cell bodies, but wide-ranging dendrites that provide large receptive fields. The axons are slender, which implies that their information is not urgent. Together with the lack of sensitivity to contrasts this indicates that they contribute less significantly to the detection of visual objects.

In those diurnal Old World primates where they have been counted the ganglion cells dealing in red/green contrasts are the most numerous by far, accounting for 70% or more of the total. Despite the fact that rod receptors greatly outnumber cones the ganglion cells which care only about luminance, not about wavelength, form only about 10%. A massive convergence of inputs at every level is the strategy for discovering contrasts in poorly lit scenes, and for revealing visual movement. Ganglion cells dealing in blue/yellow contrasts, meanwhile, constitute rather less than 10% of the total. This imbalance between red/green ganglions and blue/yellow ones reflects a similarly uneven balance among cone types. Short-wavelength receptors are not very numerous. On the other hand they are distributed evenly over the retina. The red and green receptors, in contrast, are scattered at random, with the result that there can be clumps of one or the other, the pattern varying from individual to individual.

What these ganglion cells transmit to the next stage of the visual pathway, then, is the result of a comparison, not a simple relay of photoreceptor input. Those with complex receptive fields compare the data from adjacent receptors to discover the places where it differs. BLUE-ON cells indicate that more blue light is reaching their retinal field than either red or green, and BLUE-OFF cells the opposite. GREEN-ON centre cells report that there is more green light than red hitting the centre of their field. This method of dealing with cone inputs makes up for the overlap between the sensitivities of the different types, which would otherwise cause some ambiguity.

A notable advantage is that useful colour information can be registered even when it's disguised by the current illumination. The sunlight which reaches us varies in the mixture of wavelengths it contains, according to the time of day, since more long-wavelength light is absorbed by the atmosphere when the sun is low in the sky. Comparing the proportions of each wavelength, rather than registering absolute amounts, means that colour vision can still function.

Clearly the retina filters quite a bit of useful information out of the photoreceptor input. In some species the analysis effected at this stage goes further. Rabbits have some ganglion cells which, thanks to the way the receptors are connected, register the direction of movement. So do rats and mice, and a ganglion cell has been found in mice which is wired so as to detect approaching visual objects. Something similar may exist in primates - there are some ganglion types with unknown functions, present in very small numbers.

There are some more certain differences between species. A fovea is found in carnivores as well as primates, and seems to go with having forward-facing eyes and a large overlap of the two visual fields. Most mammals have a much more uniform retina. They gather more information from the fringes of the visual field than we do - useful if keeping a watch for danger is the first priority - but uncover less detail from the centre. As already noted, many non-mammalian vertebrates have four or five different types of wave-length-sensitive photoreceptor. The retina is a nice example of something that preserves the same essential structure and organisation but evolves many variations and thereby contributes to opening up many different lifestyles.

The lateral geniculate nucleus of the thalamus

The axons of the ganglion cells join together to form the optic nerve. Curving back from each eye the two bundles of fibres meet, and at this point both divide, and the larger half of each bundle crosses to the other side of the brain. As a result the fibres carrying information about what's going on in the left half of the visual field end up in the right side of the brain, and information about the right half of the visual field goes to the left side. This follows the general, somewhat surprising principle of brain organisation: most sensory information about what happens to the right of the body, along with movement by the right side of the body, is dealt with on the left side of the brain, and *vice versa*.

In the primate the vast majority of the axons run to the two thalami in the diencephalon. They terminate on the side of each thalamus, in what's known as the lateral geniculate nucleus. This is composed of six distinctive layers of cells, interleaved with very thin layers. On their journey the ganglion axons maintain the spatial relationships of the cells from which they issue, and they connect to the thalamic cells in such a way that these relationships are still preserved. Each layer of the lateral geniculate thus constitutes a map of one half of the total visual field, so that a vertical line through the layers would encounter cells which all receive input concerning the same bit of view.

Two of the conspicuous layers contain many large cells, the other four only smaller cells, so they are distinguished as magnocellular and parvocellular. The six very thin layers are made up of very tiny cells, and have been named koniocellular, from the Greek word for dust. In the magno- and parvocellular layers the data forwarded from the two eyes remains segregated. Half the layers of each type are devoted to the contralateral eye (the one on the other side of the head), and the alternating layers get their input from the ipsilateral eye. (There's a diagram on colour page iv, and a sketch of the pathway from retina to visual cortex.)

Axons from the ganglion cells that deal in luminance contrasts, and are sensitive to novelty and movement, synapse in the magnocellular layers. Those from the ganglions that register borders between red and green terminate in the parvocellular layers. In both cases what mostly happens then is an elaboration of the operation performed in the retina. The receptive field has a centre (the size of which varies) and a surround, and a response is evoked when the reports from centre and surround differ. But here the need for contrast tends to be greater - the input from the periphery of the area covered by the cell has a stronger influence. Moreover some of the neurons found in primate lateral geniculate respond best to lines of a particular orientation, rather than just patches of contrast, and some are sensitive to the direction of movement.

The neurons of the magno- and parvocellular layers combine the ganglion inputs in many different ways, it can be assumed, so each little bit of retinal field is represented many times over in the lateral geniculate. Thus the data that has been collected in the retina is sifted further, and the contrasts identified more securely. The midget/parvocellular pathway is vital to the perception of colour, small-scale contrasts and the conglomerations of small features which make visual texture. The parasol/magnocellular pathway is important in the perception of motion and flicker. Both pathways contribute to the perception of brightness.

The reports from the small ganglion cells that register *blue* or *not-blue* - the ones that are only interested in the general distribution of shortwave light - go to the two central koniocellular layers. Most of these dust-sized cells are relatively slow at working up to an action potential, and fire at low rates. But there is more variation in their behaviour than among the cells of the magnocellular and parvocellular layers, and much of what they do remains unknown.

While only about a million axon fibres enter each lateral geniculate nucleus from the human eye, several million leave it, so we can deduce that quite a bit more information has been wrung out of the retinal data by then. The synapses through which the retinal data is delivered, meanwhile, are vastly outnumbered by those from the neocortex and from subcortical areas, all conveying neuromodulatory input. It's a fair bet, however, that the processing carried out in the lateral geniculate is nothing compared to what, in primates and other mammals with well developed visual systems, happens next.

Edges and lines

The axons carrying the output from the lateral geniculate run to the rear of the occipital lobe, to primary visual cortex, also known as V1, or - because some layers contain darker cells which give it a stripey appearance - as striate cortex. In the great blanket of cells that is the neocortex different types of neuron predominate at different levels, so that microscopists have been able to identify six layers. In the primary visual cortex of primates layer 4 is further subdivided. The magnocellular, parvocellular and koniocellular axons run mainly to different layers, and to different subdivisions of layer 4

As in the lateral geniculate, the spatial relationships of the retina are maintained, so that V1 forms a map of the retinal field. It's a somewhat distorted map, since the centre of the visual field is represented so much more fully than the periphery, and takes up far more than its fair share of space. But cells dealing with the data from one bit of retina are as adjacent as possible to the cells dealing with the nextdoor bit.

Each lateral geniculate axon connects to a large number of cortical cells, spreading its data widely. Conversely, each cortical cell collects data from a great many lateral geniculate cells, and inputs from a substantial number of them are generally needed to fire it. The principal neurons of V1 function very much like the retinal ganglions and the lateral geniculate cells, insofar as they are concerned with identifying contrasts. But here only a few cells respond to darkness completely surrounded by light, or *vice versa*, and those that do require the contrasting frame to be complete. Mostly the stimuli which produce a strong response are lines, or bars, with the optimum width varying considerably, while some cells are excited simply by edges between dark and light.

In every case the orientation of the line or edge is important. A cell that fires at its highest rate for a vertical line is much less excited by one that is ten degrees or so off the vertical. Similarly a neuron that specialises in right-slanting bars doesn't respond to vertical or horizontal ones, let alone to one that slants leftwards. In the cells that fire for edges it's necessary for the light and dark to be the right way round. In all cases a less than ideal stimulus evokes a reduced response, and one that misses the specification by a wider margin can silence the cell altogether, suppressing the normal modest rate of firing that is maintained when there is no input. A great many of the cells are also extremely fussy about the position of the stimulus on the retina.

All these neurons are linked by inhibitory interneurons, which increase the inbuilt selectivity. A stimulus that deviates a little from the proper orientation, or the proper width, or the proper position, will still provide much of the right excitatory input. But if the stimulus is such that neighbouring cells are more strongly activated then inhibitory input from them will outweigh the excitation. An orientation which falls between the preferences of two cells excites them both equally, or one a little more than the other, and it's the balance of their activities that reveals the exact slant. The correct assessment, in other words, is achieved by competition. This system for distilling more precise information out of imprecise neuronal activations is known as population coding.

Until recently the favourite stimulus pattern for testing the preferences of cells in the primary visual cortex has been a moving grating of bars. As well as being highly selective for the orientation of the bars cells make their best response to one particular width of bar. The response is also influenced by the speed at which the stimulus moves. In those neurons which get their input from the magnocellular pathway, where the ganglions are particularly sensitive to moving stimuli, the best stimulus is a moving one, and these cells are particularly good at responding to fast-moving gratings. Since the input comes from ganglion cells with large receptive fields thickish, widely spaced lines make the best impression.

The neurons which receive the parvocellular input include specialists in narrower lines, but don't cope well with fast-moving gratings. And they deal, of course, in colour contrasts as well as luminance contrasts. There are thus, in effect, two parallel pathways, one concentrating on rapid movement and matters of immediate urgency, the other on the luxuries of detail and colour contrasts.

The feature-sensitive cells were discovered by David Hubel and Torsten Wiesel, using delicate micro-electrodes to record activity in the visual cortex first of cats and then of monkeys. (The initial discovery was serendipitously assisted when one of them happened to wear a Fair-Isle sweater, and the device recording impulses in the cat's visual neurons clicked furiously every time he walked across the animal's field of view.) The cells are neatly arranged. All those that respond to vertical features are stacked one on top of each other, in a column a few tens of micrometres wide. In the adjacent column all the cells respond to features on a diagonal about ten degrees from vertical. A little further on the optimum orientation shifts another ten degrees or so. This pattern continues pretty much all over the primary visual cortex, with only an occasional discontinuity. And by the time a complete set of orientations has been covered the features that produce the excitation are being reported from an adjacent bit of retina.

The information derived from magno- and parvocellular pathways is combined in various ways in the different layers of striate cortex. Many of the neurons in the later stages of processing don't require the stimulus to be quite so precisely positioned as at earlier stages. They are known as complex cells. Some of them specialise in moving stimuli. Others seem to compensate for the fact that the eye is never quite still, so that the image on the retina is constantly shifting slightly.

Dense interconnections among cells mean that a neuron's performance can be influenced by what goes on outside its receptive field. A cell specialising in lines may fire most enthusiastically for a segment of a continuous line, while others are most effectively excited by a line or edge that doesn't completely cross their receptive field. Things are getting yet more mind-boggling as the neurologists, having worked out the basics of the system with simple stimulus patterns, study what happens when natural scenes are viewed. Receptive fields don't seem to be completely fixed, but can contract or expand under different circumstances. And with more natural stimulus patterns the responses evoked in V1 tend to be less intense.

Primary visual cortex can be described as a filter for extracting more information about contrasts from the data reported to it, but it's a flexible and dynamic filter. The data from the lateral geniculate provides only a small part of its input. A great many messages come in from all the further stages of the visual pathways, and a great many more from subcortical areas, contributing to the analysis.

The array of feature-sensitive neurons uncovers only the very basic components that go towards making up a shape, leaving much to do before anything significant is revealed about visual objects. However, it seems to constitute a pretty adequate machinery for registering visual textures. Indeed the retina on its own reveals something about texture, and the lateral geniculate quite a bit more. Some V1 cells are sensitive not only to bars of a certain width but also to the degree of spatial separation between them, which makes them ideal for registering texture. In the primate visual cortex the neurons specialising in one particular spatial frequency group together, just as those specialising in one particular orientation do.

We humans, living as we do amongst clearly outlined manmade shapes and uniformly coloured manmade surfaces, coping happily with the single lines that make the written word or a line drawing, probably tend to underestimate the importance of visual textures. In the natural world texture is frequently more important than shape. In the forest a tree is most easily distinguished from a neighbour of another species by the differently oriented edges of its leaves, and the different angles at which those edges meet; or in winter, by the angles and density of its branches and twigs. Most of nature is the same - grass, bark, fur, feathers, scales, sand, pebbles, shale.

Visual texture can play an important part in pointing the way to a potential meal, or in camouflaging a predator. It can provide clues about whether surfaces are likely to be suitable for travelling on - their horizontality or otherwise, their tactile nature, their potential for supporting weight. A texture that moves across the background is often a danger signal. A capacity for registering visual textures is probably at least as useful as the more complex machinery required to add the edges and lines together so as to decipher visual shapes, so it seems likely that ancestral species possessed only the former ability. Indeed, there may well be extant species which discriminate only textures and not objects, especially among those which are active when the light is poor.

Movement and distance

Around a fifth of complex cells specialise in identifying movement, and fire strongly only when the feature moves in the right direction. The response to a movement which is slightly off the optimum path is modest, and the cell is completely silenced by movement in other directions. The excitatory input comes from several simple cells, all activated by the same sort of feature, and spread across adjacent retinal fields. A stimulus moving in the wrong direction triggers inhibitory input. The connections are such that these neurons are not only highly selective about the direction of movement, they also have preferences as to its speed, so a burst of impulses conveys two bits of information.

The mechanism of the direction-sensitive neurons is reflected in the way they can be fooled. If a spot of light disappears and is immediately succeeded by another spot a little distance away what's usually seen is a single spot which has moved. There's a clear bias in favour of perceiving movement, rather than sudden disappearances and reappearances.

Another interesting aspect of the motion-detection system is its extreme sensitivity. We can detect movement without knowing what it is that moves. When people are asked to watch a screen which displays a moving grating of a very low contrast they become aware of an indefinable rippling effect at a contrast level which doesn't permit them to identify the bars which compose the grating. We can also be aware of a movement in the corner of the field of vision without being able to decipher what the moving object is.

The inability to discover anything about form or colour must be attributable to most or all of the input being provided by the magnocellular pathway, which supplies limited information about one, none about the other. The sensitivity must reflect the fact that the direction-responsive cell collects input from a whole row of other neurons. Each contributor may be firing too briefly and faintly to provide useful information about the nature of the stimulus, but the combined excitations add up to produce a decent rate of firing in the motion-detector.

Moving stimuli are particularly important, since they may threaten danger, or signal the potential for a collision. Judging distance is also pretty vital. There are three ways of doing it. The closeness of a very near object is revealed by the angle to which the eyes swivel to converge on it. For distant objects the only option is to make use of motion parallax, moving the head and finding out how much the relative positions of the various objects in view are altered by the change of viewpoint. The third method utilises the way the visual fields of the two retinas overlap. When we focus on something our eyes rotate horizontally until the image of the object in which we are interested falls on the fovea in both retinas. Any other object at the same distance will also be registered on exactly corresponding spots in both right and left retinas. But because of the slightly different view from the two eyes the retinal images of all objects at greater or lesser distances are slightly displaced horizontally. (This is easily demonstrated by focussing on an object a few feet away and seeing how the background shifts when you close first one eye, then the other.)

Distance is established by correlating the two slightly different views, and this is yet another task performed by those columns of cells all dedicated to stimuli of the same orientation. They are in fact paired columns. Input from the left eye goes to one of the pair, input from the right to the other. When lines or edges have been registered the results from the two eyes are brought together, and an ingenious arrangement effects a further analysis which reveals distance. It involves a whole bunch, for each site in the central retina, of what may be called distance-sensitive cells.

These become significantly excited only when the same sort of feature falls on both receptive fields. For a large proportion of such cells the double input required to fire them must come from precisely corresponding areas of the two retinas. In other words they are excited only by a feature which lies at the distance on which the gaze is focussed. Other cells, however, combine the input from retinal areas that don't quite correspond, in ways that reflect varying degrees of disparity between the sites at which the feature is registered in the two eyes. Some cells are activated by features beyond the plane of focus, with different ones tuned to different distances. Similarly there's a range of cells that's excited by features situated in front of the plane of focus.

The disparity-registering cells don't measure absolute distance, of course, only relative distance. Converting the latter to the former must involve adding into the equation the degree to which the eyes are swivelled inwards, which gives an indication of where the focal plane is. The system is effective at shorter ranges, but less helpful at longer ones, as disparities and adjustments of convergence both become more marginal. Then motion parallax becomes a better bet than binocular parallax. In normal circumstances we make use of both sources of information, and of learnt cues as well, and people who have sight in only one eye may find it possible to make fairly adequate distance judgements.

But the binocular disparity detectors alone do a very efficient job, over the range at which distance judgement really matters. Anyone who has ever looked in a stereoscope will know this. Bela Julesz demonstrated the efficacy of the system yet more powerfully with his invention of the random-dot stereogram. He created a random pattern of tiny black and white squares, and then duplicated it, but with a central area of the pattern shifted slightly to one side. The difference between the two is imperceptible until the patterns are viewed stereoscopically, one to each eye. Then the central area detaches itself from the background and appears nearer or further away, according to the direction in which the dots or squares have been shifted. It can take a little time to achieve the effect, probably because co-ordinating the focussing of the eyes through a stereoscope requires a little practice. But once established, it's very clear.

John Ross and others developed the idea further by creating stereograms of computer-generated moving dots. When the two screens are appropriately viewed, one with each eye, the shape defined by binocular disparity appears to float in front of, or behind, the rest of the display. Ross described it as looking like *a sheet of dark plastic, on which points of light move about. Although the pattern of points of light on the surface changes continuously, the square itself presents a solid and unchanging appearance.*

Stereograms have been created which contain areas of several different disparities, representing up to ten different distances. Moreover, the three-dimensional effect can be obtained even if the figure and the background both occupy the same area. If the dots of the apparently nearer plane are small and numerous they are seen as texture on a transparent surface. When they are larger and fewer the sense of depth is still created, but they are seen as isolated points. The effect of movement towards or away from the viewer can also be arranged.

There are some neurons, meanwhile, which are sensitive to both distance and movement, as might be expected since nearby movement or movement towards the self may urgently demand action .

Colour

The contrast-sensitive cells in layer 4 of primary visual cortex receive the input of the parvocellular pathway, and can be supposed to deal essentially in red/green contrasts and what they reveal about textures and edges. Some koniocellular blue-OFF cells also send axons to layer 4. But the two middle koniocellular layers of the lateral geniculate nucleus deliver most of their output to a specific part of layers 2 and 3. Treating striate cortex with a substance called cytochrome oxidase turns these layers spotty. Lots of little dark patches appear, fading at the edges into the paler surround. (Cytochrome oxidase has this effect because the cells in the patches have a somewhat different metabolism from their neighbours in the matrix.) The koniocellular axons deliver their information to the patches, and information from the parvocellular pathway is relayed there too, via layer 4. And here the neurons seem to be particularly interested in colour. In the lateral geniculate the wavelength-sensitive cells give their best response to the wavelengths which produce the maximal effect on the cone receptors. In this part of V1 cells can be found which make the biggest response to shades such as violet, orange, or lime - shades that result from a mixture of the four basic colours.

Nevertheless activity in these areas still seems to correlate with the actual wavelength mixtures that are reported from the ganglion cells of the retina. It doesn't explain the subtleties of our colour experience - the way, for example, the same mixture of wavelengths, at the same strength, is seen as orange in poor light and brown in good light. Nor does it account for the fact that snow still looks white, even on a dull day, and coal looks black even when it's reflecting twice as much light as poorly illuminated snow; or for the fact that we can distinguish between a reddish-tinged object observed in normal light and a normal range of colours seen in red light. Perception is clearly influenced by overall lighting conditions. It's also influenced by local context.

One of the leading explorers of the effect of different lighting conditions was Edwin Land (who also invented the Polaroid camera). He used displays composed of variously coloured rectangles which resembled Mondrian's paintings, so he called them mondrians. He found that whatever the mixture of wavelengths illuminating a mondrian the blues still looked blue, the reds red, and so forth. In fact it's only when a patch of colour is completely surrounded by darkness that the conscious perception corresponds to the mixture of wavelengths that reaches the eye and is recorded in the retina. (To describe short-wavelength light as blue is a convenient shorthand, not strictly accurate.) To begin to explain such mysteries the chains of excitations must be followed out of V1 into the several adjacent areas of occipital lobe that process the visual input further.

Before we go there, however, it's worth noting that the visual cortex of primates has more neurons to the square millimetre than that of other mammals, so we probably see the world in greater detail, as well as with more colour.

Beyond the primary visual cortex

V1 is at the very back of occipital cortex, mostly on the part that folds in, where the hemispheres meet. From there the next step in the pathway is V2, which similarly forms a retinotopic map, one in which practically all the neurons that cover the central visual fields are binocularly driven.

A suitable stain turns V2 stripey. There are thick dark stripes, narrower pale stripes, and very thin dark stripes. Each little collection of one thick, one thin and two pale stripes deals with one small portion of visual field.

The thin stripes get a good deal of input from the dark patches of V1, especially those in layers 2 and 3. This suggests that they are particularly concerned with colour, and indeed many neurons here respond very selectively to hue. They are neatly organised, moreover, with cells that respond best to red adjacent to others that respond best to orange, and so on, through all the colours of the rainbow, with intermediate colours, such as turquoise and yellow-green, tucked in at the appropriate sites.

The thick and the pale stripes seem to provide further processing for the information about contrasts, motion and disparity. This is where the data from the fragmentary receptive fields of V1 begin to be put together to reveal longer edges. Some cells here respond to the 'object' side of an edge but not to the 'background' side. Assembling the bits of data from V1 on a larger scale also allows different sorts of texture to be distinguished, and some cells are sensitive to a border between a plain surface and a textured one. The horizontal connections between cells in the same layer stretch about twice as far here as in V1, which suggests that the principle neurons not only cover a somewhat larger fragment of the visual field but also extend a wider influence over the operations of their neighbours.

Much of the V2 output goes on to V3, which has cells sensitive to all the same qualities as previous areas, but exactly what goes on is still unclear. Further combination of the information to discover rather larger-scale shape and movement would seem to be the logical guess.

Part of V4 is responsible for the final correlations which provide the finesse of our conscious experience of colour. There are cells here for every possible hue, including all those muddy colours like brown and khaki, olive and indigo, which combine colour with luminance. There are also cells which respond only to white. The area is organised in colour columns, with the cells in any one column all responding to the same colour.

The neurons have larger receptive fields than in V1, and it's thought that their activity is influenced by a particularly large surround, so that the mixture of wavelengths causing the excitation in the centre can be compared with both the level of illumination and the mixture of wavelengths over a much wider field. This would explain the illusions we experience at borders between light and dark or between different colours. Since our colour experience is influenced by the overall level of illumination it seems likely that there is also an input from those ganglion cells which are themselves sensitive to light, as well as combining the data from rod and cone receptor. Their information, as well as regulating the dilation of the pupil and the operation of the biological clock, must be applied here.

The neurons of V4 are generally not activated by colour alone, however, but are also concerned with form. They require a particular shape, a certain fragment of colour- or texture-filled outline. It seems that by this stage the edges and textures of V1 have been assembled into units that can become the components of visual objects, with colour and texture playing important roles. Damage to this area can interfere with object recognition and discrimination.

V4's harvest of information is relayed onwards to the lateral part of the occipital lobe and the lower part of the temporal lobe. In the monkey, damage in the latter area disrupts both colour discriminations and the ability to detect a specifically coloured stimulus. In humans it results in colour-blindness, and an inability even to dream in colour. In both species such damage also causes considerable difficulties with object recognition and discrimination. But we still haven't quite reached the stage at which whole objects are perceived.

Meanwhile information about movement goes to an area in the middle of the temporal lobe, generally referred to as MT, which specialises in visual movement. Here the cells have quite large receptive fields, and are very selective for the direction of movement, with columns dedicated to different directions. A substantial number are not selective about orientation, however. This makes sense, since moving objects tend to be seen in new perspectives as they move, and the animate sort change shape as well.

The tiny receptive fields of V1 can only indicate movement at an angle to the orientation of the stimulus. The larger fields of MT add together the information from a good many of the miniscule apertures to discover what's really happening.

The area gets input from the thick stripes of V2, from V3, and some from V1. There's also some direct input from the lateral geniculate nucleus and by another subcortical route. These latter contributions must be responsible for the fact that MT can respond very quickly to retinal input.

Next to MT is an area called MST (Medial Superior Temporal), which has even larger receptive fields. Neurons here respond to the patterns of overall movement inscribed across the retinas by the movement of the self through the environment - patterns of expansion and contraction, of rotation, and of simple flow. They also receive information about eye movements, and seem to adjust the interpretation of these patterns of optic flow to compensate for eye movements.

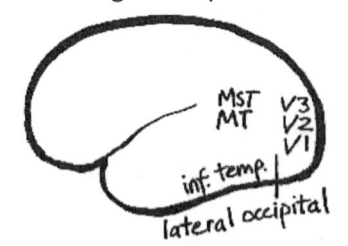

The essential strategy of the neuronal processing systems should be clear by now. Each layer of neurons through which the messages pass serves to compare outputs from the previous layer. In the retina the activity of the cones and rods within a small area is compared by ganglion cells, enabling them to sift out contrasts. The output of adjacent ganglions is then similarly correlated in the lateral geniculate nucleus of the thalamus, where the magnocellular pathway concentrates on luminance contrasts, and is most effectively stimulated by sudden or brief inputs, while the parvocellular pathway deals largely in contrasts between longwave and mediumwave light, in finer detail, and can maintain its response for longer, all of which suits it for registering stationary or slow-moving stimuli.

In the primary visual cortex the magnocellular and parvocellular axons synapse on cells which filter out information about contrasts that extend over slightly larger areas, for the most part oriented edges or lines. Further processing discovers the density at which similar features are registered, while other cells combine the inputs about orientation to reveal distance, and movement. In short, the data provided by the photoreceptors is analysed by filtering it through a series of parallel channels, each extracting a different sort of information.

At every level the contrasts are sharpened by inhibitory connections between the excitatory cells, allowing those that are most strongly activated by the input to suppress those that are less excited. This compensates for the fact that neurons are not precise on-or-off machines, but merely fire their best for a perfect stimulus, and a little less for a slightly imperfect one. It also permits the strategy known as population coding to be applied. This all adds up to an efficient means of extracting information without the need for an impracticably huge collection of components.

When the input data has been passed through all these filters, and the analysis has produced numerous individual fragments of information, the business begins of assembling the fragments to see if they add up to anything significant and actionworthy.

Recapitulation

In the retina ganglion cells receive inputs from four different types of photoreceptor and use them to measure the overall distribution of light, and to discover points of contrast between light and dark, between longwave (red) light and mediumwave (green light), or between blue (short-wave) light and a combination of both red and green.

In humans by far the most numerous type of ganglion cell reports on margins at which longwave photoreceptors are more excited on one side and mediumwave ones on the other.

Around 10% are sensitive to contrasts revealed by reports from both cones and rods. Some of these fire only when a contrast first imposes on their receptive field, others both when a stimulus enters the receptive field and when it leaves. Both sorts are therefore most effectively excited by a moving stimulus.

Some ganglions are simply excited by blue light. Others are excited by blue and inhibited by a combination of red and green, or *vice versa*.

A small number are in themselves photosensitive, as well as receiving input from both rods and cones, and report on the overall level of illumination.

The major part of all this data is transmitted to the lateral geniculate nucleus of the thalamus, where it is dealt with in two magnocellular layers, two parvocellular ones and six very thin koniocellular layers.

Information from the two eyes is kept separate at this stage, but all the data from the left side of both visual fields goes to the right thalamus, and *vice versa*.

The results of the processing in the lateral geniculate nuclei are relayed to primary visual cortex, or V1. There the reports of contrasts are sifted to find larger areas of contrast - edges, lines of differing widths, corners.

The cells of the neocortex are organised in columns, and in V1 each column deals with a particular orientation.

The results from the feature-sensitive cells are fed into neurons which compare the positions of a feature on the two different retinas to discover its distance.

Other V1 neurons receive their input from a number of cells with adjacent receptive fields and register movement.

In layers 2 and 3 the reports from the Blue-ON and Blue-OFF retinal ganglions, relayed via the koniocellular pathway, arrive at 'blobs'. These form part of the pathway which turns wavelength information into the experience of colour.

All this information about small areas of contrast, colour and distance is passed on to V2, V3 and V4, and put together as increasingly large edges, corners and so forth, with colour information contributing to the shape definitions.

A part of V4 provides the calculations which turn the data about the relative amounts of input in different wavelengths and the overall levels of illumination into the huge range of colour shades we experience, and allows us to recognise them under different lighting conditions.

Data about visual movement goes to middle temporal lobe, MT, where the inputs from the tiny receptive fields of many V1 movement-sensitive neurons are added up to reveal more about the nature of the movement.

Chapter 7

Learning to See

How are the edges, colours and distances interpreted as visual objects? The way the challenge is met is demonstrated by those ambiguous figures that turn up in so many works on visual perception. Consider this one, devised by Edgar Rubin in 1915. Is it the outline of a vase, or is it two faces? It can be seen as either - but only with much practice as both at once. Below is the even more venerable Necker cube. Probably you will see it first as a cube viewed from above. Look at it long enough and it will suddenly turn into a cube viewed from underneath. If you haven't met the Necker cube before it may take a while before your perception of it suddenly flips, but with practice you can switch back and forth between the two perspectives fairly easily.

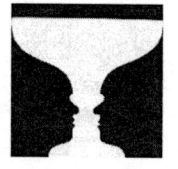

This is curious. The retinal stimulation and the information transmitted to lateral geniculate and on to visual cortex remain unchanged and yet the conscious experience alters. We see a cube from below or a cube from above, we see two faces or a vase. Clearly the conscious experience is not simply a reflection of the retinal input. It's an interpretation of it, translating it into known objects. Somewhere between the registering of lines and edges and the conscious perception decisions have been made about just what the lines and edges mean - decisions that are usually quite unconscious.

There's also a third way of seeing the Necker cube, and there's significance in the fact that it's generally ignored. It could reasonably be interpreted simply as a two-dimensional pattern, not particularly meaningful. To someone brought up in wholly natural surroundings and ignorant of cubes it would be just that. And to a man who had lost his sight in infancy but had it partially restored by a corneal graft at the age of fifty, it was just a jumble of lines. But for those accustomed to a largely right-angled world, and schooled in geometry and its diagrams, making a three-dimensional interpretation is automatic.

The influence of experience on perception was extensively explored by Jerome Bruner and his colleagues. Using a tachistoscope - a machine which projects an image for a very brief and precisely measured period of time - they established that familiar stimulus patterns are 'seen' more quickly than less familiar ones, and less familiar ones more quickly than wholly unlikely ones. In one experiment, for instance, Bruner and Leo Postman showed brief exposures of nonsense words, and found that those which were constructed fairly like English words, in that the letters followed each other in combinations commonly found in English, were identified at significantly shorter exposures than those in which the sequence of letters was not characteristic of English. Expectations, they found, could also affect perception. When subjects were asked to identify words and were told what grammatical class of word to anticipate, they recognised those which conformed to the prediction at shorter exposures than those which didn't.

An experiment with playing cards underlined the influence of both past experience and current context. The trick was that while some of the playing cards flashed on the tachistoscope screen were conventionally coloured hearts and spades others were unorthodox - black hearts and red spades. Each card was shown first for ten milliseconds, and the exposure time then gradually increased until the subject could describe the card correctly. The average exposure required to identify the normal playing cards was 28 milliseconds. The average exposure needed before an unconventional one was identified was four times as long - 114 milliseconds. Moreover in about 10% of cases the incongruously coloured cards remained unrecognised even at the longest exposure used, a full second.

Clearly the familiar pattern was being converted into a conscious visual experience much more readily than the incongruous one. But a breakdown of the results showed that the expectations put in place by the first identification also played a part. When the first card presented to a subject was an unorthodox one the average exposure needed for a correct perception was 360 milliseconds. When the first encounter with an unconventional card followed one or more experiences of a standard card the average time required rose to 420 milliseconds. But after the subject had successfully identified one trick card the average exposure needed for the next one dropped to 230 milliseconds, and for a third or fourth it shrank still more substantially, to 84 milliseconds.

Particularly interesting is what the subjects reported before they got it right. The most frequent reaction was simply to see the unconventional card as a conventional one, some people being regularly influenced by colour, others by shape, while some were equally likely to make a judgement based on either. Another response was to describe the card as a conventional one, but with the colour distorted in some way - the descriptions included brown, purple, rust, olive, and black seen in red light. And some subjects, after reporting inaccurately but fairly confidently when the trick cards were seen at the briefest exposures, became totally confused as the exposure grew a little longer, and found it hard to make any sort of guess as to what they were seeing.

As a result of such experiments Bruner and Postman suggested a theory of expectancy, or, as Bruner would also call it, of perceptual readiness. Perception they said, involves an act of categorisation. And perceptual readiness *refers to the relative accessibility of categories to ... stimulus inputs. The more accessible a category, the less the stimulus input required for it to be sorted in terms of the category*. The neuropsychologist Donald Hebb tackled the same idea using the word 'identity'. *The object that is perceived as having identity is capable of being associated readily with other objects or with some action, whereas the one that does not have identity is recalled with great difficulty or not at all, and is not recognised or named easily. Identity of course is a matter of degree and .... depends on a considerable degree of experience.*

For me both *category* and *identity* carry rather too much baggage from their everyday meanings. Things of very different appearance can share the same category, and identity is too often applied to the individual rather than the class. In company with many modern psychologists I prefer to follow Bartlett and Piaget and use the term *sensory schema*. A conscious experience, it seems, is generally created by matching current input to an established schema. If the sensory input adds up to a familiar stimulus pattern the matching is done immediately, and perception is effortless. If there's no appropriate schema a new one must be created, and achieving a clear perception takes longer.

An activity in which well-learned schemata play an obvious part is reading, and Bruner's concept of perceptual readiness has been nicely illustrated by studies of the way people read alphabetical languages. An experienced reader doesn't need to look at every letter, but skips along the line, fixating only on every fourth or fifth word. The longer, more unusual and most informative words are allotted to central vision, and the gaze slides over the little, thoroughly familiar ones - the conjunctions, articles and pronouns - identifying them without needing the sharpest focus. Recognising words involves paying particular attention to their initial letter, to their length, and to their overall shape. This can be deduced because when a mistake is made the incorrect word is usually one of roughly the right length and shape, and with the correct initial. The sensory input is matched to the wrong schema, but one that has similarities to the right one.

At the same time reading involves expectancies about meaning and grammar. A misread word is likely to be one that still makes sense in the context, and is the right part of speech. If, as the sentence continues, it turns out not to fit, in either respect, the reader is more likely to notice the mistake and go back to correct it. So reading is guided both by sensory schemata and by the expectations aroused by context. If the meaning or the grammar is very abstruse we have to read more slowly, and identify the individual words more carefully.

On the whole, though, reading can be a highly efficient process because the possible expectations are limited. There are only so many letters, the rules of the language only allow them to be combined in so many ways, the number of possible words, though very large, is not infinite, and we can expect sentences to say something we recognise as meaningful. Thus at every level there is a restricted number of schemata to be considered, along with a relatively modest number of possible hypotheses as to the interpretation of the data, so one doesn't need to register all the information available in order to construe the sentence. Thanks to these factors we can often make sense out of distorted or unconventional handwriting. Though few of the individual letters are identifiable when considered separately, the assumption that they add up to known words, reasonable grammar, and some sort of meaning, helps in deciphering them. 'Reading' the natural visual world is a very similar task, but since the natural world is not constrained by so many predictable limitations it can sometimes be a demanding one.

A great deal of the learning that produces visual schemata is done in infancy and early childhood, but even in adulthood we sometimes encounter something novel - perhaps a complicated collection of edges and corners with a three-dimensional arrangement not readily revealed by binocular disparity or motion parallax. Or we come across a familiar object in misleading illumination. We stare at confusion, and then suddenly we realise what it is. Perhaps we run our fingers over the multiplicity of edges and check the ins and outs, or we learn what the machine is doing and how the dimly lit bits in the furthest corner must connect, or someone points out a distant shadow on the hills as the fold through which the road runs. And suddenly the shapes become clearer. It's almost as if one had been looking through a vaselined lens, and the lens has been wiped clean. The whole conscious experience is subtly modified as a decision is reached as to what one is seeing.

If the wrong schema is evoked at the beginning of the process, however, confusion results. Bruner and Mary Potter showed brief glimpses of photographs projected with varying degrees of lack of focus, and asked their subjects to identify the pictures. There were three levels of definition, lightly blurred, middling blurred and very blurred, and three exposure times, 122 seconds, 35 seconds and 13 seconds. The subjects were divided into nine groups, and each group first saw the pictures at one of the nine possible combinations of fuzziness and exposure time, and then progressed through other possible combinations.

The most striking result was that subjects who saw a picture first at a very blurred level of definition and a short exposure had great difficulty in working out what it represented when they saw it under more favourable conditions. The guesses made after a brief glimpse of a fuzzy and unidentifiable picture got in the way of accurate perceptions when they viewed it for longer, or with better focus. Subjects who viewed the pictures first under the more favourable conditions made far more successful identifications than those who saw it under these better conditions only after a first viewing had led them to make wrong guesses.

The essence of perception - or conscious sensory experience - was summed up by Hebb. "We do not really see a thing until we have recognised it."

What exactly is a visual schema?

I said in Chapter 2 that what we record is essentially a pattern of relationships, and gave as an example the visual concept, as I called it there, of a cow. A cow is a collection of components joined together in a consistent way, such that the spatial relationships between some features can change, within certain limits, and that between others cannot - unless perhaps it's dead or damaged. That definition would fit most vertebrates of course. What distinguishes the various species is essentially the shapes of the various components, and the degree to which the spatial relationships can vary. Colour and size are also important, expected to vary only within certain limits. Building a schema is well described in Eleanor Gibson's words - a process of *extracting the invariants in stimulus information*. It means learning which aspects of a stimulus pattern remain consistent in all its manifestations, and which differences serve to distinguish it from other, similar patterns. A preliminary guess at an identification can then be confirmed by directing attention appropriately.

A sensory schema must always, of course, be capable of modification in the light of new experience. Piaget's account of infant learning has proved to fit extremely well to what has been discovered about the learning involved in perception. To use his terms, new information must frequently be assimilated into an existing schema, and the schema must be adapted to accommodate it.

The essence of a schema is its functionality. We need to recognise visual objects in order to know what to expect of them, and what we might need or want to do about them. Just how much detail is necessary in order to make useful predictions about a visual stimulus pattern can vary considerably, depending both upon the nature of the pattern and on the needs and interests of the observer. Schemata can therefore differ considerably in their precision and the amount of detail recorded, and in the degree to which a general schema is divided into more closely determined subcategories. What is just another brown horse to many eyes can to one perceiver be quite clearly my horse Dobbin. A coin may be just a penny or an interesting piece of metallic art. The exactitude of the schema depends on the amount of attention paid to the exemplars.

A visual schema has to be very clear about the essentials of the class of percepts it represents. But at the same time it must usually be fairly rough-hewn, for in most cases it has to embrace quite a range of patterns of retinal stimulation. Dobbin is recognisable in many different poses, and from many different perspectives. Even something as simple as a geometric figure, a square or a triangle, can be classified regardless of its size, and is still seen as a square or a triangle when viewed obliquely. Schemata which cover whole classes of stimulus pattern must be flexible enough to be fitted to a pattern of visual stimulation which qualifies as a novel example of the familiar class, or to a familiar object in an unfamiliar orientation, or to one which is transformed in some considerable but essentially unimportant way. It's not surprising that while we are pretty good at recognising visual objects we are rather bad, for the most part, at dredging up accurate representations of them out of memory, so that even the most intimately known faces and scenes often cannot be recalled in full detail.

Distinguishing between faces is a particularly tricky business, since it involves making fine distinctions between many basically similar stimulus-paterns, and also discounting the effect of constantly changing expressions, and in the longer term the effects of passing years. This makes individual faces especially hard to remember exactly, so photographs of our friends are highly valued, and the portraitist who can capture what seems to many to be the essence of a face is the greatest success. The artist most relevant to this theme, however, may be the caricaturist, who aims not so much at portraying the actual face as at summarising the features which most distinguish it from other faces.

If the conscious visual experience is created with reference to stored knowledge, or schemata, the reverse must also be true. The visual schemata are forged through conscious experience. It's the process of turning the sensory evidence into a meaningful perception that creates the memory trace which is a schema. An experiment by Deborah Chambers and Daniel Reisberg helped to confirm this, with ambiguous drawings like this one. Some people identified this as a rabbit, some as a duck, but few recognised both possibilities without prompting. The subjects were asked to memorise the figure, without their attention being drawn to its ambiguity. The drawing was then removed and the subject was shown another ambiguous figure, and this time it was pointed out that it could be seen in two ways. Then the subject was asked to think about the first drawing and say whether that too had a second interpretation. None of the subjects could re-interpret their remembered image, even after further bouts of coaching and some hint-dropping. On the other hand they could draw the figure from memory, and having drawn it could recognise its ambiguity. This supports the conclusion indicated by earlier evidence - what's stored is an interpretation of sensory input rather than a record of the input itself. It can be deduced that the work which produces perception is also the work which creates a memory trace. The conscious experience which is our sensory model of the world-out-there and the schemata which we derive from it are inextricably interdependent.

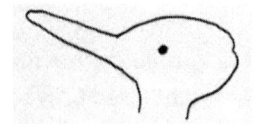

Infant learning

If experience plays such a big part in seeing, it follows that infants must have a great deal of learning to do before they can make sense of the world. They need to build up a visual vocabulary, as it were. The nature of infant interactions with the visual world was explored by Tom Bower. He discovered that the young infant treats something that stands out from the background by virtue of distance or movement as a single object. Show her a matchbox resting on a book and she will reach for book-plus-matchbox as if it were all one thing. Only when books and matchboxes have become familiar as separate objects does this behaviour change. And when Bower contrived to show infants a visual object which moved about as one unit for a while, before breaking up into several pieces which moved off in different directions, they looked very surprised. Clearly we have to learn by experience what visual arrays represent several separate objects, and are likely to be capable of coming apart.

Bower also found that young infants can be insensitive to the appearance of visual objects, but very sensitive to the speed of visual movement. When a moving object disappears behind a screen infants as young as eight weeks anticipate its reappearance very efficiently, turning their eyes to meet the exit point just as, or just before, the object emerges. Bower arranged things so that one object went behind the screen and a completely different one emerged at the other side after the appropriate interval, and infants of up to about twenty weeks continued to track this as if nothing had happened, unaffected by major changes in the shape, size and colour. Only older infants showed surprise at the transformation. (Perhaps this indifference to varying appearance is not quite as surprising as it seems if one considers that the most compelling visual stimulus patterns, people, tend to turn up differently dressed, often in quite different colours, from day to day.)

However, when an object disappeared behind the screen and an identical object appeared on the other side, but much sooner than would be expected if the original had continued at the same speed, the infants' eyes moved rapidly to the new object, went back to the point of disappearance, then jumped back to the new object again. And younger infants seemed to find this experience disturbing, and refused to look at the oddly behaving apparition more than once or twice.

Bower repeated the experiment in a new form, using an arrangement of half-silvered mirrors and cunning lighting, so that one object seemed to be transformed into another in full view. Infants of up to about four months still reacted in the same way. As long as the object continued on the same path at a constant rate, they continued to track it unconcernedly. But if the rate of movement altered the infants got upset. It was as if these very young babies *define the identity of a moving object solely in terms of movement.*

The next step was to find out what happened when a moving object simply stopped, in full view of the infant. It turned out that the gaze continued to travel along the path the object had been following, and the now stationary object was ignored. This might merely indicate that very young infants don't have very precise control over eye and head movements and cannot stop tracking exactly on cue. But frame-by-frame analysis of videotape records showed that there was in fact a pause when the object stopped, and the eye movement only continued after a few tenths of a second.

Perhaps, Bower thought, the young infant doesn't identify the stationary object with the moving one. If so, the reverse should also be true, and a stationary object which starts to move won't be identified as the same object. This idea was tested with the aid of a splendid toy train with flashing lights. It stood in one place for ten seconds, moved to another place, stopped there for ten seconds, then returned to its starting point, and this shuttling sequence was repeated several times. Infants of between twelve and sixteen weeks of age at first went on with their tracking movement after the train had stopped, but after a few trials they settled down to follow it from A to B and back, without any overshooting.

Is the infant really tracking a single object from one place to another, Bower wondered. Or has she learnt to expect a succession of separate events - a stationary object at A, which disappears and is replaced by a movement from A to B at a certain rate, then a stationary object at B, succeeded by a movement from B to A? If the second account were the true one, then the infant would be taken by surprise if the object moved off on a new track and stopped in a new location. And so it turned out. When, after trundling from A to B and back again several times, the train went off in a new direction and stopped in full view but at a new spot, the infant would ignore it, and look firmly towards the accustomed stopping location, often looking surprised and puzzled when it did not see the train there. This seems to tie in with an observation of Piaget's. He noted how his infant son's gaze followed his mother as she left the room, and then turned back to where she had been standing beside his cradle, as if he expected to see her still there.

Bower's experiments showed how little the newborn infant knows about visual objects. They suggest that hardwired elements of the visual system, such as those V1 cells which are connected up so as to be sensitive to different directions and speeds of movement, must shape the infant's earliest responses. The first expectations would seem to grow out of inductive rule-making, or conditioned learning - what has happened several times can be expected to continue happening - and are only modified as visual experience increases and more complex rules about the likely behaviour of perceived objects are acquired. No wonder the very young regard the world with such a solemn, studious gaze.

More recently electroencephalography (which records the electrical potentials on the brain's surface with external electrodes) has shown that the infant capacity for registering visual motion does indeed mature before the capacity for registering visual form - or at least most sorts of visual form. Furthermore, a study of three individuals who had been endowed with sight after being blind from birth or infancy revealed that to begin with they too could only distinguish object from background when the one moved in relation to the other. It took them between ten and eighteen months to learn to use other cues.

For infants, who can concentrate all their efforts on investigating the world around them, the learning is easier. By six months they can cope quite efficiently with objects that change their trajectory while obscured from view. Kochukhova and Gredebäck confirmed that babies of this age initially anticipate that when an object travelling in a straight line disappears behind a screen it will continue on the same course to emerge on the other side. But after two or three experiences of an object which reappears instead at a right angle to its original path they learn to look for it there, showing no signs of distress at the failure to conform with their original expectation.

In addition to developing schemata for visual objects, there are other things to learn. To begin with, we must learn how to look. The fovea, that central area of the retina which is most densely packed with receptors and capable of fine resolution, is very small, covering only about two degrees of arc. In order to study any but the smallest of objects in any detail, therefore, the gaze must move over it, focussing now on one part, now on another. The movements are known as saccades, and in adults take up about 10% of the viewing time, occurring at the rate of two to four per second and often coinciding conveniently with blinks. The fixations are usually concentrated on denser clusters of lines or edges, or conspicuous angles, areas likely to carry the most information.

The adult style of scanning takes a long time to develop, however. Newborn infants hardly scan at all; they tend to focus on just one small point, with minor excursions around it, as if just that one bit of the visual scene was enough to occupy their attention span. By two months the saccades grow rather longer. But youngsters of three or four years old still spend a lot of time focussing on the centre of a complex scene, rather than tracking round it. Over the next two or three years the habit of concentrating fixations around the outline of figures becomes established, but there are still a great many small saccades around points of detail, and the features which receive such attention are often not those which an adult would consider most informative. Finally, with increasing experience, fewer saccades are needed, fixations become briefer, and there are longer traverses between fixations.

If infants concentrate their attention on one small cluster of features at a time it may well be because some part of the neuronal machinery has to mature before information gathered from a wider field can be meaningfully assembled. But the evidence also indicates that we must work out the most profitable ways of scanning a visual object, so as to derive the maximum of information with the minimum of effort.

In adulthood each individual seems to develop their own particular scan-path for a scene, and may repeat it several times. There will be irregular excursions outside this well-used pathway, but a fairly consistent re-tracing of one pattern of saccades and fixations takes up a significant proportion of viewing time. Moreover if the same picture is shown again, a little while later, the same pattern reappears, accounting for perhaps half the viewing time. The fixations are no doubt concentrated largely on those features of the object which are essential for making an identification, or on distinguishing the current example of the class. But it seems likely that a schema is somewhat influenced by the shape of the first scanpath. If an element of pure chance helps to determine how the attention is first drawn around an object, that could be why the memories of different individuals often emphasise different aspects of the same stimulus pattern.

As well as discovering the best technique for studying an object we also have to learn some general rules about interpreting the patterns of retinal stimulation. We must learn to sort out shadows, and to appreciate their significance. We find out how visual textures relate to tactile textures, and the significance of a gleam of pure white light reflected from an otherwise coloured surface. We learn about transparency, and about the way things seen through water are slightly displaced. We work out how coloured glass affects the colour of things seen through it. We discover the properties of mirrors, and the distorting effect caused by curved mirrors, or by transparent glass of an inconsistent thickness. All this acquired knowledge underlies our adult perception of the world.

However, there do seem to be one or two important things that don't have to be learnt. One is the effect of distance on the retinal size of an object. Bower established that, trained to respond differently to a large square and a small one, an infant doesn't confuse a large square at a distance with a small one close up, even though the retinal image is the same in both cases. Equally, the infant can tell the difference between a square viewed on the slant and a trapezium viewed straight on. The inbuilt distance-measuring mechanisms serve to produce some basic assumptions about visual objects, it seems, just as the movement-detecting mechanisms do, and these assumptions point the infant in the right direction.

Occasionally it happens that someone who has been blind from birth has an operation which renders their eyes functional. They are then in very much the same situation as the young infant in trying to make sense of a visual world - but with the disadvantage that their expectations are likely to have been shaped by what they have learnt to achieve with their other senses, and they are not tackling the job with the infant's unprejudiced sense of adventure. Reports on just how much difficulty they encounter vary somewhat.

According to J.Z. Young, such a patient, opening his eyes upon the world for the first time *gets little or no enjoyment; indeed, he finds the experience painful. He reports only a spinning mass of lights and colours. He proves to be quite unable to pick out objects by sight, to recognise what they are, or to name them.* It takes a long time to learn to recognise visual shapes, and at the beginning it requires conscious concentration. *Of course, if I look carefully I see that there are three sharp turns at the edge of one patch of light and four on the other*, said a newly sighted man when confronted with a triangle and a square.

However, shapes for which there are tactile schemata may be recognised more readily. A subject studied by Richard Gregory and Jean Wallace immediately identified upper case letters, which he had learnt to read by touch, but took many months to learn lower case letters. He could tell the time by a clock too, since he had been accustomed to using a glassless watch. He could use his sight to make his way about without bumping into things, and he could judge the distance of objects he was already acquainted with by touch, but was bad at distance judgements otherwise. Many of the visual illusions experienced by the normally sighted (including the Necker cube) did not work for him, or operated much less strongly. Unfortunately there doesn't seem to have been any follow-up to discover what improvement a few years of practice made.

A more recent case was a man born without lenses in his eyes, who got his first pair of glasses at the age of twenty nine. Their assistance still left him with a far from perfect retinal image, but over the first year he learnt to recognise simple, two-dimensional shapes, and in the second year began to cope with more complex objects. He reported that at first he saw a cow as separate patches of black and white, but perceived it as all one object once it began to move, and thereafter recognised it even when it was still.

Youngsters seem to do better, judging by five subjects, aged between eight and seventeen years, who were treated for dense congenital cataracts or corneal opacity. Though previously able only to distinguish between light and darkness, and in two cases to see a bright moving light, within 48 hours of their surgery they all performed pretty well on a visual match-to-sample task, using well defined objects constructed especially for the purpose. They were also successful at matching objects by touch. But at this point they were very bad at matching a visual shape to a felt one. Three of the subjects had improved significantly on this last task, however, when tested a few days later. In another study a girl of 44 months made rapid progress in integrating tactile and visual inputs.

Illusions created by learning

Creating schemata with which to interpret sensory input is a neat way of dealing with huge amounts of data, and doing it speedily. There's a record of an archetypal form, in which some features and the relationships between them are essential and others less so. If the sensory input fits the essential pattern it can be classified swiftly. Furthermore the classification can often be made successfully even when the data is ambiguous, or when the object is only partially visible. An example of the readiness with which we do it is provided by figures such as this one (devised by Gaetano Kanisza). We are so used to seeing objects partially concealed by other objects that the interpretation is automatic.

The readiness with which we evoke well-used schemata leads to all sorts of other illusions. Many are concerned with distance and size. Our inbuilt systems for judging distance - binocular parallax for nearer objects and motion parallax for further ones - are pretty effective. But as our knowledge of the world develops we learn additional ways of assessing distance. One of the characteristics that goes into the schema for a visual object is its likely size. Using this knowledge, and our appreciation of how one object can mask another, and of perspective, we have no difficulty in translating a two-dimensional drawing into a three-dimensional scene. A painting may even look more three-dimensional when viewed with only one eye, no doubt because the effect of learned cues is not contradicted by the absence of binocular disparity. For everyone accustomed to the use of perspective in the visual arts the habit of converting two-dimensional drawings into three-dimensional percepts leads to some well-known illusions.

Psychologists have had great fun devising three-dimensional arrangements which mislead the viewer - at least if only one eye is used - by not conforming to well-established expectations. One such trick array is shown here. The subjects, observing it through a carefully positioned hole, were asked to indicate the distance of the three surfaces by aligning a rod with each in turn. They generally interpreted the display as one would the flat reproduction. But in fact the smallest square was nearest, and properly speaking neither it nor the next nearest shape was a square at all, since corners had been cut out. The accessible schema easily triumphed over the less familiar.

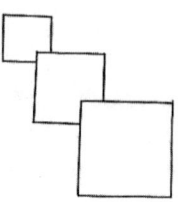

Jolliest and most bizarre of the illusions of this type is the one created by Adelbert Ames, an artist turned psychologist. He designed and built a room which, when viewed monocularly from the appropriate spot, presented exactly the retinal image that would be produced by a normal, right-angled room. In fact - you will have guessed it - it was nothing of the sort. One end of the back wall was much further away than the other. But all the features which would normally supply perspective cues - the angles between walls, ceiling and floor, the shapes of windows and doors - were adjusted so as to create the same image as a conventional room.

Ames' rather startling discovery was that when someone looked into this distorted room and saw two people, one in each of the two further corners, and one therefore presenting a significantly smaller retinal image than the other, what was usually seen was a normal room containing two people of very different size. The assumption that the room was a conventional shape was the prevailing influence on the perceptual experience. We are conditioned to expect rooms to have right-angled corners, and human beings do vary in size quite considerably, so it's perhaps natural that the latter schema should be the more flexible one. The illusion doesn't work when the people in the room are known to the observer. Nor does it work on people whose perceptions haven't been conditioned by straight-lined, right-angled architecture.

How schemata are stored

The experiments outlined above mostly date back to the nineteen fifties and sixties. It seemed reasonable to guess that there might be areas in the brain where neurons could be regarded as functioning to record schemata. This would provide a way of explaining why damage in certain areas of the brain, mostly in the temporal cortex, can result in an inability to recognise a certain class of visual percept, while leaving the sufferer perfectly capable of dealing with other types. A stroke that affects one particular area, for instance, creates great difficulty with faces - the patient may be unable to recognise even their nearest and dearest, or their own face in the mirror. Damage to another area interferes with the ability to read, and a third area is important for recognising tools.

The anatomical evidence was consistent with the idea that the relevant area might hold the schemata for these particular types of visual percept, since after the retinal input has been filtered through the various stages of the occipital cortex the results are passed forward through the lower part of the temporal cortex. It wasn't until the eighties and nineties, however, that neurons towards the front of the temporal cortex of monkeys were identified with activity which suggested they might be the physical machinery embodying the hypothesised schemata.

The first discovery was of neurons that were activated specifically by monkey faces. Some fired for any conspecific face, others only for the face of a particular individual. It seemed a reasonable conclusion that the first sort corresponded to the generalised visual schema for a monkey face, the second to a more particularised schema. Some were activated by a full-face view, while others specialised in side-views; the degree of fussiness about viewpoint seemed to vary.

Then, in the early nineties, Keiji Tanaka found cells at the occipital end of inferior temporal cortex which were responsive to more modest assemblages of features. He and his colleagues carried out their explorations by presenting three-dimensional representations of plants and animals to his monkey subjects. When they found one that activated a neuron they removed different features from the image in turn, until they found out just what combination was sufficient to produce a strong firing rate.

Later work showed that the neat columnar organisation already seen in occipital cortex continues here. Within one column cells tend to respond to roughly similar feature arrangements. An adjacent column may deal in variations on a related pattern. No doubt the vast variety of potential inputs is dealt with here, as in primary visual cortex, by population coding - most inputs will excite more than one neuron, and it will be the precise combination of excitations that conveys the information to the next stage of the pathway.

Damage in this part of inferotemporal cortex causes difficulties with shapes generally, rather than with any particular type of percept. Receiving a major part of its input from V4, it might be described as assembling fragments of pattern into somewhat larger portions. Another contributor to this process is an area between V4 and inferotemporal, the lateral occipital cortex. Neurons here are sensitive to visual objects but indifferent to textures, don't care about the size of the object, and prefer fuzzy representations to clearly detailed ones. An fMRI study showed that when people studied novel shapes and learnt to identify them activity here declined if the shapes were easily distinguished from the background, but not if they were camouflaged by a similar background. All this suggests that lateral occipital is concerned with sorting out just what edges can be interpreted as adding up to a visual object, and what's to be regarded as background - presumably with the help of distance and movement information. If the objects are already familiar, and easy to detect, it doesn't have to work very hard. The detail within the object, meanwhile, is left to later stages of the pathway to deal with.

At early stages of the pathway retinal position is still important, as might be expected since the stimulus patterns to which neurons are sensitive are generally only components of whole visual objects. Brincat and Connor carried out a study using a huge collection of subtly varied combinations of straight lines, curves and angles, and uncovered effects of relative position as well as shape.

As the visual pathway unfolds and the neurons focus increasingly on whole shapes and objects rather than lines and edges, so they have increasingly large receptive fields, with spreading dendrites that offer the potential for an ever larger number of synaptic inputs. And more and more of the receptive fields include the fovea.

From posterior inferotemporal cortex the results are relayed to the more anterior part, to be turned into information about whole objects, as in the face area. When subtle discriminations must be made among visual stimulus-patterns, with attention to detail, the neurons representing them don't fire more strongly as a result, but they become more finely tuned, and less likely to fire for stimulus patterns that only approximate the perfect one. Sigala and Logothetis created two series of drawings, one of schematic faces in which the positions of the features varied, the other of outline fish which differed in the shape of nose, fins or tail. Each series was composed of ten examples, divided into two categories, and while there were four variable features within each category only two of them were relevant to the task. Two monkeys were trained to pull one lever when they were presented with a member of one category, and a different lever when they saw an example of the other category, a discrimination which they learned to perform extremely well.   Recordings were made from 150 neurons in the anterior inferotemporal cortex, of which 96 proved responsive to line drawings, including the test patterns, and to some natural images.   Of these, 44 responded preferentially to a face which contained one particular version of a feature, or sometimes particular versions of two features.   A very similar proportion of the 65 neurons that were also tested with the fish stimuli performed in the same way. The interesting discovery was that by the time the monkeys had become adept at their task (they achieved a 98% accuracy rate) the neurons which responded to the diagnostic features had become significantly more selective than those which responded to the irrelevant ones.

In a similar experiment Chris Baker and colleagues created stimulus patterns which were based on a vertical rectangle, or stem, with various different shapes attached at either end, to form something they called a baton. Monkeys were trained to distinguish eight of these batons altogether, again being rewarded if they pulled a righthand lever for some, a lefthand lever for the others.   Then, for control purposes, novel versions of the same sort of stimulus were presented. Neurons which had developed a strong response to the learned stimuli no longer responded to other variants on the theme.   All this suggests that inhibitory influences become more powerful as learning advances, preventing the cell from firing for mere approximations to a preferred stimulus pattern.

A rather different sort of experiment by Adam Messinger and colleagues confirmed that significance can be as important as appearance to inferotemporal neurons.   Here the monkeys learnt, by trial and error, that four picture stimuli should be grouped into two pairs.   First, having fixated on a small spot, they were shown a single picture.   This was followed by a presentation of two other pictures from the set, and they were rewarded for looking at the one that was designated its pair.   As their learning progressed the stimulus patterns that were paired came to elicit increasingly similar neuronal responses.

Presumably it's by similar means that we learn that Mother is always Mother, no matter how she's dressed, and Auntie is not Mother, even if she has borrowed one of Mother's outfits.   What's learnt is where to pay attention to make a reliable categorisation, and this is what the schema-areas record.

The neuronal excitations in early stages of the visual pathway correlate to a fair degree with the pattern of retinal excitation.   Those in the later stages tend to correlate with the conscious perception.   Or, to be accurate, some ingenious experiments have shown that they correlate with what a monkey reports that it sees, and it seems reasonable to suppose that what it reports is a conscious perception.   The experiments make use of ambiguous stimulus patterns that can be seen in more than one way, and the monkeys are first trained to make one response to one of the possible percepts, another to the alternative.

Leopold and Logothetis studied what happened when different images were presented to the two eyes. Humans in this situation see first one image and then the other, an unwilled alternation that usually takes place every few seconds. Monkeys seem to have the same way of dealing with the competing patterns of excitation. In inferotemporal cortex practically all the neurons the experimenters sampled changed their firing rate just before the monkey reported an altered perception. Earlier in the pathway the activity of about half the cells changed as the perceptions switched, and even in primary visual cortex about 18% modified their response.

Bradley, Chang and Andersen used a different illusion. It's possible to create on a computer screen a pattern of moving dots which humans see as a rotating transparent cylinder. It's still interpreted this way if about half the dots are travelling from left to right, and half in the opposite direction, but a cylinder seen as turning towards the right will suddenly convert into one that turns to the left. Monkeys were trained to signal by an eye movement which way a real transparent cylinder was revolving. Then they were put in front of the illusory cylinder on the computer screen while the scientists studied the activity in MT, the area in the middle of the temporal lobe that deals with visual motion. About half the neurons here varied their activity in tandem with the experience reported by the monkey. Cells that were normally activated by left-to-right movement would fire strongly when the monkey indicated it saw rightward movement, and much less strongly when the cylinder was seen rotating to the left - despite the fact that, as with the Necker cube, the retinal input was the same in both cases.

I hope it will be apparent by now that our conscious visual experience is a thoroughly creative interpretation of retinal input. It might be described as a guess as to what the retinal input is likely to represent. The guess is heavily influenced by acquired knowledge of the world, and sometimes it means that we see what we expect to see. It can also be influenced by the current context - Jerome Bruner showed that a line and an adjacent squiggle is interpreted as a B when it's surrounded by letters, as 13 when it's surrounded by numbers. And sometimes the conscious experience is coloured by current pre-occupations and desires, so that we think we see what we're looking for, and find we were mistaken only when we look more closely. As Richard Gregory put it: *normal everyday perceptions are not selections of reality but are rather imaginative reconstructions.*

In view of its creative nature it seems quite appropriate to refer to the conscious visual experience as 'the picture in the brain'. When first used this term was roundly criticised as implying the existence of a little homunculus sitting in the brain looking at the picture. I don't believe modern readers are likely to put this interpretation on the phrase. But if you do feel tempted that way, gentle reader, please don't imagine the homunculus sitting and looking. Conceive of him as painting the picture. He's dashing about in manic haste, slapping on some paint here, filling in another bit of canvas there, rushing over to a distant corner to make a convincing interpretation of the signals he's received about that bit of the scene, and then haring back to the first area to update his work as fresh data come in. Then you turn your head, and he has to start all over again. (Terry Pratchett fans may like to think of him as performing a job rather like that of the elf inside the Discworld camera - but having to work considerably faster.)

The creativity that goes into the picture-in-the-brain is demonstrated by one particularly flamboyant bit of invention. The axons of the retinal ganglion cells all run to one spot on the noseward side, where they join together to form the optic nerve. Since the space is entirely occupied by axons there are no receptors here, and there is a small blind spot. This doesn't normally matter - the two blind spots cover non-equivalent bits of retinal field, and what is missed in one eye will be recorded in the other. But if you view the world with one eye closed there is no indication that something is missing - the brain simply fills in the gap to match the surround. You can prove this to yourself by making two conspicuous dots on a blank sheet of paper, about an inch apart. Hold the paper out at arm's length, close your left eye, and focus your right eye firmly on the lefthand dot. Then bring the paper slowly towards your nose without shifting your gaze. At a certain distance the second black spot will vanish, to be replaced by unsullied background.

The same filling-in process occurs when blind spots result from damage to the retina. This is unfortunate for people who develop glaucoma, a disease which destroys retinal cells, for they may not notice the gaps appearing in their vision until fairly significant bits of the visual scene begin to vanish, by which time the damage to the retina is well advanced. Most oddly, though, the creative cover-up can happen when the retina is functioning perfectly well and there is a genuine lack of visual input. If, in the far periphery of the retinal field, there is a texture that looks as if it should be a continuous pattern, but which actually has a small blank gap in it, the gap may quite rapidly vanish. That busy little homunculus with the paintbrush decides it's unlikely, and opts to override the data and cover it up.

The process of creating the picture might be described as a sort of proto-reasoning. It's done, most of the time, without conscious thought. The rules about interpreting the input are put to use, and the inputs are matched to the learnt schemata before the conscious experience even materialises. Even when a novel or imperfectly defined visual object is encountered the process of considering the evidence and clarifying the perception only rarely feels like something that should qualify as thought. In effect, the process must involve dealing with questions like: *is that a shadow or an edge? should greater weight be given to this feature or to that?* But the questions are not generally formulated in consciousness. Nevertheless, a juggling with learned concepts is involved in the creation of the conscious experience. The evolution of such sensory reasoning can be seen as an important step on the path towards abstract reasoning.

Recapitulation

Visual illusions show that our conscious visual perception is usually an interpretation of the retinal input, one that is greatly influenced by past experience.

Experience enables us to build up visual schemata, which are embodied in the connections of more forward areas of a pathway through the lower part of temporal cortex.

Such connections are readily activated, while visual inputs that don't correspond to existing schemata take longer to be converted into clear perceptions, or are misinterpreted.

A schema can be defined as a record of an essential pattern of relationships, and of where attention should be directed.

The firing of neurons in the schema-areas is heavily influenced by the significance of a stimulus pattern. Details may influence the neuronal response only if they are important for distinguishing the pattern from others of a different import.

This is why we are good at identifying visual objects in varying manifestations, bad at recalling their exact appearance.

There's a lot to learn before a really useful conscious visual experience can be produced. As well as building up a good collection of visual schemata, which includes knowing how to scan the most informative bits of the scene, there are all sorts of rules to master - about the effects of the direction from which light falls, for instance, and about matters such as transparency and reflection.

The neuronal activity involved in creating the conscious experience seems to be what creates the schema.

It has a lot in common with more abstract reasoning, and can be regarded as a forerunner of it.

Chapter 8

Creating the Conscious Visual Moment

A notable characteristic of the visual world is that it's a stable world. Our eyes dart about, fixating momentarily on one portion of the scene and then sliding to another, our heads turn, we walk, run and bend, so static objects are stimulus patterns that mostly move across the retinal field. Conversely, we often follow a moving object with our gaze, and the moving object remains pretty much fixed in relation to the retina. Yet what we see is a still world, with the self moving through it, and we correctly identify the other moving phenomena. This implies that some pretty dazzling calculations go into the creation of the picture.

The mechanisms responsible are found in parietal cortex. While one set of outputs from the occipital cortex goes to the inferior temporal lobe another is transmitted to the parietal lobe, to a region which is less concerned with discovering what a visual object is than with working out where it is. Here the sheets of neuronal columns don't make simple maps of the current retinal field. Eye position is brought into the calculations, and the receptive fields of the cells are determined by the direction of gaze as well as by retinal position, so they constitute a head-centred representation of the view. In later stages of the pathway input from the vestibular organs is added, allowing head movements and body movements to be taken into account. Increasingly the maps come to cover the whole potential field of view.

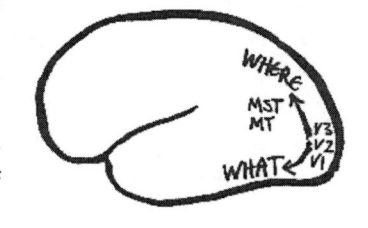

Damage to this parietal region can produce difficulties in perceiving spatial relationships, as in the case of an unfortunate man who suffered two strokes, affecting the area bilaterally. In one test he briefly saw an X on a computer screen and was asked at each trial whether it was to the right, left or centre of the screen, or whether it was in the upper, lower, or centre part. His performance on this simple task was only 70% correct. In another test he was asked to judge whether an X had appeared to the right or left of an O, or whether it was above or below, and he found this task extremely difficult and got only about half right, the result that would be expected from chance.

Shown the two letters in sequence and asked whether they had appeared in the same place or not his performance was equally bad when the relationship was horizontal, though slightly better when it was vertical. Incorrect judgements that the letters had occurred in the same place were made more frequently than the opposite error.

In addition to the calculations which anchor the picture in space there may be some help from the fact that, as Tom Bower pointed out, one's own nose is always in view - unconsidered, shadowy shape though it may be - and eyebrows too are usually detectable at the edge of vision. Most other mammals must equally be able to see their own noses, and those few species which have unusually flat faces generally make up for it with fringes of hair that hang down over their eyes, while flat-faced birds tend to have tufts of feathers similarly arranged. These intrusions may well provide an additional means of relating the position of the self to the view. Input relayed from the muscle-stretch receptors in the neck must also help to calibrate exactly where the head is pointing, and for some directions the sight of other bits of the body must contribute.

The parietal lobe seems to need this sort of help, for while its calculations about eye and body movements are very efficient at identifying spatial relationships, they don't by themselves create a wholly stable world. If someone sits in a completely dark room observing a stationary light, the light soon begins to appear to move. Moreover if, after a short while, the observer is asked to point to the light, their gesture is often inaccurate. Equally odd things happen if a subject in a dark room is shown a luminous dot in a luminous frame. When the dot moves and the frame is stationary the movement is correctly identified. But when the dot is stationary and the frame moves it's usually still the dot that is seen as moving. On the other hand, if instead of the dot there is a small luminous shape that looks like a ship or an aeroplane the observer may see it moving forward, but is less likely to be deluded into seeing it travelling backwards.

It seems that we don't depend exclusively on knowledge of eye movements and body movements to know what's static and what isn't. We also make use of the expectations derived from previous experience, and treat certain things as likely to be fixed landmarks - anchors for the 'picture' - and others as likely to move. Information from the inferotemporal pathway is used to augment the calculations in the parietal pathway. Deprived of fixed landmarks we soon become confused, which is why it's so easy to get lost in a featureless desert.

When the temporal or ventral and the parietal or dorsal pathways were first distinguished the former was described as a 'what', the latter as a 'where' pathway. However the operations carried out in the parietal lobe have turned out to extend well beyond locating the stimulus patterns that are being identified in inferotemporal cortex. The further calculations needed for preparing actions on them are also performed there. One subdivision is involved in managing eye movements, another directs the reach towards the object, a third calculates how the hand must be shaped to grasp it, and a specialist sector of the latter organises the appropriate grasp for a tool that is to be used rather than just picked up.

Just what the visual areas of parietal cortex can do has been demonstrated by a study of a patient with a lesion in inferotemporal cortex that left her unable to recognise visual objects. Nevertheless she shaped her hand correctly when she grasped something, and when she was given a rectangular object to post through a slot she orientated it in the correct way to go through the opening. Clearly the parietal cortex receives quite a lot of information about shape. Indeed, the parietal cortex is more objective about visual objects than the temporal. Our conscious experience is sometimes subject to illusions – here, for example, the circle surrounded by smaller circles looks larger than the one surrounded by larger circles, although they are actually both the same size. Even when the conscious experience is deluded by such arrangements, however, the hand put out to pick up the object gets shaped to the appropriate size.

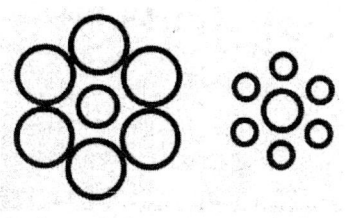

This evidence suggests that the ventral visual pathway's function might best be described as deciding whether a stimulus pattern should inspire any action, and the dorsal's as working out how the action should be performed. Establishing just where the stimulus is constitutes only one step in the latter process. However, from the point of view of creating the picture in the brain it's combining the 'what' and the 'where' that's important. The necessary connections between the two pathways are gradually being uncovered.

The unstable eye

Saccades and tracking movements are not the only motions of the eye. The fixations are not held steady; the eye regularly drifts a little, and then flicks sharply in a microsaccade - enough to shift the image over several receptors. This movement can be reduced with practice, but not wholly eliminated. You can observe the effect when you've looked at a bright light for too long and experience an after-image. Being fixed on the retina, the after-image trembles and dances along with it. The picture-in-the-brain shows something that is absolutely fixed in relation to the retina as moving.

One obvious means of exploring the significance of the mini-movements is to see what happens to perception when their effect is eliminated. Back in the nineteen fifties an ingenious device was developed to achieve this, a contact lens with a little stalk attached, to the end of which a tiny, simple line drawing could be fixed. It turned out that a stimulus thus stabilised in relation to the retina might quite rapidly vanish. Within a period of as little as two or three seconds, it was reported, the lines of the drawing would soften, reducing the contrast between line and background. Almost immediately some part of the figure would disappear, and sometimes the whole figure, the space being filled in by 'background'. The disappearance was not permanent, however. The lost bit of drawing would re-materialise within seconds, while another bit would vanish in turn.

By the nineteen nineties images could be stabilised on the retina more comfortably, with computer displays that were constantly adjusted for microsaccades, and it was confirmed that lines do tend to disappear in these circumstances. The thinner the lines are and the lower the illumination the speedier the vanishing act. Images defined purely by colour, with no luminance contrast, behave in the same way. A stabilised green disc against an orange background disappears very quickly. And something particularly intriguing happens when - as Crane and Piantanida ingeniously accomplished - the boundary between two coloured stripes is stabilised on the retina, while the outer borders are unstabilised against a dark field. The boundary vanishes within seconds, and the experimental subjects reported seeing dots of both colours, or islands of one colour against a background of the other, and in some cases a mysterious, indescribable combination of the two colours. Finally, in 2007, Michele Rucci and colleagues, using a very sophisticated technique of image stabilisation, confirmed that fixational eye movements do help with the perception of fine lines (high frequency gratings), but not with thicker ones.

Collewyjn, Martin and Steinman, meanwhile, tried a different trick. They experimented with set-ups that gave their subjects the chance to compensate for an abnormal amount of movement of the retinal image, in a way that would have made it possible to achieve perfect stability. They found that the degree of compensation chosen did not maximise stability, but maintained a certain appreciable degree of image movement.

So the eye's little drifting movements do perform a useful function. Fine detail is registered more effectively with their assistance, resulting in a better picture-in-the-brain. Further experiments by Rucci's group have shown that this is particularly useful when viewing the complex small-scale detail of natural scenes.

Part of the explanation may be that minisaccades serve to move a feature across the receptive fields of several retinal ganglion cells, several lateral geniculate cells and several of V1's feature-sensitive neurons. This increases the chances that at some point it will be perfectly centred on the receptive field of the appropriate sort of neuron and cause the maximum rate of firing. As a result population coding must work more powerfully at each level, making use of the reports from a larger number of cells than would otherwise be the case. The amount of information that can be obtained from a limited population of feature-sensitive neurons is thus increased. The firing rates of a whole clutch of V1 neurons are in effect compared over the duration of the not-quite-steady fixation, and a more reliable conclusion about the feature is reached than could otherwise be achieved.

This system should help to compensate, too, for the imperfect optics of the eye. It might be regarded as not unlike the astronomers' trick of superimposing several pictures of the night sky, one over the other, in the hope that the light which shows on all of them indicates a genuine star, while the fuzz caused by reflections off particles in the earth's atmosphere will vary from one exposure to the next and can be ignored. The drifts and microsaccades of the eye allow for dirty data to be averaged, and thereby cleaned up. Such a system might well be misled by an edge or a fine line that was stabilised in relation to the retina, and thus excited fewer feature-sensitive cells than normally happens and produced a lower overall rate of firing. From time to time such an input might be judged to be 'noise' rather than information.

In addition the constant shifting of the image on the retina means that cells which are excited only by moving stimuli, both retinal ganglion cells and complex cells in primary visual cortex, can contribute to the business of registering stationary ones. Several different types of response have been found in the cortical cells. One sort fires when a microsaccade causes a stimulus to move into, or across, or out of the receptive field, but only when the movement is in a certain direction. Another behaves similarly, except that it isn't direction-sensitive. Others, with rather smaller receptive fields, are activated only by stimuli that remained centred. The fourth sort shows a burst of activity when a stimulus is brought into the receptive field, and continues to fire at a lower rate during the period of stabilisation and slow drift. The third and fourth of these categories clearly respond optimally to stationary stimuli. The other two, especially the neurons which are sensitive to movement in any direction, must provide a means of extracting information about non-moving objects as well as moving ones, thanks to the eye's micro-movements.

From a subjective point of view it's easy to see a static stimulus pattern, more difficult to see one that's moving, and perhaps confusing the experience by changing its shape as it does so, or going too fast to be seen as anything more than movement. Stationary objects tend to provide a much more satisfactory visual experience. However, from the perspective of creating a neuronal mechanism to deal with it movement must be the easier proposition. A moving object obligingly defines itself by activating a whole row of retinal ganglion cells, of lateral geniculate cells, and of V1 feature-sensitive cells. Its passage thus provides a good deal of data at all levels of the machinery, and if some of the cells fail to operate it doesn't matter because the movement of any object of more than miniscule size will have activated a good many of them, more than enough to trigger the direction-sensitive cells of V1 and MT. (Movement-detectors have to operate pretty efficiently, of course, since they only get one crack at the stimulus.)

A stationary stimulus is another matter. Excitation in a bundle of feature-sensitive cells could indicate a stationary stimulus, but could equally be the result of a moving stimulus passing through their receptive fields. In order to discover which it is, it's necessary to establish that the stimulus being registered now is the same one that was being registered a moment ago. For some species the solution is no doubt to wait a moment and act only if an adequate level of stimulation builds up over that period. The system which compares a series of inputs as the eye drifts and then flicks back towards its original position is subtler. It exploits the necessary delay to extract finer information from the data collected. The picture-in-the-brain is thus the product of what might be called a comparing memory, whereas the animal that merely needs to accumulate a certain amount of input before making a response to a stationary stimulus uses an accumulating memory.

A parallel can be drawn between vision and the tactile sense. We learn a great deal more about an object by running our fingers over it than by just touching it. Rats and mice use a similar tactic when they are exploring - they twitch their whiskers forwards and backwards, at a rate of around eight cycles per second (a gesture known as whisking). Clearly more can be learnt from a changing input, achieved through a displacement of the receptors, than from a static one. Vision works in much the same way, except that the eye is continuously in exploratory mode, and makes use of very tiny movements as well as larger ones.

For us the benefits of microsaccades are only modest, helping us to see fine lines in poor light, or to sort out potentially ambiguous clusters of features. For species with fewer photoreceptors, fewer ganglion cells and a smaller primary visual cortex they should be more conspicuously beneficial, especially for species which undertake most of their activity in dusk or darkness. I suspect that microsaccades explain why mammalian eyes look so different from those of reptiles, amphibians and fish. For species with foveas they must be particularly useful for getting significant lines and edges centred precisely on the small receptive fields connected to that bit of retina.

Creativity takes time

Another significant aspect of the picture-in-the-brain is that it takes time to create. We seem to perceive the world instantaneously, but a measurable amount of time is needed to collect the data and turn it into conscious perception. Once created, though, the conscious experience has a life of its own, and may outlast the input. The relatively clumsy pace at which conscious vision operates accounts for the fact that a movie running at 28 frames per second is perceived as showing smooth, continuous action - though the early 24-frames-per-second looks a little jerky.

How much input is needed to produce a conscious experience, how long does it take to create the picture, and how long does it last, once created? The answers have been very revealing.

First of all, the input. The tiniest, briefest flash of light - just a few photons delivered to a dark-adapted eye - can produce a conscious experience. But experiments with the tachistoscope show that for any sort of pattern or shape to be experienced a minimum of ten milliseconds of well-illuminated input is required. And that's when the stimulus pattern is a very simple one, such as a letter, or a well-defined line drawing consisting of only two or three clearly related features, and where the contrast is good so that averaging the inputs over a longer fraction of a second isn't necessary. Considerably more time is needed for a complex pattern to be turned into a meaningful conscious experience, as some of the experiments described in the last chapter demonstrated. The time required to collect data explains why a fast-moving object is seen only as a blur - it passes across the retinal field too fast to provide an adequate sample of stimulation, and as far as the conscious experience is concerned merely interrupts the collection of evidence about the scenery behind. If the object is travelling fast enough (a bullet, for instance) it doesn't even do that.

After an adequate tachistoscopic exposure the 'picture' doesn't appear immediately, but takes a further fifty or sixty milliseconds to materialise. The fairly precise measurement is inferred from the fact that if a second stimulus pattern is registered on the retina within this time the first one doesn't reach consciousness. The deduction is that the sensory input has to be processed to turn it into conscious experience, and an interruption by further input disrupts the operation.

The conclusions reached from experiments with the tachistoscope fit well with the timing of the eye's saccades from one fixation to the next. There are generally between two and four fixations per second. The saccades in between take about 25 milliseconds each, on average, so fixations must generally last from about 200 to 500 milliseconds. On the evidence of the tachistoscope, this is about the amount of time needed to collect data, process it, and appreciate the result. Complicated scenes require fixation durations at the longer end of the range, but for dealing with simple, clearly defined and well-learned stimulus patterns such as printed words 200 milliseconds can be adequate.

What also happens during the fixation is that a brief record is established. We know this because if the scene viewed is succeeded by a blank visual field the 'picture' can linger on after the retinal stimulation that produced it has ended; on a dark visual field it can last for a matter of seconds. You may have experienced the effect under natural circumstances, during a thunderstorm on a dark night. A flash of lightning illuminates the scene that was invisible before, and the visual sensation fades slowly, lasting longer than the illumination provided by the lightning. In more usual circumstances the record must serve to promote the sensation of a static scene when the eye is saccading and what's being registered on the retina is a blur of movement. The illusion is helped by an arrangement which during a saccade automatically suppresses transmission in most of the channels leading to visual cortex.

The mechanism that creates the lingering image was named, by Ulric Neisser, iconic memory, while the image itself came to be termed the icon. The seminal experiment on the icon was done by George Sperling. He showed his subjects a tachistoscopic exposure of a three-by-three array of letters, and asked them to read out one line, but they were only instructed which line to read (by means of a high, medium or low tone) after the exposure was terminated. As long as the signal indicating which line was required came immediately after the flashed exposure of the letters they read the letters off quite easily, and reported that they were still experiencing a conscious percept when they did it.

Other workers developing the possibilities of this experiment used only a single line of letters and asked their subjects to read out just one, and this time the required letter was marked by an indicator flashed on the screen immediately after the display of letters. This turned out to work as expected as long as the indicator was just a small line underneath the space where the chosen letter had been. In the subjective visual experience this translated into an actual underlining of the letter. However, when the experimenters attempted to mark the letter with a circle enclosing the space it had occupied things turned out rather differently. The subjects saw the circle, but not the letter. The sign which was meant to indicate which letter should be read out actually made it invisible. This is how it was discovered that new input can block out preceding input if it follows too fast.

An obvious question to be asked about the icon was how it relates to the retinal image and to the big picture. Is it fixed in retinal space or objective space? Since conscious vision gives us an objective view of space the latter might be expected. A variation on that last experiment was devised to elucidate the matter. Once again a row of letters was flashed on the tachistoscope screen, and once again a circle followed, positioned so that (intentionally this time) it would obliterate one of the letters. The new wrinkle was that the subjects were trained to centre their gaze on one fixation point, and then to transfer it very precisely to another, nearby fixation point before the circle appeared. It was theorised that if the position of the icon was fixed in relation to the retina then the circle would replace the letter that now occupied the same bit of retinal field. If the creation of the icon took eye movements into account then a different letter would be masked.

As it turned out, neither prediction was fulfilled. The letter which was missing was the one that occupied the same site on the retina as the circle. But the circle did not appear surrounding that empty space. Instead it was seen framing the letter which it was expected to mask according to the icon-related-to-allocentric-space hypothesis. It must be concluded that the calculations that turn retinal input into a picture that depicts objective space are efficiently done before the conscious experience materialises, even under very strange conditions.

Other workers confirmed the finding a few years later with a slightly different experiment. The subjects saw a tachistoscopic exposure of a pattern of twelve dots. They then shifted their gaze slightly to centre on a new fixation point and a second array of twelve dots was flashed on the screen. The two collections of dots were designed so that, overlapped in one way, they would make up a regular array with one dot missing, and it would be easy to identify the gap. Superimposed in the other way they would be just a patternless jumble. The result was the same as before. The two displays were combined in the way that produced the correct relationship in terms of allocentric, objective space.

Clearly one of the operations carried out in the period before the visual scene reaches consciousness combines the results from inferotemporal and parietal pathways so as to locate what is seen correctly. Other processes are sometimes completed successfully in this brief period of time, sometimes not. If the stimulus pattern is a familiar one it evokes excitation in a clear-cut set of finely tuned neurons in inferotemporal cortex, as described in the chapter on learning to see - a process which is assumed to equate with matching a visual input to a visual schema. When that happens a clear conscious experience can emerge at maximum speed. A novel stimulus pattern, for which new neurons have to be trained, is likely to take a little longer, so that an initial fuzziness in the experience is detectable - if one stops to think about it. Wholly novel and strange objects viewed under difficult circumstances, may, as shown in the previous chapter, produce a very muddled impression until they have been thoroughly studied, and a satisfactory interpretation of the data discovered.

Another process that may be successfully completed during the initial stage of creation but may take longer is that which establishes the third dimension of the picture. We saw in Chapter 6 how the binocular disparity cells of primary visual cortex function to establish the distance of visual objects. However, there is an obvious hitch to overcome before they can do their job. For the binocular disparity cells to achieve their function the convergence of the eyes must be correctly adjusted; but before the eyes can be correctly converged, not to mention focussed, it's necessary to know how far away the object under inspection is. If the focal plane needs to be altered for the new input the problem can only be resolved by a dialogue between the focussing and convergence mechanisms and the binocular disparity detectors - in other words by a little experimentation. In most cases the binocular disparities recorded in the previous view will provide the basis for a fairly good guess about how much convergence and focus have to be altered to shift the gaze to a nearer or more distant object. But the result is only likely to be exact in familiar situations. On other occasions some further juggling may be necessary before the best possible match of the two views is achieved and the object of interest successfully engages the zero-disparity cells.

We can deduce that this dialogue generally takes place before the conscious visual experience emerges because it can be detected when, occasionally, it continues for longer. This happens when the scenery is not perfectly obliging in arranging itself so that there are no ambiguities. Where groups of identical features are clustered closely together pairing them up in the two retinal images must be tricky; if the convergence of the eyes is marginally wrong there are likely to be quite a few features which can be inappropriately paired in the receptive fields of binocular disparity detectors. On such occasions it must be necessary to sample the images through more than the usual number of convergence adjustments. Perhaps the eye's tiny drifts and microsaccades help with this process, perhaps they make it more difficult. But certainly one may stare at a complicated bit of scenery for some time before working out the exact ins and outs of it. The effort involved becomes particularly detectable when one looks at a random dot stereogram in a stereoscope. At first sight it usually appears to be a simple two-dimensional array. It's only after sustained study and perhaps conscious effort that the three-dimensional figure suddenly pops out - but once it's discovered it's easy to maintain the perception.

How the conscious experience is created

An increasing amount of evidence supports the long-held idea that the conscious sensory experience equates with an extended circulation of impulses around a complex collection of circuits. Distant brain areas become innterconnected temporarily, to work together. A certain minimal amount of input is needed to activate all the circuitry, and a certain amount of time – just how much depending on the complexity of the input, and the degree of familiarity.

In the visual system some of the circuitry must serve to link inferotemporal 'what' information to parietal 'where' information. Other connections provide for visual input to be linked to any coincident auditory and tactile input. Visual, auditory and tactile pathways all continue to the frontal lobe, where the information about the outside world can be combined with information about the inner world and the really important conclusions about sensory input are reached, namely what action it should elicit.

The various stages of the cortical sensory pathways don't just pass information forwards, however, they also send messages back to every station. Some project back to the lateral geniculate nucleus of the thalamus, where fibres delivering information from visual cortex far outnumber those carrying it in the other direction. Some connections even run back to the retina. Thus all the later stages of the pathways can influence what goes on in the earlier parts. The creation of the conscious experience can be imagined as involving a dialogue between numerous cortical and subcortical areas. The inferotemporal neurons activated by the current batch of input send messages back to the occipital lobe which serve to intensify the activity of selected neurons in earlier stages of the pathway, so as to fit with the likely interpretation of the retinal input. In the thalamus neurons fire more strongly when messages returned from visual cortex indicate that they are recording an object, rather than background.

For the most part this process makes for rapid and efficient identification of significant objects. Occasionally it leads to mistakes, making us see what we expect or hope for rather than what is actually there. And clearly it's what makes it more difficult to try out an alternative interpretation of the data once an incorrect guess has been made.

In addition to the guidance from the schema-areas there is also a strong influence from attention. For instance V4, heavily concerned with sorting out colours, is strongly activated when colour is relevant to the current task, less so when it's irrelevant. When it's the direction of movement that must be noted activity intensifies in the temporal area specialising in visual movement, MT. Even in V1 the level of firing is affected by whether what's being reported in the neuron's receptive field is of significance to the business of the moment.

Some 'top-down' communications, meanwhile, serve to clarify the data. One study investigated how the occipital responses to bars moving across a chequered background were affected when MT was inactivated. It turned out that, without feedback from MT, many neurons in V2, V3 and some layers of V1 showed reduced responses. The reductions were particularly marked when the contrast between bar and background was small, so this feedback must play a particularly important role when only limited information is available.

The same circuitry also seems to help in emphasising the ends of lines. Many lateral geniculate neurons are more responsive to the end of a bar crossing the field than to a continuous one. Feedback from MT increases this distinction. The consequent intensification of the feedforward messages presumably makes line-ends and corners more salient to primary visual cortex, and helps the ventral visual pathway with the identification of the stimulus pattern. Feedback from areas such as MT, Adam Sillito and colleagues suggested, *can inform the earlier levels about the broad borders of an object in ways that are [otherwise] inaccessible to the circuits at these levels.*

Other inputs, some facilitatory, some suppressive, come from neighbouring neurons, so that the response to a stimulus within a cell's receptive field is modulated by what's going on just outside it. McManus, Li and Gilbert suggested that the horizontal connections in primary visual cortex provide a range of possible interactions, and that connections from more forward cortical areas determine which are put to use. This helps in sorting out figure from background. As a result of all the accumulating contributions the V1 responses to a stationary, unchanging stimulus pattern evolve over a few hundred milliseconds.

The sustained circulation of impulses around a large circuit must incidentally help to iron out any problems arising from the different rates of transmission in the three separate pathways from the retina. Messages from the parasol/magnocellular system, travelling in thick axons, must arrive first, bringing information about more widely spaced features, which is in line with the fact that the overall layout of a scene is perceived first. The input from the midget ganglion/parvocellular system is needed for the details. Messages from the koniocellular blue/yellow system travel slowest, though there is perhaps some compensation in the smaller amount of processing they have to undergo on the way. But it seems possible they add subtlety to the colouring only after edges and other contrasts have been established.

Meanwhile the conversations between the various stages of the visual pathway form only part of the story. There are also multiple connections to lower areas of the brain, some of which are essential for any sort of consciousness. It's the co-ordinated flow of impulses around all these routes that creates the effect we know as sensory consciousness, a flow that can be briefly self-sustaining.

The current evolves as it goes, winnowing more information out of the original data in the process (apart from the odd occasion when misinformation results). In the chapter on the basic mechanisms of vision a set of neural arrangements was described that's often referred to as a set of filters, each extracting one particular sort of data out of the input. As knowledge of the visual system has improved it's become clear that it's a very dynamic set of filters, with feedback producing rapid modifications at every level.

By the time we are adult a great deal of what we see can be identified pretty rapidly, at least to the degree of accuracy necessary for our needs or purposes. If a stimulus pattern is a familiar one, in a likely context, the classification will be made very promptly, probably at the first fixation and before the conscious experience occurs, so the perception will immediately be clear. That recognition of familiar stimuli can be that fast is demonstrated by the tachistoscopic experiments. It's also implied by the fact that the same pattern of saccades tends to be repeated whenever an object is re-encountered, since the object must have been identified for the appropriate scanpath to be activated. The fact that we tend to be jolted into attention when the familiar scenery of our lives manifests some alteration, even though we have been scanning it for some quite different purpose, is also suggestive. The matching-to-schema process can be very efficient.

Indeed, it can be done without consciousness. Cues flashed on a screen too briefly for conscious perception can influence the subsequent interpretation of ambiguous words or scenes, or the judgement of a scene or a face. Such subliminal cues can also evoke emotions, and produce a galvanic skin response. It's only with novel or indistinct or unexpected input that the classification process takes more time than the process that leads to consciousness.

However, even the most minimal conscious sensory experience only occurs if the circulating messages reach the frontal lobe of the cortex, as Raphael Gaillard and his colleagues showed, with the help of patients who had had electrodes implanted in their brains in preparation for surgery. On a computer screen the subjects were shown first a 'mask' consisting of a row of hatch marks, then a very brief exposure of a word, then either a blank screen or another meaningless mask, the whole sequence taking only half a second. There was no awareness of the word when a mask followed it, although activity was seen in several cortical areas, including the visual cortex and parts of parietal cortex. This died out quickly without propagating further. Without the mask the word couldn't be read, but the subjects were aware of seeing it, and could even indicate whether it was a threatening word or not. The conscious sense of something there, sometimes accompanied by an emotional reponse, came with sustained activity across a larger number of cortical areas, and with a notable activation in the frontal lobe.

More recent work by Pietro Avanzini and colleagues suggests that for a fullscale conscious experience the reactivation of visual cortex by returning messages is also necessary - in other words a complete activation of the circuit. Antoine del Cul and colleagues, meanwhile, have confirmed that a fair amount of processing is done in the occipital and temporal lobes before the conscious experience emerges, and that conscious perception correlates with more extensive activity that also takes in parietal and frontal areas.

An early insight into the development of the conscious sensory experience emerged from a tachistoscopic experiment performed by Haber and Hershenson. After a very brief exposure of a seven-letter word their subjects saw nothing. When the same image was flashed again, after not less than 8 seconds, they would begin to see parts of letters, or perhaps a whole letter, on the second or third flash. Perception of the whole word might require several more flashes. But when the percept developed it was in no way hazy, and the subjects were always quite certain of it, though unable to see anything a few exposures earlier. Given what's known now we can deduce that the succession of inputs builds up the efficiency of transmission in the activated pathways so that the message eventually reaches frontal cortex.

The early findings were linked to the modern ones in an experiment by Moshe Bar. Pictures of various familiar objects were presented very briefly (for 26 milliseconds), sandwiched between two masking events, with each object appearing up to six times in the course of the presentation. The subjects were asked to press one of four buttons to indicate how well they could identify the object, while an fMRI scan was performed. Successful recognition of an object was achieved after several exposures, as activity in inferotemporal cortex gradually increased and propagated forward.

Evidence from touch

The first evidence that sensory consciousness involved expanding the excitations that resulted from sensory input into a longer activation of extensive neuronal circuits came from Benjamin Libet, and involved a different sensory channel. Back in the nineteen seventies he recorded activity in the somatosensory cortex of conscious humans, working with patients who were to receive surgery for epilepsy. (It's customary, before operating on the brain, to explore the relevant area by means of gentle electrical stimulation.) Libet found that a single weak impulse of electrical stimulation applied to the skin of the hand was sufficient to cause a tactile sensation. However, when the current was applied to the somatosensory cortex it took a half-second train of impulses, in a range between 20 and 120 per second, to produce a similar conscious sensation. Prolonging the stimulation beyond the half-second mark created the impression of a prolonged tactile experience, rather than increasing the perceived strength.

Libet also reported that there was a delay of up to half a second before the subject consciously felt a stimulus on the hand. It's been calculated that only a small fraction of a second is needed for the transmission of a signal from hand to cortex, so the rest of the time must be used for cortical processing and elaboration of the message. A further interesting discovery was that a stimulus to the skin could be either masked or enhanced by stimulation applied directly to the cortex within the same fraction of a second - the difference depending on the precise temporal relationship between the cutaneous and the cortical inputs. All this showed not only that it takes time to create a conscious experience, but also that the messages arriving from sensory receptors are considerably augmented and prolonged in the neocortex.

The time taken to set up the circuit explains why a visual input is blotted out by another that follows very fast; the second input interrupts the circulation of impulses initiated by the first. It also helps to explain another phenomenon. When someone is paying attention to one sensory event they tend to miss another one that follows very closely after. Experimental subjects asked to press a button every time they see a certain word in a rapidly flashed list usually miss an example that occurs too soon after the previous one. Attention to the first word, and to the action, can be deduced to maintain the first active circuit long enough to prevent a second one being initiated.

Taking time to create the conscious sensation is fine for static scenery, but surely it's a little dodgy when it comes to moving stimuli, which may require rapid responses? Luckily the conscious experience of movement materialises a little faster than that of stationary stimulus patterns. This is shown by an illusion. When observers watch a moving stimulus and a static stimulus is flashed briefly at the precise moment the moving one is level with it, the static stimulus is seen as appearing just after the moving one has passed. Just how far after seems to depend on the brightness and salience of the static figure.

Perhaps we shouldn't be surprised. The parasol/magnocellular pathway, geared for detecting movement, uses very fast-conducting fibres. Furthermore, collecting enough information to demonstrate the existence of a moving stimulus is relatively simple, and there are numerous neurons involved at every level to ensure it's done efficiently. Establishing the existence of a stationary stimulus pattern, on the other hand, may be supposed to take a little longer.

If our conscious perception of movement is nevertheless slightly delayed there are several factors which must combine to reduce the disadvantages. MT (the temporal area which specialises in visual motion), receives direct connections both from the lateral geniculate nucleus and from another subcortical structure that receives direct retinal input, the superior colliculus, where there are neurons particularly sensitive to moving stimuli. MT in turn communicates directly to the parietal pathways which calculate the 'where' of a stimulus pattern. So it's probable that the motor system is already being primed for use before the conscious experience has materialised. These arrangements must account for the way one sometimes finds oneself tensed and preparing to duck or to throw up a defensive arm almost before one becomes aware of the need for action.

Another helpful gimmick is that we can extrapolate the future path of a moving object, which enables the response to be calculated in relation to where the object is about to be, rather than where it is now. The calculations, based on what is consciously seen and on past experience of moving objects, take into account the delay in conscious sighting of which we are unaware.

Blindsight

One part of the cortical visual pathway that's pretty much essential to conscious vision is V1. People who suffer damage in later stages of the pathways have specific problems. They don't see in colour, or they don't perceive motion, they have difficulty knowing where objects are, or they fail to recognise particular sorts of stimulus pattern - but they do have a 'picture' even if it's startlingly deficient in some respect. People with any significant degree of damage to V1 generally no longer have a conscious experience in the part of the visual field covered by the damage. They may still have some sort of access to visual information, however, for if they are asked to guess where a light or a moving object is they may perform at well above chance levels. They can often avoid obstacles too. Some of them can also guess surprisingly well at which of two orientations is being presented, or which of two colours. The ability is generally termed blindsight, or sometimes Riddoch's syndrome, after the neurosurgeon who first studied it.

Part of the importance of V1 must lie in the fact that it provides the major route by which the output of the lateral geniculate nucleus is forwarded to later parts of the visual pathways. But there are some direct connections from the lateral geniculate nucleus to visual areas beyond V1. Movement seems to be the stimulus most likely to be detected, so the short, subcortical route to MT probably makes an important contribution to blindsight. There are also connections from lateral geniculate to V2, V3, and V4, which have been shown, in the macaque monkey, to be adequate to produce some activity in those areas.

Blindsight patients usually report that there is no conscious visual experience, they just have a sort of feeling about the stimulus. Some, however, have said that on some tests they have been aware of something - a something often described as a shadow, perhaps as a dark shadow on a dark ground. One subject who has been studied at length has said that he does sometimes have vague conscious sensations of brightness, or of visual movement. Presumably the latter is something like the sense of movement one sometimes gets at the edge of vision, without having any idea of what it was that moved. Perhaps we might deduce that a modest amount of retinal data reaching the dorsal 'where' pathway, which needs to be very sensitive, tends to have a greater impact than a similar input to the inferotemporal 'what' pathway. Sometimes, however, enough information may reach the latter to provide a hint of the stimulus.

There seems to be little consistency about these conscious sensations - the same stimulus, under the same conditions, can lead to awareness on one occasion, but none a few minutes later, for no detectable reason. Presumably it's a matter of chance whether each cortical neuron in the surviving parts of the pathways fires in response to the modest input, and consequently whether the result tips the scale sufficiently to trigger the circulation of impulses that creates some consciousness.

Blindsight is more likely to occur in people who have been cortically blind for some time, which suggested that the connections from the thalamus and other subcortical areas to parts of the cortical visual pathway beyond V1 might have become stronger than is normal. This has been shown to be the case both in monkeys and in someone whose V1 was destroyed on one side by an accident at an early age.

Recently it has proved possible to create a temporary condition of blindsight in sighted subjects by applying a brief pulse of transcranial magnetic stimulation to V1. If the pulse is presented at just the right moment it blocks out conscious perception. But these subjects too could often identify the orientation of a bar or the colour of a patch. Like the blindsight patients the reported that they were aware of having seen something, although they had no conscious experience of colour or orientation, and were only guessing at it. Transcranial magnetic stimulation to V2 can similarly prevent conscious visual sensation.

Investigations of what normally happens in the visual cortex of seeing subjects bear out the implications of blindsight. Ress and Heeger used fMRI to study activity in V1, V2 and V3 while their subjects tackled the task of detecting a brief appearance by an indistinct stimulus. They found that activity in these areas was higher when the subject reported seeing something, regardless of whether there actually was something to see. False alarms, in other words, correlated with higher activity than missed stimuli.

Possibly the early stages of visual cortex constitute not only the main means of access to the later stages but also a crucial node in the visual circuits - the place through which all the circulating messages must run if a flow adequate to create a full conscious visual experience is to be achieved. V1 and V2 are the areas with the smallest receptive fields, the ones that record all the details of the visual input, and their exact location. The receptive fields of the 'where' pathway's cells are larger than those of V1, those in the later stages of the 'what' pathway even larger. Moreover the latter pathway is concerned, the evidence suggests, with registering the essential aspects of stimulus patterns, so that they can be recognised sufficiently well for a decision as to whether they require action. Inferotemporal neurons don't seem to be very fussy about detail, except where they need to be to make useful discriminations. Indeed, there can hardly be room to record all the details of all the infinite variety of objects that may at some time or other be encountered.

V1 and V2 may therefore be the only areas that register the full detail of the scene the retina is currently reporting. As Jean Bullier has put it, maybe *the lowest cortical stages act as 'active blackboards' for integrating the results of computations carried out in the different cortical areas of the visual system.*

The binding problem

As the workings of the sensory systems have been elucidated some neurologists and philosophers have puzzled over the question of how, if the neocortex processes sensory input through all those separate channels, do the separate bits of information get put back together to create a meaningful whole? This has become known as 'the binding problem'. It may be illusory. The maintained circulation of impulses around extensive circuitry means that all the threads of evidence, even if they don't reach the final stations in frontal cortex with precise synchrony, will be brought together in a temporal overlap. This must allow the data which has been analysed in different ways through parallel channels to be reassembled into a meaningful whole. In short, sensory consciousness is an ultra-short-term memory which provides the answer to the binding problem.

This conclusion is supported by an experiment carried out by Anne Treisman and Garry Gelade, which showed that while a conscious sensation can be created out of a very brief input the minimum needed for a successful binding operation may be slightly larger. The subjects were shown very brief tachistoscopic displays of coloured letters and digits, and their attention was directed to either the letter or the digit by asking them to report its identity. Questioned subsequently about the colour of the letter or digit the subjects sometimes got it wrong. After a presentation of a red letter and a blue digit they might confidently report the opposite.

Also relevant, perhaps, to this question is the experience of the unfortunate man who suffered major damage in both parietal lobes, causing difficulties in locating objects. He also had difficulty linking form and colour. Presented with a display of two coloured letters, and asked to report the name and colour of the first to appear, he reported incorrect combinations in 13% of the trials, even when the displays lasted as long as ten seconds and his attention was wholly concentrated on the task. This indicates just how important the contribution of the parietal lobe must be in knitting the various strands of evidence back together again to build the picture-in-the-brain. The separately processed aspects of the image must be reunited largely by reference to the position they occupied on the retina. The union must be achieved by linking the parietal 'where' pathway and the temporal 'what' pathway into one circuit, and this is something else that takes a little time.

These experiments confirm that the machinery of sensory consciousness not only allows information from an assortment of different sensory channels to be correlated; it also provides the means of analysing the input of a single sense through multiple channels simultaneously, and then assembling the results. In short, the conscious moment is the manifestation of a process which brings a great deal of data together, and harvests a great deal of information from it. By means of the mechanisms of sensory consciousness the moment's information is then illuminated further by relating it to succeeding inputs, gathered over seconds or even minutes. The conscious moment is subsumed into a stream of consciousness. This process will be the subject of the next chapter.

Recapitulation

It takes a measurable amount of time to turn input at sensory receptors into a conscious sensory experience.

In the case of vision the delay serves to ensure that stationary objects are firmly distinguished from moving ones.

In addition several profitable operations are carried out during the delay.

A succession of excitations in feature-sensitive neurons is compared as the eye drifts and microsaccades, which makes for more accurate identification of poorly defined lines and edges, and allows retinal ganglion cells which specialise in detecting molvement to contribute to the discovery of stationary ones.

A dialogue between occipital cortex and the visual schema areas of inferotemporal cortex leads to the identification of familiar objects or scenes.

A very brief exposure of a familiar stimulus pattern can trigger an unconscious identification which influences the interpretation of subsequent inputs.

It takes more input to make a conscious visual experience.

Still longer study is needed to make sense of unfamiliar or ambiguous stimulus patterns.

Meanwhile the visual information has been transmitted not only to inferotemporal cortex but also to parietal cortex.

There a succession of processing stations has taken saccades, head movements and whole-body movements into account, and translated the retinal image on to a body-centred map.

Thanks to these calculations our conscious experience can show stimuli that have been registered across a succession of retinal receptive fields as stationary.

A conscious experience only emerges if the excitations are strong enough to reach frontal cortex and initiate return messages from there.

The importance of the sustained circulation of iimpulses which creates sensory consciousness is that the extended analysis of sensory inputs can be very delicate, and a great deal of information can be brought together.

Chapter 9

The Benefits of Conscious Sensory Experience

So far the discussion has centred on the single conscious moment. The focus has been on the perception of static visual stimulus patterns, over which we can afford to linger. But consciousness is an ongoing, constantly changing experience. Each instant is interpreted in the light of what's gone before, and the limited information a single moment provides is usually only meaningful if it can be fitted into a larger pattern. The most important benefit of sensory consciousness must be that the process which creates the conscious moment also creates a very accurate though brief-lived record which can be related to following inputs. Sensory consciousness is above all a stream of consciousness.

For most stationary objects we need to put the products of several fixations together to get an adequate idea of the whole. The contribution of the very short-lived record must explain how it is that our conscious visual experience manages to be so much more informative than might be expected, given the limited amount of data that can be garnered towards the periphery of the retina. Each 'picture' is informed by what was learnt in previous fixations. Having identified something as a tree we remember it as a tree, though it's now just a blur on the far edge of vision. The record means that the products of several fixations can be put together to establish the wider-scale layout of the scene. In passing it may be noted that the evolution of a retina which concentrates most of its receptors and processing power in a small central area can only have been profitable in species which possessed such an ultra-short-term memory.

The very brief, very accurate record also serves, it may be inferred, to set up a slightly longer, not quite so accurate record which is maintained in existence even after the gaze shifts to a completely new focus. This enables the new set of data to be correlated with that which has just been absorbed. As you turn from, say, the building on your left to the building on your right you can acquire a very good idea of their spatial relationship, even though you can't get them both in view at the same time. You can also make a pretty good judgement as to their relative size, their relative distance from your viewpoint, and how their colours compare. These judgements aren't quite as precise as they might be if both buildings could be fitted into one retinal image, but they're a great deal more accurate than comparisons that can only be made in a longer-term memory. This sort of record makes it possible to build up schemata for very large objects which can't be studied with one set of overlapping fixations, and to add together the succession of views seen as one moves about.

Most of our sensory schemata are similarly distilled out of a succession of sensory inputs. In us humans a particularly impressive short-term memory is the one which serves audition, enabling us to take in lengthy sentences as meaningful units. It's pretty certain that an expansion of auditory short-term memory was important to the evolution of our capacity for complex language. Much simpler patterns can be extremely significant, however. A rustle in the undergrowth which is repeated at a nearer location, for instance, is a different matter from a series of rustles moving away. Being able to distinguish between the two patterns was no doubt the sort of benefit that first made a short-term aural memory, even one of small capacity, worth having.

This memory allows us not only to capture a pattern that unfolds over a brief period, but even to replay it, for further consideration. It's been termed echoic memory. When you haven't quite caught what's being said, or when what you're hearing doesn't seem to make sense and you need to reconsider your interpretation of the sounds, you can often rewind the tape, as it were, and play the sentence again. And with luck you decipher it on the second encounter.

Other sensory channels too have a short-term memory that allows patterns to be perceived, and which provide a replay facility. We can consider a series of tactile inputs as a whole, or interpret a moving touch. That enables us to learn about a whole object (not just an edge or two) by running our fingers over it, or to recognise a letter or other simple shape traced out on the skin. For those who are both blind and deaf it makes it possible to communicate by means of a finger alphabet.

We can also retain quite a lengthy flow of visual movement, and patterns of visual movement can tell us as much as stationary shape does, sometimes more. We can often identify a familiar person by the characteristic rhythm of their walk or gesture while they are still too far away for more than a basic, overall shape to be distinguished. Just how much information can be derived from visual movement alone was demonstrated by Gunnar Johansson. He attached twelve small lights to a collaborator, at shoulders, elbows and wrists, and at hips, knees and ankles. Then he took a ciné film of the illluminated figure in a completely darkened room, so that only the lights were visible. The film was shown to unprepared observers, with unexpectedly clearcut results. *During the opening scene, when the actor is sitting motionless in a chair, the observers are mystified because they see only a random collection of lights, not unlike a constellation. As soon as the actor rises and starts moving, however, the observers instantly perceive that the lights are attached to an otherwise invisible human being. They are able not only to differentiate between walking and jogging movements but also recognise small anomalies in the actor's behaviour, such as the simulation of a slight limp.* When two people, similarly kitted out with lights, performed a folk dance, the identification was again immediate. Further experiments showed that when the actors put emotion into their movements, the observers could recognise that too.

Johansson's film provides two sorts of information. One is the rhythm of the movement. The other derives from the parameters within which the spatial relationships between the lights vary. The second ingredient is highly informative, but probably not absolutely necessary. Even the movement of a single light, I suspect, can be interpreted into a meaningful percept, or at least a reasonable hypothesis. Imagine unrelieved blackness on the cinema screen, implying a dark night. A single small light appears, bouncing up and down jerkily, and simultaneously swinging from side to side, and gradually grows a little larger. Any *aficionado* of classic horror films is surely going to suspect that this is a lantern being carried over rough ground, and will be prepared for the holder of the lantern to become visible as the light draws closer. A fairly similar erratic bouncy movement without the swing that indicates a flexible attachment between the light and its support might suggest a bicycle being ridden over a bumpy track. A light that wavered only a little as it approached would probably be interpreted as a bicycle on a smooth road - or perhaps as a flying saucer.

There are other circumstances where it may be possible to identify something by movement alone, although there's not so much as a hint of shape to be seen. When the leaves of a tree quiver and branches shake one can often guess roughly what sort of bird or mammal is causing the disturbance from the degree to which the branches bend, and from the rhythm of the movements, even though the thing itself remains invisible.

Finally, we have an accurate short-term memory for the feel of the movements we make ourselves. Immediately after making a gesture, or a sequence of gestures, we can recall the exact feel of doing it. When the movement results in stimulation to the vestibular organ the record includes that sensation as well. Patterns of rising, falling, rocking or twisting that are caused by external forces can also be replayed. Thus we come to develop schemata for the different styles of movement we can expect to be imposed on us when travelling in trains, boats, cars, escalators, or on different animals, and we have a schema for the feel of being on a swing. Having learnt to anticipate these effects we can adjust for them as they occur.

Echoic memory has a very short life, just long enough for several units from the even shorter-term memory which is the conscious moment to be put together, and appreciated as a whole. While it lasts it preserves the pattern of input from a sensory channel very efficiently. And the replay facility works even with sensory input that wasn't receiving attention at the point when it was registered. This is not only convenient in allowing you to fudge up a reply when the person you're ostensibly conversing with pauses expectantly, and you've been listening to the more interesting conversation going on behind you. It also means that attention can be switched to a stimulus pattern in retrospect, when it's only after the full pattern has been registered subconsiously that its attention-worthiness becomes apparent. (Very possibly that was how you came to be listening to the conversation behind you.) This indicates that incoming data can be preserved long enough for the pattern in the flow to be registered at an unconscious level, regardless of whether attention is being paid or not, a valuable arrangement.

It's a necessary one too, since for stimulus patterns which manifest themselves over a period of time the matching-to-schema process can only begin after a certain amount of input has been registered. The individual units of a temporal pattern must be interpreted in the context of the whole pattern, and so decisions as to the identity of an individual unit may only be made after several units have been put together. A word containing an ambiguous consonant is heard in one way in one sentence, and as an alternative possibility in a different sentence - the choice made being the one that produces a more meaningful utterance. What we hear, when we listen to a familiar language, is whole words and whole phrases, rather than a succession of individual speech sounds.

The delay involved in creating the conscious experience is thus a positive benefit in listening to language and other extended series of inputs. It allows time for several units to be registered and fed through the processing system before the conscious experience emerges. This means that there's an opportunity for the interpretation of early phonemes to be influenced by what follows. So even with the very first phoneme of a sentence, perception can be influenced by context.

There's even more scope, of course, for the beginning of a sentence to influence how the later parts are interpreted. Languages take advantage of this to reduce the amount of perceptual work that is required in listening to them. There are always rules about acceptable combinations of sounds, and about the way words can be put together. Identifying one phoneme reduces the number of possibilities for the next one. And identifying a word reduces the number of potential following words, not only because of the grammatical rules but also because a sentence can be expected to convey some likely meaning. Experimental studies have shown that listeners develop a running hypothesis as a sentence unfolds. Under noisy conditions sentences that disobey accepted grammatical rules or express an improbable proposition, and so confound all hypotheses that the listener may have developed, are less likely to be heard accurately than more conventional sentences.

Sometimes, needless to say, the first hypothesis about a sentence is wrong. The last words arrive, and they just don't fit on to the supposed meaning of the first part of the message. Then it's necessary to adjust the hypothesis and re-check the identifications made of earlier words. This is another situation where the replay facility comes into operation, making it possible to match an ambiguous sequence of phonemes to a different schema, in the light of subsequent context.

Comparing memory and accumulating memory

As with the memory that produces the single conscious instant, the patterns perceived by means of the short-term memory are patterns of relationships. Music and spoken language demonstrate this. Few of us have perfect pitch, but most of us can learn to recognise a tune. Having done so, we can recognise it again in a different register, on another instrument, or in a different key; nor will a change of tempo confuse us, provided it's not too extreme. Indeed many of us, far from being confused by a change in key, timbre or tempo, take little notice of it. Clearly what we remember most readily is the relationship between the notes, not the absolute values. When Beethoven's fifth symphony begins with its famous *da da da BOM* what we essentially register is three notes of the same pitch and duration and a fourth that is lower, longer and louder.

Listening to language, too, is a matter of identifying relationships rather than absolutes. The records stored about speech-sounds are not precise specifications but comparisons. Vowels are characterised not by a precise shape but on a flexible scale of openness; their length is measured by comparing them to other sounds produced by the same speaker. As a result we manage to cope quite successfully with different accents and different speeds of delivery. We can understand both the very refined lady who says in clipped tones *Ai don't think ai quaite laike thet* and the Australian who drawls *Ah'll tike a plite 'v 'stike 'n chips*. When we encounter a new accent it takes a little while to understand it easily because we need a fair-sized sample of speech in order to make comparisons and work out the parameters of the sounds as they are used by this speaker. But once we have done that we are unfazed by the departures from whatever we happen to regard as normal pronunciation.

So, just as a visual object is a collection of features of various relative sizes, relative colours and relative distances in a certain spatial relationship, a tune is a sequence of notes measured in terms of higher and lower, shorter and longer, louder or softer. A movement, either seen, felt or performed, is a sequence of ups, downs, lefts, rights, nearers and furthers, along with the assorted compromises which constitute diagonals, all in relation to a fixed reference and combined with a sequence of variations in speed. Absolute values exist, obviously, but on the whole it takes training to register them accurately and remember them, whereas the patterns of relationships are learnt more easily. The schemata we build may well include limitations on absolute values, but the essence lies in the record of relative values.

Hardwired stimulus patterns can also be patterns extended over time. The flashes emitted by fireflies, for instance, are signals by which the females attract mates, and each species has its own distinctive pattern of flashing, so that where two or three different species inhabit the same area interbreeding is avoided. A comparable system exists among many species of frog. In the mating season male vocalisations summon the females, and the latter move only towards calls uttered at the correct pulse rate, distinguishing them both from the signals produced by other frog species, and from those produced by a male engaged in a disagreement with another male.

For the stickleback shape and colour is transformed into a mating signal with the addition of a special pattern of movement, and various other species use the same trick. The pigeon, for example, bobs along and bows repeatedly in his attempts to woo a female. Distinctive displays fulfil the same function as a change of male plumage in other species. In some birds the signals may be emitted by both sexes, and they work themselves up to the act with a sort of ritual dance.

Such extended signals imply something that could be termed a short-term memory. The efforts of the flashing firefly, the croaking frog and the posturing bird must work as a whole pattern, not just as a series of moments. But where the signal that must be registered is an innate one (as essential intraspecific signals always must be), the construction of a suitable memory is not too difficult. Imagine a species with a male mating call that goes *da da da BOM*. For the female to register it an auditory neuron which responded to *da* would be required, and another which responded to *BOM*. The *da* neuron would connect to an integrator neuron, which would fire only after receiving three inputs. The *da* integrator and the BOM neuron would in turn feed into a master neuron, and there would probably be inhibitory connections to prevent the excitation of this master neuron if the inputs arrived in the wrong order.

The master neuron might require several complete signals to be registered before the female began to move in the direction from which the sound came, and more calls would be needed for the approach to continue. If another male started to call and his call registered more strongly - whether because he had a louder voice or because he was nearer - she would divert towards him. And when a male and a female came together further signals would no doubt be exchanged, of a tactile or olfactory nature (no species can afford to be cavalier about mating matters) before the female accepted the male.

The neuronal wiring which registers such mating calls has in fact been worked out in several sorts of insect and various species of frog. It's invariably designed to require very precise timing of both the sounds and the pauses between them. If the rhythm is slightly wrong the integrator neuron doesn't fire, so there is little chance of a female being misled.

Such arrangements for registering extended signals exemplify an accumulating memory. The point to be stressed is that this sort of neuronal device works only for the genetically ordained pattern, and all other sequences of input pass it by. The memory that (within its capacity limits) can record any sequence of input the sensory channel is capable of receiving, and which is used to make comparisons among different patterns, is quite another matter. That sort of temporary storage arrangement clearly requires far more complicated neuronal engineering. And for mammals the neocortex provides this.   Animals whose neocortex has been removed can make simple auditory discriminations but cannot learn to distinguish changes in the temporal pattern of a tonal sequence. Damage to human auditory cortex may produce a similar deficiency.

Another piece of evidence that points to the same conclusion was provided by Sharpless and Jasper. They discovered that neurons very early in the cat's auditory pathway habituate to a given sound, but can be kicked back into full action if it changes in intensity. Neurons at the next level habituate similarly, but can respond to a change in pitch. Only in the neocortex does a repeated pattern of input produce habituation, which is relieved when the pattern varies.

We must conclude that the sustained activation of interlinked circuitry which creates the conscious moment persists to provide a longer-lasting record that preserves a succession of moments. This is what allows us to perceive very accurately an infinite variety of patterns of input that are extended over a brief period of time – to experience a stream of consciousness.

Echoic memories and sensory intelligence

The essential sensory schema - one that is developed without conscious effort - must deal in the sort of pattern that can be encompassed within the span of the accurate, replayable short-term memory. Larger echoic memories mean more complex schemata, and that in turn means a larger number of schemata, because more distinctions can be made. There must be considerable differences between the endowments of different species. We humans have an auditory echoic memory of particularly large capacity. Other species which produce long and variable vocalisations, such as various cetaceans and some songbirds, must be similarlyl equipped.

Monkeys and prosimians, meanwhile, must have especially large short-term memories for patterns of visual and performed movement, and for patterns of vestibular input. These provide the knowledge they need for their speedy and acrobatic travels through the trees, allowing them to match their jumps to the space that must be covered, maximise the propulsion to be gained from bendy branches, and tune their landings to the solidity or flexibility of the landing place. Youngsters must learn partly by experiment, partly by observing their elders, absorbing the rhythms and shapes of their movement patterns, and developing a vocabulary of what's possible. Similarly, those carnivores which pursue swift-running prey that jink and swerve must learn to time their final pounce exactly. We can deduce that they too have pretty good short-term memories with which to perceive the phraseology both of their own movement and of their prey's. The means of perceiving a fair-sized chunk of movement-pattern can be deduced to be essential to learning strategies of pursuit.

Then there are those species which are active mainly after dark, or live largely or entirely underground. It's likely that they have extensive short-term memories for whisker input - to which a great deal of their sensory cortex is devoted - since it's largely by means of this specialised tactile sense that they find their way around.

As well as major differences between species in the capacities of the various sensory memories, there must also be some variation between individuals of the same species. Amongst humans the guess must be that musicians, poets and actors are likely to have above-average auditory memories, painters above-average visual memories, and dancers particularly good memories for self-movement. Such artists can create patterns of a complexity which challenge the rest of us very slightly. They can also imitate patterns a little more easily than the rest of us. Learning a complex tune, or a series of ballet steps, or painting a realistic portrait, has to be that bit easier if slightly larger chunks of relationship can be perceived at one go.

Are such differences genetically determined? Probably they are, to a great extent. One of the benefits accruing to a social species is that all members of a group can profit, in many cases, from the special talents of a few, so many varied talents can flourish, and this is of course particularly true of *Homo sapiens*. But learning and practice are required to make the most of an individual's genetically determined potential. And since we generally enjoy practising the things we are good at more than struggling with the activities in which we cannot compete succesfully, the bedrock of natural talent is difficult to distinguish from the effects of practice.

In any case, there are all sorts of difficulties about measuring these sensory short-term memories. They hinge on the vital question: what sort of units of measurement do we use? Some of the earliest experiments, concerned with the auditory memory, were done by the nineteenth-century psychologist Wilhelm Wundt. He used the simplest and most elementary of stimuli, the clicks of a metronome, running at around one per second. A subject would listen to a series of clicks, then, after a brief pause, a second series, which might be of the same length, or one click longer, or one shorter. It transpired that a series of eleven clicks could readily be distinguished from a series of ten. Counting wasn't necessary - it was just immediately perceptible as different. One might conclude that the short-term memory is sufficient for recording a minmum of ten or eleven separate stimuli; or perhaps that it can hold at least twenty odd seconds' worth of input. However, Wundt found that it's possible to impose a subjective pattern on the clicks, hearing them as groups of two, or groups of four. Doing this increases the size of the stimulus-trains which can be discriminated as similar or different quite substantially.

Using echoic memory to register a string of assorted stimuli is more demanding. Experimental subjects asked to repeat an arbitrary sequence of letters usually manage around seven, with the variation plus or minus two. But it's possible that reciting the list interferes with the echo.

However, if the subjects are asked to repeat an arbitrary string of words the score is still somewhere around seven, even though the number of speech-sounds is now noticeably greater. Moreover if the words make up a meaningful sentence the length of the input that can be perfectly reproduced increases considerably. It seems that if the stimulus patterns can be readily identified - matched to an existing schema - the amount of information that can be preserved in echoic memory is very much expanded.   The count is the same for words as for single letters, and if the words are meaningfully linked it can be made in concepts rather than stimulus patterns. Perhaps echoic memory must be measured in matching-to-schema units. The conclusion to be drawn may be that the real capacity of echoic memory is always much the same, and its apparent capacity is determined by which stage of the sensory pathway is involved, whether it's the early stage that identifies speech-sounds, a later one that identifies words, or the complex of brain areas that makes sense of whole sentences.

The experience of learning to hear a foreign language provides a natural experiment in the use of echoic memory. First we have to build up schemata for the speech-sounds the new language employs and our native tongue doesn't, before we can begin to hear words properly. And only when we can hear the words clearly do we begin to decipher whole sentences, rather than guessing at their general drift. If it takes all your concentration to identify the unfamiliar speech-sounds then the useful capacity of echoic memory may be one phoneme. If you can identify the odd word, but only with effort, then the capacity is one word. When you attempt to replay a heard utterance many of the sounds remain ill-defined, and only odd words or fragments of words turn out to have been captured. This implies that when there are no schemata to which to refer the input the capacity of echoic memory is very small. Only when all the words are familiar can echoic memory encompass a normal-length sentence. The fact that there *is* a normal maximum length for sentences must say something about the typical echoic memory capacity in humans.

Measuring the span of time or input across which similarities attract notice might be one way of assessing echoic memory. How wide can the gap between two stimulus patterns be if a match is still to be immediately noticeable?   For auditory memory a good clue may be provided by poets' use of rhyme.   Clearly rhyme only works if the second occurrence of the sound arrives while the first is still echoing in the mind.   Line 27 of a poem may rhyme with line 8 but nobody will notice. Rhyming couplets, on the other hand, are certainly unmissable. Rhymes that are two pentameters apart are pretty salient. Those separated by three pentameters, as in some sonnet forms, attract notice more delicately. Leaving a longer interval is rare, and the effect tends to be more of a hint than an attention-getter. Once the rhyme has been registered, however, it may recur five, perhaps even six lines away and still be noticed. But by then the repetition is not wholly dependent on echoic memory for recognition, since the first occurrence has given the sound a toehold in a slightly longer-term memory.

Relating the inputs of different senses

A further important benefit of sensory consciousness lies in enabling us to link the inputs of different senses. It's something we can do regardless of whether there is anything to be gained by it - not just when one input functions as feedback concerning the effect of responding to another slightly previous one. We can work out, for instance, that the song must come from the bird with the moving beak, or that the delicate scent goes with the shape and colour which is the rose. Multisensory schemata can be created simply for the interest of it. In many species associations between stimulus patterns registered by different senses only seem to be learnt if there is some profit to be gained, by way of predicting the appearance of a desirable or undesirable. Our ability to correlate different sensory inputs must depend on the existence of neocortical pathways for the senses concerned, on connections between those pathways, and on a potential for sustaining a circulation of impulses around those connections.

But it is of course the intersensory correlations which relate directly to the wellbeing of the self that are most important. With appropriate neuronal connections an animal can learn that an input in one sensory channel is frequently followed by a desirable or a noxious input to another sense. There's reason to think that the short-term record which equates with sensory consciousness extends the possibilities for discovering such predictive cues. Humans can learn to blink at just the right moment to avoid a small puff of air to the eye without being aware of either the puff of air or the blink, but only if the signal which warns of the unwelcome stimulus lasts long enough to overlap with it. If there is a brief gap between the warning cue and the puff of air only those people who notice both sensory events learn to make the avoidance response. This suggests that the process which produces the conscious sensory experience also sets up the record which allows the cue to be related to the unconditioned stimulus.

There's one form of learning, meanwhile, which clearly involves a conscious feedback, as I noted earlier. From the subjective perspective it's obvious that we humans can plan actions on visual objects because both movement and visual object are conscious experiences. We can tell whether the movement was successful by means of conscious vision and conscious touch. If it isn't, we perceive how it needs to be adjusted with the aid of conscious vision. We learn most, in fact, through our failures, and it's the picture-in-the-brain which provides a means of measuring the degree of failure. If the hand didn't quite reach the ball, or didn't quite achieve a lasting grasp, what went wrong? Here the classical arrangement is reversed. Instead of a contact sense providing feedback about the effects of responding to a stimulus registered by a distance sense, a distance sense provides feedback on an attempt at making a contact.

Other distance senses too can provide information about consequences, as when, for instance, an infant bangs two objects together and discovers they make an interesting noise - and promptly proceeds to do it again. Most importantly, it's the conscious auditory experience which enables an infant to compare the sounds she herself makes with the speech she hears around her, and gradually to adapt her vocal productions to the same style. We can conclude, in short, that sensory consciousness is what turns distance senses into sources of feedback.

Modifying motor behaviour through learning, meanwhile, requires not only feedback from distance senses about the results of an action. That needs to be correlated with information about muscle deployment, joint position and tactile inputs. The senses which register how limbs and joints are currently disposed (proprioception) must be long-established. But where all varieties of movement are fore-ordained by the nature of the muscles and the neurons that control them the proprioceptive data only needs to be applied with equal automaticity, to keep track of the position from which any new movement must start. Species whose actions are organised in this way begin life already capable of very efficient movement, or needing only the briefest of 'running-in' periods for neurons and muscles.

Promoting proprioceptive information into the arena of consciousness must have been a vital factor in turning species with a predetermined repertoire of very effective hardwired actions into animals capable of more flexible, adaptable movements, which can be honed to suit a wide range of tasks. This was the development that led to species where the young start out with wobbly, uncertain movements, and must learn to perfect them, developing their motor potential by practice and experimentation. The repertoire of movements is no longer limited, and the animal can learn to shape its movements to suit varied goals.

First, though, a young animal may have to learn how to use the feedback. The role of the picture-in-the-brain may possibly be relevant to one of the most enigmatic aspects of vision. A cause of much early pondering was the fact that the image formed on the retina is upsidedown, and reversed left to right. The maps in visual cortex also maintain this relationship to reality. Does this matter? From a modern perspective the answer might be assumed to be no. Neither the orientation of the retinal image nor of the brain's retinal maps would seem to be important to the complex processing of the information they provide, or to the parietal calculations about the location of the objects registered.

However, towards the end of the nineteenth century a classic experiment was performed to discover what happened if the image on the retina was the 'right' way up. G. M. Stratton devised a pair of goggles with prisms instead of lenses, which inverted the pattern of light entering the eye. Heroically, he wore these all day and every day for a period of weeks. And he saw, of course, an upsidedown world, not to mention an inverted self, with hands that descended into view from above. With practice he managed to adapt his movements to this new perspective. But then, after several days, he found that he was looking at the world the right way up again, despite the goggles. His picture-in-the-brain had changed to match the reality. Furthermore, when he finally left the prisms off his view of the world turned upsidedown again, and he needed a further period of readjustment.

The experiment has been repeated since, sometimes with up/down inversion, sometimes right/left, with varying results. Not everyone's conscious experience adapts. When it does it seems likely to be the result of relating visual input to the eye movements and arm movements that achieve desired goals. This implies an intriguing possibility - that a considerable element of early learning may go into the functioning of the 'where' pathway, just as into that of the 'what' pathway. Very young infants concentrate on very small bits of the visual scene - one fixation's worth at a time - and don't seem to deal in larger spatial patterns, so if they lack a proper sense of orientation it probably doesn't matter. The earliest eye movements and arm movements are stereotyped enough to indicate they are hardwired; it can be deduced that they are controlled by subcortical circuits, and can be performed without reference to the conscious 'picture'. Maybe the infant has to learn how eye movements change the visual input. And perhaps, as he looks at an attractive object and experiences proprioceptive feedback from the arm that is flung out towards it, he learns to construct his picture-in-the-brain in the correct orientation. In a species which uses consciousness to bring many strands of sensory evidence together, working out just how to do that may conceivably be the first step.

A fair amount of evidence supports the proposition that always seemed likely: as far as we mammals are concerned the sort of fullscale sensory consciousness we experience depends on the neocortex. (The neocortex can only work, of course, with the extensive support provided by its subcortical foundations.) The neocortex is a massively complex network, even in small mammals, and especially so in primates. There are numerous parallel pathways and complex interconnections among them. That makes it a very efficient machine for making correlations - enormous numbers of correlations. It's also very thoroughly connected to all the other areas of the brain. There are complex and lengthy pathways around which the neuronal impulses can circulate.

The sustained currents of impulses that create consciousness must be necessary to knit all the diverse activity together. At the same time modifications can be made to the circuitry at each pass, making the correlations increasingly delicate and more accurate. The benefits of these arrangements in extracting subtle and reliable information from the sensory data generally outweigh the disadvantage of extending the interval that elapses before a decision to act is made.

The process which creates the conscious moment, I've suggested, can be regarded as an ultra-short-term memory. The record it creates is incorporated into a more capacious short-term memory, which allows a succession of sensory inputs to be assembled into a single percept, and equates with the stream of consciousness. That in turn allows the more definitive aspects of stimulus patterns to be preserved, as experience accumulates, in longer-term memory.

The information is stored in the later stages of the neocortical sensory pathways, in assocation areas of the cortex, or schema-areas. What's recorded are those aspects of the pattern which have attracted attention, which are generally those which will prove useful for recognising it again, or for recognising something of similar significance. Further encounters with similar stimulus patterns will, however, be able to modify the schema.

The record, reciprocally, helps to shape later conscious experience, as we saw in chapter 7, *Learning to See.* A two-dimensional arrangement of straight lines looks like a cube because we are familiar with cubes. An Ames room looks like a conventional set of walls joined at right angles because that's what we've learnt to expect. In short, sensory consciousness functions to bring together large amounts of sensory information, both information that comes from all sorts of different sensory channels and information that is spread out over time.

Harry Jerison's account sounds even better now than it did in 1973. *I regard the mind and conscious experience as constructions of nervous systems to handle the overwhelming amount of information that they process.* Or as Francis Crick and Christof Koch put it, with reference to vision, *the biological usefulness of visual consciousness in humans is to produce the best current interpretation of the visual scene in the light of past experience .... and to make this interpretation directly available - for a sufficient amount of time - to the parts of the brain that plan possible voluntary outputs of one sort or another.*

Operating without sensory consciousness

A modern animal that probably doesn't have much conscious vision is the frog. The study of this species' visual abilities began with a seminal paper by J. Y. Lettvin and colleagues in 1959, *What the frog's eye tells the frog's brain.*

Lettvin and co. found that the frog's eye is similar to ours in basic design, but the receptors are distributed evenly across the retina, and most of the ganglion cells receive input from a great many receptors, so that there's no area of special acuity. Furthermore the majority of the ganglion cells are geared to respond to moving contrasts, with preferences as to the direction of movement.

One type of ganglion is excited by moving stimuli with small, dark, tightly curved outlines. A stimulus pattern of this sort following a zigzag path nearly always means a flying insect, which equates with a potential meal. When such a stimulus is registered the frog unrolls its long tongue at ballistic speed, and with luck catches a fly on the sticky end of it and rolls the prize back into its mouth. Another variety of ganglion is sensitive to a larger darkness crossing the visual field, or a sudden dimming of the overall light level. Activation of these cells prompts the frog to make a jump, and avoid being trodden on by an animal which was oblivious to its presence. (Maybe you too have been startled, in fading light, when a bit of the ground you were about to tread on suddenly leapt away.)

What we have here, then, is a pair of stimulus patterns with specifications that are hardwired in the retina. Both of them depend on movement, which is a strength in one way, since plenty of things might be mistaken for a stationary fly, but not much moves like one. But a fly that stays still is safe from the frog's tongue, and if you notice a frog before its threat-sensitive ganglions register you it's fairly easy to creep up on it. Identifying movement in the outside world is pretty straightforward for the frog, incidentally, because no allowance has to be made for eye movements. The eye doesn't swivel in its socket, except to adjust for movement imposed on the animal by external forces, as when the lily pad he's sitting on is lifted by a ripple. Another sort of ganglion cell provides the information about global movement that allows the view to be stabilised.

The axons from the ganglion cells that respond to flying insects and to danger run not to the thalamus but to the optic tectum (a structure known as the superior colliculus in mammals). The target neurons here conserve the retinal relationships and form a map of the retinal field, in the usual fashion, so the trajectory of the stimulus is reflected by the neurons it excites. And from there the onward route to motor neurons which trigger the fly-catching action or the escape jump is short. Consequently the response to the appearance of a fly can be very fast - which is just as well, since the fly has even shorter neuronal pathways from eye to muscle, and can manoeuvre very fast to escape. The rapidity with which the tongue-flip is initiated no doubt helps to explain why frogs are so much better at catching flies in midflight than we are. Visual consciousness, in this context, could only make for disastrous delay.

Frogs don't seem to be so good at navigating their way around. On several occasions I've seen one jump straight into a solid fence or wall. After bouncing off once the policy seemed to be to try again. When that failed they made a right-angled turn and set off in a new direction. However, frogs do possess a type of ganglion cell that responds to stripey textures, which provides some guidance for navigation. One species jumps towards horizontal lines, away from vertical ones, while another, which lives in grassland, prefers to jump towards vertical texture. Again, the axons run to the midbrain, and again, it's only moving stimuli which can engage the mechanism, which must be a little inconvenient for a species which gets around by jumping rather than running. However, frogs are light, tough, and bendable, so they don't damage easily.

There's yet one more type of ganglion in the frog's retina, not very numerous, which records the general pattern of light and darkness in terms of the blue wavelength. This seems similar to the mammal's blue-responsive ganglion, also uninterested in contrasts. It must be what enables the frog to head for shade when it needs to cool down, for sunshine and blue sky when it needs to warm up. This ganglion sends its message to the thalamus, and the hypothesis, of course, is that applying its information may perhaps involve a modicum of consciousness.

More recent research has focussed on toads, which have basically similar visual systems, but with variations. While frogs sit around waiting for their food to come to them, toads are hunters, seeking out worms, caterpillars and beetles, with the aid of retinal ganglion cells which are sensitive either to small blobs or to long thin horizontal objects. It takes movement to activate either sort of cell, but toads can seize a prey-sized object after it has stopped moving. They can also identify a motionless object revealed by their own movement. And although they have an escape reaction to larger moving objects they can learn in time to associate the moving object which is the experimenter's hand with the delivery of food - which must make them more convenient animals to maintain in the laboratory.

There are four or five different types of ganglion in the toad's retina, with different-sized receptive fields and different preferences as to the speed of movement, and the connections from retina to thalamus and optic tectum are much the same as in the frog. In addition, an area has been identified which contributes to distinguishing prey or predator from background, and movements in the external world from the effect of the animal's own movement. There are circuits for managing orientation towards the prey, and approach towards it, and a loop which seems to function to suppress the prey-seizing action when appropriate.

Many mammals also have some retinal ganglion cells with complex requirements. The rabbit retina, for instance, includes quite a lot of ganglion cells sensitive to the direction of movement. In some species even orientation is identified in the retina, and only a rather modest proportion of ganglions is responsive merely to small, unoriented contrasts, or to the presence or absence of light.

In many cases, moreover, the larger proportion of the axons from the retina goes not to the thalamus but to the superior colliculus. In rats many of the fibres divide, so there is a considerable overlap; over 90% of the ganglion axons project to the superior colliculus and about 30% to the thalamus. And the arrangement is much the same in the hamster, with the interesting twist that the upper part and the periphery of the visual field (the locations most important for spotting impending danger) are particularly well represented in the superior colliculus.

A likely deduction is that hardwired stimulus patterns are reported to the superior colliculus. Such patterns are pretty certain to include those that portend danger and require a rapid response. For some species they will also include the stimuli that represent food. And for many they seem likely to include a particular combination of edges and perhaps colours, moving in a characteristic rhythm, all adding up to an adequate definition of a conspecific. The superior colliculus is organised in layers, and the number of layers varies considerably between species. It's possible that the number of layers is related to the number of visually directed behaviours that are hardwired.

Some mammals appear to be seriously handicapped by the removal of the superior colliculus, but get by quite well without a visual cortex - at least on the tests that experimenters have thought to try them with. It may be that the data from the ganglion cells with the less demanding requirements is reported to the lateral geniculate nucleus, and thence to the visual cortex, where it can be put together in flexible ways so that varied sorts of information can be extracted from it. That information can be used to guide the behaviours that can't easily be triggered by innately determined stimulus patterns, the most obvious of which is acquiring geographical knowledge of the environment. For many species the information that goes to the visual cortex is quite limited, and likely to be useful only for a very broad-scale mapping. There must be considerable support from senses such as touch and olfaction.

The rat's visual cortex, for instance, is reported to be largely concerned with overall patterns of light and dark, and much less concerned with featural detail than a monkey's or a cat's. There are orientation-sensitive neurons, but vertical and horizontal orientations are apparently more strongly represented than obliques. In the mouse the cortical neurons sensitive to the orientation of lines and edges are distributed randomly, but orientation-tuned cells are also found in the superior colliculus, where they are arranged in columns as in the primate visual cortex. One might guess, perhaps, that rodents are pretty good on visual textures, not so good on visual objects. They're probably more dependent on their whiskers for learning about shapes.

In us only a modest proportion of the optic fibres go to the superior colliculus - which is probably why it was at first thought to be responsible only for co-ordinating information relevant to eye movements, and for registering the sort of sudden movement that always attracts visual attention. But we have larger eyes than rats and rabbits, more densely packed with receptors and ganglion cells, so it's possible that the number of ganglion axons conveying information to the superior colliculus is no lower in us, merely overwhelmed by the vastly increased number communicating to the lateral geniculate.

Recapitulation

The process which creates the moment's conscious experience also establishes a brief, very accurate record of it, so that a succession of inputs is interpreted as a unit. The products of several fixations can be assembled, or the effects of a changing viewpoint, or the sensations that result from running one's fingers across an object.

This makes it possible to deal with the many sensory stimulus patterns that are inherently evanescent - with the flow of auditory input, and with movement, either seen or felt.

We can replay the sequence preserved in this short-term memory, even if the input only triggers attention after the full sequence has registered. It's consequently termed echoic memory.

Most sensory schemata are created out of such a succession of inputs. The echoic memory allows us to discover the relationships between the various components of a stationary or changing stimulus pattern. It can be termed a comparing memory.

The capacity of echoic memory is hard to measure. It no doubt differs somewhat in different sensory channels. It must vary considerably between species, and somewhat between individual members of a species, though it can be expanded by training. Its capacity must determine the maximum degree of complexity possible for a sensory schema.

Humans have evolved a particularly capacious auditory short-term memory as part of their language ability.

The fact that the conscious experience takes a good 50 milliseconds or so to create means that even the first syllable of a sentence can be interpreted in the context of what follows.

An important function of sensory consciousness is that it allows the percepts made in different sensory channels to be related to each other, regardless of whether there is any immediate profit in discovering the correlation.

A notable consequence is that distance senses can come to function as feedback senses.

Conscious visual feedback and a consciousness of muscle stretch, joint position and so on make it possible for some species to be born with a very rough-hewn repertoire of movements, and learn how to shape them into a wide variety of subtly differentiated ones.

Stimulus patterns spread out over time can be hardwired, however. This is particularly likely to be the case with the signals that bring the sexes together for breeding. In both vertebrates and insects such signals tend to involve very exact timing.

The frog has a hardwired visual response which enables it to catch moving flies, something which it can do more efficiently than us because - we can be pretty confident - it doesn't have to wait for a conscious visual experience.

Chapter 10

The Internal World

The input from our eyes and ears gets more extensive processing than that from other senses, and provides much more complex information. Being so well equipped for sight and hearing we tend to overlook the fact that elaborate distance senses are something of a luxury. Knowing what's going on at a distance can be very valuable, but it's not as vital as knowing what's happening to the self. Sight and sound give warning of what's approaching, what might usefully be approached, and what should be avoided. Boundary senses register actual encounters - things that are already in contact, or beginning to penetrate. Internal senses report, equally vitally, on the current state of the organism. The essential function of distance senses is to provide forecasts of what might come to activate these more intimate sensory channels, desirably or otherwise.

Furthermore, the data provided by peripheral and internal senses, derived from actual contact with the stimuli, tends to be less ambiguous than that from the distance senses. The inputs require less processing to turn them into useful information. What needs to be discovered about a tactile stimulus is essentially revealed by how many receptors it excites, how powerfully it impacts, how long the pressure lasts, and whether it's repeated. The same goes for the skin's temperature-sensitive receptors. As for taste, the shape of the food molecule determines whether it keys into the taste receptor, and this is a strong clue to its identity, so the message is an inherently informative one. Internal senses, meanwhile, have only a finite and pre-determined range of possibilities to record.

In contrast, the visual and auditory channels require a good deal of sophisticated neuronal machinery before they can reliably discover more than a few significant facts. A pattern of light and shade has to be registered in some detail before it provides as good a guide to an object's significance as does the feel, or the chemical composition. Any sudden noise can easily be received as a warning signal, but a capacity for analysing the sound is needed before much is conveyed about the nature of the threat. The fact that something is hard or soft, hot or cold, sweet or bitter, is generally a great deal more useful to know than the fact that it reflects green light, or emits a rather high sound, qualities that are likely to be shared by a great many other very different entities.

Distance senses, in short, have to evolve a far greater degree of sophistication than contact ones before they can gather comparably significant information. The exception is perhaps olfaction, which exploits the way soluble or volatile molecules detach themselves from the substances of which they form part to drift through water or air. The molecule which activates an olfactory receptor is fairly likely to deliver useful information about the nature of its source.

Boundary and internal senses are thus well suited for providing reliable reports on the consequences of responses to distant stimuli. It was when neuronal connections began to evolve that enabled inputs from these senses to provide feedback about the results of such actions that an elaboration of distance senses and an ability to register more than just the stimulus patterns that elicit hardwired responses became increasingly profitable. The capacity for this form of learning provided a means of dealing with the infinite variety of the external world. It will have been particularly valuable in the context of locomoting towards distant targets since this involves more effort than responding to nearby stimuli and there are considerable energy savings to be made by exercising discrimination.

With adequate feedback arrangements in place there is no real limit, in theory, to what can usefully be learnt about the external world, though there may be practical limits on how much data can be collected and how much time can be spent processing it. Given feedback, it's worth collecting all sorts of distance information, on the chance that it reveals something of importance to the health of the self, or to the survival of the genes, or saves wasting energy on avoiding something that isn't really a threat. So it was the development of feedback channels that led to species which do a great deal more analysis on the input of distance senses than on that from contact senses. The advent of connections that turned the inputs of contact and internal senses into potential feedback about behaviours prompted by distance senses must be regarded as among the most seminal occurrences in animal evolution.

There's something of a conundrum here, however. Why should sensory abilities have increased before there was any learning ability to make such an evolutionary development profitable? And why should learning ability have evolved before there was the sort of sensory ability that would make it useful? Prior to the evolution of feedback channels there would have been nothing to be gained by collecting sensory information beyond that which evoked the innate behaviours, so how could the extra information which would allow hardwired behaviours to be modified be acquired?

Part of the answer must be that quite a variety of events can function usefully as Pavlovian signals, and they don't have to be very closely defined. It was always likely, therefore, to have been worth acquiring some new way of registering, in any sensory channel, anything that suggested a sudden disturbance in the neighbourhood, since sudden disturbances often imply danger. This probably accounts for some of the earliest additions to sensory abilities.

Another factor is that occasionally a new variety of olfactory, visual or auditory receptor must have proved useful in adding to the definition of one of the hardwired stimulus patterns. In particular the signals by which potential sexual partners identified each other must have become more complex as the number of sexually reproducing species increased. Sometimes that will have meant that new types of sensory receptor evolved which responded to aspects of the species that had not previously functioned as a signal. Sometimes the input to existing sensory receptors must have been combined in new ways. As we have seen, some hard-wired responses can require a combination of features to trigger them, and this is probably particularly true of the systems which ensure that animals mate with appropriate partners. Any signals which could bring potential breeding partners into proximity, increasing the chances of an actual encounter in which the more reliable contact senses could be put to work, would be very selectionworthy.

It's a reasonable surmise that the evolution of a feedback system relating the input from a distance sense to that from a boundary or internal sense involved, in the first place, combining the data from the distance sense in new and flexible ways. Once a feedback system was in place all sorts of extensions to the distance sense might prove to be worthwhile - additional receptors, new types of receptors, or new analytical machinery. Data that would have been superfluous before could be put to good use. Additional means of collecting sensory data would in turn mean scope for further new learning abilities to be profitably employed.

The first expansion of vision?

Some expansion of sensory systems involved applying the data they produced to new purposes. This may have been the case with vision. The earliest form must have served merely to register light or darkness, enabling an animal to head towards one or the other according to the time of day, or its current needs, or the current state of the sea. The photoreceptors were probably maximally receptive to blue light, and the data no doubt went straight to the diencephalon, to what may have been the origin of the simplest koniocellular pathway. Or perhaps the very earliest form of guidance came from the same photoreceptors that regulated the biological clock.

The first addition to the visual eye could have taken various forms. Perhaps the most probable is another sort of photoreceptor, receptive to a different wavelength. Or the innovation might have been a neuron connected to several photoreceptors in such a way that it fired in response to an edge between light and darkness. Either could serve to warn that the animal was close to a rock or other hard surface and in danger of a damaging collision. The response needed was the same twist away from the threat that would be prompted by an actual collision. It would be most profitable, therefore, if the data went to the brain region in which escape responses were already being triggered by tactile input, rather than to the front of the brain. This, we can speculate, is how a visual centre in the midbrain, in what would become the tectum, came into being - the structure that in us is best known for its role in directing eye movements, and in frogs triggers the escape movement elicited by a looming visual stimulus.

An apparently odd fact is that in the modern eye the photoreceptors that connect to OFF bipolars (the ones that register darkness) do it via ionotropic receptors - the fast sort that would be desirable for making rapid escapes. The connections to ON bipolars, on the other hand, are the rather slower metabotropic kind. That would make sense if the latter cells evolved in the first place to serve the less urgent business of guiding locomotion.

In early days the evolving distance senses can only have reported on objects in the immediate environment, revealing things that were not quite in contact with the animal but close. Eyes will not have acquired lenses for focussing contrasting edges on the retina before there were contrast-sensitive neurons to benefit, so when such neurons appeared the only features detected, to begin with, are likely to have been nearby ones.

The fishes' lateral line sense similarly responds only to fairly local disturbances in the water, and the first benefit must have lain in detecting the disturbances in water currents that occur around rocks, providing another cue for accelerating away from the risk of being dashed up against a hard surface. We can suppose that the information went to the same motor system that responded to alarming tactile input and - if the hypothesis is correct - visual input indicating a nearby darkness or a nearby texture. Whether the visual or the lateral line warning system came first must be a matter for guesswork.

Olfaction seems likely to have been more concerned with desirables. In the beginning it may have functioned simply to tell an animal when it was already close to something desirable, such as food or a potential mating partner. Olfactory inputs, all genetically determined by the nature of the receptors, probably prompted a deceleration of locomotion, enabling an animal to deploy touch and taste more efficiently or more cautiously.

Even at such an early stage feedback connections should have been well worth having, linking odours with tastes, or visual and lateral line inputs with tactile ones. And once they evolved it would have become more worthwhile to discover what was going on at a distance, so they will have contributed to the evolution of more far-reaching distance senses. Correlating visual and olfactory reports with information from the internal senses will also have been very important. There are reasons for suspecting that feedback connections from internal senses were among the first to appear.

Internal senses

Internal senses first evolved, of course, as a means of regulating the state of the organism. The earliest multi-celled animals, as filter-feeders, no doubt simply absorbed food as and when they could. Slightly more sophisticated feeders were probably still in no danger of overeating, and when there was no food about could go into a state of suspended animation, as bacteria do. But increasing talents for finding food, and the more complex structures into which it had to be converted, tended to mean that it was useful to pause from consumption from time to time, even if more food was readily available, and to rest while some digestion went on.

There were enormous benefits to be reaped, meanwhile, when the ability evolved to maintain the internal temperature as close to an ideal point as possible, making use of the warmer and cooler areas of the local environment. Most of the chemical reactions that constitute the processes of life work at maximum efficiency within a fairly modest thermal range. At low temperatures metabolism slows, and getting too cold is fatal, as is getting too hot. Early forms of animal life won't have acquired the various adaptations which allow some modern species to live in extreme temperatures. An animal which could control its temperature by avoiding excessive heat would have a better chance of survival, and one that could move to warmer places when it got too cold could do a lot to avoid dormant periods. That would mean it lived faster. It could spend more time eating, and, most importantly, it could reproduce more frequently than animals which were at the mercy of chance for finding themselves in regions of suitable temperature. The species which first evolved a means of monitoring its temperature, and moving along a temperature gradient to concentrate its activities in thermally optimal areas, must have had a powerful competitive advantage over its rivals.

In short, the most successful animals ceased to be entirely the slaves of fortune and acquired the means of directing their behaviour towards achieving homeostasis in all sorts of respects. They acquired receptors that measured things like internal temperature and stomach fullness, oxygen, blood sugar and fluid levels. In vertebrates some of the receptors that monitor such things came to be situated in the hindbrain, and some in the hypothalamus, a structure sited, as its name indicates, just under the thalamus, and slightly in front of it. The thalamus is indicated by stripes in this cross-section through the medial part of the right hemisphere, the position of the hypothalamus is shown by the solid area. (The dots indicate the axons which connect the two hemispheres – the corpus callosum.)

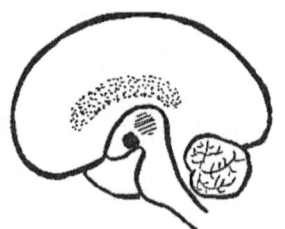

Gradually many of the senses that report to the hypothalamus developed connections that allowed their data to be applied as feedback. This would mean that not only would a non-optimal condition prompt action to rectify it, but now the achievement of an optimal condition - something like a full stomach or a satisfactory internal temperature - could be related to the circumstances in which it was achieved.

It seems likely that a feedback arrangement of this sort was the first of an intersensory nature to evolve. The internal conditions monitored by the hypothalamus don't change rapidly. It takes time to fill up an empty stomach, or to warm up or cool down, and then the effects last for a while. And to take advantage of a desirable temperature or a plentiful supply of food an animal must linger in the place where it's found. That means there's plenty of time to register the ongoing inputs which represent the characteristics of the place. When they must be registered by primitive sensory systems which could be misled by too small a dose of evidence this is likely to be an important factor.

Evolving the means of learning about such longlasting coincidences must be an easier proposition than connecting two brief inputs in different sensory channels. And it must certainly be more straightforward than the developments needed to relate transient forms of feedback, such as taste or a brief impact on nociceptors, to stimuli which have vanished or been distanced as a result of the response made to them. Learning about places that have provided a good meal or a comfortable rest doesn't require the means of maintaining a record of information from inputs which have ceased.

Olfaction is a likely candidate for the first sensory channel to provide contextual information which could be linked to an internal report, and the most probable information to be related to an odorous context is perhaps digestive repleteness. Some forms of vegetation must have provided more extensive meals than others. A means of learning to distinguish the odour which offered the best promise of a full gut should have been very valuable.

Vision would also come to contribute. My guess is that at some point those new cells in the eye, excited perhaps by some other wavelength than the aboriginal receptors, perhaps by dark/light contrasts, began to send their data to the diencephalon as well as to the rear of the evolving brain. They would provide modest information about the prevailing colour or the visual texture of the environment, which could be associated with information from the proto-hypothalamus about gut fullness or satisfactory temperature. There would then be two dividends to be reaped from the new retinal component, a cue pointing to reward as well as a warning of danger.

A diversion on the benefits of temperature control

A hardwired behaviour can provide only a limited advantage in temperature control. A rule such as *head for the light when too cold and for the shade when too hot* can be effective. But the place where the sun shines brightly can also be the place where an icy current is at its worst. Similarly, a genetically determined practice of retreating into a dark hole as dusk draws in or as dawn approaches can do quite a lot to promote survival, since dark holes generally provide protection from extremes either of heat or cold. But not every dark hole is suitable. The species which acquired the means of learning what sort of input recorded by a distance sense tended to coincide with an optimum internal temperature would benefit considerably. Vision seems the most likely distance sense for providing helpful information, and it may be imagined that even a very primitive visual system, capable only of discriminating a few basic textures, could help somewhat in the identification of desirable thermal environments. A feedback link which created the possibility of conditioned learning on this subject should have given the species which first acquired it a noticeable competitive edge. So I suspect that the hypothalamic monitor of internal temperature may have provided one of the earliest forms of feedback.

A species which could mark out thermal information on a cognitive map would benefit even more strongly, for temperatures can vary considerably over small areas. In the sea cold currents can slice through warm ones, creating quite a sharp thermal boundary. On land erratically blowing breezes can confuse the picture. In short, finding a suitable place for warming up or cooling down can be a serious challenge. A further hypothesis, therefore, is that thermal information may have been marked on the cognitive map at an early stage. Knowing where suitable sites for temperature modification are to be found, and how to return to the location which has proved a satisfactory hideaway in which to spend the inactive part of the twenty four hours, would seem to be particularly valuable. A trait that constitutes a very novel piece of neuronal engineering (and is therefore unlikely to evolve very often) must have a better chance of getting established if it brings a really big advantage, and the potential for efficient temperature control by behavioural means seems to be where one of the biggest benefits of cognitive mapping lies.

Reptiles and amphibians certainly use behavioural means to regulate their temperature. There is also evidence that at least some species of reptile possess cognitive maps and that thermal information is registered on those maps. As noted earlier, crocodiles bury their eggs on dry land, and then return to the site to ferry the hatchlings to the adults' watery habitat. Remembering where the nest is implies the possession of a cognitive map, as does the fact that the same site is used again and again. And the fact that the eggs must be buried in ground that will keep them within a very small temperature range if they are to hatch successfully suggests that the place must be marked on the map as a site of appropriate warmth. Similarly, the way certain snake species return to the same cave to hibernate year after year implies a cognitive map, and thermal information may well be recorded on the map, since the most important function of a site for hibernation is to provide shelter from the worst of the winter cold.

Mammals and birds, of course, have evolved the ability to keep their internal temperature at a fixed level by burning energy derived from food, but they couldn't have acquired this trait if they weren't efficient at finding their way to suitable thermal environments. Becoming warm-blooded has meant becoming more vulnerable to excessive heat or cold. Even with the aid of heavy insulation in cold climates and assorted cooling devices in hot ones, the metabolism which maintains a consistent temperature is not so efficient that it doesn't have to be supported by behavioural means. The more complex mechanics of homothermic animals only work within a smaller temperature range. The essential systems of mammal and bird cannot be put into the sort of cold storage that can be survived by reptile or amphibian - even if some mammals can wind down their metabolisms sufficiently to survive extreme cold by hibernating in suitably insulated dens. It can safely be assumed that the capacity for maintaining a constant blood temperature couldn't have evolved in animals that weren't already well equipped for regulating their internal warmth by other strategies.

Delayed feedbacks

Establishing a record of extended coincidence between a rewarding experience and the visual or olfactory environment in which it occurs seems relatively straightforward. Learning to make an association between two transient sensory inputs would seem to require niftier machinery. The challenge is perhaps not all that great where the two sensory inputs overlap in time, even if only briefly, so that the input to the boundary sense begins before the signal to the distance sense terminates. As already noted, when this is the case we humans can learn to blink in response to a cue without being aware of the cue, the puff of air that we blink to avoid, or the blink itself.

Learning to associate a sensory input with one that follows less closely is a different ballgame. A record of the input that presages something undesirable must be preserved until the noxious event occurs if similar inputs are to function as warnings in the future. When the stuff that has just vanished into the mouth turns out to taste bad it's necessary to have a record of what it looked like in order to associate the taste with the appearance and avoid trying anything visually similar again. Often it's only necessary to know that it was some of the stuff that is mostly still there, available for further study, but even that requires a modest amount of memory. (Linking the taste of food to its odour might not pose the same problem, since the odour is still available while the food is in the mouth - at least in those species where nose and mouth are connected.)

Correlating two sensory events that don't overlap in time, so that the second provides feedback on the result of a response to the first, clearly requires a short-term memory, of the same order as that which enables successive visual inputs to be correlated. The record is surely maintained in a similar way, by a sustained circulation of impulses around the relevant brain areas. Perhaps that means that an ability to make a connection between two slightly separated sensory inputs implies sensory consciousness - or perhaps that's a guess too far. But certainly conscious pains and pleasures are a far from simple matter. The specific feedback connections between two sensory channels which provide for conditioned learning merely serve to encourage or suppress an essentially hardwired response. Conscious pains and pleasures are something else.

Conscious Pains and Pleasures

Just like our conscious perception of the external world, our conscious pains and pleasures are influenced by past experience. All sorts of other factors also play a part. The same pattern of stimulation can produce different effects at different times. Food gives pleasure when we are hungry, but becomes repulsive if we overeat. Being too hot or too cold is unpleasant, but going into an overheated place for a while can be agreeable when one's cold, and *vice versa*. Activity and relaxation can both be enjoyable, but too much of either can lead to discomfort. All this clearly reflects a subtle system for keeping the organism operating as efficiently as possible, and promoting homeostasis. The effects produced by inputs to the boundary senses are blended with information from the control systems of the hypothalamus to create our conscious pleasures and discomforts.

And this is only half the story. A fair degree of hunger can be quite agreeable if there's a good meal in prospect, but the same level of emptiness is thoroughly unpleasant if you find yourself stranded miles from the nearest source of food. Stretching out in the sauna with a cold shower waiting beyond the door is enjoyable. Being stuck out in a broiling desert with no means of escape is an entirely different matter, even though the temperature is not so high.

Moreover pleasure and discomfort can be greatly influenced by past experience. Sweating in the sauna may not be wholly agreeable the first time, but becomes so once you know what to expect. Even a sweltering desert heat can become quite addictive for some, and people who are brought up in conditions of extreme heat or cold are generally not bothered by them - a fact which reflects mental adaptation as well as physical. The subjective experience of extreme conditions can also vary considerably according to whether one is undergoing them by choice, or whether they have been forced upon one.

Pain, which might seem to be a simple matter of how much pressure is being applied to how many nociceptors, is similarly subject to a variety of influences. People raised in cultures where demonstrativeness is the normal rule seem to feel more pain that those brought up to value a stiff upper lip. The man who thinks he has indigestion seems to suffer rather less than the one who's worrying about cancer or heart attacks. Yet more tellingly, soldiers can be badly wounded in battle, but feel no pain until later, when the dangers of the battlefield are past. Moreover an adequate theory of pain must take into account the paradox of masochism. And people may willingly choose to undergo pain, for instance in the acquisition of tattoos and piercings, though they would resist strongly if someone inflicted the same degree of pain without their consent.

Clearly our conscious physical pains and pleasures don't reflect a simple measurement of sensory input. They're the end result of a series of judgements about both internal and external conditions, heavily influenced by past experience and by values acquired by imitation. Our bodily sensations constitute a model of the internal world in the context of the prospects offered by the external situation, just as our perceptions of the external world constitute an assessment influenced by the internal situation. Indeed, while current needs and emotional associations do a great deal to colour our perception of the external world, the influence in the other direction is perhaps even stronger.

One particularly interesting and mysterious form of learning is the business of learning what not to eat, when the feedback is provided by internal discomfort and perhaps vomiting, which may occur hours after the food was eaten. We share the capacity for this sort of learning with other adventurous omnivores, notably the rat. The feedback connection which links channels reporting on alimentary distress to the gustatory system seems particularly sophisticated, since the feedback arrives so long after the action-evoking stimulus, so it may be a relatively recent innovation. Just how the record of what was eaten is preserved, so as to be linked with the unpleasantness, remains unclear.

What's interesting is the way this sort of learning manifests itself in us. When we become ill after eating something we may apply reason to the question of what was responsible, pondering, as we recover: *was it the meat pie for lunch that did it, or was it the curry at dinner?* But we may also find that any distinctive flavour eaten within a few hours before the illness has ceased to be attractive. Whatever the mechanism behind this useful arrangement it doesn't operate with much precision. When I was about five or six I had one of my favourite puddings for lunch, chocolate blancmange. That evening I developed the first symptoms of a nasty bout of tonsillitis. I never nursed any illusions of a connection between the two events, but chocolate blancmange never tasted the same again. The flavour had become vaguely repulsive, and remained so for many years afterwards. Some readers, I imagine, will have had a similar experience.

The altered conscious sensation when a stomach upset turns a previously liked flavour into an aversive one could be said to parallel the case of the Necker cube, and other ambiguous visual stimulus patterns. A different experience is obtained from precisely the same sensory input. In the case of taste, however, the change in the perception is considerably longer-lasting.

This sort of learning applies only to distinctive flavours. We don't develop aversions to bland and boring food. If we become ill after eating a ham sandwich we go off ham but not bread. The other side of the coin is that we treat particularly distinctive flavours cautiously when we first encounter them - but they are often also the flavours which we come to find especially rewarding. *It's an acquired taste* we say, of pheasant, or curry, or whisky.

Rats behave very much like us, requiring only one upsetting encounter to put them off a distinctive flavour. And when they meet a new taste they eat only a small amount to begin with. If no ill effects follow they come back for more, and make it a slightly larger portion, and the next time they will eat yet more confidently. This is obviously a usefully adaptive behaviour, which allows a novel substance to be introduced into the diet with a minimum of risk.

Both rats and humans can also learn by the experience of others. Bennett Galef flavoured standard laboratory food with cocoa, cinnamon, aniseed or marjoram, and fed each flavour to a different group of rats. He then put pairs of rats of different gustatory experience together for thirty minutes, after which he returned them to their home cages and gave them a choice of two flavours of food. He found that a rat would select the taste that its recent cage mate had been fed on over another novel flavour, and would even prefer it to the one it had already tried. Rats are happy to try something new if they have evidence that it seems to be safe, and presumably the cage-mates still smelt, to another rat's nose, of what they had just eaten. Rats can also learn what to avoid if they encounter a conspecific which has been ill after eating something with a distinctive smell.

We can use other senses to the same effect. I once encountered some vomit in which lots of little bits of chewed-up tomato skin featured. The tomatoes could scarcely have caused the stomach upset, but it was about a year before I got back to feeling any enthusiasm for the fruit. Any sort of vomit, of course, tends to put one off eating while one is within sight or smell of it, a healthy tendency if more of the stuff that caused the sickness is likely to be found in the vicinity. And if you're of a delicate disposition this discussion may have prompted the feeling that perhaps you're not, after all, quite hungry enough to eat just yet.

Our conscious sensations of physical pleasure, pain and discomfort can be seen as the result of weaving together multiple strands of information. The result points to the sort of behaviour that is appropriate in the circumstances. The weighing and balancing of evidence which lies behind these sensations may be totally unconscious, but nevertheless it constitutes a sort of proto-reasoning - just as our perceptions of the external world do.

And just as it's necessary to learn how to create a model of the external world out of the information provided by eyes and ears and touch, so we seem to learn how to make use of the reports from the inner world. The evidence for this is slender, since normal childhoods provide enough experience for the essential discoveries to be made. But the occasional unusual case is suggestive. The wild boy of Aveyron (subject of a film by François Truffaut) was apparently abandoned in the forest as a child and lived like a wild creature until he was discovered and captured at the age of about eleven. Accustomed to going naked he at first refused to wear clothes, and tore them off when they were put on him. Later, however, he became extremely fussy about the temperature of his bath water. Presumably by then he had learnt to appreciate the possibilities of controlling one's temperature by external means.

The implications of the limited human evidence fit with the results of an experiment in which puppies were reared in total isolation, with no opportunities for educational interaction with the environment. As grown dogs they put their noses into candle flames in an exploratory fashion, and didn't know enough to withdraw again quickly. The lesson seems to be that the complex neuronal wiring which can create models of the world and of the self, if not properly programmed with experience, can end up as a liability.

Thanks to the neocortex we humans can link up evidence from many assorted senses, allowing us to make all sorts of predictions. Inhaling a flower scent we can invoke the appearance of rose or lilac, jasmine or lavender. Seeing a fruit from a distance we can guess if it's likely to smell bad. An experienced musician can often tell by the appearance of an instrument what sort of quality of sound it is likely to produce. But some intersensory connections seem a good deal more intimate than others, functioning without any need for attention or volition. The most notable example, perhaps, is the way gustatory and olfactory input are combined to create the conscious sensation of flavour. Both these chemical senses can provide warnings about substances it would be dangerous to consume. No doubt the close link between them is part of the machinery that enabled us to become omnivores.

Another notable partnership between two senses, though not quite so close, is that between sight and touch. We instinctively put out a hand to check whether something has the texture its appearance leads us to expect. (That's why museums need all those *do not touch* notices.) And if odd lighting or short-sightedness makes the exact form of a complicated object ambiguous, we check by touch on which are the convex curves and which the concave, what's solid and what's shadow.

Evidence of just how close the connection is comes from people who have been blind since birth. Their complete lack of visual experience doesn't prevent them from producing drawings which can be recognised by the sighted. They even make use of common artistic conventions. Clearly shape-concepts developed by touch alone are entirely comparable to those distilled from both visual and tactile experience. This suggests that visual schemata and tactile schemata are pretty closely integrated.

The essential link between the two senses must be pretty ancient, if one of the earliest uses of visual information lay in avoiding collisions. It's an interconnection that must have been very much strengthened in the early days of mammals. As I observed earlier, fur not only provides insulation it also extends the reach of touch, turning it into something that can provide advance warning of a potential contact with the skin. Even in us humans, so sensitive are the receptors attached to the base of our hairs that a small spider landing on one's head may make itself felt. Fur is for landlubbers, one might say, pretty much the equivalent of the lateral line system for aquatic animals.

In modern mammals the extension to the tactile system is especially valuable for navigating in the dark. In species which frequent the dark a large proportion of the neocortex may be devoted to processing whisker input. Many subterranean tunnellers like to build their tunnels to fit rather closely, providing a maximum of input to whiskers and fur, and when above ground they prefer scuttling along by a wall to venturing out into the open. Naked mole-rats, committed almost entirely to a life in underground tunnels, are not entirely naked but have a few hairs dotted around their skin to provide subtle tactile input. Harbour seals, which hunt their prey through murky waters, make considerable use of the information provided by whiskers. Bats have a few hairs extending from their wings, which provide information about air currents just as a bird's feathers do.

Unfortunately the fossil record is unlikely ever to reveal just how long ago it was that whiskers came into being. But we can be sure that tactile information must have been of great importance to early mammals, in helping them to move around when other animals were inactive. The visual information available in low light levels is often ambiguous, especially for an animal making its way through vegetation - dim shapes of sky at the end of a dark vista, and sometimes a reflection from something shiny, which may move or vanish as the moon moves on. In the early days of mammals one of the developments that opened the way to exploiting twilight and darkness must have been the evolution of neocortical pathways which correlated visual and tactile information more thoroughly.

The outstanding feature of the neocortex is that it brings together a wide range of sensory information, from internal, external and boundary senses, so that almost any sensory input can be correlated with any other. No longer are the neuronal connections limited to those that are inherently relevant to a behaviour.

The conscious sensory experience, whether of the external world or the internal self, reflects all the analyses. It's a judgement both on the current potential of the external world and the current needs of the self. The feedback that made conditioned learning possible has been elevated into something much more complicated.

Physical pains and pleasures are not the only sort, however. There are other, less obvious forms of feedback which have an enormous influence on mammalian behaviour.

Recapitulation

Internal and boundary senses are inherently more reliable than distance senses, which must deal with the infinite variety of the external world.

Some of the most seminal events in animal evolution must have been the evolution of feedback channels which made it possible to learn to associate stimulus patterns recorded by distance senses with effects recorded by boundary senses such as touch and taste, or by the hypothalamic sensors which monitor internal conditions.

This meant that any improvement to a distance sense, or any extension of the range at which it could operate, was likely to be worthwhile.

Perhaps one of the first feedback connections to evolve was one that linked longlasting sensory inputs from the external world with longlasting effects on the internal situation.

Connecting relatively brief inputs to the distance senses with relatively brief effects on the boundary senses would be a greater challenge, especially for two events which do not quite overlap in time so that a record of the first must be preserved.

It would seem to require at least some of the neuronal machinery that enables a succession of sensory inputs to create a stream of consciousness.

Our conscious experience of the inputs to internal and boundary senses that can function as classical reinforcements is just as judgemental as our conscious sensory experience of the external world.

It reflects not only the current state of the self but also the current potential offered by the outside world in relation to the self's physical needs. And it's greatly influenced by past experience.

Chapter 11

Subtler Forms of Reinforcement - Cognitive Emotions

The learning which is concerned only with whether to respond to a stimulus or to refrain from responding seems relatively straightforward. Desirable input to internal or boundary senses, if there is wiring which turns it into feedback about the effects of the response which preceded it, can strengthen the connection which triggers the response. (Just how that happens we'll come to later). Undesirable effects can bring inhibitory connections into operation and suppress the response. Learning about environments which tend to coincide with a satisfactory internal condition and are worth approaching can be supposed to work along the same lines.

However, it's clear that not all learning depends on this sort of feedback, for sometimes there are no immediate rewards or discouragements in the shape of physical effects on the organism. Some learning, most notably that involved in constructing the cognitive map, produces longer term benefits. The animal which explores its territory doesn't necessarily benefit from its discoveries at the time of exploring. In fact encounters with immediate rewards seem to distract from the business of exploration. Similarly, when young animals play they discover their own physical powers, and they learn something about the world they play in, but they don't gain any immediate material profit.

Activities like exploration and play must have some other form of motivation. Indeed, we know from subjective experience that exploring a new terrain can be enjoyable for its own sake, and play is, by definition, done 'for fun'. The reinforcements which power this sort of learning are clearly not simple reports from peripheral and internal senses, but something much more difficult to pin down. Just what is 'fun' in scientific terms?

Mysterious they may be, but these non-physical forms of reinforcement shouldn't surprise us. The neuronal machinery involved in creating the conscious experience, and in the process building sensory or motor schemata, constitutes a major investment, and one that is metabolically expensive to maintain. It couldn't have survived the pressures of natural selection if it didn't incorporate a device to ensure that its possessors made the further investment of effort which is necessary if the benefits are to be reaped.

Donald Hebb noted the importance of this form of reward and went so far as to define pleasure as *fundamentally a directed growth of development in cerebral organisation. It is thus a transient state of affairs in which a conflict is being reduced, an incipient disorganization being dissipated, or a new synthesis in assembly action being achieved.* In other words a new schema is being created, or an established one extended or modified.

We derive reinforcement, indeed, not only from spending energy on play, and on creating cognitive maps, but also from the less conspicuous efforts of mere everyday perception. For us humans the conscious sensation of seeing or hearing functions as the reward for the unconscious work that goes into creating it. We like to put our advanced sensory powers to work - to have something to look at, and something to listen to. We get painfully bored when there is nothing to activate our senses. Indeed, sensory deprivation experiments show that in the total absence of sensory input we tend to start suffering from hallucinations, as if our perceptual systems couldn't bear to be unemployed. It seems there is a built-in drive to make the fullest possible use of our sensory apparatus. Back in 1947 Robert Woodworth wrote: *perception is always driven by a direct inherent motive which might be called the will to perceive. Whatever ulterior motives may be present from time to time, this direct perceptual motive is always present in any use of the senses ..... to see, to hear - to see clearly, to hear distinctly - to make out what it is one is seeing or hearing - moment by moment, such concrete immediate motives dominate the life of relation with the environment.*

The urge to achieve the best possible perception is present from a very young age, as objective experiment has demonstrated. Infants from one to three months old were shown a fuzzy motion picture, and provided with a dummy, or pacifier, which was connected to the projector in such a way that their sucking affected the focus. They readily developed sucking strategies that achieved a sharp picture. No doubt it's for the same reinforcement - the reward to be obtained from turning a fuzzy experience into a clearer one - that infants learn to control the focus of their own eyes.

We can deduce that all species which possess the neuronal apparatus for conscious sensation have the same urge to put it to use effectively, derive positive reinforcement from doing so, and suffer negative effects if they don't get the opportunity to give proper employment to whatever schema-building abilities they have. It's long been remarked that many caged animals look as if they suffer acutely from boredom. The impression received solid backing when it was shown that laboratory monkeys would learn to perform quite demanding tasks when the reward was simply the opening of a shutter, allowing them to look out of their cages at scientists and laboratory technicians going about their business. (It's to be hoped that laboratory animals are now routinely provided with diverting perceptual fare - both for their sake and because animals suffering from depression are unlikely to provide useful evidence on any subject other than depression.)

Subjective experience also suggests that there's positive reinforcement to be obtained from the possession of schemata once they're created. The lack of schemata, meanwhile, can produce aversive emotions. Not knowing where one is, or how to get to where one wishes to be, is at the very least annoying, and sometimes quite unsettling. Similarly, registering a sensory pattern that can't be related to a schema can be disturbing if it seems like something that might have significant implications and one doesn't know what they are.

In short, there is both an urge to put the elaborate machinery that creates conscious sensory experience to work, and two sorts of reward to be gained from having done so - the experience itself and the knowledge gained. The ability to derive these reinforcements from conscious sensory experience must have been crucial to the evolution of a brain that could produce it.

Familiarity and unfamiliarity

The ability to build a sensory model of the world carries an important corollary. When some patterns of sensory input can be matched to a schema those that can't must be classified as unknown, with unknown significance. In contrast, the unknown is something that simply doesn't exist in the context of hardwired responses to genetically determined stimuli. An animal makes a response when one of the predetermined stimulus patterns is registered - perhaps performing an actual action, perhaps just becoming alert and extra-ready for further sensory input - and that's it. Other, irrelevant events may excite a few sensory receptors but make no further impression. The situation is the same where some of the hardwired responses can be modified by conditioned learning, so that they're inhibited by some variations on the basic stimulus pattern and strengthened by others. The sensory machinery registers only relevant bits of data, and anything else simply slips past, leaving the animal unaffected by what it doesn't know.

When the sensory apparatus grows elaborate enough to register large amounts of data, and the data has to be extensively analysed to work out which combinations signify something important, the situation changes. It becomes necessary to distinguish between the familiar and the unfamiliar because the unfamiliar is worth studying, to see what implications it may carry.

The sensations of familiarity and unfamiliarity are such a primary element in our conscious lives that they receive little notice. Familiar stimulus patterns usually don't get much attention, and so neither, as a general rule, does the sense of familiarity. We stop to think about it only when something seems familiar that, as far as conscious memory serves, ought not to. Then we talk about a sense of *déja vu*. Unfamiliarity receives a little more thought because the wholly new is always likely to trigger attention. Occasionally there's a conscious internal debate about whether something is familiar or not, whether it's perhaps an echo from the distant past or merely something with a partial and irrelevant resemblance to a known stimulus pattern. But mostly we can make instant categorisations.

Moreover we can sometimes identify the fact that there is a novelty in the environment without knowing quite what it is. We find ourselves intrigued and alerted by a niggling sense of *something's different here*, but it takes time to work out that a friend has a new hairstyle, or that the furniture has been re-arranged. Similarly, that sense of *déja vu* is a feeling that something is familiar although we can't remember why. In other words a sense of familiarity can be experienced without being accompanied by conscious recognition. It seems, therefore, that the sensations of familiarity and unfamiliarity are created by unconscious processes. The business of matching current sensory input to sensory schemata is an activity which runs constantly in waking life, and automatically.

Presumably the feeling of familiarity is produced when the neuronal impulses resulting from a pattern of sensory inputs readily find their way into a clear-cut circuit, and unfamiliarity is the effect when a new circuit has to be created. The two sensations can be seen as reports on the state of the schema-creating systems of the neocortex, just as conscious physical sensations function as reports on the state of the body. Hunger, thirst, tiredness, indicate that more food or water or sleep is desirable. Unfamiliarity indicates that more data is needed. And generally it mobilises attention, so that more data can be gathered. Familiarity and unfamiliarity are perhaps the most basic sensations of a cognitive life. They're the most essential of what we might call the cognitive emotions.

Lacking an established schema to which to relate current input means not knowing what sort of significance for the self may be implied, and not knowing what sort of response might be required. Since the unknown may be dangerous it's best treated with caution until more information is collected. Where there's an appropriate schema there's a record of what sort of consequences for the self, if any, have accompanied similar stimulus patterns in the past, and a means of making predictions about current probabilities. Even a danger is less frightening when it's understood than when it's a mystery. So on the whole the unknown is tinged with insecurity and therefore a little bit aversive, and the familiar is reassuring. On the other hand, the unknown may turn out on investigation to contain all sorts of desirable things, and at the very least the insecurity of strangeness can be banished. So venturing to explore it is attractive too, and perhaps exciting. Thus familiarity and unfamiliarity, like other sorts of feedback, function as reinforcements which can be either positive or negative according to circumstances. Just as food is rewarding when we are empty and aversive when we are already overfed, so the familiar is comforting and desirable when we are tired, ill, or frightened, and the unfamiliar is worrying. On the other hand, when we are full of energy, and especially when we are young and healthy, the unfamiliar beckons.

However, the attractions of the unfamiliar are limited. We like things that extend our existing schemata, or can be related to them in a meaningful way. We don't like things for which we possess no relevant schemata whatsoever, so that we have no guide-lines about what to expect from them. If there's a suggestion of possible danger about such a novelty it's particularly frightening. (The great thing about horror films is that they offer novel sources of terror within the safe and understood context of the cinema or television screen.) Even when something offers no sort of threat an inability to make sense of it can still produce an edgy reaction, or at least a mild distaste. Mathematicians pore over equations with delight while those to whom the figures are mere gibberish turn away from them. Cricket or soccer scores are fascinating to fans but not to the uninitiated. New knowledge needs to be fitted into an established framework.

This taste for novelty, but not too much novelty, is particularly marked in infants. They enjoy encountering variations on familiar themes, which can be related to something already known, and they turn away from the wholly strange, looking upset. There's pleasure in extending an existing schema, but confronting the totally unclassifiable is aversive.

Cognitive emotions and territorial behaviour

Territoriality is obviously closely linked to the evolution of cognitive mapping ability. If an animal is going to stick to the patch of land it knows, defending it against competitors, and if that patch is larger than it can survey at one time by means of its distance senses, then it needs a means of identifying what it regards as its territory. And of course an animal capable of mapping its patch of environment, and learning where useful things are to be found, will benefit from staying on the ground it knows. The ability to construct a cognitive map makes territoriality both possible and desirable.

If we may judge by appearances the creation of the map and the territory is motivated by cognitive emotions. A young animal which must leave the area in which it was born to set up its own home, being healthy and full of energy, sets out with a cocky, purposeful look, as if it finds reward in the adventure of exploring wholly new ground. Once it has established its domain it's likely to leave only reluctantly, if pushed out by a rival, or if the food supplies become inadequate, or something else occurs to make the territory unsuitable. We may suppose that it's motivated by a sense of the advantage of remaining in the known and predictable environment. In some species the male has to range over a more extended area at breeding time to find a mate. But otherwise, once settled, the animal can usually give up exploration and substitute the less demanding business of checking over the territory regularly and noting the effects of the seasons and any other change.   (No doubt such species, like us, do not consciously have to look for change but are hit by a '?' feeling when something in the environment doesn't match up to the stored schema.)

The initial exploratory behaviour, meanwhile, must be helpfully shaped by the principle that new information is attractive only if it can be comfortably added on to existing schemata. An aversion for the wholly strange should function to preserve an exploring animal from venturing into habitats to which it's not suited. When a young adult establishes its own territory it's likely to choose the same sort of terrain as that in which it grew up, the sort in which it knows the rules and knows what to expect. Only under pressure will it venture into a different kind of landscape. The forest-born creature explores readily in the forest but is super-cautious about venturing out into the open plain, we can guess, and this is generally not so much because it knows that the open plain is dangerous for it as because the possible dangers are unknown.

The territorial animal's map could probably be drawn in contours of familiarity: the densest colouring in the heartland, where rest and sleep are taken and the young are born; a slightly lighter shade covering much of the rest, with perhaps yet lighter patches here and there, representing areas of territory that are not investigated very often because they don't have much to offer; a still paler area beyond the borders, known only by means of distance senses, or through occasional forays taken when the neighbour isn't looking; and beyond that the bleak white of *terra incognita*.

All this implies that building a cognitive map involves a certain minimum of sensory consciousness. What would it be like to belong to a species that used consciousness only for cognitive mapping, with perhaps only one type of reward-site marked on its map? Probably the map would be pretty blurred by our standards, with just a few salient landmarks. We might imagine its possessor moving around its territory with a fuzzy consciousness focussed wholly on reaching the goal, and on the purpose for which it is travelling, and perhaps experiencing something a little like confusion and alarm if it gets lost. Just as its sensory model of the world is very hazy compared to ours, so, we may suppose, are the feelings which relate to the map. They can afford to be, since the animal doesn't have any other conscious feelings from which they might need to be distinguished, or which might distract its attention. Just as we blink, or digest our food, or produce antibodies against invading bacteria, without having to make any conscious decisions, so this animal performs nearly all of its activities without giving them conscious attention or having any conscious feedback about the results.

In contrast, our range of conscious feedback is extensive, and our place schemata are correspondingly complex. They are hung about with all sorts of feelings which reflect the values each place has for us, and the experiences we have had there. The place where a proposal of marriage was made and accepted has a special magic - or possibly a special pain or a rueful regret, depending on how things turned out. The feelings attached to a supermarket, for me at least, are a muddled compound of the hopeful promise of good eating, the boredom of regular shopping, and the frustration when they've changed all the sectors around again and I can't find what I want.

More generally, we learn to judge landscapes and cityscapes for their potential, usually without thinking about it, and they acquire an emotive flavour from the assessments. Does the new place offer interest or boredom? Is there a possibility that danger may threaten, so that it's important to stay alert, or is it safe to relax? Is there shelter from the elements if it's needed, and are food and drink available, or are there prospects of discomfort? Are there other people around, and if so do they seem to be friendly? A blend of all these elements colours the record of the place that's established in memory.

A sense of physical comfort or discomfort, and perhaps particularly of temperature, often seems to play an important part. Was one regularly hot and sweaty there, or cold and shivering, or drenched with rain, or braced against a gale - or was it deliciously comfortable? Canny writers know that the most effective descriptions of places colour them with these subjective qualities. Landscapes are bleak and forbidding, or calm and fruitful. Skies are lowering or smiling. Townscapes are awe-inspiring or tawdry, bustling with life or eerily silent. The feel of sweat dripping and humid air pressing on the skin is evoked, or the tingle caused by cold. And the best descriptions convey not only the emotive flavour of a scene, but also the smells - more evocative than any visual description.

The instinct for attaching an emotional assessment to a place has often been expressed by awarding distinctive sites their own deity. Mountain, stream, grove and forest have been conceived of as being inhabited each by its own spirit, powerful or minor according to the nature of its domain, to which visitors were well-advised to pay respect. The idea must have been a useful one, encouraging people to give proper attention to the nature of the terrain. For instance, thinking of a mountain-god as a powerful figure, readily angered, should convey to the inexperienced that mountains can be particularly dangerous places.

A scientific account of fun

Play is exploration applied to subjects other than geography. It's the experimentation by means of which a juvenile animal investigates its own physical potential, simultaneously discovering what can and can't be done with the components of the landscape around it. Whereas exploration enables an animal to plan where to fulfill its needs, play provides the knowledge that allows it to plan how.

If young animals are studied closely it's clear that the rewards of play come from the development of new skills and the discovery of new features of the world about them. New motor patterns are practised until they can be performed without thought, and then appear less frequently. Objects are thoroughly investigated by whatever means the species possesses, and then put aside. The theories formed as a result of the investigations may be checked again by further experiment, on future occasions. But once the knowledge is secure attention will be turned to other subjects, or to new variations and extensions of the established skills. In other words, the fun comes from forming schemata about what the self can do, and about the objects that things can be done to or with.

Making connections between simultaneous experiences in different sensory channels is clearly part of this. The feel of the arm making a hitting motion, the sight of the hand striking the toy and the feel of the impact all become connected to the noise that results; and the sight of the hand missing the toy can be related to the absence of the usual noise.

Where larger brains provide the means of building a greater variety of schemata more abstract discoveries can also provide fun. Psychologists of the sixties and seventies, studying learning in human infants, began to realise that their subjects often looked as if they were as interested in working out the rules according to which rewards materialised as in obtaining the rewards. They seemed to be trying out strategies, and as the experiment progressed showed expressions of concentration, surprise, pleasure or disappointment. It looked as if the experiments devised by the psychologists were treated by their young subjects as if they too were carrying out experiments, with similar motivation.

Sometimes the apes involved in learning experiments give the same impression. A psychologist told a story of realising during a research session with a chimpanzee that it might have to be curtailed because he hadn't provided a large enough supply of the grapes he was using as rewards. But when the grapes ran out the chimpanzee continued performing the tests, and started returning a grape to the experimenter after each was completed.

Jerome Kagan, pondering the significance of the smile in infants, observed that it often seemed to occur *when the infant matches stimulus to schema - when he has an 'aha' reaction; when he makes a cognitive discovery*. For instance, when a regular representaton of a human face is presented to a very young infant *there is a short period during which the stimulus is assimilated to the schema and then after several seconds, a smile may occur. At eight months the face is recognised immediately but there is less likely to be a smile*. Now, however, a smile may be elicited by a picture of a distorted face. This and other experiments led Kagan to conclude that stimulus patterns which require *an active process of recognition* are a source of pleasure. But there's no such reward from input that's too easily assimilable, or too difficult.

At every age, in fact, we like something that is a little bit novel, but not too novel to make sense of. Setting up new neuronal connections somehow results in a reinforcing feedback. This feedback must play a large role in the smile-evoking capacity of jokes, which usually give a novel twist to a wellknown idea. Jokes too might be said to involve *an active process of recognition*. So, indeed, does looking at pictures, reading books, watching theatre or films. To entertain there must be novelty, or a degree of complexity such that there is always something new to discover; but total incomprehensibility repels.

Do humans ever quite grow up?

The light-hearted business of play is usually confined to juvenile animals, still living under the protection of mother and provisioned by her in one way or another. When maturity arrives play is generally replaced by the more serious, adult satisfactions to be obtained from exploration. Geographical exploration can't usefully start earlier, for the neuronal wiring that is essential to mapping must first mature. What helps to keep the youngster from trying it, though, must be the urge to stay close to the reassuring familiarity of mother, or the den. The mapmaking facility is ready for use by the time the animal has grown to adult size, has discovered and developed its own physical abilities and investigated the more interesting aspects of the world around it, and is feeling capable and confident. Now pleasure can be derived from putting this knowledge to work. And in adulthood the demands of maintaining a territory, finding food, finding a mate and producing young, generally combine to use up all the available energy. If there is any energy to spare the best policy may be to conserve it, in case it's required for a sudden emergency.

In some species play may continue into adulthood, however. When play is observed in adult animals it's usually in a social species, and the benefit is assumed to be that social bonds are thereby strengthened. In other words when individuals derive pleasure from their interactions the group functions better as a co-operative unit. Large-brained species are also more likely to indulge in adult play than smaller-brained ones. This probably reflects a correlation between brain size and the reinforcement to be derived from play, but the picture is confused by the fact that large brains make their possessors more efficient at acquiring food, so there's more likely to be time and energy to spare. Captive apes and monkeys spend more time in play (provided they are kept in suitable social groups and in large and adequately furnished enclosures) than do wild animals, and more time in experimentation with whatever objects are available. Spared the dangers of the wild and needing to spend no energy on finding food and water (maintained, in other words, in much the same protected and leisured conditions that characterise childhood), they have plenty of leisure to fill. We humans are just the same, of course. We play when there's time and energy left over from doing the things that seem more essential to survival; and even in adulthood there is often time to spare.

Some of it may be devoted to sport and games. But many of the more serious activities undertaken in adulthood can qualify as sophisticated forms of play, for they're obviously similarly motivated by the reinforcement we gain from the development and maintenance of schemata. Our devotion to this sort of reinforcement is probably the most defining characteristic of *Homo sapiens.* It's what drives artists, scientists and inventors, philosophers, lawmakers and the creators of religions. It's responsible, too, for the fact that some people put much energy into decorating their homes, while others go in for elaborate cooking, or devote considerable attention to dressing stylishly.

In fact humans often seem to obtain a higher degree of reward from the pursuit of new ideas than from more vital and long-established forms of reinforcement, cheerfully postponing the latter to pursue creative activity. Children don't want to come in for tea until they've finished their game. Teenagers are capable of using their lunch money to go to a film, or to buy a lipstick with which to create a new face. The artist may be willing to starve in a garret, the scientist not only fails to look for a more profitable career but stays up all night to watch experiments, and the explorer is even prepared to risk his life. The intellectual reward can take precedence over mere matters of food, drink or sleep, at least until the need for sustenance or sleep becomes really pressing. It's in the neuronal mechanisms which produce this avid pursuit of new schemata that the essential key to the evolution of reason must lie.

There's reward in the commissioning of a new neuronal circuit, whether it embodies a piece of wholly original thinking or an idea that has been conveyed from outside, having been born in another brain (which luckily means that artists, scientists and other creators don't always have to starve in a garret). Establishing a significant new schema provides a great buzz. A good story, or a scientific theory that makes several previously disparate facts hang together, or a painter with a new and distinctive vision, or anything else that expands the mind and the understanding, can be a source of delight.

The ability to obtain reward from acquiring another individual's ideas must have been important in the evolution of large brains. It would increase the probability that when someone invented a new tool or a new technique the idea would be passed on to their offspring. If beneficial ideas died with their inventors there would be little selective pressure to promote the survival of the genes that make for inventiveness. The idea itself, meanwhile, stands a better chance of lasting survival if others not only imitate it but gain an intellectual pleasure from it, as well as the physical reinforcement obtained by its application. Once the ability to reap pleasure from a new schema was firmly in place ideas which didn't lead to immediate physical rewards could also be appreciated. In a society in which cultural transmission was well developed large and inventive brains will have begun to produce seriously valuable dividends.

The capacity for cognitive pleasure varies from individual to individual, of course, and is influenced by both genes and nurture. An animal whose juvenile period coincides with a time of severe food shortage or constant danger, so that it lacks the opportunity for play, won't discover the joys of that sort of learning. The same goes for a human infant raised in an environment that is lacking in stimulation, or one whose curiosity is constantly discouraged. Conversely, the more a juvenile experiences of the pleasures of investigation and experimentation the more likely he or she is to pursue them in later life.

But the underlying strength of the urge to explore and investigate is determined by the genes, and some of the genetic variations that make for exceptionally curious individuals have been unveiled. Since curiosity is a trait which leads to risk-taking, and since the fellow members of a group can benefit from the discoveries of the few, populations should tend to thrive best if there is a fair degree of variation here. A downside of the trait is that when individuals with a strong need for cognitive reinforcements don't get the chance to learn how to satisfy that need in ways that are useful to society they are likely to look for other means of obtaining excitement.

Humans have extra opportunities to become hooked on the pleasures of creating new schemata thanks to the efficient means of conveying schemata from one brain to another provided by language. The possibilities have been expanded further with mathematical language, and devices such as diagrams. The possibility of transferring ideas easily means that the individual encounters far more schemata than they could possibly create on their own. Such frequent encounters (as long as they are not too demanding and therefore aversive) must tend to intensify the addiction to establishing new neuronal circuits.

It's worth remembering, though, that today's world of constant change and innovation is not typical of the cultural environments to which humans were exposed even in the comparatively recent past, let alone in more distant times. Large brains and a taste for new ideas must have been around a long time before they altered human society to such a degree that new ideas could spread quickly and widely, and thereby interbreed easily with other new ideas. Among ancient hunter-gatherer groups the invention of a new tool, perhaps even a new story, must have been a rare event. Or perhaps one invention sometimes sparked a few more, and there were bursts of creativity followed by long pauses. The archaeological record shows that there can have been no ongoing tradition of looking for new ways of doing things. The imitation that sustains culture is its strongest element, and the normal situation has been that the vast majority of people assume things will go on the way they apparently always have done. Despite the possibilities opened up by large brains, leisure and playfulness, only the occasional genius introduced something new. In small societies with restricted horizons the very idea of pursuing novelty would have been novel. We can be pretty certain of this because it's the attitude more likely to promote survival. The old ways of doing things have been tried and tested, and only occasionally cease to work, while innovation often involves risks.

However, it's clear that the human brain, in an encouraging environment, is capable of creating schemata in enormous numbers. Humans also seem to be alone among species in being capable of becoming dedicated to an idea to the extent of sacrificing physical wellbeing.

Recapitulation

The evolution of cognitive brains, brains that create conscious experience out of sensory input and in the process build schemata about the world, must have been accompanied by the evolution of a drive to use that machinery and a capacity to obtain reinforcement from doing so.

The investment in such complex neuronal machinery could not have paid off otherwise.

Where there's conscious sensory experience there's a satisfaction in that experience (sometimes even in exploring painful experience).

There's also reinforcement to be had from the investigatory behaviour that sensory consciousness requires to make it profitable, as well as in the resulting knowledge that makes useful predictions possible.

Conversely, it's unpleasant to be deprived of sensory experience, or to lack the knowledge which would support predictions.

These pleasures and displeasures have been called cognitive emotions. They are what motivate exploration and play.

Other cognitive emotions are familiarity and unfamiliarity, which function as feedback about the relation of current sensory input to the contents of the schemata store.

Cognitive emotions, like physical pleasures, pains and discomforts, are somewhat ambivalent. Whether familiarity and unfamiliarity are pleasurable or otherwise, and whether exploration is attractive, depends on the current state of the organism.

A uniquely highly developed capacity to obtain reinforcement from establishing new neuronal connections and discovering new ideas seems to be the most important distinguishing characteristic of *Homo sapiens*.

Language has played an important part in promoting this trait.

Chapter 12

Social Emotions

It was only after sensory consciousness and cognitive emotions were well established that the mammalian (and avian) practice of continuing to feed and protect infants after they have become mobile can have become practicable.

One benefit of sensory consciousness is that it enables distance senses to be exploited for feedback purposes, just like the internal and contact senses. This is what makes motor learning possible. It must also underpin the business of keeping mothers and infants together. Where mothers must feed infants that are capable of locomotion it's essential that mother and infant learn to recognise each other. Looking after infants uses up energy and time which the adult might otherwise spend on promoting its own wellbeing, so natural selection ensures that animals only make the sacrifice where they are investing in the future of their own genes. The care must be given to the right infants if the genes that produce the behaviour are to be passed on. And the young, of course, need to identify the female that is prepared to feed them.

How is this peripatetic togetherness managed? Some early students of animal behaviour, focussing on mammals, leapt to the conclusion that infants grew attached to their mother because she provided them with nourishment. The idea is instantly refutable by reference to infant birds of the precocial sort (the ones that leave the nest almost immediately after hatching), for they learn to recognise mother very rapidly, and follow her around with great devotion, although they feed themselves. Moreover, keeping close to mother takes priority over eating - if they become separated from her they rush frantically up and down looking for her, and seldom stop for anything else. In contrast, altricial birds (the sort that are fed by their parents until they are fully grown) seem to remain vague about identifying their benefactors for some time, at least by visual means. As late as the first day or two out of the nest they will accept food from any source, if it's offered in the right way. (Perhaps this isn't altogether surprising, since what the nestling mostly sees of a parent must be the inside of a food-laden beak.)

The idea that an infant primate's attachment to mother might be motivated by her role as supplier of food was put conclusively to rest by Harry Harlow, who took newborn monkeys from their mothers and housed them in cages furnished with two dummy figures. One was made of wire, and equipped with a nipple from which milk could be obtained. The other, similarly shaped, was covered with soft terry cloth, but had no nipple. The infant monkeys would venture on to the wire figures when they needed nourishment, but spent a great deal of time clinging to the padded one, and would rush to it when alarmed. Thus they demonstrated an innate response to soft tactile input rather than to the source of food.

Precocial birds, however, have provided the most convenient subjects for studying mother-infant attachments. In the farmyard it has long been common practice to put orphaned clutches of eggs under any conveniently broody female, and the hatchlings learn to follow a foster mother of a different species as readily as other hatchlings follow their natural mother. Konrad Lorenz brought such adoptions to the notice of the scientific world with his account of a family of goslings which he raised himself. They waddled along after him wherever he went, and showed signs of great distress if they were separated from him. The perceptual learning which allows an infant bird to adopt a substitute mother when the real thing isn't available Lorenz termed imprinting.

Imprinting became a popular subject of research. Many experiments were done to find out just what newly hatched precocial birds look for in a mother-figure. It was determined that, in the absence of the real mother, infant ducks, geese, chicken and turkeys, not to mention coots, quails and moorhens, would rapidly learn to approach instead all sorts of unlikely objects - cardboard boxes, footballs and flashing lights among them. Given a choice, though, they invariably preferred something more like the real thing, an appropriate shape to an inappropriate one, the right colours to those of another species, and so forth. Gradually the experimenters worked out just which features of an adult bird operated most powerfully to elicit a reaction from a naïve infant. A head of the correct shape, eyes, and a rolling, two-legged gait generally proved to be among the most powerful factors. It seems that the innate specification is adequate, under normal circumstances, to fix the attention on mother, and only the details need filling in. Precocial birds are programmed to follow the right sort of mother-figure, just as monkeys are programmed to cling to the right sort of texture. But the urge to do it is so strong that almost any salient object will be adopted as a substitute if it's the only thing available.

The most revealing fact about imprinting may be that it must happen (in precocial birds) during the first two or three days of life. Within this period an infant which has become attached to an inappropriate stimulus pattern will desert it for a female of its own species. After that it will be too late for a transfer of affections and the unsuitable attachment will persist. On the other hand, having once met a proper mother-figure at the impressionable age the infant will never become attached to any other sort of stimulus pattern. Significantly, this critical period is also the time at which the newhatched infant's visual system is first exposed to patterned input. If the chick is reared in the dark for a few days, or fitted with goggles which allow only diffuse light to penetrate, imprinting can still take place when the visual world is revealed.

The infant's following response is elicited by mother's voice too, and under normal circumstances learning about her voice comes first, for it begins in the egg.

Eckhard Hess observed that mallard ducklings begin to make a high-pitched, peeping sound while they are still in the shell, twenty four hours or more before they hatch, and the mother duck clucks quietly in response. The peeping increases as hatching time draws nearer, and comes to coincide more with the mother's vocalising, and when a hatchling breaks out of the egg the mother gives a prolonged burst of rapid clucks, lasting a minute or two. It's usually somewhere between sixteen and thirty two hours after all the eggs are hatched that the duck leads the ducklings out of the nest, and then her clucking sounds begin to turn into quacks.

By the time the ducklings leave the nest they have therefore had at least a day and a half in which to get used to the pattern of their mother's voice. Mallard ducks all produce much the same sounds, but the calls of different individuals can be distinguished by their characteristic rhythms. It seems likely that the distinctive rhythm of mother's voice provides the main cue which allows the infant to distinguish her from other females. Certainly, if an infant strays too far away from mother it's the sound of her voice that must provide the most practical means of locating the correct female - and mothers act in accord with this assumption, and call persistently when an infant goes missing.

Distinguishing between two mallard females by visual means is hard at close quarters, and would seem to be impossible at a distance. Moreover the ducklings don't seem to get much chance to make the comparisons which would let them learn to do it, since the species of birds which produce precocial young, though generally sociable at other times of year, tend to go off and find isolated spots for their nests. This is a practice which makes things more difficult for predators, which at least can only pick off one brood at a time. It also has the advantage that the hatchlings are not in much danger of exercising their following instinct on the wrong female. When the young are led out of the nest, the mothers still tend to keep away from conspecifics - the little flotillas on river or pond are well spaced out. So while the infants may hear the distant calls of other mothers it seems likely that they don't get much opportunity to make useful visual comparisons until they are quite well grown.

The importance of the visual stimulus pattern in keeping precocial chicks close to their mothers must lie in the fact that, in contrast to mother's utterances, it's a continuously available signal, which doesn't require any effort from mother. It's also one that can be located with precision. Her intermittent calls serve to remind the infants, from time to time, to pay attention to her, and must be particularly important when she moves off after some stationary foraging.

What's interesting is that the sight and sound of mother not only constitute stimuli which the infant is pre-programmed to approach, they also provide a reinforcement that rewards that behaviour. The lost infant works hard to find mother again, ignoring all other distractions. In a laboratory experiment, chicks which had become imprinted on a flashing lightbulb devoted themselves for long periods to jumping on the pedal that switched it on. And the infant that has located a lost mother visibly obtains reassurance as he approaches her.

How do we explain the mysteries of imprinting - the fact that a visual stimulus can function as a reinforcement, and the fact that an infant precocial bird which doesn't encounter its mother within its first three days of visual experience will thereafter reject her in favour of a perhaps quite unsuitable substitute? I think the questions are answerable if we assume that the infant is a cognitive animal, using conscious sensory experience in its relationship with mother. This assumption seems reasonable, in view of the subtle distinctions the infant must make in order to distinguish its mother's voice from that of other females. Identifying an individual animal's call means distinguishing a rhythmic series of varied pitches, in other words a collection of relationships distributed over time. This implies a short-term memory which enables the relationships to be registered and the pattern perceived. The short-term memory which allows such comparisons to be made equates - if the theory is correct - with sensory consciousness.

Working on this assumption we can deduce that the visual and auditory stimulus patterns which are mother function as positive reinforcments because they very rapidly become familiar stimulus patterns. In the natural way of things mother is always going to be the most familiar stimulus in the precocial hatchlings' young lives. While they're still in the nest she's the most salient stimulus, visually and aurally, and when she leads them out into the big, wide world she must continue to be the most attention-demanding stimulus, simply by virtue of being a moving object and near enough to provide a large retinal image. In addition the hatchlings must obey the dictates of their hardwired response to this stimulus and follow her, which means their attention is pretty much fixed on her. Thus she virtually monopolises their visual attention in the first days of their lives. Similarly she emits, at close quarters, sounds to which the infants are programmed to pay attention. Inevitably, mother is the subject of the first visual and auditory learning the infants do, and they do it extremely thoroughly.

Mother thus becomes the one familiar stimulus pattern to an infant which has no other experience of the world. Once familiarity is firmly established for this first stimulus pattern all other salient stimulus patterns must register as unfamiliar, and therefore anxiety-arousing. Many of them will become familiar in due course, but the first-learnt pattern is the one that provides comfort and security most powerfully. For the infant which has become imprinted on a cardboard box a real live female is a strange and therefore alarming phenomenon, and can no longer fill the role of stimulus-to-be-followed. It feels safer to stick close to what's known and comfortingly predictable.

Imprinting can be regarded as cognitive learning grafted on to an old foundation. All species must have the means of identifying conspecifics, and in all species where there is any sort of parental care of offspring the infants must be equipped with an innately wired representation of the essential features of the adult which will enable them to respond to the adult's care - even if this is only the basic skeleton of a pattern, like the gull's yellow beak with a red spot on it. Such genetically determined stimulus-response programmes are obviously still at work in precocial birds, and since they are hardwired they must be able to function without any need of consciousness. It can be deduced that the newly hatched bird's initial following of mother doesn't depend on the neuronal system which creates conscious vision. But following mother means that visual attention is focussed on her for considerable periods, and so she's the subject of the first visual schema to be developed. She rapidly becomes a familiar stimulus pattern, known in detail. The familiarity makes the sight of her rewarding, in a still unfamiliar world. And every time the infant is soothed by the sight of mother or by her vocalisations it will come to associate her even more firmly with reassurance, and the mother-schema will be strengthened.

The bit of the avian brain that allows the infant to learn how mother looks has been identified. If it's removed before the chick becomes imprinted imprinting doesn't happen. If it's removed afterwards the chick no longer recognises the mother-substitute it became imprinted on. However, the lesioned chick still shows a preference for a stimulus with the right sort of head and neck arrangement, demonstrating that the hardwired reflex is still intact. And it can still learn a visual discrimination for a different sort of reward, confirming that perceptual learning about the mother-figure is a special sort, with its own special wiring.

In time, of course, youngsters learn about the world they live in. They begin to know what to expect of the various percepts encountered, and as the realm of the unfamiliar shrinks there is less need of comfort from the mother-figure. They also grow larger and more physically competent, so mother looks less and less like a figure that can provide a superior defence against danger. The juvenile is prepared to stray further from her, and eventually feels ready for independence.

Mammals, like birds, can be divided into precocial and altricial species. Some infants tumble out of the womb ready to gather themselves together and stand on their feet, others are born relatively or completely helpless, and must be nursed in a sheltered place, or carried, until they grow sturdier. The talent common to them all is an inbuilt programme for finding a teat and sucking on it.

Mammalian infants have the opportunity to hear something of their mother's voice while they are still in the womb, just as birds do in the egg. The quality of the sound may be a little distorted, but the rhythms and patterns of intonation are discernable, so mother's voice must already be to some extent familiar when the infant emerges from the womb. Mother is also, again, the first and most salient visual stimulus the infant sees, and the one that becomes outstandingly familiar. In altricial infants the learning may be delayed until the sensory systems mature, but it's still focussed on mother, the major attention-demanding stimulus to appear in the sheltered den.

Smell is also important in the attachment between mothers and infants. Even in us humans, with our relatively limited olfactory sensibilities, mother and infant rapidly learn to distinguish each other's scent, and to take pleasure in the familiar smell. On the infant's side the odour is, again, something that will already have become familiar in the womb, while for the mother the learning must be easy, since the infant's odour must be closely related to her own.

For altricial mammals tactile input must generally play an important role in mother-infant relations too, as Harlow's experiment showed. In many species infants spend a good deal of early life snuggled up to mother for warmth. In us primates, where infants are regularly carried around by a parent, this sort of contact soon becomes rewarding, so that a good cuddle can be a cure for many minor hurts or frights.

Like birds, infant mammals rapidly build on the innate responses and the familiarities that began to be established before birth, and learn to distinguish mother from other females by means of the distance senses. And they rapidly learn to take pleasure in the sight and sound, as well as the smell and feel of her. If they get lost, her voice can act as a beacon to indicate her whereabouts, and all four sensory channels will contribute to calming the refound stray.

A non-classical reward

Clearly mother and infant function as reinforcing stimuli for each other. But it's a distinctly non-classical sort of reinforcement that doesn't depend on physical reward. For much of the time infants and mothers are happy if they are within sight and sound of each other. Furthermore it's an ongoing sort of reinforcement, which requires an ongoing response. More conventional learning - the type investigated by the Behaviourists - involves a stimulus which usually disappears when the response is made (having been consumed or successfully avoided) and a reinforcing event which is generally brief, or at least finite in duration. The sequence may be repeated many times, but each bit of it is transitory. Mother, in contrast, is a stimulus which must be kept in view but for most of the time not actually acted upon, and produces a rewarding effect which is continuous.

In this, the business of keeping mother within a certain distance is more like the business of keeping within a territory - something which is also defined by familiarity. And, like keeping an eye on where one is geographically, keeping in touch with mother is something that has to be combined with other occupations. There are other things to be done simultaneously. Young precocial birds have to feed and preen themselves. Young mammals play. And they all have to study the world about them. The rewards gained from these activities have to be interleaved, as it were, with the reward gained from the presence of mother.

It's difficult to imagine an inbuilt, unconscious stimulus-respose mechanism that could cope with doing two things at once. This sophisticated infant behaviour is made possible, I think, by sensory consciousness. Because the sight and sound of mother are conscious sensory experiences they can be checked up on occasionally, in between paying attention to other matters, allowing the infant to juggle two disparate concerns simultaneously. Conventional means of deciding between two different stimuli can't apply here. It's not a matter of an approach-mother reflex that always takes priority over everything else, or the infant would never do important things like investigating its environment except when mother was absent - the opposite of what actually happens. Nor is it a case of the stimulus pattern which evokes the strongest response in neurological terms getting the behavioural response. The further the infant strays from mother the smaller the retinal image she will present, and the weaker her effect on other sensory systems, but the infant's motivation to return to her becomes stronger. Conscious visual sensation must be what enables the infant to respond to the diminishment of the retinal image, and the conscious sense of confidence he obtains from mother is what prompts him to remain in her proximity. Sensory consciousness and emotional consciousness between them allow the infant to weigh up the desirability of competing forms of behaviour.

Even when mother isn't the source of nourishment the benefits an infant obtains from staying close to her are several. One obvious advantage is that mother is big enough to protect him against predators for which he represents a worthwhile meal. Another is that he can observe mother's actions and imitate them, learning what she eats, where she finds it, and the techniques she uses to obtain it, what sort of place she considers good for sleeping in, or for escaping from danger.

By the time extended maternal care and social emotions came into being conscious sensation of basic internal physical states was surely already in place, earning its keep as a means of marking useful features on the cognitive map. Such sensation was therefore available to play a part in the education of the young. The young hatchling or the weanling mammal, following its mother around, learns to associate the sort of places where food is found with the comfortable feeling in the stomach which food produces. If it needs to learn what to eat it does so as a result of the good taste experienced when it tries out what mother is eating.

In the more sophisticated species mother may even take an active role in her offsprings' education. Both cats and meerkats start to provide their young with living but disabled prey as they grow older, and then, as skills improve, with undamaged prey, so the kittens can practise catching and killing animals relatively easily before they begin stalking food for themselves. Chimpanzee mothers similarly give their offspring opportunities for learning. In the Taí National Park in the Ivory Coast the chimps regularly crack nuts on a wood or stone anvil, using a stone or a dead branch as a hammer. Mothers with youngsters who show an interest in this process may leave a nut or two near the anvil, along with a hammer, while they go off to find more nuts. Individuals without infants don't leave nuts or hammer unguarded. Mothers may also let infants take over their hammers, or take nuts from their store. On rare occasions they may even demonstrate how the hammer should be held or the nut positioned.

The greatest benefit of prolonged maternal care, however, is probably the chance it provides for infants to learn from mother's reactions rather than her actions. It's by observing mother that the infant discovers which of the many novel stimulus patterns that bombard its senses should cause anxiety and provoke retreat, which should be treated with caution, which approached, and which ignored. In other words, mother provides the first part of the education which turns an unfamiliar world into a familiar one, about which value judgements can be made. And in the interim she provides protection against the unknown.

The implication is that not only must the infant be a cognitive animal in order to recognise its mother, it also needs to recognise its mother because it's a cognitive animal. It needs help with all the learning it has to do. In particular, it needs guidance when it confronts the unfamiliar. Providing evaluations of the stimulus patterns the infant encounters is a vital part of a mother's role, and a strong motivation for staying close to mother is that the infant needs her authorative opinion about the significance of novel stimulus patterns. It also needs her comfort when a novelty is overpoweringly strange and upsetting.

The larger the brain the greater the capacity for perceiving the complications and dangers of existence. Social lifestyles add another layer of challenges, and infants may need not only guidance on how to behave but also some protection against other members of the group. Mothers may have to make some sacrifices to provide such protection, as well as in guarding their infants against other dangers. What inspires them to do this is of course the phenomenon known as mother-love, and seeing it at work provides the offspring with a sense of security, and inspires reciprocal love.

Facing life without this sort of guidance and this sort of comfort is very disturbing for a cognitive infant. That's why children who are deprived of proper maternal care, or who are separated from their mothers in childhood, frequently grow up to suffer from depression, as eloquently documented by John Bowlby. Such individuals are likely to know the world as a place where certainties are hard to come by, where the good and the bad have to be distinguished by hard experience, and where there is no comfort when things go wrong. Even if the child's fortunes improve later, such foundations are hard to build on.

A corollary to all this is that brains capable of dealing with vast amounts of sensory input in flexible ways are only likely to evolve in species where the young keep company with a parent for a while, and can learn from their example.

Other special forms of learning

The rapid learning which allows mothers to be sure of directing their care to their own infants is heavily influenced, in mammals, by the hormones which flood the body at the time of parturition. As well as triggering the expulsion of the infants and the commencement of milk production they elicit appropriate behaviour. Something similar must happen in birds, for a broody duck can readily be persuaded to sit on another bird's eggs, but will reject them if the developing infants start making noises sooner than her own would have done. In some mammals the infant's odour also seems to be important to initiating maternal responses. Sheep farmers know that if they want a bereaved mother to adopt an orphan they must wrap it in the skin of the dead lamb.

The rewardingness and the caring activities serve to focus attention on the infants, so mother rapidly learns to recognise them, and the learning is continuously updated as the offspring develop. But by the time they are grown the maternal hormones are ceasing to flow, so generally mother and young go their separate ways. Where there is a period of juvenility, however, the rewards with which the young have come to be associated create a more enduring bond.

The mechanisms which keep mother and offspring together find some parallels in those which keep sexual partners together, in species where they share a territory and co-operate in caring for the young. As a first step they must learn to recognise each other, since they must be able to distinguish mate from territorial intruder. Familiarity may provide part of the reinforcement that maintains the relationships, especially where other conspecifics are kept at a distance.

The forces which make social lifestyles possible, meanwhile, must be closely related to those which tie mothers and infants together. It seems, indeed, safe to assume that the one grew out of the other. In social species the urge towards togetherness and the ability to be rewarded by the presence of familiar company persists into adulthood. We know that the presence of other members of the group functions as a reward because animals of social species will work at demanding tasks in order to be reunited with their fellows, just as infants will put a great deal of effort into looking for their mothers. Moreover, keeping in touch with the group, like keeping within sight of mother, is something that has to be done simultaneously with the pursuit of other goals.

Most of the special forms of learning dicussed in this chapter take place only at a specific time. While precocial infants learn to recognise their mothers almost immediately after they are born, altricial infants take longer, but the learning is in place by the time it's needed. The maternal hormones dictate the timing of the mother's learning. For many species pair-bonding takes place at the beginning of the breeding season, and is similarly influenced by hormonal condition. Only the business of learning to recognise other members of a social group is a continuous process. The infant must learn to identify its mother first - quite quickly if it's mobile and there are a lot of other females in fairly close proximity. Gradually it will become familiar with all the other group members too; and in adulthood it will be necessary to continue learning the new infant additions, the occasional incomer from another group, and the slow changes wrought in all the fellow group members as time passes.

There are two or three other forms of learning which must happen within a certain period of life if they are to happen at all. In some of the numerous species of songbird the song is innately programmed and invariable, and males grow up to sing it perfectly, regardless of whether they have ever heard it. For other species only the rough outline is pre-determined, and the male develops his own individual song by creating variations on the basic theme. These variations are influenced by what he has heard other members of the same species singing. In some species a bird that doesn't hear an appropriate model produces only a partial and distorted version of the usual song, and in others he doesn't sing at all. In those species which have been studied, the exposure to the model has to take place during the first year of life, even if the bird won't start to sing until his second summer. In what looks like a close parallel, on the rare occasions when human children have been raised without exposure to language before the age of about seven or so, they have never properly mastered it thereafter. With careful tutoring they can come to make some use of words, but they don't become fluent, and never master all the intricacies of grammar.

There's a likely hypothesis as to why these particular varieties of learning are only done in infancy. Songbirds, for instance, produce altricial young, and when the fledglings leave the nest they hop around in the undergrowth, bleeping regularly, which lets their parents know where they are so that food can be delivered. As they grow bigger they venture on longer excursions, and there must be a danger that they will stray off the parental territory and end up in the domain of another pair, where the bleeping signal may be out of earshot, and where their parents may anyway find it difficult to feed them. It seems possible that the distinctive pattern of father's song has by then become familiar to the fledglings, and that, as well as advertising his territorial claims to rivals, it serves as a beacon for his offspring. If they hear the familiar tune from far away they can head back towards it. Father's song thus (perhaps) functions to keep the infants close to their parents in much the same way as the call of the mother and her appearance function in precocial birds. And maybe it provides comfort and reassurance too.

If this hypothesis is correct, then innate response mechanisms must first cause the nestlings to pay attention to father's song, so that by the time they leave the nest its detail has become familiar. And when it's familiar it becomes rewarding, so that the fledgling continues to pay attention to it. Neither factor is at work by the time the youngster has become adult, and that may be why it's then too late for song-learning.

It may be for somewhat similar reasons that human language learning must apparently begin before the age of seven or so if language is ever to be fully mastered. Learning to communicate offers several rewards to the infant. One is the fun of interacting with carers in new ways, sometimes even influencing what they do and exercising a sort of power over them. Another is that simply uttering the appropriate word turns out to provide a more effective means of obtaining material rewards than was previously available. A third is that language provides a powerful way of obtaining help in making sense of the great big mysterious and sometimes alarming world, and of checking on hypotheses about it.

A child who grows up without language has to manage as best they can without these rewards. They don't discover the pleasures of socialising through language, they grow big enough to obtain what they want for themselves, and their schemata develop without the aid of information that could have been supplied through language. When a certain stage of self-sufficiency is reached there is no longer an adequate incentive for the hard intellectual concentration involved in mastering grammar. And meanwhile the innate tendency to pay attention to human vocalisations has faded, in the absence of reinforcement. An added complication is that without language in early life the child is unlikely to develop the capacity for the sort of thinking that requires language to express it. (However, it should be stressed that it is very rare for a child not to be exposed to language in infancy, and in the few known cases where a child has been locked up in isolation, or abandoned in the wild, the possibility that they were mentally disadvantaged in the first place can't be ruled out.)

To birdsong and human language must be added the arts of social interaction. Harry Harlow demonstrated that infant monkeys brought up with their mothers but with no other company never learnt how to get on with other monkeys. Those separated from their mothers but allowed the opportunity to play with other juveniles didn't do quite so badly. The few recorded cases of wild children suggest that this is also true of humans. Socialising can only be properly mastered if the observation of fellow beings is begun at an early age.

In this case we may guess that the strange stimulus-pattern doesn't fail to attract attention but attracts too much. And to a mind that has become familiar with only a limited range of undemanding stimulus patterns a conspecific - moving, autonomous and unpredictable - is quite frightening, not to be assimilated as anything other than frightening.

In short, it may be hypothesised that some rewards can work only if they are brought into action at the appropriate stage of life, and gain the pleasing sheen of familiarity. If the stimulus patterns that ought to be rewarding are encountered after this critical period they don't function as rewards, and may even acquire a negative import. Such are the perils associated with cognitive ability.

Learning by emotional infection

The most important advantage of keeping company with mother, I've suggested, is that the infant learns how to evaluate its world - what's worth approaching, what's to be avoided, what's insignificant. In addition it learns where to go to satisfy various needs, and (in species where this is necessary) what's good to eat. Part of the education can be acquired simply by following mother and doing what she does - eating what she eats, resting where she rests, and experiencing the satisfactory effects which follow. The most valuable lessons, however, must be the negative ones, the ones about what to avoid. By learning from mother youngsters can acquire this vital knowledge without experiencing pain or suffering damage. And they learn not so much from mother's actions as from her reactions.

It's a particularly tricky bit of education, since when danger threatens mother may not be setting an example of approriate behaviour by running away, but doing her best to stand between it and her offspring. Observation suggests that the way this particularly valuable and particularly tricky bit of information transfer works can be appropriately phrased in subjective terms. The infants learn what's frightening or anxiety-arousing as a result of seeing mother frightened or anxious.

Judging by our own subjective experience, this transfer of vital evaluating judgements is achieved through conscious emotion. One of the most notable characteristics of emotion, and particularly of emotions such as fear, is that it's highly infectious. The sight or sound of someone else's fear triggers at least a nasty twinge of apprehension in the observer, even if rational analysis of the situation rapidly indicates that there's no good cause for it. Anxiety too is catching. A display of anger, meanwhile, arouses reciprocal anger, or fear, or at best discomfort, in the person at whom it's directed. The emotional reaction is automatic. It can be triggered by events on stage or screen, as well as by those which have actual practical significance. And the effect of a horrifying event on screen is largely determined by the emotion displayed by the actors. When the hero maintains his cool under all circumstances we are entertained more than upset by the threats of world-scale disasters.

Laughter and the positive emotions are infectious too, though perhaps less reliably so. (All that canned laughter in the television shows is designed to get the audience at home laughing, and assure them that the jokes really are funny.)

Human infants, it can readily be observed, are even more sensitive than adults to emotional currents. At the slightest sign of sadness or irritation on the part of a carer they soon begin to grizzle in sympathy. In humans this ability to read even quite subtle emotions plays a vital part in learning. Language would hardly be possible if the infant couldn't appreciate the feeling expressed when an adult utters a word such as *good, nasty,* or *naughty,* and thus deduce the sort of judgement it conveys. And until it's possible to communicate adequately by means of language, persuading an infant to refrain from dangerous or undesirable actions must be done by means of intonations suggesting disgust or fear, and appropriate facial expressions. Such is the sensitivity of the audience, however, that the expressions mustn't be too realistic or they'll cause distress.

Since human infants grow up to experience all these emotions consciously, and since there is no sign of a point in development where emotional behaviour seems to look as if might switch from an unconscious mode, it's reasonable to deduce that what looks like conscious feeling in an infant actually is. And when other mammalian infants show all the signs of emotions such as fear, and appear to be sharing them with mother, it similarly seems reasonable to imagine that they too have a conscious experience of the emotion.

That consciousness is involved seems the more certain in that when an infant learns from observing its mother's reactions it has quite a complex job to do. The stimulus pattern that provokes the response may be a complicated one, and the infant needs to learn enough to be able to recognise it again. But before it can do that it must establish what has evoked mother's emotion, by ascertaining the direction of her gaze, or at least which way her head is pointing. That means saccading between mother and the stimulus pattern. If the mechanisms of consciousness serve to bind together events that are slightly separated in time, it would appear that they are needed here.

It must have been extended maternal care which made the evolution of emotional consciousness particularly profitable. In turn it made extended maternal care more useful, and must surely have been a factor in its further development. The variety of conscious emotions probably grew in tandem with lengthening periods of juvenility, which brought more time for education, and coincided with the brain expansion which meant that more could be learnt.

Many non-mammals seem to reveal emotions by the rhythm of their movements and vocalisations. Mammals have added to these means of expression. Complex and adaptable musculatures give them a wide range of postures, and there's usually a fair range of vocal calls. Some mammals have evolved expressive faces. Moreover the variety of postures, calls and facial expressions tends to correlate with larger brains.

In primates there are particularly suggestive developments. The divided upper lip of prosimian muzzles fuses into a continuous one in monkeys and apes, which, as W. E. le Gros Clerk observed, allows a wider range of facial expression. In apes and humans, furthermore, the hindbrain nucleus which determines spontaneous, emotional expressions is larger than would be expected on the basis of a mere scaling up, while the density of face neurons in motor cortex, which controls voluntary movements, is exceptionally high.

Expressiveness and the ability to learn from mother's reactions no doubt evolved together. Where social lifestyles developed they became even more valuable, since there were more models to learn from, available throughout life, meaning that adult learning could be passed on. Here they will also have tended to promote co-ordination of group activities.

But ....

But what exactly is this conscious experience we call an emotion? A century or so ago William James proposed that what we experience as strong emotion is actually a consciousness of physical reactions. What we know as fear is what happens when the body prepares for running away or for self-defence - rising blood pressure, shallow breathing, tensing muscles, and as James dramatically put it, *quickened heart-beats, trembling lips, weakened limbs and visceral stirring*. Anger involves some of the same effects (preparing the body for action and for coping with possible damage) along with some contrasted ones. *Can one fancy the state of rage*, asked James, *and picture no ebullition in the chest, no flushing of the face, no dilation of the nostrils, no clenching of the teeth, no impulse to action, but in their stead limp muscles, calm breathing, and a placid face?*

Antonio Damasio in particular has insisted that James *seized upon the mechanism essential to the understanding of emotion and feeling*. Firm evidence has come from a man who had to have both adrenal glands removed. He is still fully capable of recognising dangerous situations, but he no longer experiences any sensation of fear.

The difficulty I felt when I first encountered the idea arose of course from the question of temporal order. The natural assumption was that one must consciously judge an event to be frightening, or worthy of anger, or whatever, before the bodily reactions take place. It had to be the emotion that prompted the tensing of the muscles and the raising of the blood pressure, not *vice versa*. Now we know that conscious experience takes a little time to develop. and that sensory inputs which are not consciously perceived can have significant physical effects. It has been shown, for instance, that sensory events that don't make it to consciousness can nonetheless influence skin conductance. It's easy to imagine, therefore, that where the significance of a stimulus pattern is already known – or where it's alarmingly strange - physical responses to it can be triggered while conscious perception is still developing.

Long before any species was possessed either of conscious emotions or of a conscious means of judging the need for action there were hardwired behaviour patterns for escaping danger and for combating rivals. These responses to genetically determined stimulus patterns were accompanied by adjustments to the animal's physical state which allowed them to be performed with maximum efficiency. When the capacity for conditioned learning evolved, and previously neutral stimuli could come to function as signals warning of a potential need for action, these same neuronal and hormonal effects would be triggered by the conditioned stimulus. And as consciousness evolved the internal responses would have to be part of it, so as to provide the essential valuations on the inputs to external senses, and be incorporated into choices and plans.

In social mammals, especially larger-brained species, more complex social emotions seem to emerge. There are signs of feeling that might roughly correspond to respect, shame, ambition, affection. The subtlest emotions, though, must develop only with extensive introspection and the ability to communicate with language. Our feelings can get inextricably entangled with the perceptions and thoughts which coincide with them, so that previous experience influences the current sensation.

In the more sophisticated human activities the role of autonomic feedback is modest. People whose brains are intact but whose autonomic systems are disabled at peripheral levels are only slight disadvantaged in gambling tasks. But in patients implanted with electrodes for medical reasons it's been shown that a little covert stimulation in one of the nerves by which information about what's happening in the body is conveyed to the brain improves the choices made. It can also make for more efficient learning.

A balancing act

I've already hinted that conscious feedback and social emotions don't function only as a means of education. They also provide a means whereby conflicting needs or conflicting drives can be weighed against each other. In a species which works entirely on genetically ordained programmes the possibilities for sorting out such conflicts must be limited. Responses to danger take priority, and when there are no such distractions internal sensors indicate whether eating, drinking or resting is the next most important thing to do. At the appropriate season sex hormones begin to flow, and as their strength grows push sexual activity towards the top of the agenda. In species capable of conditioned learning the order of precedence can probably be modified by the effects of past experience.

But when the reports on the physical state of the organism are expanded into conscious sensations a more delicate balancing act is possible. For instance, the need for food can be weighed against the need for water in the light of the potential of the current environment, and an estimate of the amount of effort that will be required to fulfil each need. Consciousness, we might say, constitutes an arena in which many different factors can be brought together.

This is important to the evolution of prolonged maternal care because that's a tricky business. Mothers have delicate judgements to make. The name of the game is preserving the genes into future generations, so natural selection has favoured mothers that take good care of their offspring. To some extent a mother must put the infants' welfare before her own. But infants without mothers have a severely reduced chance of survival, whereas mothers who lose their infants may live to breed again, so self-sacrifice musn't go too far. A mother with infants in tow must frighten off a predator if she can, where if she were on her own she might flee. But if the predator is too big there's no sensible alternative but to run, rather than lose both offspring and her own life. The danger can't be treated with a simple reflex response, but requires a judgement of the whole situation. A mammalian mother of helpless infants may also have to balance her own need for food against the interests of her young. Leaving them unguarded may be risky, but if her milk fails they are doomed anyway. Judgements must be made that take into account the time likely to be needed for finding food, any signs of imminent danger, and the capabilities of the infants.

Conscious emotion, introspection suggests, functions quite effectively as a means of gathering all the elements of the equation together, so that mother-love can be balanced against hunger, and the result will depend on the degree of hunger and on the infants' precise degree of vulnerability. Or mother-love can be weighed against fear, and since the degree of fear will be related to the size of the threatening predator, or the perceived degree of risk, a sensible decision is probably reached.

Then the offspring grow bigger and stronger and the costs of caring for them become greater. The pleasure of suckling them is lost. If they are being provisioned with solid food their appetites grow bigger and bigger. Moreover they gradually cease to resemble the stimulus pattern that evokes and rewards mothering behaviour, and the maternal hormones fade. The patience with which the small offspring were treated is gradually replaced by a tendency to irritation when the juveniles' demands become excessive. Finally mother's patience is exhausted and self-interest takes over again, while the juveniles become self-confident and no longer value mother's support as much as the prospect of independence. Thus maternal care may draw to a close slowly, rather than simply ceasing abruptly at some signal such as the termination of the milk supply, or the maturation of the youngsters' locomotive ability. The balancing act of emotional consciousness supports finer judgements than can be achieved by a hardwired hierarchy of priorities, just as finer discriminations can be made about the external world as a result of consciousness in the external senses.

Emotions might be regarded as the conscious result of calculations carried out at an unconscious level, and they generally tend to produce judgements that promote survival. Since the more ambitious applications of reason and the building of complex societies often require that emotions be controlled or even suppressed, reason and emotion have often been classified as opponents. But emotion can be seen as a sort of proto-reason. Fully-fledged reason makes it possible to put more information into the equations and reach subtler judgements. But the mechanisms of conscious emotion provided an earlier means of weighing up several sorts of evidence, and reason builds on that beginning.

Recapitulation

Parental care of mobile infants demands that parents be able to distinguish their own infants, and that infants can identify the adult or adults that are prepared to provide care.

It also requires parents and infants to stay in contact while simultaneously giving attention to other activities.

This is possible, I suggest, because input from distance senses is translated into conscious sensory experience, and a consciously experienced sight, sound or odour provided by the one party can function as a reward for the other.

For the infant the rewardingness derives from the fact that mother rapidly becomes a familiar stimulus pattern, in a world where nothing else is.

This could only happen in a species equipped to do more than refine responses to genetically determined stimulus patterns, one that explores the world, inspired by cognitive emotions, and needs to distinguish familiar from unfamiliar.

We may deduce that the evolution of prolonged maternal care has been closely intertwined with the evolution of conscious emotion, and of the conscious sensory experience which makes fine distinctions possible.

Prolonged parental care offers scope for learning from mother, and therefore for evolving the sort of brain that could learn by following her example, both in actions and in emotions.

The talents that became useful where extended maternal care evolved also permitted the development of social lifestyles that hadn't been possible before. Living in social groups offers much the same sort of benefits as maternal care, but on a larger scale.

Chapter 13

Social Lifestyles

Sociable lifestyles have evolved several times, in different forms. The simplest is founded on an innate behaviour pattern which causes conspecifics to stick together. It's perfectly exemplified by the small fish which move about in shoals, in which a change of direction often runs through the whole crowd like a travelling wave. If one or two fish turn aside, prompted by a hint of danger or a prospect of food, its neighbours follow, more follow them, and so on. The bit of brain that produces this impressive co-ordination has, in one species, been identified. Removing it produces a freak, an independent fish that goes its own way, and becomes the regular leader of the shoal as a result.

A major advantage of shoaling lies in improving the chances of escaping predation. A predator generally needs to spend longer over locating a shoal than is needed to collect a satisfactory meal when the prey are dotted about at intervals. Sociability also means that there are many sensory systems at work to spot approaching danger, and with a large shoal the hunter has to meet the challenge of focussing on just one moving stimulus among the flashing myriads. The most telling factor in favour of the genes which promote togetherness, though, must be the fact that when predators attack it's the individuals which push into the centre of the group most successfully which are most likely to survive, which makes for a strong selective pressure bearing directly on the herding instinct.

Various species of bees, wasps, ants and termites go in for a different style of social living - huge, highly organised societies in which the members play specialised roles, or a succession of roles. Queens are elected by the accident of being born at the right moment, when season, weather, and/or the size of the colony prompts the workers to feed some of the larvae with a special diet. Interactions between the members of the society are a matter of genetically determined responses to automatically produced signals. Thus a large number of brief-lived individuals combine to function as what could be called a superorganism.

Among mammals there are ways of life that carry echoes of both these types of socialising. Some species of herbivore herd together in enormous numbers, in a way that suggests a simple urge for pure, unstructured togetherness, though only when migrating between winter and summer pastures. The naked mole rats of East Africa, meanwhile, organise themselves rather like the social insects do. They live in colonies of a hundred or so, in a great warren of underground tunnels which give access to the tubers it eats. Only one pampered female in the group gets to breed, and only one or two males get to mate with her. The members of the colony are consequently closely related. But now and then a strong, super-sized and presumably frustrated male braves the dangers of the world above ground and sets off to find another colony, on the rather chancy bet of achieving a consortship there.

However, the parallels are not very exact. Within the migrating hordes of wildebeest or reindeer there are social units and social relationships - females accompanied by their offspring, males which have achieved status by their battles with other males, and subordinate males which have yet to make a successful challenge. In other words there's a good deal of social structure. And the naked mole rats differ from the social insects in that the top positions can be achieved by competition.

Dominant rank in a mammalian society usually goes with breeding privileges, to a greater or lesser degree. Sometimes being top dog means being the only individual who gets to reproduce, sometimes it merely confers superior opportunities. In the animal world competition for breeding opportunities, usually between males, is a pretty standard procedure. It means that the strongest and most successful individuals make the biggest contribution to the next generation's gene pool. In social mammals, and in some bird species, the practice has provided a foundation for a system of social organisation. Group leadership is decided by the same means as breeding opportunities, usually by combat, and once established the dominance is maintained as far as possible by displays of threat.

Consequently group structure is flexible. Leaders can be deposed and lose their power when younger, stronger individuals take over. And in both mammalian and avian social groups there is often not just a leader but also a hierarchy of power and influence, or sometimes an intermeshing pair of hierarchies, one male, one female. Position in the hierarchy can be influenced by parentage, by relationship to the leader, by competition, by the forming of alliances, or by sheer power of character - the forces at work grow subtler in the larger-brained species. The result is a social structure which can accommodate a variety of differentiated individuals, in contrast to the insect societies with their pre-determined collection of stereotyped roles. All this indicates that social mammals and birds must be capable of some fairly subtle perceptual learning.

In addition to the potential defence against predation, for those vulnerable to it, other factors have favoured social lifestyles for some species. An important one is diet. Animals whose food is spread widely and sparsely across the terrain can only feed well by working alone. Those whose food is distributed patchily, in rich but widely separated clumps, are likely to keep coming together as they seek it out. Indeed, for these species there's a lot to be said for sticking with the crowd and congregating where other individuals are, since they're probably there because they've found food. We can suppose that many species have had sociability thrust upon them. As their ancestors evolved the perceptual abilities necessary for reliably seeking out the sort of food that's found in widely separated clumps, along with the digestive systems necessary for specialising in it, they were obliged, if they were to be successful, to evolve the means of getting along in the company of conspecifics.

The small fish that congregate in massive shoals mostly live on plankton, particles of nourishment that tend to drift in clouds. Many individuals were therefore always likely to be attracted to the same area, moving on when the supplies were exhausted. In less unbounded habitats than the open ocean it's often worthwhile for an individual or a breeding pair to defend an area of resources only adequate for the sustenance of one or two animals. But when there are richly stocked but widely separated patches of terrain that, for limited periods, provide enough to feed many individuals, then it's not practicable or profitable for one or two individuals to try and monopolise them. On the other hand it may be worthwhile for a group which can make full use of the supply to defend it against another group. This is probably the factor that played the primary role in establishing the habit of relatively permanent grouping among fructivores, and other consumers of patchily distributed and seasonal foodstuffs, species which notably include many primates.

Sociability has also evolved several times among predators. There are species of fish which hunt in shoals, coping with those seething masses of smaller fish by ganging up to cut off their escape routes or to split them into more manageable targets. Some species of dolphin use the same tactics. On land, co-operation serves to bring down larger prey than an individual could manage on its own. This isn't necessarily a direct benefit, since there are also more individuals to consume the prey, but it may mean that a larger number of fairly similar-sized predator species can co-exist successfully on the same terrain.

But if diet has the biggest influence on sociability, habitat is important too. Congregating together in huge numbers is fine in open sea, where usable space stretches away in three dimensions and there are plenty of escape routes. Close to a rocky shore or reef it's better to be in a smaller crowd, or to have a means of hiding from predators. On land, a two-dimensional habitat, the potential for huge numbers is restricted. Herbivores of the open plains and savannahs live in fairly large companies, but only join together in dense masses when they are migrating. It takes a lot of grass to make an adequate meal, and they need room to start running if the need should arise. Group size for the consumers of patchily distributed foods, meanwhile, is not only constrained by the amount of food found in a typical patch, but also by the fact that such food is often found in forests, or other terrain where it would be difficult for a large group of animals to keep together. For creatures of the air things are different again, and various species of birds and bats mass together to roost or to breed, homing in on the safest sort of terrain for the purpose and gaining the protection of company at the same time. But the sociability doesn't necessarily extend beyond the roosting place.

In species whose food is thinly spread mothers must see grown offspring as rivals for the food supply, and be ready to drive them off, if necessary, as a sensation of hunger, in the context of her knowledge of the food potential of the environment, comes to take priority over the remembered rewardingness of their infant forms. But where talents evolved for finding the sort of food that occurs in widely distributed clumps, mother and offspring would tend to find themselves eating in the same places, and would only occasionally have reason to argue over the division of the food. Siblings and cousins would accumulate, choosing to stay in company as long as there was enough food available, or until there was an urge to find an unrelated mating partner. Used to being in company, they would be loath to break the habit, though groups would fission when they grew too big for the typical food clump. And as time went on adaptations that made the togetherness easier, and more profitable to the individual, would be selected.

It's easy, at least, to see how this might have happened among fruit-eaters and other vegetarian species, and genetic evidence suggests that primates did indeed progress from loose, unstable associations, as in some lemurs, to more structured groupings. The evolution of sociability among carnivores is more puzzling. Perhaps it could be attributed to accidental discoveries of the possibilities of cooperative hunting among groups of maturing youngsters, and to brains which could conceive of developing the idea. It seems appropriate to regard mammalian sociability as something which began as a cultural habit. Genetic adaptations which made an individual particularly likely to be successful at group living would then tend to be selected, establishing the practice more firmly. It has certainly evolved several times, for many social species have non-social sister species.

What it takes to be a social mammal

The talents needed for the mammalian social lifestyle are essentially the same ones required for prolonged maternal care of the young. Infant and mother need to be able to distinguish each other from other individuals, and members of a group must similarly be able to recognise each other.

Sight and sound provide the most practical means of identifying individuals at a distance, and this may be reflected in the fact that the more closeknit and complex societies are mostly found among primates and predators, species which for other reasons need good stereo vision. However, even sheep, though not obvious visual specialists, are up to the challenge of distinguishing betweeen photographs of different sheep faces. For many primate species the task must be easier because the individual faces are more conspicuously differentiated than in most mammals. Voices too have become more variable.

At least some non-human primates are not only capable of identifying each other, but also recognise the relationships among their fellows. Ranking in the hierarchy is often heavily influenced by parentage, which implies that the members of the group are thoroughly aware of family connections. This was confirmed by Dorothy Cheney and Robert Seyfarth, who observed wild vervet monkeys over many years. When they played the recorded alarm call of a youngster to a group of monkeys the mother of the apparently frightened youngster looked towards the sound, while the other mothers looked towards her. In a similar experiment baboons heard recordings suggesting a disagreement between two females, one more dominant than the other. When the threat-grunts were in the voice of the more subordinate female, the fear screams in the voice of a higher-ranking one, the hearers paid greater attention than when the exchange was the more likely way round.

A consequence of a leadership-by-contest system, meanwhile, is that the leadership can change any time a rival makes a successful challenge. Group hierarchies can change too. Social animals therefore have to be able to do a lot of re-learning about the social power structure. Yesterday's juvenile grows into a tough and ambitious young male, battles his way to leadership of the group, and may survive to become an unimportant old has-been. Both apes and monkeys can clearly cope with these changing situations.

For successful group living it's also helpful to be able to read the emotions of others. The evolution of increasingly expressive faces and voices, especially among primates, will have made social living easier. Not only are fellow members of the group readily identifiable, but it's usually fairly easy to discern their current mood.

The benefits of sociability

The benefits to be gained from social living are pretty much the same as those to be gained from growing up with mother, writ larger, and continuing into adulthood. Mother can protect small offspring against predators that are bigger than them but pose less threat to her. Social adults can gang up against a predator from which a solitary animal would be obliged to flee, and maybe drive it off, or convince it that they constitute the sort of meal that's not worth the trouble. And several pairs of eyes, ears and nostrils are always better than one when it comes to spotting the approach of danger.

Living in a group also means that the juvenile has many models from which to learn, instead of just a parent or two. For those vegetarians whose diet is found in widely scattered clumps this is particularly useful, and especially so for fruit-eaters and other species whose sources of food are not only widely scattered but also seasonal. There are several older adults to carry the vital knowledge of where the food is to be found, and which sites are currently likely to be worth visiting. The youngsters tag along and learn the routes, and in their turn will be able to guide the next generation. The older members of a group may have other important sorts of knowledge too, such as where water was found in the previous drought, a few years back, and may perhaps be found again. In various primate species an elderly group member, not the dominant animal, occasionally leads the group's expeditions, sometimes to a great distance, apparently for such purposes. In groups of hamadryas baboons which include two adult males it may be the elder who chooses the direction of travel, even if he is no longer the dominant male in other respects.

Among non-specialist eaters youngsters can learn from the older generation not only where to find food but also what makes good eating. Having a number of models to observe rather than just one increases the possibilities for useful learning. Moreover, if one member of a group discovers a new food the other members can benefit from the discovery.

Tool-using techniques too are spread through imitation, so the potential for establishing such ideas is greatest in social animals. Learning by imitation, strengthened by an inclination to conformity, leads to distinctive cultures. Different practices are found among the several groups of wild chimpanzees which have been observed over long periods. Some eat oil-palm nuts, others don't, though they live in areas where these nuts are equally available. Two groups have been observed to crack nuts by hammering them on an anvil, but in one stones are always used for both hammer and anvil, while the other group uses stone or wood for either role, and sometimes just a hard bit of ground for the anvil. When ant-fishing the Gombe chimps use a long stick, collect the ants off it by sweeping the stick through the fingers, and transfer them to the mouth *en masse*, while a couple of West African groups are in the habit of using shorter sticks, and picking the ants off one by one with their lips. The traditions of a group observed in Guinea are subtler - the length of the stick used is related to the species of ant and the chances of being bitten.

For termite-fishing one group regularly uses the midrib of a leaf, a tool which has not been seen in any of the other groups that have access to termites. Another group uses a stick that has been frayed at the end to form a sort of brush, which can hold a larger number of termites. Savanna-living chimpanzees seem to be particularly adept at making and using tools, employing them both for hunting and for digging up roots.

There are different group habits outside the field of technology too. In some groups individuals who are being groomed often stand with their hands clasped above their heads, in others this behaviour has never been seen. The leaders of seven major long-term field studies pooled their observations and listed sixty five distinguishable sorts of tool use or cultural habit, of which only seven were found in all groups. After discounting those techniques which were not applicable in the environment of some groups they found that a distinct cultural profile could still be produced for each group, with none using precisely the same range of techniques. Even among neighbouring groups practices may differ.

By way of illuminating the imitative talents and tendencies of chimpanzees an experiment was carried out with captive animals, using a specially devised apparatus from which food could be extracted by two different methods. One high-ranking female from each of two groups was trained to use it, one in each technique. They were then reintroduced to their groups, where all but two of thirty individuals learnt the new game, while none mastered the apparatus in a third group which lacked a demonstrator. Most of the chimpanzees copied the practice used by the tutored animal, and even those that discovered the alternative technique eventually switched to the one more widely used in their group.

Evidence is accumulating of distinctive cultures among wild orangutans, which also use tools for assorted purposes and have other ingenious practices. (Orangutans brought up among human models can be even more enthusiastic tool-users than chimpanzees.) Gorillas, as leaf eaters, don't need tools for obtaining food, but they have techniques for breaking through painful plant defences - for extracting the pith from nettles, for example.

Cultures can also evolve among monkeys, as shown by the example of the juvenile Japanese macaque who discovered how to clean first grain and then sweet potatoes by washing them in a brook or in seawater, practices that were subsequently copied by other group members, and were still in use generations later. The experimenters originally put out food to lure the monkey troop down from inaccessible cliffs to where they could more readily be observed. Eventually the population grew uncomfortably large, and the amount of provisioning was reduced. At one point, perhaps by way of compensation, the monkeys took to foraging for marine prey such as fish, octopus and limpets - a practice which continued for about six years, and then faded out again. This, too, must have been an innovation which spread by imitation, as, indeed, was the business of venturing into the sea in the first place. Macaques can swim quite effectively, but don't necessarily think of trying it. However, another Japanese population, living in central mountains, took to warming themselves up during the winter snows by bathing in the hot springs there, a practice which seems to have been prompted by seeing humans immersing themselves in the warm water.

As described in chapter 3 varied sorts of tool use have also been observed in other macaque populations, and also among the various species of capuchin monkey. Capuchins -along with the closely related squirrel monkeys - have the largest brains, for their body size, of any non-human primate. In captivity they have always seemed particularly curious-minded and creative.

In a social group a young animal can also continue learning about situations which cause fearful reactions in older and more experienced individuals after it has left mother's side. Awareness of potential danger can thus be passed on even if no member of the society has actually succumbed to it within living memory. Furthermore, fear, caution and anxiety, reactions that can be passed on particularly easily, can rapidly unite the group in response to a threat. And when a member of the group does fall victim to accident or disease their display of pain functions as a warning to their fellows to try and avoid a similar experience.

The capacity for imitation leads to practices which can be handed down from generation to generation without any need for an alteration in the genetic recipe which shapes behaviour.   And in contrast to random genetic mutations new cultural practices are likely to appear only where they suit the potentials of the environment.

Balancing benefits and costs

Living in a group may bring substantial benefits, but it has its costs, for the interests of the individual don't entirely coincide with those of the group. Since the individual has a much better chance of surviving within the social unit it's worth making some sacrifices in the interests of the unit, and complete selfishness is unlikely to be a successful policy. On the other hand the prospects for passing on the genes are likely to be endangered by an excess of altruism. The trick is for the individual to get as much out of the group as possible, without damaging its viability as a group.

Primates devote a good deal of time and energy to the upkeep of their social bonds, by way of mutual grooming sessions and ritual displays of submission to more dominant members of the group. Despite all this a certain amount of time and energy goes on squabbling. Weighing the advantages of the social life-style at the species level, the effort that's saved by belonging to a group (more efficient food-finding and better defence against predators) must be counterbalanced by what's expended on promoting social relationships, and on intra-group conflicts.

There are also balancing acts for the players of particular roles within the social structure. Weaker members of the group benefit from the protection of the stronger ones, but when food is scarce the more dominant individuals get a larger share. Being top male carries all sorts of privileges, including the vital one of best access to females, but it involves expending energy on maintaining the position. Challenging the top male, meanwhile, is risky. The would-be challenger needs to calculate his chances accurately, which is likely to mean comparing himself with the established male for size, for current physical condition, and for psychological condition. Success is most likely if he can weigh all these factors and get the timing of his bid just right.

Such judgements must take into account several variable factors, often including an assessment of the whole contextual situation. They are possible, I suggest, thanks to the capacity for conscious emotion which evolved in conjunction with the development of prolonged maternal care. Being bullied by the dominant animal is distressing, anxiety-arousing, and sometimes painful, and losing food to him is at the very least tiresome. But having him around when a predator appears, or when a rival group is on the rampage, is reassuring. For many group members the prospect of escape by leaving the group and facing the world alone - something they are totally unaccustomed to doing - is no doubt just too frightening to be thinkable. Even when the boss is a monster, at least he's a familiar monster. For the young male, though, self-confidence increases as the body grows and muscles strengthen, and eventually comes to outweigh caution. Then he may leave the group to roam for a while on his own, or join a gang of other young males (according to the traditions of the species). Similarly, the up-and-coming young male will make his challenge to the boss of a group at the point when his confidence in his own strength and energy outweighs his respect for the opponent's.

Emotions are likely to provide a reasonably good assessment of this sort of situation - not good enough to make the outcome anything like a certainty, but good enough, generally speaking, to prevent an animal embarking on behaviour which has little chance of success. This is how they work for us. Our emotional reactions reflect, in a rough-and-ready but generally quite effective fashion, the prospects for benefit or harm to the self, direct or indirect. They are heavily influenced by past experience of reward or unpleasantness. They can take kinship into account, and familiarity. The immediate emotional reaction to someone also tends to vary according to whether they seem to qualify as a fellow group member or a non-member - despite the multiplicity and complexity of our groupings.

Many social animals, especially among primates, can be inferred to have a greater capacity for making judgements than is involved in merely sticking together. They go in for co-operation, and sometimes for apparently altruistic acts. For example, when one individual in the group spots a predator it utters a warning cry, giving its fellows a chance to escape too, and taking the risk of drawing the predator's attention to itself. This may not be as altruistic as it seems, since in some cases the only animal running for safety might be the one that caught the predator's attention, but it can still be a risk. Apart from the fact that many group members are relatives, it should be worth taking for two reasons. One is the possibility that on another occasion some other member of the group will be prepared to take the same action, from which the current signaller will benefit. The other is that if there are benefits in belonging to a group it's just as well not to let the other members get picked off if you can help it. The warning cry hardly counts, therefore, as truly altruistic behaviour. It's the sort of act that's worthwhile if you have the means of judging the costs and benefits involved. Those primate species which are not constantly social certainly do, for individuals wandering through the forest alone keep quiet when they see a predator.

It seems, too, that giving a warning cry is not always risky. Meerkats do a lot of digging for their food, and can't keep a good look-out at the same time. Frequently one individual stands guard on an eminence which commands a good view, emitting a regular call which tells the others they are there. A lengthy study has shown that the guard generally takes up a position within five metres of the burrow, and is among the first down it after giving an alarm. In over two thousand hours of observation no guard fell victim to a predator. So it seems that doing look-out duty carries no great risk, so far as meerkats are concerned.

An individual was more likely to do guard duty, moreover, when it was already well fed. First thing in the morning the job was quite often neglected, but individuals to whom the observers slipped some hard-boiled egg did significantly more look-out duty than usual. Otherwise adults shared the task pretty evenly, including the occasional incomer not related to other members of the group. Only the dominant female, who generally produces around 75% of the group's offspring, did less than others. And look-outs were posted a great deal more frequently in habitats where predators were common than in those where they were fairly rare. All this suggests that a meerkat can judge the likely dangers of the surroundings, assess the risks of eating when there is no look-out on duty, and weigh these factors against the current degree of hunger.

Among primates mutual grooming is another instance of co-operation, and equally one where the pay-off may be delayed. A grooms B today and B grooms A tomorrow. The grooming role is not without immediate rewards, since the groomer usually eats the parasites removed from the groomee's fur, and gives the impression that they constitute a tasty snack. But high-ranking individuals get more grooming than they give, so we can assume that to receive is more valuable than to give.

In a more clear-cut example of delayed reciprocation, individuals may give each other assistance in intra-group conflicts, or in manouevres which provide one of them with some advantage. One young male provides a diversionary distraction, perhaps, while the other copulates surreptitiously with a receptive female, evading the notice of the dominant male. Most often the co-operators are siblings, or otherwise closely related, so the helper is advancing the interests of their own genes.

Apes and monkeys have the business of co-operation quite well sorted out. Favours are fairly regularly returned. An individual who has received support in a fight is likely to support the helper in turn on a subsequent occasion. With grooming the picture is complicated by the fact that lower-ranking animals tend to do a little more than their fair share and higher-ranking ones a little less, but individuals tend to belong to regular grooming partnerships, taking turns at the two roles. When the Gombe chimpanzees gang up to kill a monkey the individual who ends up holding the prey is more likely to give a share of the meat to one who has previously given him meat. The repayment of debts also extends to retaliation. A monkey worsted in a fight may attack the victor's sibling an hour or two later.

Biological theorists have worked out the principles an animal should follow in order to maximise the benefits it gains from co-operating with its fellows and minimise the costs, or in other words, to avoid being cheated. This exercise involves what is known as game theory, and the use of computer models which allow the effects of a long series of interactions to be calculated. The simplest profitable strategy is tit-for-tat. An animal should co-operate at the first encounter, but not oblige again if the beneficiary fails to reciprocate when the opportunity presents itself. Subtle variations on this rule can modestly increase its efficiency. And of course, since gene survival is what's at stake, kin can be allowed somewhat more leeway in keeping up with their debts than non-kin. The results of the computer modelling largely coincide with observed behaviour.

The repayment of debts implies a means of keeping track of who owes whom, as well as who is related to whom - which might seem like quite a demanding intellectual requirement. However, introspection shows that we humans can do the job without necessarily making use of intellect. Again, emotion provides the means. We tend to behave nicely to our kin, on the whole, because we feel happy to make them happy. We feel well disposed to people who do us a good turn, and inclined to return the favour, and we feel badly disposed towards those who do us a bad turn, and tempted to return the injury. Conscious calculation doesn't necessarily come into the matter. We just know that we feel friendly towards X, or don't quite trust Y, even if we would have difficulty enumerating the incidents that led to these feelings.   It's reasonable to suppose that the social behaviour of other primates is similarly based.

Moreover, time plays a role in our feelings which is paralleled in monkeys' behaviour to their fellows. An animal's behaviour to a fellow member of the group is often predictable on the basis of their recent interactions, much less so on the basis of more distant history. Our feelings too are much more influenced by yesterday's good turn or yesterday's insult than they are by a similar incident of a year or two ago, especially if the ancient memory has since been overlaid by other experience of the individual. This fading influence of all but major debts makes for flexibility. Fences can be mended, new alliances forged, new leaders can establish themselves.

But research shows that most of us do like to punish those who offend, even if it costs something. The subjects in these experiments may even be prepared to make some modest sacrifice to punish an offence which didn't affect them personally. No doubt this is why we're prepared to pay taxes for financing law courts and prisons. In this sort of experiment with adults, however, it's impossible to distinguish between genetically determined predispositions and the influence of upbringing and social pressure. It's therefore interesting that infants as young as eight months seem to prefer characters who behave on tit-for-tat principles, not only acting nicely towards nice individuals but also showing unfriendly behaviour towards those who behave badly.

Sociability and brain size

It's clear that considerable sophistication is needed for the mammalian style of structured sociability, especially as practised by some primates. Advanced perceptual systems are necessary for recognising fellow group members, and subtle learning abilities for remembering their current status. Furthermore there must be a means of weighing selfish advantage against the benefits of belonging to a group, a talent which in some species has been sufficiently expanded to support reciprocal assistance between individuals. Advanced perceptual systems depend on extensive cortical networks, and the conscious emotions which can be supposed to provide the evaluative judgements and the means of remembering debts similarly imply a generous amount of neuronal processing. So this sort of lifestyle is found only among species with largish brains.

The probable intellectual demands of social living were noted by Michael Chance and Allan Mead in the nineteen fifties, and again by Alison Jolly in the sixties. Another decade on, Nick Humphrey proposed that learning about their fellows, and how to interact with them, is *both the most difficult and the most important task to which social animals must turn their minds. No other class of environmental objects approaches in biological significance those living bodies which constitute for a social animal its companions, playmates, rivals, teachers, foes .... In these circumstances the ability to model the behaviour of others in the social group has paramount survival value.*

There are some confounding factors in the correlation between sociability and large brains, given that social species tend to be consumers of patchy resources. They therefore need particularly good cognitive maps, and usually it's also helpful to be able to predict seasonal effects on the target features of the map. Fruit-eaters, and specialists in fresh young leaves, usually possess especially refined colour vision - red-sensitive photoreceptors as well as green and blue, and the additional analytical machinery which is required to make use of the extra data. All that takes up cortical space. So there are at least two reasons, in addition to the demands of sociability, for their larger than average brains.

It's among the consumers of patchy and fluctuating resources that the most flexible social systems are found. Chimpanzees form quite large groups, but do most of their foraging in smaller units, and mature males sometimes go off for long treks on their own. Orangutans may spend much of their time alone, or as partnerships of mother and infant, or mother and juvenile, while a dominant male exercises control over a large territory that contains several females and some immature males. But in richer habitats several individuals are often discovered feeding together, and behaving in a sociable fashion. Clearly orangutans are capable of adapting their lifestyle to suit the availability of resources. The same flexibility is found in some monkeys. One study compared howler monkeys and spider monkeys living in the Barro Colorado forest in South America. For howlers, which live in permanent groups, fruit makes up just over 40% of the diet. But the larger-brained spider monkeys live in more fluid societies, breaking up into smaller units for much of their foraging, and manage a diet of over 70% fruit.

Such adaptability is not confined to primates. Coyotes living in high country which is covered in snow in winter survive that season mostly by scavenging on the carcasses of large herbivores which die in the cold. Families tend to remain together, and to defend the territory which contains these valuable resources. Coyotes which live at somewhat milder levels can eat rodents in winter, as in summer, tend to live in pairs, and don't defend a territory. The typical pride of lions contains maybe half a dozen adults and hunts animals such as zebra and wildebeest, but a pride of thirty individuals has been observed, which makes a habit of hunting elephants.

A really flexible social structure, meanwhile, is exemplified by dolphins, which may swim together in units of very varied size. When two or three dolphins come across a really large shoal of fish they have a way of signalling to others to come and join in the fun. They then herd the fish into a spiralling mass and take it in turns to grab at the prey, or drive them towards a shallow beach where there's no escape. The ability to adapt the grouping to meet the current potential of the environment implies an ability to make very acute judgements, and correlates convincingly with large brain size.

Being sociable requires a large brain, then. But equally, social lifestyles make it worthwhile to be large-brained. Living in the protection of the group makes an extended period of juvenility practicable, which allows for more play. Greater opportunities for play increase the chances of making discoveries which give rise to useful cultural adaptations. And sociability increases the likelihood of discoveries and inventions being passed on, since the whole group can copy them, not just the inventive individual's offspring and their descendants.

Social living also creates circumstances in which it's profitable to evolve a new kind of talent. It becomes usseful to conceive of the self as a distinct individual, and of fellow individuals as other, parallel instances of the same sort of entity, whose behaviour can be predicted by considering how one might behave oneself in the same circumstances. This implies an ability to hold in mind simultaneously both a knowledge of one's own feelings and behaviour and the schemata that have been developed about the behaviour of conspecifics.

Awareness of the likely motivations and perceptions of fellow group members means that new sorts of plans can be dreamed up, based on their predicted behaviour. An appreciation of their emotional responses is probably the most valuable component of the new talent, but there are also benefits in deducing the different perception provided by a different viewpoint. One consequence is that an individual can practise deceit, and promote the interest of their own genes at the expense of other individuals. Another is that the individuals on whom deceit is practised can learn to recognise the habitual cheat, and defend themselves. The delicate balance to be achieved between the interests of the individual and the interests of the group becomes yet trickier.

The ability to appreciate the feelings of others also opens the way for the development of social practices that make co-operative efforts more likely. It must be noted, however, that just as the idea of using a stone as a tool must be discovered by some imaginative individual before it can be passed on as part of the culture of the group, so must the idea that individuals are more likely to expend energy for the common good if there is something in it for them.

Such an insight on the part of some percipient individual may, conceivably, account for the contrasted hunting habits of the chimpanzees of Gombe in Tanzania and those of an Ivory Coast troop which has also been observed over a long period. The former collaborate in the hunt with a fair degree of efficiency, but once the prey is caught there is much squabbling over the meat. Most frequently it ends up in the possession of the dominant male, who often consumes much of it himself, while doling out bits to his most prominent supporters, or to a female who begs very persistently. In the Ivory Coast group the spoils are shared out more fairly, and it's tempting to think that this may help to explain why this group hunts quite regularly, while the Gombe chimps, after a few hunts within a relatively short period, seem to lose interest for a while. The Ivory Coast gang also co-operates more effectively than the other, which means it can catch older monkeys, not just juveniles, and there is more meat to share out.

A possible interpretation is that the dominant animals of the Gombe troupe follow the guidance of their emotions, and part with meat only to those to whom they are feeling particularly friendly. Getting help from colleagues in a hunt perhaps doesn't produce the same direct emotional effect as getting assistance in a quarrel, or being groomed, since attention during a hunt is likely to be directed to the quarry rather than the helpers, and since the enjoyment of the meat is somewhat separated in time from the business of obtaining it. Add to this that a young monkey constitutes a good meal for one chimp, and only modest snacks if divided among all the hunters. And meat is a major attention-grabber that perhaps doesn't leave much room for social thinking, or looking forward to the next hunt -

just as the sight of a bunch of bananas seems to shut out the possibility of thinking about an innovative use of tools. The emotional response that usually provides a fairly good way of weighing debts and striking bargains over quite a range of activities may not work too well here. But perhaps some dominant individual of the Ivory Coast group had the insight to realise that it was worth suppressing the usual responses, rewarding his underlings more generously and inspiring them to work more frequently. Thus a more profitable tradition was set up..

That may be just a fairy tale, of course. What must be certain is that the imaginative insight which makes it possible to strike bargains that promote willing and effective co-operation was a major factor in hominid evolution. Groups which could achieve this trick would tend to flourish better than those that couldn't, as Darwin noted. Within the group, meanwhile, those individuals who had some insight into their fellows' behaviour would generally be most successful at promoting both their own interests and those of the group. A talent for well-managed co-operation would have been particularly valuable to the hominids who left the shelter of the forests for the savannah. It's a talent that no doubt facilitated larger groupings, and since larger groups have an advantage when there's inter-group competition for resources there would have been strong selection for it.

The capacity for emotional infection, so useful in the education of the young, must provide the basic underpinning for the understanding of others. It's clearly the foundation of empathy, which in turn must go far to account for the phenomenon of altruism. If one shares the emotions one observes there's a motive to try and reduce the pains, and pleasure can be obtained from increasing someone else's pleasure. Where one individual inflicts harm on another an observer is likely – if no relatives are involved - to empathise with only one, often the recipient rather than the agent, which no doubt accounts for the willingness to inflict punishment on offenders, even at some expense, regardless of the absence of personal injury. Here lie the emotional underpinnings of a sense of justice.

A well developed concept of other individuals as conscious, feeling beings, essentially like the self, must have been in place before language developed. Language in turn would extend the potential for comprehending the feelings of others, and of perceiving that other individuals were performing similar calculations in their social relationships. It no doubt contributed to an expansion of the capacity for empathy, which seems to be greater in humans than in other primates.

Both language and the concept of self in a society of other, similar entities must have been widely established before larger, more complex units could be organised. Language made it possible to frame rules about appropriate behaviour, and to create stories to illustrate or justify the rules. A highly developed capacity for empathy has no doubt contributed to our ability to conceive of the idea of justice. But since the interests of the individual often don't coincide with those of others an innate sense of morality isn't sufficient to support large groupings successfully. The most successful groups would be those that promoted cooperation by establishing rules of behaviour, rules which would work best if presented with a strong rationale phrased in the concrete terms available in the language of the time. We can guess that the roots of myth and religion lay in meeting this need - and then offered scope for satisfying an increasing desire for explanations about other aspects of life. The rules and the myths serve to bind the social group into a sufficiently coherent whole to operate effectively. The major religions of the world were clearly inspired by the need to meet the challenges of ever larger groupings.

Survival and expansion were more likely if the rules were fitted to the environment – for instance resisting the temptation to eat pork was a sound practice in hot dry terrains, even if pigs do breed much faster than other meat animals of similar size. Often it hasn't mattered too much if the rules and practices of one group seemed silly, wrong or revolting to another group. Nor has it mattered if the relationship between neighbouring groups has sometimes been hostile, since competition over limited resources tends to lead to the survival of the fittest. And while, if a group is small, interbreeding between groups is desirable, it's often also desirable that it should largely be confined within the geographic area to which the genetic recipe is adapted, with only limited outbreeding. In the past a distrust of conspecifics who seemed too strange will have tended to promote the survival of genes. Now we are armed with weapons which could make the world uninhabitable, and that world has truly become, as Marshall McLuhan put it, a global village, we clearly need to acknowledge and control this instinct.

Only in a social species, then, could the large brains evolve that might eventually lead to the potential for abstract reasoning. Living in a group provides the opportunity for large brains to be well educated, and for any innovations they produce to be readily disseminated. Furthermore the faculties which have made social living possible - advanced perceptual systems, conscious emotion, the ability to imitate, complex means of communication - are also the foundations of reason.

The ability to calculate in rational, verbal terms the costs and benefits of social living allows us humans to form groups in which most of the individuals are strangers to each other. And such extended societies have, in their turn, provided the opportunity for a more extensive practice of abstract thinking, fuelled by increasingly extensive communication among the thinkers. But the architecture of those large brains has been very largely shaped by the challenges of living and procreating successfully within a less sophisticated sort of society.

Social thinking still comes much more easily than abstract thinking. A logical problem phrased in terms of human action is solved by many people who fail on a version that involves objects. And the earliest attempts to explain life, the world, the universe and everything have generally involved gods and unseen powers which behaved in largely humanlike ways, and were often envisaged as bearing at least some similarity to human form. This is an area in which our thinking has certainly been shaped by our evolutionary history.

Recapitulation

A preference for togetherness has evolved among many species, and is likely to emerge in any species which specialises in the sort of food that comes in clumps adequate to feed several or many individuals.

In some species the togetherness is managed by genetically determined responses to genetically determined stimuli, but mammals are different. Mammalian societies apply the same talents needed for mammalian motherhood, on a larger scale.

Social living gives the young wider opportunities for education, through play and through imitation of the actions and reactions of their elders.

Living in groups has favoured the survival of animals which can judge how best to forward their own interests without sacrificing the benefits of belonging to the group.

The costs and benefits of social living are represented in the form of social emotions, and the balance of such emotions mostly supplies a reasonable guide to the most profitable behaviour.

Larger primate brains have the means of correlating the emotions experienced by the self with the emotional displays of other individuals, and developing sufficient psychological insight to make some useful predictions about the likely behaviour of their fellows. This was an essential precondition for the evolution of language ability.

Language clearly made it possible to develop such insights much further. That was one of its earlier benefits, and in turn made it possible to form increasingly large groups, in which new ideas could be passed on with more reliability, and could interbreed more widely with other new ideas.

The demands of living successfully within a society have been perhaps the most vital factor shaping our brains in more recent evolutionary history, and this is reflected in much of our thinking.

Chapter 14

Attention

A notable aspect of sensory consciousness is that it is, as it were, graded. As one examines a picture on the art gallery wall the picture is a vivid experience while the out-of-focus surroundings are only just present in awareness. In a crowd buzzing with conversation we listen to one voice and the other voices meld into a blur. Enough information reaches consciousness for the blur to have some degree of shape - we usually hear enough of the overall patterns of intonation to know if the conversations are mostly in our own language, and whether any emotion is being expressed; but only the voice to which we listen is analysed into a pattern of words. The background information intrudes on awareness just enough to indicate its availability, as it were, so that we can divert attention to it if we choose. And every now and then some sensory event thrusts itself into the spotlight of full consciousness whether we will or no, imperiously demanding attention. But much of what tickles the sensory receptors and is processed in ways that make it potentially available to consciousness makes only a hazy contribution to the sensory experience, as a context for the main show. So clearly attention plays a very big part in shaping that experience.

A selective attentional system is essential for any animal which can collect large amounts of sensory information, for that can only be worth doing if there is also a means of ensuring that the urgent bits don't get lost in the crush. George Miller wrote: *We are constantly swimming through oceans of information, far more than we could ever notice and understand; without some effective way to select what is important, we would surely drown.* Complex sensory systems clearly require a means of focussing their power where it's needed.

We might begin by noting that attention has two distinct aspects. I shall call them behavioural and cerebral. Cerebral attention is of course the mysterious phenomenon which produces the focus of conscious experience. For a good deal of the time it goes hand in hand with the behavioural sort, but it can be divorced from it, and covertly directed elsewhere. The eyes can be focussed straight ahead, the mind on something that is going on in the periphery of the visual field. We can face someone and give a reasonably good impression of listening to what they are saying while actually attending to the discussion going on behind us. And it's not inconceivable that you're sitting there with your eyes tracking back and forth across the page, and thinking about what to have for dinner ....

Behavioural attention is demonstrably common to a good many species. A salient stimulus in any external sensory channel causes the current behaviour to cease, and all sensory apparatus to be pointed towards the direction from which the stimulus came. The head is turned, the gaze directed, the ears pricked and twisted, the nostrils flared - as many of these movements are performed as the species is capable of. Focussing all the distance senses towards the site of a possibly significant event makes for the best possible chance of picking up any further information that may be available from that direction.

A species without moveable sensory organs would have to turn the whole body towards the stimulus – something which has to be done anyway before an approach is made. A stimulus not strong enough to attract a fullscale response would have prompted a turn, positioning the animal for the actual behaviour if more stimulation arrived. Behavioural attention grew more elaborate as natty new arrangements such as rotatable eyes and flexible necks evolved, so that it wasn't always necessary to turn the whole body towards a stimulus for a preliminary examination of it.

Cerebral attention might reasonably be assumed to have grown out of the mechanisms of behavioural attention, and this is what the accumulating neurological evidence indicates. What's particularly interesting is the ability to divorce cerebral from behavioural attention, for this is obviously a necessary prerequisite for the business of thought, which involves turning attention inwards. Even the simplest form of thought, such as deciding what to do next, entails consulting the contents of memory to access knowledge of the possibilities. Planning requires the ability to exercise recall, rather than mere recognition, and recall can be defined as an inwardly directed form of attention.

In Chapter 2 I defined three levels of consciousness, which must have evolved in ascending order of complexity. They might be redefined in terms of an increasing ability to turn attention inwards. Simple sensory consciousness, which allows the individual to make predictions about how to achieve desirable effects and avoid undesirable ones, entails paying attention to the records stored in memory, but only to the extent of putting them to use to make simple plans. The second level of consciousness is the one required for developing a concept of the self which can be used to predict the behaviour of others. This involves turning attention on to the records of emotional and sensory experience and reflecting on the way that experience shapes behaviour. By this means a schema for the self can be constructed, correlated with the way conspecifics behave, and used to understand their perceptions and motivations, and to predict their responses in familiar situations.

A yet finer focussing of inward-directed attention makes it possible to see the self as a thinking being, with thoughts separable from the self and communicable to others. That in turn facilitates an appreciation of words as signs that refer to objects and actions but belong to a different order of reality. Thus the evolution of the capacity for abstract reason might be seen as an ever-growing talent for contortionism on the part of the attentional system.

Some reliable attention-grabbers

Some things can be almost guaranteed to thrust themselves into our awareness. They are the stimuli which have always been worth noticing, the ones that it's prudent for an animal to check for possible significance, the ones for which a larger response may well have to follow.

These automatic attention-grabbers can be classified into groups. One type of intrinsically salient stimulus is any novelty in the environment - a sudden noise, a new visual movement, a new smell, a sudden tactile input. The appearance of a visual object out of nowhere will do it too - as long as the eye is not saccading at the time, in which case it may be missed. Anything like this can intrude on consciousness to interrupt the current focus of concentration; equally, any of them is likely to evoke a detectable attentional response in other species. Experiments have shown that many species are alerted by even a slight change in an ongoing stimulus pattern. The reason is obvious: any such unexpected event may herald approaching danger - a predator perhaps, or some inanimate threat such as a falling tree or rock.

We humans don't necessarily make the behavioural response and turn towards the novelty, we can exercise choice about it, making use of experience to decide how much attention is desirable. But not paying behavioural attention is often as much a matter of suppressing a response as of not choosing to make it. Sometimes the eye quivers, even if there is no saccade towards the movement at the periphery of vision. If the sudden stimulus is a powerful one we don't generally respond with instant flight, but hardwired and very rapid connections prepare our muscles for such a response, and may tauten them so effectively that we give a visible jump.

However, the effect of a novel stimulus fades rapidly. A small sound in the midst of silence calls for consideration, but a much louder noise is ignored once it has become a regular occurrence. A small twitch in the grass commands attention but not the constant swaying caused by the wind. A smell which seems overpowering at first soon becomes almost unnoticeable. And sometimes the novelty that attracts attention is the absence of stimulation - a sudden silence, a movement that suddenly stops, or something that vanishes. Even a mere reduction in the level of sensory input can impinge on attention.

Habituation can no doubt partly account for this sort of salience. Neurons at many levels of the nervous system fire strongly when they haven't been prodded into activity for a while, but less energetically as the input is repeated. It seems probable that the more powerful firing of neurons responding to new input can trigger the attentional system. Even the response to sudden silence can be explained in this way, since neurons have been found in auditory cortex which respond specifically to silence.

The habituation explanation cannot hold, however, for one especial class of novelty - a pattern of sensory input which is not merely novel in the current context but wholly unfamiliar and unclassifiable. That must receive attention until its implications become clear, and it can be judged to be harmless.

Another class of stimulus patterns with inherent salience is other human beings. A human voice always calls for notice, unless it's masked by a chorus of other voices. Indeed it can be difficult to ignore. The human figure in the landscape - actual or painted - usually holds the eye longer than other components.

This seems to be a hardwired attentional response that has been strengthened by experience. The gaze of newborn infants is attracted to faces, and particularly to eyes. Given a choice between looking at a roughly circular shape with dots where eyes would be and a similar shape with some other form of decoration they look longer at the former. Real faces are even more compelling.

Infants also give more attention to biological motion than to other sorts. Two-day-old infants spent more time looking at point-light displays of animal movement (animations in which movement is revealed, on a dark screen, by lights where joints would be, as in the Johanssen experiments described in chapter 8) than at other sorts. This must be another arrangement which ensures that infants pay attention to mother, and to any other humans around. As they learn to distinguish among faces and styles of movement they also learn how important other humans are to their welfare, and how unpredictable they can be, so that it's worth continuing to pay attention to them.

Many mammals seem to share our sensitivity to the stimulus pattern which is a pair of eyes, if the eyes are looking in their direction and thus presenting a perfectly circular bull's-eye pattern. Herbivores such as horses and cows become visibly nervous when stared at. An animal with two forward-facing eyes is likely to be a predator, and if both pupil/iris patterns are fully visible the animal in view may be being considered as potential prey, so a hardwired response to this stimulus must be very valuable to prey species.

The response may be more ancient than this implies, or have evolved more than once, if the theory is correct that eye-spots on butterfly wings serve to discourage birds that would otherwise prey on these species. Certainly nesting birds of species that are subject to predation are often frightened off the hidden nest when one looks directly towards it, but put up with a nearby human presence otherwise. And one might hypothesise that the peacock has been able to evolve its seemingly impracticable tail because the array of eye-like shapes, along with the loud rattle when the tail is shaken, can frighten off most of the predators that would find a peacock a satisfactory meal.

There's reason to suppose, then, that the response which directs the human infant's attention towards a face is long-established, and perhaps somewhat ambiguous, having been co-opted for a new purpose. As we progress from infancy we come to dislike being stared at, especially by strangers. A brief gaze is acceptable, a prolonged one uncomfortable - unless there is a sort of prior agreement to make it acceptable, as when the audience watches the actor, or lovers gaze into each others' eyes. It would appear that some sort of maturing process makes the stimulus pattern less thoroughly rewarding.

This seems to be a process which goes too far in autistics, who generally intensely dislike being looked at, and avoid looking at the eyes of others. However, the normal reflex is still at work in the first months of life.

Human eyes are particularly conspicuous, since the pupil and iris are surrounded by a white sclera, rather than being distinguished only by their shiny texture, as in other mammals. When they are seen at close quarters the constant saccading movements must also help to draw attention. One benefit may be related to the exceptional expressiveness of human faces. Whereas other primates tend to express their emotions vocally as well as facially, thus drawing attention to themselves, humans often reveal a great deal in their faces without making any sound. This trait can only have been worth evolving if the communication was pretty sure to be noticed. The white sclera also makes it easier to see where the eye is looking, a skill which plays an important part in the education of human infants but which seems to be less thoroughly developed in other primates.

The inbuilt tendency to turn towards biological movement also looks like an attentional device which was in operation long before maternal care began, and came to have additional value. Other animals are of vital relevance to survival for individuals of any species, as competitors or breeding partners, prey or predators, or sometimes just as giant moving objects which can cause disruption.

"Pop-out" stimuli

The stimuli discussed so far command attention fairly powerfully and can often distract it from its previous focus. A third class doesn't divert attention quite so readily, but in the absence of stronger competition is pretty certain to get noticed. As Anne Treisman put it, a certain sort of stimulus can 'pop out' of a scene, so that we are immediately aware of it whether we are looking for it or not.

Treisman investigated just what qualities make for this automatic salience, She used a tachistoscope to measure just how long an exposure was needed to discover a single different stimulus amidst a collection of identical ones. The results confirmed what subjective experience suggested. A single blue figure is immediately obvious amidst an array of red ones, as is an X among a crowd of Os, or a diagonal line among verticals. In a crowd of moving stimuli all going in one direction the unique stimulus moving in a different direction can be picked out at once. So can the figure which is nearer or further away than the others.

The rule seems to be that when the target figure differs from the 'background' in one simple quality, such as colour, or the orientation of one of the component lines, or stereoscopic depth, or when curved lines are contrasted with straight, then it's detected very rapidly. Moreover the time taken to discover the target stimulus isn't affected by the number of distractors - it stands out vividly among few or many. The simple qualities correspond to those registered by the feature-sensitive neurons of the visual cortex (counting colour and direction of movement as features) so the rule could be expressed as: a target figure will stand out if it excites at least one sort of feature-detector not excited by the background figures, or excites one of them more strongly. Consonant with this interpretation, Treisman found that a target figure is more readily distinguished by the presence of a feature than by its absence - a Q is detected among Os more easily than an O among Qs. Similarly, a single slightly longer line is discovered faster than a slightly shorter one, a slightly darker line more rapidly than a slightly paler. But a single stimulus which is distinguished only by a unique combination of qualities completely fails to 'pop out'. In an array of red Vs and blue Os, for example, a red O or a blue V is not immediately obvious.

```
OOOOOOOO
OOOOOOOO
OOOOOQOOO
OOOOOOOO
OOOOOOOO

QQQQQQQQQ
QQQOQQQQQ
QQQQQQQQQ
QQQQQQQQQ
QQQQQQQQQ
```

The same rule governs the salience of visual textures. A figure distinguished by its texture stands out readily if it excites at least one set of visual features that the background does not. It's also easily noticed if its texture is denser or sparser than the background - another instance of the same rule, since the visual cortex has cells which are sensitive to the density of textures. But if the figure excites the same sets of feature detectors as the background, and is distinguished only by the fact that the features of the two textures are arranged in different ways, then it takes time and study to spot the difference between the two.

When the odd-man-out stimulus is not of the sort that automatically grabs attention the exposure required for its discovery lengthens as the number of items increases. The array has to be scanned carefully until the target figure or the patch of different texture is reached. The automatic singularity-detecting mechanism, it seems, works only to identify either single instances of a feature within the visual field, or patches of texture distinguished by containing a feature that is not present in the surrounding texture. It can't deal with unique combinations, nor is it activated by the absence of a feature, or by a feature that is distinctive only by being smaller than others.

Exceptions to this rule are found when one of the qualities of the target stimulus is either movement or stereoscopic depth. A unique blue moving stimulus is salient among those which are blue but stationary and those which are moving but not blue. It's also inherently noticeable if the array occupies several distinct distances and it's the only blue stimulus in its plane. Movement and distance are so vital that they operate on special principles.

The rapid response to singularities must depend partly on widespread inhibitory connections in early visual cortex. One of the many types of inhibitory neuron in V1 must link excitatory neurons which respond to the same sort of feature, so that when many of them are activated the firing rates of all are somewhat subdued by mutual inhibition. When only a few are excited - by a single edge or line or a patch of similarly oriented lines or edges – there's no inhibition from their fellows, so they fire more strongly than neurons with other sensitivities that are more widely activated and are, in some degree, inhibiting each other.

Certainly the V1 response to a line or edge is less strong when it's surrounded by similar features at similar orientations. This, presumably, is what makes a texture a texture - that is, an experience of repeated, densely crowded, oriented features which are not perceived as separate entities. However there are also other factors at work which we will come to later.

What's true for visual stimuli is also true of sounds. If attention is to be paid to one particular voice it must differ in some way from others that are being registered simultaneously. It must have a different pitch, tone, timbre or rhythm, be noticeably louder, or come from a different direction. We don't distinguish among the chorus of sopranos all singing in tune. But the solo instrument can be balanced against the sound of the whole orchestra.

The mechanism that emphasises singularities is obviously valuable. The texture or colour that doesn't quite match up to the background may be a leopard lurking in the dappled shade, or, from the leopard's perspective, the next meal. The bit of landscape with a different texture from the rest may be the place to go for a drink. The sound that's distinctive amidst the background noise could be a danger signal. Turning the attention on such singularities would be useful even in a species that possessed only limited means of perceiving patterns in aural input or deciphering the visual scene - one that didn't have the means of emphasising continuous lines, for example.

Another aspect of the matter is that it's very difficult to focus attention on any one bit of sensory input without some little touch of singularity to help. For attention to be maintained on one particular visual object there has to be some way of distinguishing it. Position can be an adequate defining feature, even among rows of identical objects. But when there are many stimulus patterns of the same shape, colour and size, all moving with the same rhythm, and all constantly changing their spatial relationship to each other, there's no way of fixing attention on just one of them for more than a brief moment. This, no doubt, is one of the reasons why shoaling and flocking can be a useful defence against predators. Even though the predators' reactions are very fast we can guess that they have some difficulty in lining up their sights on one individual in the gleaming, shifting mass of fish, or on one bird amidst a tumbling flock.

Conditioned attention

The attentional response to novelties, singularities and conspecifics must depend on innate neuronal wiring. Other attentional responses are learnt, for attention can readily be conditioned. Stimuli which become associated with emotional experience are always likely to attract notice, especially those associated with serious pain or fear. But associations don't have to be bad to be powerful. One tends to react to the sound of one's own name, even if it occurs in a conversation one wasn't listening to. Hammered into memory by all sorts of reinforcements over the years, it penetrates consciousness as few other words can. The phenomenon has been illustrated by experiments in which subjects hear two different stories, one in each ear, and are instructed to listen to and report on only one of them. The sound of the subject's name reliably deflects attention to the 'wrong' ear. A relatively quiet utterance of a sleeping person's name can also be more effective than any other stimulus of similar intensity for waking them up.

A major part of sensory learning clearly consists of establishing where to pay attention. Learning to classify a visual object means deciding which are its essential features, the ones that will prove to be permanent, and filtering them out from such ephemera as current viewpoint and lighting effects. The schema is a record of these essentials, and when a visual pattern reappears which looks as if it might fit the formula these are the features that must be checked. If the stimulus pattern is a familiar one the checking process will be so well practised as to be automatic. The pattern of saccades which occurred at the first encounter is likely to recur again, which suggests that the input from the first fixation functions as a conditioned stimulus which triggers a conditioned response - the sequence of saccades which is behavioural attention.

A brief tachistoscopic exposure is usually sufficient for the identification of simple, familiar objects, so it may be deduced that for an adult operating in a familiar environment, making use of the expectations prompted by context, the first fixation or two is similarly sufficient. We also recognise the objects that form the well-known background of everyday life without paying them any heed at all, and are only jolted into attention by an unanticipated absence or alteration. The saccades and fixations with which we explore many stimulus patterns are not to confirm the essential identity of the visual object so much as to check up on the details of its current manifestation.

The conditioned learning that directs attention for much of the time explains how we recognise familiar objects so easily, without stopping to think about it - and do have to puzzle consciously over unfamiliar ones. It also accounts for the way we sometimes register a stimulus pattern as entirely familiar when in fact it differs in some small but important detail from what we have met before. Experience has not yet taught us to pay attention to that particular detail.

If experience teaches us where attention can most usefully be directed it also teaches us what isn't worth attending to. A sensory event that occurs frequently and is never accompanied by any significant effect soon comes to be ignored. If that sound or sight should then begin to be followed by something pleasant or unpleasant - something that qualifies as an unconditioned stimulus - it takes much longer for the familiar event to become established as a signal presaging something significant than it does for a novel sensory input. Once something is familiar it's discounted, and perhaps isn't even noticed. Other species have the same problem. Reversing 'learned inattention' takes longer than learning to respond to a new signal.

Learned inattention plays a significant role in perception. We automatically extract from the pattern of retinal stimulation the data that allow us to identify familiar objects, and we automatically ignore the reflections, the shadows, and the interactions between colours. We learn in childhood how to use highlights and shadows as information about three-dimensional shape, and not to be fooled into thinking of them as tangible phenomena. For the most part we then read the implications of such effects without taking conscious notice of them.

Perhaps the most dramatic example of learned attention and inattention is provided by language. Human infants are born with the ability to make well over a hundred different speech sounds, and with the potential to learn to distinguish all of them. But no language uses all the available sounds - no doubt the strain of having to make so many subtle discriminations would interfere with rapid communication. Initially the burbling infant experiments with all the possible speech sounds, but gradually the sounds that aren't heard are omitted, and the infant conforms with the linguistic environment.

Then, as the growing child works out how the sounds relate to meanings she discovers that some distinctions between quite similar sounds are important to her native tongue and some aren't. In English it doesn't matter just how you pronounce the initial sound in *kin*, your meaning will be understood, and the *t* sound in *otter* can be pronounced in the same way as that in *cater*. The young Scot, in contrast, finds that there is a meaningful difference between *lock* and *loch*, and the young Italian must learn to discriminate between *cita* and *citta*. European infants find that you need to shape your tongue precisely when pronouncing *rock* or *lock*, and pay attention to the difference when listening to somebody else; many oriental languages have only a sound which is somewhere between *l* and *r*, so their speakers tend not to notice the distinction when listening to English. One of the problems of learning a new language, probably the major one, is learning to hear the distinctions that have previously been discounted as insignificant.

Attention must be given, furthermore, not only to differences that have previously been regarded as unimportant, but also to sounds not used in the native language. Examples, from the English speaker's point of view, are the breathy *r* in French, and the several guttural sounds in Arabic. Sometimes it's necessary to take notice of sounds not previously classified as speech sounds, or phenomena not previously relevant to semantic meaning. Students of some African languages have to learn to pay attention to clicks, and students of Chinese to intonation.

In short, learning to hear a new language involves re-training the attention, and reversing various bits of learned inattention. Until the attentional system works fast and efficiently, without conscious effort, only the occasional word can be matched to its schema with sufficient speed to create a clear conscious experience. We don't hear the unfamiliar consonants properly without practice.

Certain stimuli are inherently salient, providing an alert that can usually divert attention from its current focus. Others acquire a similar power as a result of conditioned learning. Yet others we become conditioned to ignore. But for much of the time we do feel as if we are exercising choice over where we direct our attention. But just how great is the element of free will? Certainly attention is a wayward thing. It needs chivying along or enticing, it needs reinforcement of one sort or another to keep it going. If one needs to look out for something not inherently attention grabbing it's all too easy to be distracted by something more interesting, or by irrelevant thoughts. Practice is necessary for a not naturally salient stimulus-pattern to become automatically noticeable, preferably augmented by some regular reinforcement.

For us humans the greatest reward comes from the satisfaction of curiosity, the chance to extend our experience. We can concentrate on a book or a film for hours, as long as we find it interesting and want to know what happens next. But once we get bored it's very hard to keep the attention focussed. Off it goes on another track, and we have to keep trying to drag it back. This instinctive hedonism is tough for the student trying to absorb a dull text. But the attentional system evolved to make sure that we deal with matters that are relevant to survival, as defined by the pressures of natural selection, and it pursues its own priorities.

The reinforcements which sustain attention can work well and powerfully, however, and the attentional system does a remarkable job, meeting some very demanding challenges in selecting just one strand of input amidst the huge flood. Following the voice of a single speaker in the midst of a sea of conversation, for instance, involves focussing on one timbre and rhythm among many, and the distinctions to be made may be subtle. We derive more clues from lip movements than we tend to realise, and in natural circumstances the location from which the sound comes can also help to distinguish the particular train of input one wants to hear. Listening to films and television must usually be done without this latter cue, however. The considerable demands of operating the selection process are perhaps reflected in the fact that it becomes more difficult as age advances. (If only modern cinematographers would allow for our deficiencies.)

Retrospective Attention

We've seen that a sensory input too brief to be turned into a conscious experience can nevertheless be processed and classified sufficiently to influence the perception derived from a following input. This is also true of the unattended stimuli which are just part of the blur of background noise. For instance, in a dichotic listening experiment subjects were presented with an ambiguous sentence in the attended ear, and in the unattended ear a sentence relevant to one of the possible interpretations. In one ear they might hear *he put out the light to signal the attack*, and in the other *he extinguished the light*. Although the subjects reported that they didn't hear the unattended sentence it nevertheless influenced their interpretation of the ambiguous utterance. Another experiment showed that if the words in the unattended ear became a logical continuation of the attended message they might push their way into consciousness, in place of those reaching the designated ear.

That the unattended message doesn't go unprocessed can also be inferred from the operation of echoic memory. The utterance is deciphered sufficiently to indicate that it's interesting, and recorded well enough to be 'replayed' a moment later. Here the unattended input reaches full consciousness, but only after a slight delay. There's a partial parallel to those tachistoscopic experiments where the subject sees a brief exposure of three lines of letters and after it ends can read off whichever line is asked for - but only one line. All three lines must have been preserved, but only one can be promoted sufficiently to be deciphered.

We saw in chapter 8 that sensory input only reaches consciousness if the sustained circulation of messages it prompts gets as far as prefrontal cortex and prompts a return message. The various attentional processes must be a factor in determining whether that happens. Inherently attention-getting stimuli get a boost at an early stage. Conditioned attention kicks in just a little later, and is sometimes sufficiently delayed to mean that a replay in echoic memory is needed. Once the message reaches the frontal lobe then prefrontal cortex can direct where attention goes - the process that feels like exercising choice in the matter.

Attention and the consciousness it underpins are graded values, however. Sensory events which are inherently salient, or of learnt importance, and those which seem ambiguous or have the potential to add to our knowledge in interesting ways, are processed very thoroughly, with full involvement of prefrontal cortex. As a result details are painted in and the colours brought up to full brightness, as it were. Much other input is processed less thoroughly - just enough to create that fuzzy awareness which functions as background and context for the input that does warrant attention. The implication is that only a limited amount of sensory input can be subjected to the full treatment. Generally speaking, of course, only a modest portion of the current sensory input really needs it.

The idea that only a certain fraction of sensory input can receive the full-scale processing offers an explanation for the phenomenon, already mentioned, known as the attentional blink, revealed in experiments where subjects are asked to detect a stimulus that appears amongst a rapid stream of other stimuli. The detection of one target substantially reduces the chances that a second one occurring within a few hundred milliseconds will be reported, although the evidence provided by brainwaves, as measured by electroencephalography, indicates that the missed target does get registered. It seems that the operations involved in noting the appearance of the first target stimulus and despatching a motor command for the response occupy prefrontal cortex for long enough to prevent the second one being promoted into consciousness.

Attention can be characterised as a constructive delay which, in any species, often occurs between a stimulus and a behavioural response. The delay may be needed for reorientation, or for collecting more input about an ambiguous stimulus. Where there is sensory consciousness there is also a delay while an initial analysis of the current input is carried out. With luck the first circulations of the data reveal which parts of it might repay further analysis, and these are boosted into the special sort of circulation which constitutes attention.

But clearly when there's a lot of incoming data only a portion of it can be treated in this way. Cerebral attention is limited, just as behavioural attention is. We can only look in one direction at once, only sensitise our auditory system to sound from one direction. Cerebral attention evolved to match behavioural attention, even if it has become divorceable from it. Consequently we can only achieve one perception at a time in any sensory channel. We can't listen to two different conversations simultaneously - we can only interleave attention to one with attention to the other. On the other hand we can drive a car and have a conversation at the same time, using visual attention for the one and auditory attention for the other - just as we use different sets of muscles. However, this works well only when neither task demands too much attention, when the driving involves familiar sensory input and automatic instructions to muscles, and when the conversation is not emotional. The limits on paying cerebral attention also apply to thinking. We can only juggle with a limited number of ideas at one time.

So how does all this happen? The last few chapters have been concerned with analysing the factors contributing to consciousness, and just what it is that has to be explained. Now it's time to buckle down to the neuroscience.

Recapitulation

Attention clearly makes a vital contribution to sensory consciousness, rendering some parts of the sensory input vivid, leaving the rest as fuzzy background.

The attention which affects the conscious experience can be defined as cerebral, and distinguished from the behavioural sort.

Cerebral attention is of particular interest to the subject of reason, since it can be turned inwards, to the content of memory.

It can be supposed to have evolved, however, out of the machinery of behavioural attention, and the beginnings of that must be very ancient.

An animal must cease from anything it might currently be doing before making a response to a new stimulus. It must be correctly orientated before it can act on the stimulus or make a successful escape from it. And if it has several relevant sensory systems it's worth putting them all to work on an ambiguous stimulus before spending energy on a fuller response.

The stimuli that capture attention can be divided into different categories.

Any sudden new input will do it.

Any visual stimulus pattern which stands out from its background by virtue of a single feature or a unique colour, or by movement, or by moving in a unique direction among other moving stimuli is another attention-grabber. Indeed it is only possible to pay attention to a visual stimulus if it is distinguished from others around it by some unique feature, or if it stays put in one position.

Similar considerations apply to auditory stimulus patterns.

Fellow members of the species exercise an inherent salience, thanks to innate behavioural responses which are strengthened by experience.

Attention can be conditioned, and the process forms a major part of our perceptual education. We learn where to pay attention in order to identify the important consistencies of the external world, and what we can afford to ignore.

Attention can also be directed by choice. This works best when curiosity is involved, when we're finding out things we want to know, and obtaining cognitive reinforcement.

A long-tailed macaque in the Khao Sam Roi Yot National Park in Thailand using a stone hammer to chip an oyster off a rock.
Photo by Mark MacEwen.
Nature Picture Library

A macaque using a piece of string to floss its teeth, at the Monkey Temple in Phra Prang Sam Yot, Thailand.

Photo by Mark MacEwen
Nature Picture Library

An adult bearded capuchin using a 2kg. stone to crack a palm nut.
Photo by Elisabetta Visalberghi of the EthoCebus Project. (www.ethocebus.net)

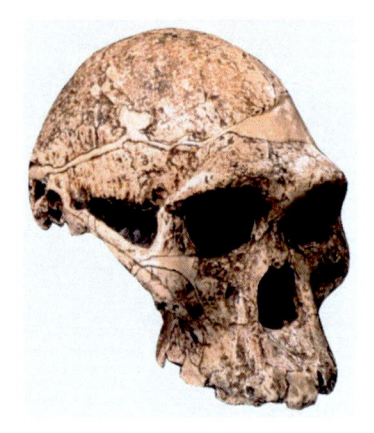

Fossilised skull of an *Australopithecus africanus*

Science Photo Library

Footprints of upright-walking hominins, who walked across wet volcanic ash around 3.6 million years ago at Laetoli in Tanzania.

Science Photo Library

A hagfish
Photo by Andrew Murray
Nature Picture Library

A lamprey
Science Photo Library

A diagram of the retina. The receptors are dark blue, the rod a little lighter than the cones. Bipolar cells are orange, ganglion cells pink, the horizontal cell green and the amacrine cell light blue.

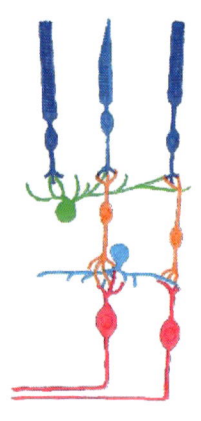

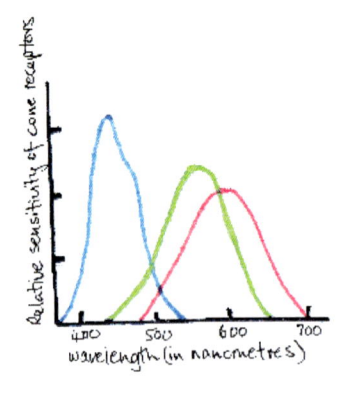

The wavelength sensitivities of the three types of cone receptor.

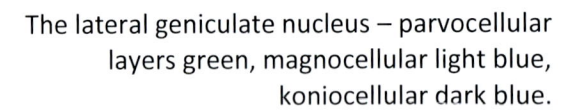

The lateral geniculate nucleus – parvocellular layers green, magnocellular light blue, koniocellular dark blue.

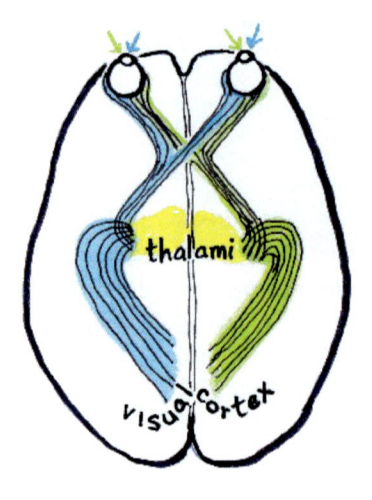

The axons from the ganglion cells on the left side of the retina, carrying information about the right side of the view, run to the lateral geniculate nucleus of the left thalamus, which connects to left visual cortex. Information about the view on the left is similarly conveyed to right visual cortex.

The basal ganglia and closely connected structures. Dark blue is the striatum, dark green the final output stations, interior globus pallidus and substantia nigra pars reticulata. Yellow is the thalami, grey the subthalamic nuclei, red the dopamine-deploying substantia nigra pars compacta. The higher brown patch is the laterodorsal tegmental nucleus, the lower one the pedunculopontine tegmental nucleus. The front-to-back plane is somewhat compressed, but not sufficiently to show the nuclelus accumbens, which occupies a more forward position, with the higher striatal area curling forward and down in a long thin tail to end close by.

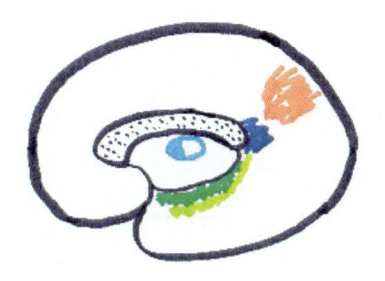

Areas involved in creating the cognitive map: the hippocampus, dark green; parahippocampal gyrus, light green; retrosplenial cortex, dark blue; anterior thalamic nucleus, light blue; parietal sectors, orange.

A chimpanzee known as Gremlin – a member of the troop studied by Jane Goodall in the Gombe National Park in Tanzania - fishing for termites while her infant watches.
Photo by Anup Shah   Nature Picture Library

An orang utan washing fabric at Camp Leakey rehabilitation centre, Kalimantan, Borneo, watched by her infant.
Photo by Andrew Murray Nature Picture Library

Chapter 15

The Neurology of Movement

It's easy to forget that complex sensory apparatus, conscious perception and cognitive abilities have all come into being as a means of putting muscles to work in the ways most likely to promote survival. Perception and cognition are what seem to be particularly relevant to reason, and they can be studied without much reference to the organisation of movement. For a long time, moreover, the neuronal machinery behind action seemed even more daunting than that which deals with sensory inputs, so I hoped I could write this book without much reference to motor matters. However, I finally realised that the subject is vital to understanding how brains evolved. And happily, its mysteries are beginning to clear.

The situation is particularly complex in mammals because in most cases each of the many fibres which make up a skeletal muscle is innervated by its own motor neuron, whereas in other vertebrates a single neuron generally controls a number of fibres. The mammalian arrangement allows for great flexibility in the deployment of the whole muscle. We can also learn to employ our muscles in many different combinations. These factors underlie our potential for using them in a great variety of ways, for many different purposes.

Voluntary actions are directed from the neocortex, from a sector at the rear of the frontal lobe, just in front of the central sulcus which separates it from the parietal lobe. There it's conveniently sited between the three lobes that process sensory input and the frontal area which weighs up all the sensory evidence and decides what action is desirable. Each bit of neocortex has its own distinctive structure, but motor cortex is particularly differentiated, in that the largest excitatory neurons here are found in the layer from which most long-distance axons leave, whereas in sensory cortex the largest neurons are in the input layers.

There are several divisions. Primary motor cortex is supported by premotor cortex and supplementary motor cortex, all of which are involved in formulating directions to muscles that move body parts. Another group of sectors is concerned with eye movements. Somatosensory cortex is close by, to provide tactile feedback. And there are dense connections to motor areas from the parietal sectors where retinal images are converted into body-centred locations and calculations are done about how actions on objects must be shaped.

Some of the neurons in primary motor cortex despatch straightforward messages to muscles, but others seem to code complete gestures or other body movements, defined by the results rather than the individual muscle contractions.

Half a second's worth of artificial stimulation in a certain part of a monkey's motor cortex, for example, produces a movement of the hand towards the mouth and an opening of the mouth, regardless of where the hand is when the current starts. Stimulation in another area causes the hands to be positioned centrally, as if there was something to be examined visually, in other areas it triggers reaching, or a defensive movement. Much of motor cortex, in short, seems to deal with movements in terms of their functions. It's organised, Michael Graziano has suggested, *along ethological lines*, with different subregions *emphasising different categories of behaviourally useful actions*. One might say that it holds the schemata for movements, just as the temporal lobe stores perceptual schemata. A perceptual schema enshrines the idea of a cup, a tree, a voice, defined by its significance. Motor schemata represent ideas like reaching, grasping, running, and relate to the purpose of the action.

Clearly the commands sent out to muscles from motor cortex must be adapted to current circumstances, such as the position from which the movement begins, and the degree of challenge presented by any objects acted upon. And all sorts of feedback are needed. Connections carrying the proprioceptive and somatosensory sort have been found in primary motor cortex. For the visual guidance of movement premotor cortex may be more important. It has a significant proportion of neurons which respond to visual input and is particularly active during visually guided movement. Much of it is concerned with reaching and grasping movements. Some of its neurons fire at the sight of any object which requires a certain sort of grasp, regardless of what it looks like, or whether the movement is actually made. Others have receptive fields centred on visible parts of the body, such as the arm. When the arm moves the visual receptive fields centred on hand, wrist and arm move with it, so that the neurons continue to respond to any stimulus which is close to arm or hand. This part of premotor cortex apparently came into being with the evolution of primates. The abilities it confers, it's been suggested, may have been what made it possible to harvest fruit and young leaves from the ends of branches, adjusting a movement-in-progress when either the target or the branch supporting the animal swayed.

The supplementary motor area seems to be especially concerned with voluntary but well-practised movements. Artificial stimulation in this region can trigger movement, but it takes rather more current than in primary motor cortex. Experiments with epileptic patients preparing for brain surgery have revealed that a level of stimulation not strong enough to produce actual movement produces an urge to move, described as resembling a compulsive desire.

The evidence suggests that supplementary motor cortex is involved in the parsing of sequences of movement - something that has to be learnt. It's also implicated in the switching of responses when the demands of a task are changed. Switching a response means, of course, suppressing the one that has been practised up to then, as well as adopting a new one. Lesions in the supplementary motor area sometimes lead to the production of automatic responses in circumstances where they aren't appropriate.

Movements can be elicited by artificial stimulation of premotor as well as supplementary motor cortex, but again a higher level of current is required than in primary motor cortex. Moreover the activity preceding a movement usually begins earlier in the former two areas. So although there are some pathways which descend directly from premotor and supplementary motor areas the more important route for their influence seems to be via primary motor cortex.

In order to plan a movement, meanwhile, the motor cortices clearly need to know just where the object-to-be-acted-upon is in relation to the self. The information comes via connections from the parietal areas which transform a retina-based view into one that takes account of the direction of the gaze, the orientation of head and body, and even the position of arm and hand. It's also parietal cortex that allows a hand to be shaped to fit the object it's about to grasp.

Artificial microstimulation in a certain part of the lamprey telencephalon can evoke specific eye or body movements, but there is little sign of motor learning in this species, or indeed in reptiles or amphibians. It seems reasonable to deduce that an area which once served only to release hardwired actions has been expanded in mammals to support the use of conscious feedback for guiding actions. Something similar seems to have happened in birds, though with a rather different form of neuronal organisation. Thus both mammals and birds have acquired, to varying degrees, the ability to experiment, to develop a knowledge of what the self can do, and to elaborate on inbuilt motor patterns, adapting them to suit particular purposes and varying circumstances. The size of the cortical motor areas must surely correlate with the amount of motor play indulged in by the young of a species.

Confirming the idea that these areas serve to develop and store action concepts, or motor schemata, activity has been measured there when actions are merely imagined. It ran at about 25% of the level recorded when movements were actually made. And in monkeys premotor neurons were found which fired not only when the animal performed an action but also when it saw the same action carried out by another animal. These have been termed 'mirror neurons'. Some of them can also be activated by the sound effects that suggest an action being performed out of sight, or by the implication of an action, as when a hand reaches into a box known to contain food and can be supposed to be grasping it.

Mirror neurons have also been found in the lower part of parietal cortex. They probably account for the way newborn humans sometimes reproduce the facial expressions they see, although they have had no chance to compare their own facial expressions with those of other people. Infants of quite a few species seem to learn from parental example, though equally unlikely to have a consciously developed rationale for doing so. It seems likely that the earliest form of mirror neurons evolved in parietal cortex, and simply functioned to release an innate action.

We may guess that in infant mammals a report from visual cortex when an action is observed and a report from the proprioceptive system as it's imitated converge on motor cortex to inaugurate the mirror neurons there. The sensations resulting from the automatic imitation inspire study of the model. As the brain matures the hardwired connection which triggers the imitation is overwritten, and is replaced by intentional experimentation with the performance of the gesture. Thus the concept of the action becomes established.

In apes and some monkeys the mirror neurons of premotor cortex no doubt became especially numerous, underlying their capacity for learning to use tools. In humans they must be yet more numerous, allowing for more exact imitations. In brains which can reflect on perceived and performed actions the inbuilt capacity for imitation must form part of the foundation for the concept of others as beings like the self.

The evolution of motor systems is reflected in the complex way the motor cortex exercises control over muscle deployment, which involves several different routes to the neurons which synapse on muscles. One set of connections from neocortex to spinal column runs down through telencephalon, diencephalon and midbrain, through a series of structures known collectively as the basal ganglia. Another goes via various hindbrain nuclei to the cerebellum - that "little cerebrum" that bulges out conspicuously from the back of the brainstem.

In both cases the messages eventually arrive at nuclei in the hindbrain, situated in a network of neuronal fibres known as the reticular formation. From there they are despatched down the spine through the reticulospinal tract. Another pathway, the vestibulospinal, carries directions from the vestibular or balance organs which allow an animal to maintain a satisfactory relationship with gravity, and can prompt a rapid readjustment of muscle tensions when there's a danger of overbalancing. Both the reticulospinal and the vestibulospinal tracts are of ancient evolutionary origin.

The rubrospinal tract, originating in a structure called the red nucleus, is not quite so long-established. The instructions it carries go mainly to extremities, and it's thought to have evolved in association with lateral fins. (It's missing in the two sorts of fish which lack such fins.) With the appearance of this first bilateral body-accessory it will have become useful to have rather more independent management of the two sides of the body, and the red nucleus perhaps emerged to deal with this complication. In line with this theory it has been lost in certain snakes, and in caecilians, amphibians which similarly get around without legs.

With the evolution of neocortex came another innovation, the corticospinal pathway, which provides a direct connection to spinal neurons and interneurons. This pathway too mainly carries instructions for limbs and especially their furthest portions. Damage to either corticospinal or rubrospinal pathway affects the control of the more distal components of limbs, and damage to the corticospinal pathway particularly compromises finer movements - the independent use of fingers or the exact placing of a foot on uneven ground. Predictably, monkeys have vastly more corticospinal fibres than cats. The ratio of corticospinal to rubrospinal neurons in macaques is estimated at around a hundred to one.

In the more sophisticated primate species there are also some axons which run all the way from motor cortex to the actual motor neurons. These are concerned in the control of the finest finger movements. Normally the corticospinal pathway and the direct one work co-operatively, but if the former is temporarily silenced finger movements gradually regain their dexterity, and if it's unsilenced again there is a further period of readjustment. It's clear, therefore, that the fast direct pathway deals with movements that have to be learnt, and perfected by practice - the sort of movements that are mostly guided by visual feedback.

The basal ganglia

The pathways through the various structures which constitute the basal ganglia illustrate the important role inhibition plays in neuronal machinery. The messages from neocortex to the input station of the basal ganglia, the striatum, are excitatory while the output neurons at the end of the chain are inhibitory. The latter keep up a fairly steady rate of firing, until an excitatory message from the neocortex activates intervening inhibitory neurons and suppresses that output, releasing the targeted neurons from the suppressive signal. The excitatory neurons are shown as dashed lines in the diagrams, the inhibitory ones as solid lines.

Some of the inhibitory output from the final stations goes to two centres where artificial stimulation can elicit locomotion, the diencephalic locomotor region in the thalamus and the mesencephalic locomotor region in the mid-brain. Both project to the hindbrain, and both are maintained in an inhibited state, apparently by input from a variety of structures, including the basal ganglia, until that inhibitory input is itself silenced, allowing locomotion to be initiated. The connections have only been traced so far in the lamprey, but are likely to be essentially similar in other vertebrates.

Locomotion must be the oldest form of brain-directed movement in vertebrates, so it's reasonable to suppose that this arrangement has very ancient origins, and that the basal ganglia have evolved through elaboration of the early form of locomotion control. As new body parts and new forms of movement evolved the system expanded to manage those too.

In non-mammals the basal ganglia are strongly and conspicuously inter-connected with the thalamus, which must provide much of the sensory information which guides locomotion. In mammals that circuitry is overshadowed by extensive connections which lead on to the neocortex. There are three main circuits, incorporating various thalamic nuclei. The one already described, which suppresses the inhibitory output of the final station, freeing the related part of the thalamus to communicate with the cortex, is known as the direct pathway, shown on the left.

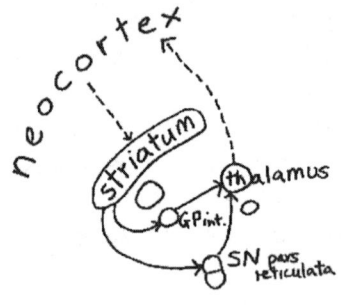

The indirect pathway (right) takes a slightly longer route to the two output stations and intensifies the tonic inhibition, so that the activity of thalamic neurons is more firmly suppressed.

In the third pathway, discovered after the other two and known as the hyperdirect, an excitatory message from neocortex to the subthalamic nucleus evokes another excitatory message from there to the ouput stations, which strengthens the inhibition very rapidly.

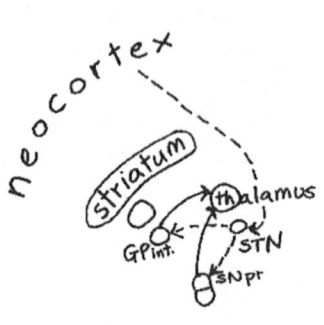

The neurons of the indirect and hyperdirect pathways spread their influence widely at the output stations, while the axons of the direct pathway terminate in quite small patches. The total effect, therefore, is that inhibition on the thalamic targets of the basal ganglia can be lifted quite selectively, and a focussed message goes to the neocortex.

(A more elaborate diagram of the basal ganglia, with complete labels for the components, can be found on page vii of the colour section.)

The three pathways, which are linked by quite a few interconnections, pretty certainly deliver their messages to the output stations in rapid alternation, so as to organise very precise communications to the neocortex and very precise timing of muscle contractions. Deficiencies in various components of the basal ganglia pathways produce illuminating effects. Inadequate inhibition in one area results in wild, uncontrollable movements - a flinging about of the limbs known as hemiballismus. In a different area it causes Tourette's syndrome, in which the symptoms are repetitive tics and hyperactivity. Huntington's chorea, which manifests itself in frequent sudden twitches, is attributed to synaptic alterations leading to cell deaths mainly in the indirect pathway. In some of these motor disorders, as well as Parkinson's, there may also be twisted and uncomfortable postures caused by simultaneous activation of antagonistic muscles, which suggests that a major part of the problem may be in the timing of commands. Recent work has shown that the fine timing of a complex movement depends on learning accomplished in the input sector of the basal ganglia.

Parkinson's disease is the result of too much inhibition. The symptoms can include a difficulty in initiating voluntary movements, very slowed-down movement, and tremor. Even more exaggerated are the effects of Encephalitis lethargica, the sickness described by Oliver Sacks in his book *Awakenings*, which often entailed lengthy periods of complete freezing. In this illness voluntary movements seemed to be more strongly affected than well-practised responses. Sacks found that sometimes, if he suddenly threw a ball towards a motionless patient, they might fling out an arm to catch it, although they couldn't make the same movement by conscious will. Parkinson's sufferers also have more difficulty with chosen than with cued movements, but here it's the learnt movements which could once be performed automatically that seem particularly badly affected.

In both Parkinson's disease and encephalitis lethargica the problems stem from a shortage of the neurotransmitter/modulator dopamine. This is produced and deployed by neurons of a structure adjacent to one of the basal ganglia output stations. Because of their dark colour the two were initially named as a single unit, the substantia nigra. The GABA-deploying output station is now known as the pars reticulata, the dopamine producing part as the pars compacta. The substantia nigra pars compacta sends many axons weaving through the basal ganglia, with an exceptionally dense distribution of contacts, exercising a strong influence. The complication is that it operates through two different types of receptors, with opposing effects. In the pathway which inhibits action it makes neurons less likely to fire, so when dopamine is in short supply they fire more readily. In the pathway which selectively interrupts that inhibition dopamine operates through a different type of receptor and facilitates firing. A reduced supply therefore means that the inhibition is less likely to be lifted. Consequently a subnormal supply of dopamine weakens the message that leads to action and strengthens the suppressive force. Probably this is not only detrimental in itself but also upsets the timing of the messages in the three pathways. Certainly firing patterns become abnormal when the dopamine supply is inadequate.

The influence of dopamine is perhaps particularly important at the input stage of the basal ganglia, the striatum. In contrast to the neurons at the output stage, which fire more or less continuously at a fairly high rate, the neurons of the striatum maintain a strong negative charge and need a good deal of input to get them going. They have an extremely large number of input synapses, as many as ten thousand to one cell, and require excitation at a good many of them to overcome the barrier. The first inputs merely lift a neuron from a 'down' state to an 'up' state, so that it becomes more likely to fire. This shift from 'down' to 'up' state happens less readily when the dopamine input is less than normal.

The basal ganglia are richly supplied with neuromodulators. Apart from dopamine the most conspicuous is acetylcholine, which comes from clusters of large cells scattered through the area. Intriguingly, both dopamine and acetylcholine seem to contribute to steering. An imbalance of dopamine between the two sides of the brain results in an animal that turns constantly towards the deprived side. With an uneven acetylcholine supply the animal also veers towards the side of the loss, but less tightly. The effect seems to be mediated in the nucleus accumbens, part of the striatum situated in a more forward plane than that shown in the diagrams, and close to the inward end of the part shown here, which curves forward and downward. Injection of either dopamine or acetylcholine imitations into the nucleus accumbens on one side of the brain causes the rat to turn towards the other side. Both dopamine and acetylcholine can also influence the speed of locomotion.

This suggests that the basal ganglia machinery evolved to control locomotion. Then, as body-plans grew more complicated and new potentials for action were added to the repertoire, it expanded to co-ordinate all movements, releasing only compatible behaviours simultaneously. And when, in mammals, muscle control grew more complex, with independent connections to individual fibres providing an increasing degree of versatility, the basal ganglia came to play a major role in learning how to exploit the new potential.

However, the sort of learning first supported by the basal ganglia must have concerned not how to carry out an action but when. Or perhaps it would be more accurate to say that it concerned when not to act. The basal ganglia machinery is clearly well suited for suppressing a hardwired response when experience has shown that the addition of a certain feature or features to the genetically specified stimulus pattern makes action inadvisable.

Before there was a neocortex the sensory cues which trigger genetically determined responses can only have been registered in older parts of the brain, such as the optic tectum (the sensory centre of the midbrain), or the thalamus. The additional sensory information which indicated that the response to a particular version of the cue should be suppressed must have been relayed by the same route. And the older circuits which link thalamus and basal ganglia no doubt supported the learning about which versions of the cue repaid action, and when the hardwired response should be inhibited. Such connections can be supposed to support conditioned learning, as defined earlier, and they must still operate in mammals. The evolution of the neocortex allowed subtler analysis of sensory inputs, finer distinctions, and thus a more sensitive application of inhibition, but the old system was extended, not superceded.

Motor learning must involve releasing movement on a much finer scale - not just a complete action at one go, but a sequence of carefully timed muscle contractions. The learning process must involve discovering the precise sequence of instructions which is required to achieve the desired purpose. Once a movement is well-learnt an initial consultation between motor cortex and basal ganglia presumably serves to evoke the appropriate set of instructions.

On a larger scale the basal ganglia are certainly involved in organising longer series of movements. Neurons have been found that consistently fire for the first movement in a regular sequence, or the second or third - regardless of what the component movement or the whole suite of movements might be. When rats groom themselves they typically carry out a series of actions in a regular sequence, and during an action performed in the course of this habitual sequence many basal ganglia cells fire which don't do so when it's performed on its own. Lesions of the basal ganglia interfere with the execution of the well-ordered routine.

The cerebellum

The cerebellum has long been known to be involved in the fine control of movement. It's not necessary to the initiation of voluntary movements, but damage to it can lead to movements of erratic size and direction, slurred speech, clumsiness, and problems with postural stability - all the symptoms of excessive alcohol consumption in fact, suggesting that the cerebellum is particularly susceptible to the effects of alcohol. Its functions include co-ordinating the efforts of different sets of muscles, and organising the adjustments necessary to take account of outside forces working against the action of the muscles. It's much involved in that essential skill that is learnt in early childhood and thereafter is usually automatic, the business of balancing upright on two feet, with the aid of feedback from the vestibular system. The fact that when we're drunk it takes a detectable amount of time to respond to messages from the vestibular system, whereas when we're sober we remain upright without having to think about it, probably gives a good idea of what the cerebellum normally achieves.

Imaging experiments show that when a task requiring fine motor control is being learnt there is considerable activity in both basal ganglia and cerebellum, as well as in premotor and parietal cortex. As the skill improves, and movements grow smoother, premotor cortex becomes less active, primary motor cortex and supplementary motor cortex more so. The activity in cerebellum and basal ganglia, meanwhile, becomes more focussed.

What happens as we learn a skill, Ohyama, Mauk and colleagues pointed out, is that we no longer need feedback, which can only tell us what has gone wrong. We come instead to use feedforward information, which provides the means of forecasting what will perturb a movement, and can be used to make appropriate adjustments. The cerebellum, they suggested, is a device which learns from feedback information how to apply feedforward, and thus how to shape movements successfully. The cerebellum keeps us upright, for example, by dictating the minute muscle adjustments that prevent a departure from the perpendicular, as well as the rapid larger adjustments necessary if we are unexpectedly tipped off-balance. Jessica Brooks and colleagues confirmed the idea. Predictions about feedback are compared with actual feedback – much of which comes direct from the spinal column, as well as from the nearby vestibular nucleus, and thus arrives very fast.

The cerebellum (little cerebrum) consists of a bunch of nuclei wrapped around with an extensive cortex. This has a simpler structure than the neo- or cerebral one, but the neurons are smaller and much more densely packed. (Its exterior has delicate horizontal wrinkles, in contrast to the big blobby gyri of the cerebral cortex.) In humans the cerebellum accounts for only a tenth of the weight of the brain, but various estimates suggest that it contains as many neurons as the whole of the rest of the brain, or perhaps even more. In further contrast to the cerebral cortex the cerebellar one is very uniform in its microstructure.

Each small slice of cerebellar cortex is connected to a specific sector of the neocortex, and to specific portions of the spinal cord. As well as despatching messages to the spinal cord via the reticulospinal route the cerebellum is also, like the basal ganglia, linked back to the neocortex by pathways through the thalamus. Cerebellar activity seems to be more closely linked to that in the supplementary motor area, the region particularly involved with well-learned movements, while basal ganglia activity correlates more strongly with that in primary motor cortex. These connections further support the idea that the cerebellum is essentially concerned with 'on-line' adjustments to movement, the basal ganglia with advance planning and choice.

Some co-operation between the two must be required, and there are direct connections between them, as well as a pathway with only one intervening synapse, in a thalamic nucleus. Both basal ganglia and cerebellum are linked to the red nucleus - the origin of the rubrospinal pathway.

As well as providing a means of learning new skills the cerebellar machinery is also responsible for some sorts of conditioned learning. The most extensively studied conditioned reflex is the eye-blink. Both humans and rabbits blink when a puff of air hits the eye, and if the puff is regularly preceded by a signal, such as a tone, both species can learn to protect their eyes by blinking in anticipation. The unconditioned stimulus (the puff of air) is reported to the cerebellum and leads to a modification of connections in the cerebellar cortex, followed by similar alterations in a cerebellar nucleus. That this is the basis of the learning is confirmed by the fact that after lesions in the cerebellum the learning is no longer possible, or if the reflex has already been established it's abolished.

In rabbits the learning only takes place if the unconditioned stimulus arrives before the warning signal has ceased. In humans the response can be learnt in these circumstances without conscious awareness of either conditioned or unconditioned stimulus, or of the blink, so it's assumed that this learning similarly depends on the cerebellum. However, despite the temporal limitation the cerebellum is very sensitive to the length of the interval between the beginning of the signal and the arrival of the puff of air, and triggers the blink at just the right moment.

The business of learning to produce a response to a sensory input which indicates that something nasty is about to happen is not all that different from learning to adjust an action as it unfolds. Both involve discovering the cues that presage an undesirable event, and responding rapidly and with very fine timing. In both cases the learning and its application can be carried out without reaching consciousness. Indeed, a readjustment to an ongoing movement often needs to be performed too fast for conscious direction. Since the ability to learn to avoid a noxious input must be an older talent than the ability to learn how to shape actions, it may be deduced that the latter skill has grown out of a new application of the machinery underlying the former one.

The cerebellum is pretty ancient, but not quite as old as the basal ganglia. It's not present in identifiable form in one of the two families of fish which lack bilateral fins, and is only just detectable in the other. It's reasonable to suppose that it grew out of an arrangement for avoiding harm, promoting larger responses to undesirable inputs than could be managed by local connections between sensory neurons and motor neurons. An early function certainly lay in managing the relationship with gravity. It seems likely that the structure evolved further, in collaboration with the red nucleus, as a means of controlling fins, and co-ordinating their deployment with the muscular contractions of the body. It also has a role in eye movements, which may perhaps have developed even earlier.

Central Pattern Generators

What I've been describing so far are quite advanced forms of motor control. But muscles have been around a great deal longer than brains, and the control systems that were in place before brains appeared are still there, so instructions from the brain are not necessary for every type of movement. Even in the most sophisticated species actions that need to be very fast, such as withdrawal from a harmful stimulus, are achieved by means of a direct sensory-neuron-to-motor-neuron connection. An example is the rapid tightening of leg muscles triggered when we tread on something that starts to tip and threatens to cause a fall.

Other forms of action are organised by a device known as a central pattern generator. This may involve neurons with leaky input synapses and charge-restoration gates so well balanced that there is continuous steady firing, except insofar as modulatory inputs increase or slow the firing pattern to suit current circumstances. Or it may be composed of neurons linked in a pattern of reciprocal inhibition, so that they fire in a regular cycle. Or the two mechanisms may be combined. Rhythmic activities such as breathing and chewing, or propelling food through the gut and blood around the circulatory system, are controlled by pattern generators. But the oldest one, at least so far as vertebrates are concerned, is that which governs locomotion, which is situated in the spinal cord.

The chordate phylum, it may be remembered, is defined partly by the notochord, a stiffener along the back, which vertebrates have replaced with a more effective stiffener, either of bone or cartilage. As noted earlier, the best idea of what an ancestor of the vertebrates might have been like is probably provided by the lancelets, more officially *Amphioxus* or *Branchiostoma*. In *Amphioxus* larvae there are thick, chevron-shaped blocks of muscle which are contracted first on one side, then on the other, to produce a clumsy threshing movement. As the larvae mature a staggered pattern of alternated muscle contractions along the more posterior part of the body is added to this central flexion, for somewhat more stylish swimming. Fishes improved on the design, with narrower blocks of muscle. Along with the improved stiffener this makes it possible to produce an undulating ripple of movement, growing more expansive towards the tail, which is both more graceful and more efficient.

Dorsal fins, which help to counteract the sideways force produced by the undulations so that they can be more efficiently converted into forward movement, evolved at a fairly early stage. When lateral fins had evolved some species developed them further to provide a good deal of their propulsion. Others stiffened the backbone, and depend largely on waving a powerful tail. Many species combine both tactics.

The controls which refined the output of the locomotory pattern generator became ever more delicate, as fins and then legs appeared. But the device is still in use in terrestrial vertebrates. It can be seen at work when a baby's legs wave in the air, with one raised first, then the other. We don't have to learn the basic pattern of alternating leg action; we just have to learn how to put it to work effectively, how to speed it up and slow it down, to steer, and to adapt it to the demands of gravity. Things are complicated in tetrapods, however, in that we have separate pattern generators for each limb, while the original, central pattern generator functions to co-ordinate them.

The essence of a pattern generator's counter-balanced circuitry is that it operates spontaneously and continuously - as is obviously necessary in powering devices like hearts and lungs. It's less obvious that the locomotory pattern generator should work along similar lines. But water, especially the sea, is almost constantly in motion, so that some degree of muscular effort is frequently needed just to stay in one place. Consequently the pattern generator seems to have evolved as something that needs to be suppressed when it's not needed, rather than prompted into action when it is.

In vertebrates the chain of control is complex, of course. The hindbrain nucleus controlling locomotion is under ongoing inhibition, which can be interrupted by inhibition from either of those two brain areas mentioned earlier in the chapter, the diencephalic locomotor region (in the thalamus) and the mesencephalic locomotor region (adjacent to the midbrain area which controls muscle tone). Both, as we saw, are subject, in turn, to control by the basal ganglia. The inhibitory connections to the mesencephalic centre come from the substantia nigra pars reticulata, those to the locomotor region in the thalamus from the other main output area of the basal ganglia.

Quite why there are two systems to organise locomotion is not yet clear. It may be significant that the substantia nigra pars reticulata has an important role in the control of eye movements, so perhaps this pathway deals with locomotory movements that contribute to orientation. Maybe it's also responsible for retreats from threat, which can be triggered rapidly through the basal ganglia. The input to the diencephalic or thalamic centre, meanwhile, originates in a wide variety of areas. Possibly it's concerned with the less urgent matter of locomotion to more distant goals.

If the earliest multicellular animal was a simple bag-shape, capable only of withdrawing from noxious contacts and orienting its proto-mouth towards the richest concentration of food, it was the evolution of a pattern generator that turned it into something which could use its contractile cells to propel itself through the water or across the ocean floor. The withdrawal movements made in response to noxious inputs were thereby organised into patterned alternations of contraction.

Eye movements and other forms of behavioural attention

Eye movements are controlled by circuits that are largely analogous to those for other skeletally attached muscles. But since there are only three pairs of opposed muscles to rotate the eyeball, plus one odd one, the mechanisms, though not exactly simple, are a little less complex than those that direct the deployment of body muscles. As with locomotion, there's a central pattern generator, but in species which have evolved foveas and the sophisticated gaze control that goes with them its patterns are constantly overridden and barely detectable.

In mammals eye movements are directed from the frontal lobe, from regions known as the frontal eye field, the supplementary eye field and the prefrontal eye field, just in front of the areas which direct body movements. Each has two sectors, one for promoting saccades, the other for 'smooth pursuit', when the gaze follows a moving object.

For saccades some messages are despatched direct to the hindbrain centres of eye muscle control. Others go to a structure which has been studied for a long time, and is somewhat less wrapped in mystery than the corresponding midbrain and hindbrain relays for instructions about locomotion and other corporeal action. The superior colliculi (mentioned earlier) are bumps on the back of the midbrain. They receive direct input both from the retina and from the visual cortex, in addition to that from the frontal eye fields, and they relay their own instructions to the eye-muscle command centre. In addition, they co-ordinate head movements with eye movements, sending instructions to neck muscles, and contribute to orienting the whole body towards a stimulus and directing locomotion towards it. In animals with twistable ears, or whiskers, the superior colliculi seem to participate in directing their movement too. They might well be regarded as the seat of behavioural attention. It's eye movements, however, that are best understood.

The superior colliculus has several layers, each mapping the contralateral retina. Retinal input arrives in the upper layers, where a target to which a saccade is to be made is represented by firing in the appropriate map location. The instructions for the saccade are issued from the lower layers, where the excitations occur just before the saccade is effected. Each neuron codes for a particular direction, and as with the orientation-sensitive neurons of the visual cortex, population coding is employed. Quite a number of cells fire, and the average of the whole bunch determines the direction of the saccade.

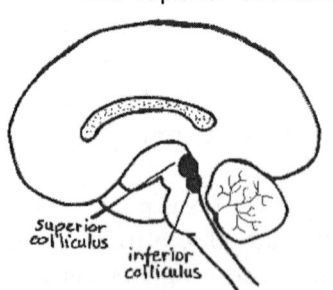

In addition to the direct signal from frontal cortex the superior colliculus also receives an indirect, inhibitory one via the basal ganglia. The GABA connections mostly run to the same-side (ipsilateral) superior colliculus. They normally fire steadily, thus inhibiting the superior colliculus neurons. When some of the neurons in the basal ganglia output station are themselves inhibited a small patch of superior colliculus neurons becomes free to fire in response to visual input.

Transmitting the decision about which bit of the current visual scene will best reward viewing is thus something of a belt-and-braces affair - a common situation where new and more sophisticated neuronal machinery is superimposed on older, simpler arrangements. The direct route from retina to superior colliculus is clearly older than that from the neocortex. It must be responsible for the way a young infant's gaze is captured by the densest cluster of features, or by other salient stimuli. The inhibitory connections from basal ganglia to superior colliculus must serve, among other things, to suppress such automatic saccades, as the infant learns what's worth looking at and develops profitable habits. The prefrontal, frontal and supplementary eye fields provide the potential for choosing what to look at, operating through the basal ganglia pathway as well as the direct connection to the eye muscle control centre. If the superior colliculus is damaged eye movements do, in time, recover to a great extent, though they are no longer as fast. And if the frontal eye field is destroyed the eventual outcome is much the same, probably because a thalamic centre concerned with eye movements can take on an extra share of responsibility.

The inferior colliculus does a similar job on auditory input, registering what might be worth turning visual attention on, and then transmitting the result to its neighbour. Tactile information too is relayed to the superior colliculus, so that head and gaze can be turned towards the source of an unexpected touch.

A further contribution to saccade control comes from the cerebellum. In addition it plays a major role in two complementary skills. One is the business of keeping the gaze fixed on an object while the head is moving. The vestibular nucleus provides information on how much and how fast the head is moving, so that eye muscle action can be adjusted accordingly. Equally sensitive muscle control is needed to keep the object of interest in the foveal field when it's the object that's moving. Both skills require that the cerebellum registers what the object of interest is, though no doubt it doesn't have to be defined in terms of its precise nature. It seems likely that it's the superior colliculus which provides that information to the cerebellum. Both tasks must depend on very rapid feedback, or better still feedforward information, as to whether the operation is succeeding; guidance provided by the lengthy neocortical pathway would arrive too late.

An interesting thought is that lampreys, the finless, jawless fish possessed of a very small cerebellum, do have rotatable eyes, so it's possible that co-ordinating eye and body movement was one of the first functions of that structure.

Recapitulation

Organised movement existed long before brains. A bunch of interconnected neurons firing in rhythmic alternation - a central pattern generator – dictated the opposed contractions of body muscles that propelled early chordates through the water.

Pattern generators still play a part in our locomotion.

Similar devices came to govern activities such as pumping nutrients around the body, breathing and chewing.

The control of locomotion grew increasingly complicated with the advent of fins and then legs.

The oldest pathways to the spinal motor neurons run from the hindbrain and are called reticulospinal. They mostly control body muscles.

The vestibulospinal pathway carries the commands for responses to changing inputs about the pull of gravity.

The rubrospinal pathway (running from the red nucleus) seems to have evolved about the same time as fins, and is mostly concerned with the operation of fins or limbs.

In mammals things became more complicated with the evolution of motor cortex in the rear part of the frontal lobe. This governs voluntary movements.

Motor cortex, divisible into three sectors, seems to be organised largely in terms of ends to be achieved. Visual, tactile and proprioceptive feedback reaches these areas.

Motor cortex contains some neurons which fire not only when the self performs an action but also at the sight of someone else doing it. These might be supposed to embody the motor schemata which allow the shape of an action to be planned.

Some mirror neurons are also found in parietal cortex.

The motor cortex has direct connections to the spinal cord, via the corticospinal pathway. The instructions conveyed by this route mostly concern the operation of limbs and especially their extremities.

In primates and especially in humans this pathway is well developed, and in addition there are some neurons which run direct from cortex to motor neurons.

A bunch of structures known as the basal ganglia provides another route to the spinal column's motor control arrangements.

In mammalian brains a much more conspicuous circuit links the basal ganglia back to the neocortex through some thalamic nuclei, circuitry which is important to motor learning.

The cerebellum plays an important and complementary role in motor learning. Another, older function is to organise finely timed avoidance movements when noxious tactile input threatens.

Eye movements are also overseen by a neocortical motor area, in combination with the midbrain structure known in mammals as the superior colliculus.

Both basal ganglia and cerebellum are involved in learning how eye movements can most usefully be deployed.

Chapter 16

The Neurology of Attention

In us attention is a complex matter, but its original, simple function is obviously as a prelude to action. An animal interrupts whatever it's doing and reorientates itself to a different stimulus, focussing whatever moveable sensory organs it may have in that direction. This provides the opportunity to gather more information, if possible, and decide whether a larger response is necessary. As could be expected, much of the neuronal machinery of cerebral attention clearly derives from the machinery of behavioural attention.

Among the devices involved are some that seem merely to provide a general alert, by means of a burst of impulses. This constitutes a message to the rest of the brain indicating that something has just come up which may require action, without being specific about what it is. The same mechanisms also function to govern an animal's general state of alertness, by their ongoing, or tonic, activity. In sleep their impulses cease, or are much reduced. But if either external or internal sensory receptors report something disturbing, it's the alerting system that kicks a sleeping brain into wakefulness.

Some of the centres that contribute both to attention and wakefulness are low in the brainstem, close to the beginning of the reticulospinal pathways which despatch motor instructions down the spinal cord, and also to centres which regulate things like blood pressure and breathing. The nucleus gigantocellularis and the adjacent nucleus paragigantocellularis have widespread dendrites weaving among the bundles of axons which carry data upwards from contact and internal senses. Small branches sprout off from the axons to synapse on these dendrites. A burst of firing in any of the ascending sensory pathways can thus trigger activity in these hindbrain neurons, even as the message is on its way to higher parts of the brain. The resulting boost to a low baseline firing rate spreads through the interconnected neurons, allowing the effects of several inputs occurring within a brief period, and perhaps in different parts of the network, to be summed together. The result is a generalised message to the neocortex, where the sensory signals which provoked the alert are meanwhile arriving in the appropriate area.

Just above these nuclei and receiving input from them is a more conspicuous component of the alerting system, a structure conveniently distinguished by its colour: the locus coeruleus (blue place, as in cerulean blue). It deploys a distinctive neuromodulator, norepinephrine (sometimes called noradrenalin). The locus coeruleus is small - in the rat it contains only about 1,600 neurons, in the average human maybe 16,000. But these cells have long, fine, unmyelinated axons which weave all through the neocortex and thalamus, branching repeatedly and contacting numerous cells. Other norepinephrine fibres from adjacent structures project to pretty much all of the rest of the brain, and synapse most densely in subcortical areas particularly associated with attention. Some go to the hypothalamus and other centres of the autonomic nervous system, which may in turn send messages to augment the effect of those already despatched from lower in the brain, increasing the response to an alarming input. The sketch indicates the distribution of norepinephrine from the locus coeruleus to the neocortex.

One of the effects on neocortical cells is to modify the flow of ions through a channel by which negative charge is restored to the neuron after an impulse, which means that another impulse can follow more readily. Spontaneous firing is repressed, but the neuron becomes more sensitive to both excitatory and inhibitory input. Since the influence is wielded through metabotropic receptors it can last for some time. However, there are at least four different types of norepinephrine receptors, and there are two distinct types of locus coeruleus cell communicating with the cortex, one with a more widespread axonal arborisation than the other, so the full story is by no means clear.

Activity in locus coeruleus neurons is suppressed during sleep, and contributes to the recovery of consciousness after anaesthesia. When an animal is awake but nothing much is going on, the cells fire at a low rate, usually around one to two pulses per second. But when something attention-demanding appears, such as food or a threat of danger, the rate speeds up. In a cat, impulses were observed to run at about one per second in relaxed circumstances, increasing to an average two and half at the sight of a dog, and accelerating to as much as ten per second if the dog behaved in a threatening manner. Activities like eating and grooming, which interfere with vigilance and are undertaken when an animal feels secure, coincide with a low rate of firing in the locus coeruleus. A burst of firing occurs when anything prompts an interruption of such activity and a turn towards the new stimulus. If the stimulus turns out to be unimportant the norepinephrine response habituates rapidly, but if it seems worrying there's no habituation. For instance, stump-tailed monkeys showed a strong and continuing norepinephrine response to the sort of vocalisations made by unfriendly monkeys.

In short, locus coeruleus responses occur for all the sensory events that can be classified as automatic capturers of attention - unexpected tactile input, conspecifics, predators, sudden novelties in the environment, excessively high rates of sensory input in any channel, or internal distress. They can be triggered by input direct from the spinal cord, or by messages from higher brain areas. The latter inputs mean that a response can be prompted by the conditioned stimulus that foreshadows the important event instead of the event itself. And if a conditioned stimulus fails to evoke a burst of locus coeruleus firing the behavioural response usually doesn't materialise either.

As this suggests, the attention-arousing function of the norepinephrine system is important in learning. Damage to it has serious effects on learning ability, in a way that suggests the animal has difficulty paying proper attention to the stimuli. Artificial stimulation of the locus coeruleus during the training period also disrupts the learning process. This tallies nicely with the introspective feeling that learning is easiest when attention can be coolly focussed on the relevant factors, circumstances which would seem to correlate with bursts of locus coeruleus firing that occur only for relevant input. Artificial stimulation must produce ill-timed firing which parallels the effect wrought by external distractions.

Similarly, selective attention seems to work best against a background of fairly low norepinephrine activity. Four Cynomulgus monkeys were trained to release a lever when they saw certain visual stimulus patterns, and then put in an situation where these rewarded stimuli occurred only infrequently, among many distractor stimuli. They performed the task most efficiently when the basic locus coeruleus firing rate was around two pulses per second. When the rate went up there were more lever releases for unrewarded stimuli.

The norepinephrine system arises in the area from which the neurons of the reticulospinal pathway depart, and sends axons down the spinal cord as well as upwards to the rest of the brain. The descending neurons influence the locomotory pattern generator and the spinal connections to motor neurons, and it seems likely that they evolved first. One might guess that in a primitive vertebrate a sudden strong tactile input triggered not only a reflex withdrawal of the part contacted, but also a message to a norepinephrine centre. Signals from there to the locomotory pattern generator would cause an acceleration of effort to take the animal away from the threat. As sensory systems grew more complex new norepinephrine pathways went upwards, functioning to ensure that the new machinery was focussed on the really important matters, thus helping to make it worthwhile for such complexities to evolve.

Long axons that branch extensively and weave all through the neocortex as well as most of the rest of the brain, influencing it with a distinctive neuromodulator, also arise from another hindbrain area. Here too neurons may fire at a slow rate when nothing is happening, and produce a burst of activity when new or interesting sensory input is registered. But serotonin is implicated in - among other things - suppressing the sort of neuronal activity that interferes with sleep, and in the comfortable feeling of reduced alertness that follows a good meal. It's also involved in reinforcement and reward, so I'm leaving it for the next chapter.

The first structures the alerting signals reach are the pedunculopontine tegmental nucleus and the laterodorsal tegmental nucleus - which we met in the previous chapter as targets of descending basal ganglia pathways. Both nuclei receive messages from various higher areas, as well as from the basal ganglia. They project back, meanwhile, to the locus coeruleus and raphe nuclei, providing a route by which sudden visual inputs or other alerting signals deciphered at higher brain levels can evoke the same sort of excitation as raw data reported directly from spinal pathways. The pedunculopontine nucleus plays a part in the control of muscle tone as well as locomotion; it's no doubt thanks to the direct input it receives from the alerting centres that our muscles can tense ready for action even while we're still registering a sudden unexpected sound or touch.

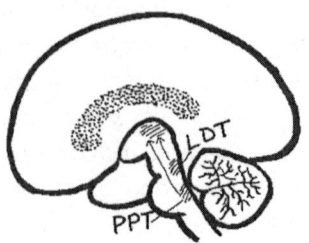

The alert is carried upwards from these two nuclei to the thalamus by neurons using acetylcholine as transmitter. We encountered the thalamus earlier as the site of the lateral geniculate nucleus, through which visual input is processed on its way to the visual cortex. Auditory and tactile input is similarly relayed, via other sectors. These form only a small part of the mammalian thalamus, however. It contains several larger, more important-looking nuclei, with some smaller ones - the intralaminar nuclei tucked in along the centre, and the midline nuclei strung out along the inner side. The largest acetylcholine input goes to some of the intralaminar nuclei (which also receive a heavy innervation from the locus coeruleus) and to two large nuclei at the front of the thalamus.

This combination of alerting systems is important to consciousness. Anaesthetics usually target either the areas from which the messages arise or the pathways by which their message is relayed to the neocortex. In humans damage to the relevant intralaminar nuclei of the thalamus may lead to coma, just as cutting off the input from the alerting systems does.

A third contribution to wakefulness and alertness comes from histamine, dispersed to a wide variety of brain areas from a nucleus in the hypothalamus. This is why anti-histamines make one sleepy, unless they're formulated so as not to cross the blood-brain barrier.

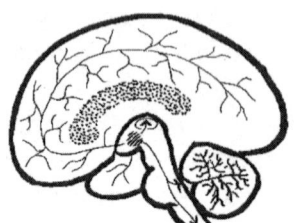

The spatial aspect of attention

Alertness is unfocussed, a prelude to attention, but the thing itself, when it's directed at the outside world, is aimed towards some particular point in space. Overt attention obviously involves movements of eyes, head, maybe even the whole animal.

As we saw, an important component of the system for managing eye movements is the superior colliculus, which despatches instructions to the motor centre that actually orders the movements. Some neurons here fire when a saccade is about to be made to the stimulus in their receptive field. The same neurons also fire when it's necessary to take note of what happens at that spot but no saccade is made. A monkey may be required to press a lever to report when a spot of light in the visual periphery starts dimming, for instance, while keeping its eyes steady on a central fixation point. Paying attention to the peripheral stimulus activates the machinery that would be involved in preparing for a saccade. Furthermore, very gentle artificial stimulation of the superior colliculus neurons representing the appropriate retinal field, at a level not sufficient to trigger a saccade, has the effect of rendering the monkey more sensitive to the dimming light, and reduces its reaction time. The same neurons are also active if the task involves making a delayed saccade to the spot where the peripheral target was, when cued by a signal that occurs after its disappearance. In other words they function to hold a record of the location.

In these circumstances the superior colliculus is activated by 'top-down' instructions from the neocortex. They come largely from the frontal eye field, another structure we met in the previous chapter. Cells here similarly fire when the stimulus in their receptive field is receiving covert attention but the saccade is not made. And here too a low level of artificial stimulation to such neurons can improve the efficiency with which change to a stimulus in the receptive field is perceived in peripheral vision. But in the frontal eye field the receptive fields of the neurons are somewhat flexible, and can shift towards the point in space which is about to be fixated. The frontal eye field maps current visual attention, it seems, rather than the current visual field.

A saccade is obviously directed at a limited portion of space, and it seems likely that superior colliculus and frontal eye field have much to do with the limitations on how much can be squeezed into the focus of attention. If humans are asked to prepare for a saccade to an upcoming peripheral stimulus they are likely to miss an unexpected stimulus elsewhere in the retinal field. This suggests that preparatory activity in a frontal eye field or superior colliculus saccade neuron improves attention to events in the receptive field, but leaves less attention available for elsewhere. This must be why quite startling events can go unnoticed when attention is already occupied.

The superior colliculus projects to the reticular activating system, and no doubt helps to direct attention to some of the inherently attention-grabbing stimulus patterns. Notably, it contains neurons which are highly sensitive to movement, without being fussy about the direction of the movement. Moreover, their activity is much increased when the cells around them, representing adjacent bits of visual field, are reporting something different.

Close to the superior colliculus is a nucleus which communicates closely with it, and which similarly receives reports from more than one sensory channel.

Here the major neurons are linked by inhibitory neurons which exercise an unusually widespread influence. One sort of cell fires very strongly for a visual or auditory stimulus within its receptive field, as long as no other cells are excited by the same type of stimulus. The advent of similar stimuli elsewhere in the vicinity can cause it to turn off abruptly. Other cells are also much affected by competition from elsewhere, but their responses decline gradually as those in the other cells increase. Both types of neuron send their axons to the superior colliculi, using acetylcholine as their transmitter. The first type can be held responsible for identifying some of the 'pop-out' stimuli that tend to prompt eye-movements, and the second for measuring relative saliences more delicately.

The nucleus is known as the parabigeminal in mammals. It also receives the sort of inputs from the superior colliculus which indicate a looming visual stimulus, in other words something approaching fast enough to imply danger in one form or another. But much of the investigation of its role in salience has been carried out on the equivalent structure in owls, and it's not yet clear just how many varieties of mammalian 'bottom-up' attention can be attributed to it.

Also important in attention is the part of the parietal lobe where eye position is correlated with head and body orientations. Some neurons here give a quick burst of firing when there's any salient change in visual input, or when salient inputs occur in other senses. Others are active when a saccade is planned, and some of these, like those in superior colliculus and frontal eye field (areas with which there are strong links), continue to fire while the location of a vanished stimulus is being remembered. In the adjacent sector which relates visual input to arm position similar activity is found when a monkey has been trained to reach for an object at a signal. In these parietal areas the rate of firing evoked by a stimulus pattern can be influenced by the size of any reward associated with responding to the stimulus.

People who suffer a lesion in this area may cease to be aware of what happens on the side of the body contralateral to the lesion, and consequently ignore it. Asked to mark the midpoint of a horizontal line they put the mark well to one side of centre. They may read only one side of a page, eat only one half of a plate of food, perhaps even fail to get dressed on the neglected side. Any or all of the sensory modalities may be affected - vision, touch, audition, proprioception, smell. A stroke patient who suffers paralysis of the limbs on one side may ignore even that, and deny that there's a problem. The information that escapes attention does get registered at an unconscious level, though, for a stimulus-pattern presented to the affected side of the visual field can influence the interpretation put on one subsequently presented to the other side.

The definition of the neglected space is sometimes relative. Both stimulus patterns may on occasion be in the same half of the visual field. One half of an object may be faded out if it's seen on its own, but if it's perceived as part of a group it will be one side of the whole group which is lost. And just which stimuli are ignored may be affected by eye direction, head position and body posture. Moving the arm across the body may make it possible to perceive a tactile stimulus on it that otherwise goes unnoticed. Some stimuli, moreover, are less easily ignored than others. Faces, especially emotional faces, may attract attention to the normally ignored side when they are in competition with less interesting stimulus patterns. Furthermore, a stimulus on the neglected side may be noticed if there is no competing stimulus on the favoured side. It's also been found that attention may be guided towards the neglected side by a predictive signal.

This suggests that the problem lies as much in disengaging attention from the sensory input handled by the unaffected half of the brain as in directing it towards the opposite side of the scene. It looks as if left and right are in competition, linked by mutual inhibition, and when one side is damaged it not only works slowly but fails to exercise adequate inhibition on the other and maintain the normal balance of power. In an experiment with normal subjects transcranial magnetic stimulation was used to incapacitate the parietal lobe on one side and the detection of stimuli which appeared briefly in the periphery of the unaffected visual field actually improved.

The crux of the matter may be that it takes longer for a sensory event to build up into a conscious experience on the affected side of the brain, so information reported to the undamaged side can make its mark first and block off attention to what's happening on the damaged side. Patients asked to judge whether a visual event on the left preceded or followed one on the right took about half a second longer, on average, to become aware of stimuli reported to the lesioned hemisphere. But a cue in advance of the visual stimulus could speed things up. The function performed by this bit of neocortex seems to lie in shifting the attention to a new part of the environment. (Sustaining the focus on the target is attributed to another parietal area, a little higher up.) Thus if one side works slowly it presumably has little chance of catching up with the other, unless it gets some assistance. The effects of a lesion in actual neocortex seem to be less drastic than the effects of damage to the white matter underlying it - the axons which connect it to distant areas. It must take longer to set up a maintained circulation of neuronal messages when this link in the chain is missing.

Lesions in the right parietal lobe are often followed by neglect of sensory input on the left, in one or more modalities, but comparable lesions in the left hemisphere seldom lead to neglect of inputs on the right side. When they do it's likely to be in a subject who is left-handed. It seems that the right parietal usually plays a larger role in directing spatial attention than the left. This is in tune with the general principle that the right hemisphere, where the principle neurons tend to have longer dendrites than those on the left, spreading over wider areas, seems to concentrate on the larger scale patterns, the left on detail. Locating the interesting small-scale patterns in the first place, before they can be studied in detail, has to be the job of the large-scale analyser. Overall spatial attention is best dealt with in one place, it must be supposed, rather than divided between the two hemispheres.

Thalamic influences

Also important in visual attention is an area of the thalamus known as the pulvinar complex. Humans with lesions here, like those with lesions in the parietal attentional centre, have difficulty in directing their attention to targets in the contralateral visual field. It's particularly hard if the target is one among many stimulus patterns. PET scans of normal subjects show raised activity in the pulvinar when this sort of task is performed. And in monkeys the ability to perform discriminations of colour and form against a background of distractors is seriously reduced when the pulvinar is inactivated, although the same discriminations can be made efficiently without distractors. So clearly it's important to selective attention, and particularly to attention which is being directed by choice, by top-down channels rather than bottom-up ones. In line with this role, its cells fire strongly for a visual stimulus that is currently significant and requires either overt or covert attention, but the same stimulus provokes little response when it's not relevant. (There's a diagram of the thalamus on page 399.)

The pulvinar complex is a tantalising structure, in that it constitutes a very substantial sector of the human thalamus - much bigger than the lateral geniculate nucleus, the visual station intervening between retina and cortex, which is just a small bump on the side. It's only conspicuously large, however, in primates, and its size seems to correlate with the degree of complexity in visual cortex. The primate pulvinar can be divided into four sectors on the basis of cytoarchitectural appearance (the pattern of cell shapes and sizes).   Differences in the chemical signatures of the cells suggest that further subdivisions are possible.

Between them the sectors receive numerous afferents which look as if they provide all the information a centre for directing visual attention should need, with broader scale information predominating in some areas, finer scale in others. At least two sectors receive axons from the superior colliculus. And each major division connects with a different selection of cortical areas, some parts with primary visual cortex, others with more advanced regions of the visual processing pathways, both temporal and dorsal. There's a particularly heavy dialogue with the area concerned with visual motion, MT. Auditory and tactile stimuli are reported too, and contributions come from areas that deal in emotional matters.

In various primates the pulvinar has been shown to contain at least two and sometimes three maps of the visual field, as well as sectors which aren't topographically organised. Neurons in the former divisions have varied properties. Some are selective as to the orientation of a stimulus, some as to the direction of movement, and some are fussy about both orientation and movement. Some cells respond to the direction of a moving object, rather than just the direction of individual moving features, thanks no doubt to the input from MT. Others are selective about colour.

In the maps where the cortical input comes from the primary visual areas, the receptive fields of the cells are fairly small and the responses fairly rapid. The sector which communicates with 'association' areas of the cortex and which gets news about other senses as well as the visual, contains cells with larger receptive fields and slower responses. In an experiment with monkeys about 40% of the cells in this area showed an increase in activity just before the monkey produced attentional behaviour, such as making a saccade to the stimulus in the receptive field. Activity also occurred when the monkey was about to reach towards the stimulus in the receptive field but had been trained not to move its gaze.

It's tempting to think that the evolution of a particularly large pulvinar in primates might have something to do with the need to expend more careful consideration than most other species on stationary objects, rather than concentrating mainly on moving ones. Animals which swing or leap from branch to branch must pay attention to the thing to which they are about to entrust their weight. Fruit-eaters benefit from being able to make a rapid identification of a meal that is ready for eating, before a competing consumer does. Perhaps the pulvinar helps with this sort of thing. The idea receives some support from the fact that it is also relatively large in squirrels and tree shrews.

Certainly the pulvinar seems to be involved in the creation of the conscious visual experience, at least in primates. Melanie Wilke and her colleagues trained two monkeys to make a response to a spot of light, under circumstances which in humans would from time to time render it invisible. They found that neuronal activity in the lateral geniculate remained the same whether the monkeys reported seeing something or not, while that in the pulvinar was distinctly higher when a perception was reported. The pulvinar has quite a strong influence on the responses of neurons in primary visual cortex, V1, so a dialogue between the two probably contributes significantly to consciousness.

Focussing attention on one thing means reducing the amount of processing done on other inputs. Wrapped around much of the thalamus like a blanket is the thalamic reticular nucleus. This receives input both from the rest of the thalamus and from the cortex, while the output fibres run only into the underlying sector of the thalamus, carrying an inhibitory message. It serves to modify the strength of the excitatory signals sent from thalamus to neocortex. Neurons in the part adjacent to the lateral geniculate nucleus, for instance, are influenced by visual input, their receptive fields roughly similar to those of the parvocellular and magnocellular neurons. When the stimulus in the receptive field of one of the reticular neurons is the subject of attention it fires less strongly. This allows the lateral geniculate neurons to which it connects to fire more strongly, so that their messages to visual cortex become more potent.

Control over the reticular nucleus is exercised by prefrontal cortex – the forward part of the frontal lobes – which connects to it with inhibitory neurons. When attention must be paid to visual stimuli a message from the frontal lobe suppresses the inhibition in the reticular nucleus overlying the lateral geniculate nucleus. If it's an auditory signal that attracts attention the suppression is exercised above the thalamic sector that relays auditory information.

The reticular nucleus is also involved in switching attention to sudden new sensory inputs. In this case the inhibitory message that allows a thalamic input to the neocortex to be strengthened is likely to come from a subcortical source. This filter between thalamus and cortex must surely play a major part in that mysterious business of making one part of any current moment of sensory consciousness notably more vivid than the rest. The stuff which remains more or less present to awareness, but only as background, is presumably the information which has not been prioritised by a reduction of the reticular nucleus inhibition.

Forebrain Influences

How is attention guided when it's not captured by a stimulus which (thanks to the design of the neuronal system registering it) is inherently salient? The decisions are made in prefrontal cortex. This is where all the sensory pathways finally lead, including those that deal with rewards and with pain. The monitoring centres of the hypothalamus report here too, as do the parietal 'where' pathways. In short, all the information needed for decisions about attention and action is made available here, conveniently close to motor cortex.

As with the parietal cortex, damage to one prefrontal area can lead to a neglect syndrome, and again, lesions on the right produce this effect more commonly than those on the left. The right hemisphere is also more important than the left in sustaining a high level of attention. Subjects asked to monitor changes in a vibrating touch on a toe showed, in a PET scan, greater activation of right prefrontal and parietal cortex, regardless of the side of the sensory input. The same areas are activated when a change in the visual scene is detected, not when it slips by unnoticed. Patients with lesions of right prefrontal do badly on tasks requiring sustained attention, while those with lesions on the left don't show impairment. In patients whose corpus callosum has been cut, destroying most of the communications between the two cortical hemispheres, the right side of the brain does significantly better at such tasks than the left.

A likely interpretation of parietal neglect syndrome is that it's caused essentially by destruction of the connections linking prefrontal and parietal, the routes by which parietal cortex informs prefrontal of inherently salient inputs and prefrontal cortex indicates to parietal which bits of the current input most need attention. These connections, it can reasonably be hazarded, form a particularly important part of the circuitry around which neuronal impulses must be maintained in order to create a conscious experience.

It's not only forebrain pathways which can be recruited to the circuitry of attention by prefrontal cortex, however. Most notably those alerting centres in the lower brain which are directly activated by significant input from spinal pathways can also be prompted into action by messages from prefrontal. An interesting finding is that after lesions of the right hemisphere, but not the left, norepinephrine levels are reduced on both sides of the brain. This suggests that the locus coeruleus gets its major cortical input from right prefrontal. When that area is not in a position to recruit a norepinephrine input, voluntary attention is underpowered - though the bottom-up sort should be unaffected.

In the previous chapter we met the basal ganglia, and its inhibitory pathways to the thalamus. These too make a contribution to attention. A large proportion of the routes through the thalamus run to prefrontal cortex. As with the other circuits, there's a constant flow of inhibition from the basal ganglia to the thalamus, punctuated by brief selective releases from inhibition. This pathway is important both to prefrontal control of attention and to learning where attention is worthwhile. Dopamine wields a significant influence here, as in the motor circuits.

The many-faceted importance of prefrontal cortex to attention is demonstrated by the effects of lesions in some subdivisions. Damage to one area causes severe distractibility, so that many tasks become difficult or impossible. Patients with major lesions in another sector may be completely apathetic, simply not interested in paying attention to anything.

The area which controls voluntary attention and makes the decisions that feel voluntary rather than automatic, is of course an area of particular interest in the context of the evolution of reason. Prefrontal cortex will get a chapter to itself later on, so I will leave the subject there for now.

Another structure which contributes to selective attention is a centre in the lower part of the forebrain, to which axons from the nucleus gigantocellularis reach. Like other players in the attention game it sends out many long, dividing axonal fibres, and these communicate with all of the neocortex. Some deploy acetylcholine as transmitter, some GABA. Known as the nucleus basalis, or the basal nucleus of Meynert, it contains, in humans, around 200,000 neurons as against the 16,000 or so of the locus coeruleus. Particularly suggestive is the fact that each acetylcholine neuron innervates a relatively small area of neocortex, something between one and a half and three square millimetres, in contrast to dopamine, norepinephrine and serotonin axons, which spread out over much wider fields.

Another thought-provoking aspect of the nucleus basalis is that it's especially large and complex in *Homo sapiens*, to a degree rivalled only in cetaceans. Lower centres, in contrast, seem to have changed little over a long stretch of evolutionary time; they are fairly comparable in size (proportionate to the rest of the brain) across a wide range of species. Human neocortex, moreover, is more densely innervated with acetylcholine fibres than that of any other species. It seems a reasonable deduction, therefore, that the expanded nucleus basalis is a significant component of what makes us human. One might hazard the guess, for instance, that it's important to our ability to use neocortex, and consciousness, for a wider range of purposes than other species.

When attention is being deployed in a demanding task, such as picking out a stimulus from a flow of distractors, the level of acetylcholine in the neocortex is higher than when there's no distraction. Depleting the supply of the transmitter leads to deficiencies both in selective attention and in sustaining concentration. A drug which increases the amount of acetylcholine available, on the other hand, enhances perceptual learning. And artificial stimulation of the acetylcholine fibres has been shown (in mice) to improve the performance on a visual discrimination task. The effect is achieved through modulatory synapses.

This acetylcholine system is one of the brain components conspicuously damaged in Alzheimer's disease, with the degree of damage paralleling the degree of dementia. Deficits in attention are detectable before the worst effects develop. Degenerative changes here are also found in other conditions which produce some degree of dementia or confusion. A final intriguing consideration is that the nucleus basalis is slow to mature, and becomes fully functional only in adulthood. Unfortunately its extreme development in humans means that only limited illumination can be obtained by studying monkeys, let alone rats, so it remains in many ways frustratingly mysterious. However, one or two interesting facts are emerging, as we shall see later.

The centre receives inputs from many areas, among them the locus coeruleus, the dopamine system and the raphe nuclei, and it responds with increased firing to any novel, arousing, rewarding or stressful occurrence. There's also a great deal of input from the neocortex. Nucleus basalis neurons can be conditioned to fire for a stimulus that signals a rewarding or an undesirable event, and some have been observed that change their pattern of firing during the conditioning process, and again if the reinforcement ceases to materialise. Some vary their response to a food stimulus according to the desirability of the food and the current degree of hunger.

The effects of acetylcholine on the neocortex vary, for once again there are several different types of receptors - at least one ionotropic, or current-initiating, and five metabotropic, or modulatory. The effect of acetylcholine is inhibitory in cortical layer 4, but facilitatory elsewhere. It's accepted, however, that the nucleus basalis contribution provides yet another way of making incoming messages clearer. It also allows a maintained sensitivity to continuing input and thus, where the input is important, helps to counteract the habituation of neuronal responses that might otherwise occur.

A clue that encourages attributing major importance to the system is that acetylcholine activity in the neocortex correlates with the degree of desynchronisation of the electroencephalogram record which occurs when the brain's neurons cease the gentle, more co-ordinated pulsing that characterises idleness, and get to work. A further bit of evidence is provided by the second of the distinctive wiggles that appears in the electroencephalograph trace after a new sensory input, and which is thought to correlate with conscious attention. If the acetylcholine receptors are blocked, by means of a drug, this wiggle takes longer to appear and is reduced in size, or it may disappear altogether. Such evidence provides grounds for suspecting that this acetylcholine system plays a significant role in creating full-strength consciousness.

As with norepinephrine, too much acetylcholine produces a sharp rise in responses to 'wrong' stimuli in a discrimination task. It seems that efficient perception may depend on having the proper amount of both transmitters, not too much and not too little. This ties in with the effects of various drugs in humans. Drugs that block the cortical acetylcholine receptors can cause hallucinations.

Adjacent to the nucleus basalis are several other structures which also exercise an extensive influence by means of spreading acetylcholine fibres - the whole collection is referred to as the basal forebrain. Whereas the nucleus basalis axons cover the neocortex the other structures project to other parts of the brain. Destruction of the whole has worse effects, as one might expect, than the loss of the nucleus basalis alone.

It don't mean a thing if it ain't got that swing

Finally we must consider not a structure but a process. In the major neurons of the forebrain, with their myriads of input synapses linking them into complex networks, the intracellular membrane potential is typically in a constant state of flux, as excitatory inputs arrive that occasionally add up to an action potential, and the pumps that remove the positively charged ions work to restore the negative charge again. Just how the two effects balance out depends on the precise selection of gates and pumps the neuron possesses, and what metabotropic influences are at work. In many cases, and most notably in the principal neurons of the neocortex, the membrane potential fluctuates continuously, causing a low level of spontaneous firing, readily distinguishable from the volley of action potentials brought about by meaningful input. Both the subthreshold fluctuations and the random firing are co-ordinated over quite wide areas, producing the 'brainwaves' that can be detected from outside the skull by electroencephalography. The effect seems to be intrinsic to neuronal networks, at least of certain sorts of cells, for it's been found that a cluster of as few as forty interconnected neurons, cultured in isolation, produces synchronised oscillations.

Part of the explanation has to be that a neuron producing a spontaneous action potential influences all the neurons to which it connects, which in turn may influence those to which they connect, and so on. However, there is also an arrangement which can intensify the synchronisation. Many varied types of inhibitory interneuron weave among the excitatory neurons, synapsing variously on dendrites, cell body or axons. One type wraps around the cell bodies of the principal neurons, where inhibition can have the most powerful effect, encasing them in a way that has inspired the name basket cells. These get their input from the principal neurons, which means that every time an excitatory cell fires it prompts an inhibitory return. But the basket cells are also extensively interconnected with each other, both by excitatory and inhibitory synapses. The excitatory synapses are electrical, or gap junctions, through which the current flows through to the next neuron without pause, so the excitatory message spreads very fast through all the interconnected basket cells, causing a GABA input to many neighbouring principal neurons. The effect is terminated when the inhibitory message transmitted through the slightly slower chemical synapses, catches up. Consequently neighbouring basket cells tend to fire together, imposing a pulsed inhibition on the local patch of excitatory neurons and accentuating their rhythm.

Different brain rhythms accompany different modes of behaviour. Fast rhythms go with active exploration, less fast ones with occupations like eating and grooming, slower ones with relaxation, the slowest with quiet sleep. Several subcortical brain areas have distinctive rates of oscillation, which are thought to influence the neocortical rhythm. In many areas two or more rhythms may co-exist, the faster a multiple of the slower and nested, as it were, inside it. The slower rhythms tend to be more widespread, while the faster ones are necessarily quite local, since the time to conduct neuronal impulses over longer distances puts limits on how far the co-ordination can extend. However, the slower, widespread rhythms can modulate the evolution of the more local, faster ones.

Input that arrives at dendritic synapses when the neuron is on the upswing of an oscillation is more likely to trigger an impulse than the same input arriving when the negative potential is increasing. This suggests that co-ordinated oscillations are very helpful to a large neuron which receives numerous inputs. It takes excitatory inputs at hundreds or even thousands of synapses to fire such a cell, and in many cases the inputs all have to arrive within a very small temporal window to have their full effect. If all the cells in the neighbourhood are increasing their sensitivity at much the same time it must improve the chances that the outputs of each layer of processing will be passed on to the next stage in unison, so that a sensory stimulus pattern will be analysed with maximum efficiency. Investigations have shown that synchronisation of oscillations does indeed occur where it might be expected to be helpful - over several columns of primary visual cortex, for instance, which means that information about orientations in the bit of retinal field represented there can be analysed to best effect.

It has also been demonstrated that such synchronisation improves perceptual efficiency. Monkeys trained to detect a change in one of two visual stimuli performed best when there was strong synchronisation in the appropriate V4 receptive field both before and after the change, and responded more slowly when the synchronisation was higher in the receptive field covering the other stimulus. This might mean that the change was detected faster because the animal was paying attention to the correct stimulus at the moment when the change occurred, and another experiment supports the idea that synchrony correlates with attention. Three monkeys were trained to press a key when one of three white squares dimmed slightly, and to perform a tactile discrimination. Then they were trained to switch between the two tasks, performing the one indicated by a cue, with both visual and tactile stimuli present. Recordings made from pairs of neurons in secondary somatosensory cortex showed that a larger proportion of them fired in synchrony when attention was being paid to the tactile input. Yet other experiments show that synchronisation can continue when a stimulus has to be remembered, or occur in advance when a stimulus is expected.

Another experiment implied that synchrony correlates with conscious experience. Cats were shown leftward movement in one eye, rightward movement in the other, and rewarded for reporting what they saw. A higher level of synchrony in direction-sensitive neurons in MT (the bit of temporal cortex concerned with visual movement) corresponded with the direction reported. Other dichoptic experiments also indicate that it's the co-ordination of neuronal firing, rather than the rate, which provides the best clue to conscious experience. This ties in neatly with the fact that the synchronisation can last for a significant fraction of a second, a period roughly comparable to the time spent on a visual fixation, or the time taken to create a conscious experience.

Basket cells exert their influence over a fairly small area, and for most of the time the rhythms recorded in different neocortical areas are unco-ordinated. But two quite widely separated bits of neocortex can become synchronised when they are dealing with different aspects of the same stimulus pattern.

Synchrony between visual, parietal and motor areas was found in cats performing a visuomotor co-ordination task, synchrony which was most precise when the cat was expecting the appearance of the stimulus, or attending to the cue which indicated which movement was to be made. In another study cats which had learnt a *go/no go* task involving visual stimuli showed synchrony between visual and parietal cortex, and it was more intense on the *go* trials than the *no go* ones. In humans, meanwhile, the electroencephalograph recorded wider-reaching synchronisation when an ambiguous drawing was perceived as a face than when it was seen as a meaningless shape.

It's been suggested that synchronisation may underpin the answer to the neurophilosophers' 'binding problem' - the interesting question of how several disparate bits of sensory information get linked together to form one perception. At the input level some fine temporal co-ordination seems to be important for efficient analysis of a stimulus pattern through all the various pathways of the visual cortex. Human subjects were less successful at discovering a line of similarly oriented bars hidden in a texture when the elements of the display were presented in a temporally staggered way, though the temporal difference was so small as to be undetectable. Continuing synchronisation also seems to be required. The experiment showing that a shape isn't always successfully linked to its colour if the exposure of the stimulus patterns is very brief probably indicates that it takes time to establish synchronisation across distant areas.

Moreover it's notable that we don't learn to associate stimuli in two different sensory channels unless we are paying at least a little attention to them. We only notice, for instance, that a bird's beak is moving in the same temporal pattern as one particular chain of sounds if we have singled out both visual and auditory stimulus patterns from the background. It seems likely that an accurate assessment of the temporal patterns in the two channels depends on synchronising the activity of the two areas. And if it's correct to deduce from the experiment with cats described above that long-range synchronisation equates with conscious experience this provides support for the idea that the answer to the binding problem lies in sensory consciousness. Further evidence comes from a study of anaesthesia in humans. As one of the most widely used anaesthetics takes hold co-ordination between different areas breaks down, and networks become fragmented.

But how might distant areas of neocortex become synchronised? One obvious suspect is the cortex/basal ganglia/thalamus circuitry. The constant, regular pulses of inhibition and release from inhibition imposed on the circulating information would seem to provide a means of transmitting the same rhythm to all the areas involved in a currently active circuit. Some of the GABA projections from the basal forebrain have also been implicated. And in visual cortex there seems to be a contribution from the pulvinar, for when it is deactivated the oscillations are less co-ordinated.

Synchronisation is proving to be important in the motor circuits of the basal ganglia, and in their co-operation with the cerebellum. But desynchronisation is important too, and it may be that producing a planned action involves cutting a smaller circuit out of the universal one by giving it a different rhythm. Synchronicity seems to ebb and flow a little as an action is performed, reaching its peaks at the point when the cue is received and when the movement is actually triggered. Dopamine seems to be important in achieving the co-ordinated rhythms, and it looks as if disruption of the normal patterns of synchronicity may be one of the factors at the root of Parkinson's disease, and of the various other motor disorders caused by abnormal dopamine influence in the basal ganglia.

Consciousness, the theory goes, is produced by a maintained circulation of impulses around neuronal circuits activated by the current input. Attention can be supposed to equate with the synchronisation of the rhythms in these circuits, which allows them all to be linked into one interconnected network. The co-ordinated firing means that at each stage of a circuit the excitations arrive in unison, and can thus have the most powerful possible effect. It can reasonably be deduced that the time taken to build up the synchronisation, and achieve this degree of efficiency, is the time taken to create the conscious experience.

The wide scope of attention's influence

As we've seen, sensory input is relayed forward through the neocortex, along complex pathways which sport reverse connections at every stage. Only if the impulses reach prefrontal cortex and prompt return messages, completing the biggest of the potential circuits, does a conscious sensory experience occur.

The top-down connections from prefrontal cortex explain how it is that neurons all along a sensory pathway fire more strongly when attention is paid to the stimulus they're registering. When the task a monkey must perform to gain a reward involves directing the attention elsewhere the same stimulus in the same place gets a lower response than when it guides the current action. The effect of attention is greater in more forward reaches of the pathways, but some magnification of the response is found in the earliest stage of the cortical visual pathways, V1, and even in the retina.

Imaging experiments with humans tell the same story. When a subject is shown a display of moving, coloured stimuli and asked to attend to the colours, the brain area particularly concerned with colour is more active than when the task involves discriminating between movements. When movement is the focus the relevant area of middle temporal cortex becomes the more active site.

Anticipatory attention can have the same effect. If the subject knows where an upcoming visual stimulus will appear, a higher level of primary visual cortex activity is found in the quadrant of the visual field where it's expected than in the other three quadrants. More intriguingly still, the same sort of heightened activity occurs in primary visual cortex when the subject is asked to visualise something. Neurons in the visual pathway can fire at significant rates when a stimulus pattern is merely being imagined. These excitations must be the result of inputs from more forward sectors of the brain. They represent attention that is turned inward, as in thinking and planning.

Attention not only makes neurons fire more strongly, it also makes them fire more selectively, as an imaging experiment with humans has shown. The sharpening of the response reflects increased activity in the inhibitory neurons which mediate competition among the big excitatory cells. An increasing amount of evidence indicates that this is achieved through topdown connections to the inhibitory neurons.

In vision this must be how attention increases the sensitivity to contrasts, making a very faint contrast more visible. The number of primary neurons in the synchronised circuit is restricted by intensified GABA activity, under the influence of prefrontal cortex. When one listens to a single voice in a crowd the buzz of other conversations which forms the background must be the effect of less constricted firing in less efficiently synchronised circuits. And some sensory input doesn't get incorporated into a sustained circuit at all, and remains completely unconscious, but may nevertheless influence the interpretation of something that is consciously experienced.

Recapitulation

Centres low in the brain influence the general degree of alertness, among them the locus coeruleus, using norepinephrine in its widespread axons, branching throughout the brain.

Another important contributor is acetylcholine, from the pedunculopontine and laterodorsal nuclei a little further up.

There are descending pathways from these two nuclei, and from a centre adjacent to the locus coeruleus, which influence the motor systems of the spinal cord, pathways which probably predate the ascending ones.

Neurons in the centres involved in the control of eye movements, head turns, and other forms of behavioural attention are active during cerebral or covert attention.

A thalamic area called the pulvinar complex, particularly well developed in primates, plays an important part in selective attention.

So does the reticular nucleus, wrapped around much of the thalamus, filtering its exchanges with the neocortex.

Also important is the nucleus basalis in the forebrain, with acetylcholine neurons that run to relatively small patches of neocortex. Its exceptionally large size in humans suggests a significant relationship to the size and the extended functions of human neocortex.

The overseer which directs voluntary attention is the prefrontal cortex, the end-point of all the sensory pathways.

Return connections cover the whole of the brain, and their top-down messages, targeted to whatever pathways currently seem important, increase the efficiency of processing there. Sharpening local inhibition is part of the effect.

In most neurons of the brain the membrane potential is constantly oscillating, and the oscillations tend to be synchronised over small areas, which must make the processing of sensory inputs more efficient.

Synchronisation of the slower oscillations can unite fairly distant brain regions when they are all dealing with the same events. This synchronisation seems to coincide with attention, and with the focus of sensory consciousness.

Chapter 17

The Neurology of Reward

An early discovery in the exploration of the brain was that tiny pulses of electrical current delivered to certain precise spots could trigger various sorts of behaviour. Depending on just which bit of the brain was targeted the animal might be prompted to start eating, drinking, or exhibiting sexual behaviour, would suddenly bristle with aggression or cower fearfully, or maybe would lie back and relax, and even go to sleep. In 1953 a neurologist called James Olds was engaged in mapping such areas when one of his rats reacted in an untypical way. Instead of showing one of the expected behaviours it sniffed around the corner of the enclosure where it happened to be when it received the microstimulation, and repeatedly returned to that corner, as if hoping the same thing would happen again. Olds tried giving it a pulse of electricity every time it went there, and it kept going back for more. Later the same rat learned to run a T maze without error, purely for this reward when it reached the goal. Excitation in a certain brain area, it seemed, could produce a strong attraction to the place where it occurred.

The logical next step was to see whether rats would learn to press a pedal for the same sort of reward. It turned out they would sometimes work harder for it than for food or water. A number of sites were discovered where artificial stimulation produced this result. With the electrode in one site a rat might press the pedal ceaselessly, at a rate of more than one press per second, as long as the stimulation kept coming. The rats seemed to be prepared to starve rather than desert the pedal. Self-stimulation in other areas would be more leisurely, and accompanied by low levels of other activity.

A number of sites were also identified where artificial stimulation had an aversive effect. An animal would press the pedal a few times, or perhaps only once, and then keep well away from it. Or if stimulation in one of these areas could be terminated by pressing a pedal, then that trick would be learnt.

The areas which can evoke either dedicated pursuit of the artificial stimulation or firm avoidance are small and clearly delineated. If the micro-electrode is placed in any other part of the brain the rat shows neither aversion nor addiction to pedal-pressing, but merely tries it now and then, in the course of its explorations. Moreover the rate of pressing remains the same regardless of whether current is delivered or not.

H. J. Campbell, who mostly used rabbits in his experiments, described the typical reaction of an animal pressing the lever for the first time and receiving its first experience of artificial stimulation in a 'reinforcement' area: ...... *always the animal does what might be called a double-take. It becomes instantly more attentive to its surroundings, moves about quickly in various directions, sniffs, licks and touches everything again.* Such behaviour continued until the animal established the connection between pedal-pushing and stimulation, but most of his rabbits learnt the trick within three sessions. They then started to come forward as soon as they saw the apparatus and bend their heads down of their own accord while the plug was inserted into the socket bolted to the skull. Over the next two to four days the pressing rate *rises quite rapidly, while the animal discovers how to press the lever with the greatest economy of movement.* After that it settled down at a steady level.

A willingness to work for a reward of artificial brain stimulation has been demonstrated in all sorts of species, from monkeys to pigeons to goldfish. And the sites for both reinforcement and aversion have been mapped out in detail. Many are in the hypothalamus, the structure that monitors internal conditions and determines when essential behaviours such as eating, drinking and copulation should be pursued. Self-stimulation here must mimic the messages that are received when the animal actually does one of these things. For instance, there is an area in which neurons are activated when the animal has food in its mouth, close to the sites which prompt eating to begin or to stop. When the stimulating electrode is here the intensity with which the animal depresses the pedal depends on such factors as blood-sugar levels, the fullness or otherwise of the stomach, and body-weight. Presumably a hungry animal receiving stimulation here doesn't stop to eat because it already seems to be achieving the initial effect of eating – without, of course, reducing the level of hunger.

Another reward centre is similarly related to drinking. A rat with an electrode in both centres and a choice of pedals presses the one that relates to its current need. A third reward centre is clearly concerned with sexual matters, since the degree of self-stimulation that is applied there can be reduced or increased by manipulation of the sex hormones.

The hypothalamus is by no means the only place where self-stimulation can be obtained. There's a strong appetite for stimulation in the cable of dopamine axons which ascends from the midbrain to spread its fibres all over the neocortex. The cell bodies are found in an area known as the ventral tegmentum, close to the other major dopamine-deploying centre, the substantia nigra pars reticulata, which distributes its influence mainly in the basal ganglia. With the stimulating electrode nestled amidst the bunched axons rising from the ventral tegmental area the self-stimulation trick is learnt very quickly and the rate of pedal-pressing is high, up to two hundred per minute. In between pressing sessions there is energetic exploration of the environment.

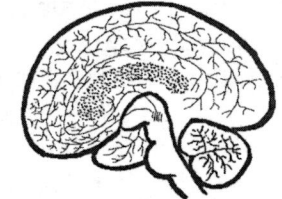

The connection between activity in these various areas and reinforcement is confirmed by the fact that self-stimulation works well as a reward in conventional learning tasks. If an animal is allowed to give itself a pulse of brain stimulation after each correct response, learning proceeds apace. The same stimulation adminstered by the experimenter, though it does work to some degree, is not so effective, while continuous or random stimulation of a reward centre interferes with learning. The temporal relationship between action and reward must be fairly precise for learning to occur.

Artificial brain stimulation has been tried out on human subjects, with the co-operation of people undergoing brain surgery. The results haven't been as enlightening as might have been hoped. In many cases various sorts of pleasure have been reported; sometimes there's just a general sense of well-being, sometimes sexual pleasure is experienced, and sometimes the stimulation provokes laughter. But the reports were often confused, particularly when the stimulation site was in the hypothalamus. Given the chance to administer the micro-current themselves some patients would operate the switch with an intensity not unlike that seen in some animal experiments, but when asked why they were doing it seemed vague. One man claimed he was doing it because the doctor wanted him to - but since the enthusiastic rate of pressing had only started after the current was switched on the explanation was less than convincing.

One possibility, of course, is that some subjects were embarrassed to admit that they had been obtaining sensations of sexual pleasure. Another is that some sorts of feedback or reinforcement may operate to promote a behaviour without being accompanied by a conscious sensation of pleasure. The ability to experience conscious pleasure must have evolved on the foundations provided by the machinery which turns sensory inputs into reinforcements, but the ability to learn to repeat an action is far from carrying any implications about consciousness.

Dopamine

The fact that self-stimulation is readily learnt in the ascending cable of dopamine fibres is particularly interesting because this system is also heavily implicated in drug addiction. All addictive drugs interact with components of the dopamine pathway in one way or another. Amphetamine increases the amount of dopamine released and inhibits production of an enzyme involved in the recycling process. Cocaine blocks the transporter mechanism which normally grabs the neurotransmitter from the synaptic gap and returns it to the interior of the cell for re-use, so that the dopamine is left floating around in the synaptic gap for longer than usual, and can continue to trigger the post-synaptic receptors. (These drugs also affect the motor system, making users somewhat hyperactive, so they are grouped under the general term psychostimulants.) Alcohol, nicotine, opiates and marijuana all have an indirect influence, modulating the rate of firing in excitatory, inhibitory or neuromodulatory neurons that influence the release of dopamine. In short, addictive drugs provide a chemical form of artificial stimulation which can fool the brain in much the same way as electrical self-stimulation. The compulsive behaviour of the drug addict or the alcoholic parallels the dedicated pedal-pressing of the self-stimulating rat.

Much of the power of addictive drugs lies in hi-jacking the machinery of conditioned responses. People trying to kick the habit find that the urge to relapse grows very strong if they revisit the place where they used to buy their drug, or see someone else in the act of taking it. Dopamine seems to be particularly involved in the conditioning process. Destroying the dopamine neurons largely prevents an animal from acquiring a taste for alcohol, but doesn't affect consumption in one that has already acquired the habit.

The picture is far from simple, however. Dopamine is by no means the only neurotransmitter affected by addictive drugs. Furthermore, addiction has several facets - the 'high' when the drug is taken, the compulsion to keep taking it, and the physical withdrawal symptoms which make abstinence miserable. Not all drugs produce all the effects. Some drugs continue to provide a pleasurable high, others cease to. With cocaine, for instance, the dopamine response decreases as consumption increases. But the compulsion to take an addictive drug doesn't necessarily fade when it no longer provides pleasure.

A complication is that people vary considerably in their susceptibility to addiction. Estimates of the proportion who are particularly vulnerable range between 10% and 25%. Similarly, only a modest percentage of rats develop a drug habit of their own volition. Many prefer saccharine to cocaine.

The genetic variations which lie behind this varying susceptibility are increasingly being uncovered, but experience and emotional state also play a part; unhappy or bored people are more likely to become drug-dependent. The same story emerges from animal experiments. Monkeys housed in individual cages readily learnt to obtain a shot of cocaine. When they were re-housed in groups of four and a social hierarchy emerged, the subordinate animals continued to self-administer cocaine but the dominant animals lost interest. (Whether dominance led to the lack of interest or the characteristics which made the animal drug-resistant also made it more likely to achieve leadership is an interesting question.) Mice housed in standard laboratory cages, with no entertainment, form a preference for a location in which they sometimes receive a shot of morphine, but mice in an 'enriched' environment generally don't. Moreover mice which had developed addiction-like behaviours no longer showed them after being housed for thirty days in a drug-free but enriched environment. However, different strains of both mice and rats vary in their susceptibility to addictive drugs.

The capacity for forming habits, though it's put to unproductive use by addictive drugs, is clearly valuable, particularly for species which can exercise some degree of choice about their actions. If eating, and the business of looking for food, are not entirely dictated by the genes they need to become a habit, especially for mammals, which need a regular supply of calories to maintain their body temperature. There's the added complication that as they grow up they have to abandon the method of feeding determined by an inbuilt stimulus-response mechanism, and switch to a completely different and more demanding one. Weanling rats, it's been shown, have to learn to look for water when they need to quench their thirst. It seems likely that in mammals (not to mention altricial birds) the adult action pattern for obtaining food is genetically determined, but the sight of mother performing the action helps to release it, and the resulting reward is needed to encourage regular performance. In those species whose diet is not widely and continuously available this mechanism may be particularly important.

There are probably several stages to learning about food. By the time a mammal is weaned it knows about the feel of nourishment in the mouth, and the rewarding effect on the internal senses which monitor stomach fullness and blood sugar levels. When it starts experimenting with the adult style of eating it may have to learn the appearances and odours that go with desirable food. What has been most studied, however, is dopamine's role in learning to recognise cues which predict the availability of food, something which may be the last stage of a mammal's education in nourishing itself.

Dopamine also contributes to attention, behaving in some ways like the norepinephrine system. Dopamine neurons pulse away slowly when nothing much is going on, but produce a burst of firing for any salient or interesting new stimulus, with a particularly strong response for the important stimuli that attract behavioural attention, including of course the stimulus patterns which represent potential rewards.

However, when a stimulus of no inherent interest is regularly followed by something rewarding the dopamine response gradually transfers from the desirable sensory event to the signal that precedes it. This happens by way of an increase in the number of cells responding to the signal, accompanied by a reduction in those responding to the unconditioned or hardwired stimulus. A significant dopamine response to something inherently rewarding, such as food, occurs only when its advent is unexpected. Moreover, if an expected desirable fails to materialise there is a detectable pause in the normal, spontaneous rate of dopamine firing. Dopamine plays an important role in learning, and neuroscientists like to describe the burst of firing as a prediction signal and the inhibition of the normal, tonic firing as an error signal, indicating that some more learning is needed.

The prediction signal is graded to indicate the size of the expected reward. When monkeys learnt to recognise five different stimulus patterns as indicating different amounts of fruit juice, dopamine neurons made no response to the cue when no juice was anticipated, and produced an increasingly widespread response for increasingly large expecations.

The dopamine flush can also reflect the probability of a reward. When a conditioned stimulus indicated that delivery of juice was certain there was a burst of firing for the predictive stimulus and none for the juice. If the odds were in favour of a reward materialising, though it wasn't certain, there was a greater response to the predictive stimulus than to the juice; if they were against it the balance was the other way around. And if the indicated chances were fifty-fifty the response was split pretty evenly between the two events. When it was uncertain whether juice would materialise at all there was a sustained but modest dopamine response to the cue, which lasted until the point at which the reward arrived, or could have been expected to arrive.

The dopamine system also acknowledges advance information about the size of a reward. Given the opportunity to obtain a forecast as to whether a reward would be large or small by making a saccade to the correct stimulus, monkeys showed a strong inclination to earn this information. And the dopamine neurons responded to the cue indicating that the information was available in the same way that they responded to the cue signalling the physical reward.

Dopamine neurons are particularly busy when an animal is still learning, when it hasn't yet got the hang of a situation, and doesn't quite know when a reward can be expected, so that attention must be paid to everything. Once it has become clear what must be done to obtain a reward, and attention can be concentrated on the relevant factors, the dopamine input too becomes more focussed. So the system must be particularly important to discovering just what correlations can be made between actions and reinforcements.

In line with this, blocking dopamine receptors prevents the learning of new associations and new responses, but doesn't immediately interfere with established habits, though they may fade eventually. Dopamine's part in learning clearly involves linking the reinforcement to the action which obtains it. Dosing healthy subjects with a dopamine antagonist (which blocks the dopamine receptors) reduces their ability to choose the most rewarding option in an experimental task, while dosing them with the precursor from which dopamine is made increases it.

The neuromodulator is also engaged when the prize is being pursued. It's released as a rat makes its way to a distant reward, in a quantity that relates to the expected size of the reward. In addition, the amount generally increases as the rat draws nearer to the goal.

If dopamine activity can be correlated with conscious pleasures it looks as if it must be with the pleasures of anticipation, the prospect rather than the actual achievement of material reward. It goes with finding a restaurant, and the promise held out by an interesting menu and delicious smells of cooking, rather than the actual taste of the food or the comfortable feeling of repleteness after eating. Its influence may be important when a reward can only be obtained with some effort, or when an extra dollop of attention is required. Dopamine levels were significantly increased in rats which had to press a lever ten times for a dose of brain stimulation, compared to animals which had to make only one lever press.

There are high dopamine levels during investigatory activity, which makes it tempting to credit the neuromodulator with the pleasures of exploration and play, and the satisfactions of developing new circuits and creating new schemata. It can be deduced that the pleasure obtained from having something to pay attention to, and making full use of a sophisticated sensory system, is attributable to dopamine.

In short, this neurotransmitter seems likely to lie at the root of cognitive emotions, and the capacity for making a habit of activities that don't bring any immediate physical reward - the pleasure an infant derives from turning a confused mass of visual input into a coherent perception, or from mastering the art of grasping an object. It seems more than likely that it powers the elation a scientist feels at the creation of a new and powerful theory, or the artist at the moment of inspiration. These can be powerful pleasures which at their strongest seem to compare with what the druggers call a high.

The fact that rats with depleted dopamine levels are completely without curiosity supports the idea. They don't explore a new environment, and may even settle down to eat without first checking out unfamiliar surroundings. They are also unlike normal rats in being very reluctant to try new foods. Even normal rats vary quite a bit in their reaction to a new environment, however. Some explore much more energetically than others.

The ventral tegmental area innervates the input regions of the basal ganglia pathways very heavily, especially the nucleus accumbens. Other axons, after travelling upward in the medial forebrain bundle, spread out to form a fine dense network throughout the neocortex, with an especially lavish distribution in the frontal lobe. Single neurons in the network may have half a million to a million synapses on cortical neurons, so their influence is widespread. Some connections, meanwhile, go to subcortical structures implicated in emotion, reward and attention.

Substantia nigra pars reticulata and ventral tegmental area were once seen as firmly distinguished from each other, both by the more extensive spread of the latter's axons and by their functions, the former concerned with movement and the latter with the rewards to be obtained by movement. They are coming to be seen more as a continuum, one area blending gradually into the next, with some central overlap both of distribution and function. Some authorities like to define the central area as a third entity, the retrorubral. A recent study indicates that in the developing embryo dopamine neurons emerge in three stages, with slightly different chemical profiles. The differences may account for the fact that substantia nigra neurons are more vulnerable to Parkinson's Disease than those of the ventral tegmental area.

The dopamine neurons are in fact proving to constitute a fairly heterogeneous collection. They can be subdivided into several groups, which get their inputs from different sources, project to different targets, and have different firing patterns. A modest proportion of them, moreover, are activated by aversive inputs rather than desirable ones. But then aversive events are just as attention-demanding as rewarding ones, and sometimes, if they are not too extreme and are alleviating boredom, can be not unpleasurable.

The effects of dopamine on the cells it's delivered to are not straightforward. As we saw in the chapter on the neurology of movement this neuromodulator can be either facilitatory or inhibitory. There are at least five varieties of dopamine receptor, each with a distinctive pattern of distribution in the brain. All of them are the metabotropic sort, modifying the internal state of the cell rather than leading to clearcut messages. D_2, D_3 and D_4 types have been regarded as facilitatory, while D_5 receptors can be classed with the D_1 sort, which have been thought to be inhibitory. But there's evidence that in some cases the effect may depend on the state of the receiving neuron. Moreover in some areas dopamine doesn't operate in the classical way but is merely released into the extracellular fluid to reach receptors not contained in synapses.

Serotonin

Dopamine seems to be connected with the more exciting forms of pleasure, and with the actions performed to acquire desirables. For the more placid satisfactions of actual consumption or enjoyment of a reward there's a likely candidate in another neurotransmitter/modulator, serotonin, or 5-hydroxytryptamine (5HT). This too is distributed by a fine network of fibres that spreads through most of the brain, including the thalamus and the cerebellum. It originates in a clutch of nine nuclei low in the brainstem, in the reticular formation. Most of the forebrain innervation comes from the two largest, the dorsal and median raphe nuclei.

Let's face up to the complexity first with this one. There are at least fourteen types of serotonin receptor, of which one opens an ion channel while the rest are metabotropic. In addition there are two distinct sets of serotonin fibres innervating the neocortex, with different patterns of distribution. This neuromodulator thus wields an exceptionally large variety of influences.

A major function seems to lie in modifying various sorts of brain activity to fit the current needs of the organism. An early discovery was that it's released into the brainstem as an animal grows sleepy, and plays a part in shutting down the waking operations of the brain. It's also released after a meal, which may account for the sleepy feeling that a large meal produces. Drugs which increase serotonin production or its release suppress food intake, and those which reduce the amount of serotonin available or block serotonin receptors prompt an increase in eating. This effect depends on the $5HT_{2C}$ receptor, for mice which lack it, or have certain mutations in it, eat more than normal animals, and grow unnaturally heavy. They also suffer epileptic fits, which suggests that this receptor influences the general excitability of the brain - not unreasonable, since a lively brain is needed for finding food but an animal with a full stomach can generally afford to relax.

As this implies, serotonin has significant effects on mood, and variations in its functioning are associated with various mental health disorders. Depression is associated with a reduced number of $5HT_{1B}$ receptors at the cell surface, and one anti-depressant medication works by increasing production of the protein which anchors the receptor in place. Alcohol also has a strong influence on mood, of course. Mysteriously, mice in which the $5HT_{1B}$ receptor is absent are endowed with a strong resistance to alcohol. They are prepared to consume, on average, twice as much as those without the mutation, and after a matched intake are steadier on their feet. Both mice without this receptor and those with a mutation in the gene for it are also particularly susceptible to cocaine addiction. Anxiety and depression in humans, meanwhile, frequently seem to correlate with a particular variation in the promoter region of the $5HT_{1A}$ receptor gene.

The significance of serotonin in mood is underlined by the fact that Prozac, the drug that relieves depression and stress in most of the people who take it, is a re-uptake inhibitor. That is, it blocks the recycling system and leaves more transmitter floating around in the synaptic gap, just as cocaine does with dopamine. However, the effects are not totally consistent - which is perhaps not surprising since serotonin fills so many varied roles. Rats may become bolder under the influence of Prozac, and less inclined to knuckle under to a dominant animal. In just a few humans the effect goes further and they are made aggressive by the medication, rather than happier.

The neuromodulator also influences sexual preference, and it has an important role in maternal behaviour, for female mice which lacked serotonin in the central nervous system made very poor mothers, taking such inadequate care of their pups that many died. Some serotonin neurons, on the other hand, promote patience in waiting for a reward.

Other serotonin fibres descend into the spinal cord. Some influence motor networks, others connect to sensory pathways and have a suppressive effect on messages being relayed from nociceptors - the messages that lead to the perception of pain. In addition, serotonin plays a part in the control of blood pressure, breathing and temperature, and it's involved in triggering nausea and vomiting. Just how far the serotonin influence on mood is due to direct effects on brain neurons and how much to the indirect effects of feedback about what it does in the body remains an open question. Discovering the precise correlation between ascending and descending messages should be usefully illuminating.

Serotonin is a particularly ancient neurotransmitter/modulator, in widespread use among very varied phyla, and has clearly had plenty of time to evolve its large variety of receptors and acquire a great many functions. Since some of the functions seem contradictory perhaps it's comforting to know that the situation is no clearer in invertebrates. In crayfish a flip of the tail plays a large part in battles for dominance. In victorious individuals the excitability of the neurons which control the tail-flip is enhanced, and in the losers it's suppressed, but both modifications are achieved by means of serotonin, acting through two different receptors. Clearly here it's contributing to a form of learning, and it has also been shown to be crucial to other sorts of invertebrate learning.

In mammals it seems to be essential to life. Mice which were genetically manipulated to lose their serotonin systems, through a mutation of the gene for the reuptake mechanism, died within days of birth.

Some other relevant neuromodulators

The list of neuromodulators is by no means exhausted with serotonin, dopamine, norepinephrine and acetylcholine. There are dozens of others. Many of them also serve other functions elsewhere in the body, for instance as hormones. Some are small chains of amino acids known as neuropeptides, and these are often produced in the same neuron as another transmitter and released with it in the same vesicles.

Of particular interest here are three neuropeptides that are classed together as opioids, because they work through receptors that are also activated by drugs derived from the opium poppy, such as heroin and morphine. Opiate drugs are used to suppress pain, which implies that this is one function of the natural version. In addition, opioids can enhance the pleasantness of sweet tastes, and decrease the unpleasantness of bitter ones. They seem to be another part of the machinery involved in the actual enjoyment of rewards.

Opioid receptors are strongly clustered in an area where neurons are sensitive to the voices of conspecifics, and especially to mother's voice or that of offspring. That suggests opioids might play a part in social relationships, and the idea is confirmed by infant mice which lack one of the opioid receptors. They don't form such strong attachments to mother as other mouse pups, and emit fewer ultrasonic calls when they are separated from her, though they complain just as vociferously as normal pups when exposed to cold. In monkeys, meanwhile, the level of one of the opioids increases when an animal is being groomed, and when the level is low it's less likely to groom others and more likely to solicit grooming. Very low levels are accompanied by extreme irritability. Perhaps we can deduce why a soothing voice or a nice maternal cuddle can function to reduce irritability or pain.

As Paul Maclean speculated, it seems likely that some opiate addicts use the drug to obtain a substitute for social warmth and support which is otherwise missing in their lives. But the soothing effect of brain opioids doesn't necessarily have to be provided by mother or someone close. A doctor in a white coat adminstering an injection of "painkiller" which is actually placebo can, in some individuals, have a strong pain-reducing effect, and increase blood flow in forebrain areas that are rich in opiate receptors. Moreover painkilling injections administered covertly tend to be less effective than those of which the subject is aware.

Another neuropeptide implicated in emotion is oxytocin, which doubles as a hormone involved in mammalian parturition, prompting uterine contractions and milk production. As a neuromodulator it affects maternal care. Female rats in which brain oxytocin is diminished are not nearly so devoted to the care of their young as normal rats.

Oxytocin also plays a significant role in the establishment of pair bonds. Prairie voles, unlike meadow voles or mountain voles, bond for life, and the male helps to raise the infants. If oxytocin is injected into the cerebral ventricles of a prairie vole it speeds up the pair bonding process. The treated individual becomes dedicated to the partner with which it's caged even before they have mated.

Miranda Lim and colleagues carried out a genetic manipulation which increased the number of oxytocin receptors in male meadow voles. The result was that these normally promiscuous animals tended to prefer the company of one female, just like prairie voles. Different behaviour patterns in closely related species can rest on quite modest foundations, it seems.

In another experiment oxytocin was administered to humans before they began playing a game (much beloved among social psychologists) in which one player receives a certain sum of money, and has the opportunity to increase it if they entrust it to another, unseen player - provided the unseen player co-operates. The subjects treated with the neuropeptide proved much more trusting than the controls treated with a non-active substance. This might be taken as demonstrating that oxytocin contributes to social bonding, as well as pair bonding, and it seems that the avian equivalent of oxytocin can indeed serve such a function in sociable bird species. Female zebra finches became much less sociable when the appropriate receptors were blocked.

Both oxytocin and another neuropeptide, vasopressin, also seem to contribute to learning to recognise another individual - something which is obviously essential to bonding with a partner. Male rats normally recognise a female again after only one encounter, but take much longer if the oxytocin system is malfunctioning, or if vasopressin receptors in the olfactory bulb are blocked.

Other neuropeptides are associated with pain, with stress, with the body's current store of energy in the form of fat, and with other factors directly or indirectly related to reinforcement. There are also a couple of neuropeptides classed as cannabinoids, since they activate receptors which also respond to marijuana and its relatives. One type of cannabinoid receptor is thought to influence the amount of transmitter released at glutamate, GABA and dopamine D_2 synapses.

None of the neuromodulators can properly be considered in isolation, however. Things are much more complicated than that. Neuromodulators influence both excitatory and inhibitory neurons, and some of them feed back on to the neurons which produced them to wield an influence there. Furthermore they interact extensively, and often reciprocally, so that just how much of one neuromodulator is released is often regulated by input from another. There is a complicated web of interactions to be sorted out.

The basal ganglia again

Since dopamine is heavily implicated in shaping behaviour to obtain rewards it will be no surprise that the basal ganglia are too, just as they are involved in learning where to direct attention. The striatum, where the neocortical messages arrive, also receives input from all the areas known to be concerned with the sort of sensory input that can function as feedback, and there are dense connections with the prefrontal area that is particularly focussed on rewards and discouragements. Much of this feedback or reinforcement information is directed to those patches of cells in the striatum described in the basal ganglia section of Chapter 15. Known as striosomes, they in turn provide a strong input to the dopamine centres, and they are sites in which animals readily learn to give themselves artifical stimulation.

The arrival of a goody is greeted by strong activity in ventral striatum, the input station of the basal ganglia that's especially concerned with desirables. Other sensory events evoke a response in just a few neurons. But when it becomes apparent that a sensory input functions as a cue that foreshadows a potential reward the number of neurons responding to it increases substantially – just as with the dopamine response. Some of them are activated sufficiently far ahead to suggest they influence the choice of movement, while others increase their firing immediately before the movement is begun, or after it's completed but before the reward arrives.

If an animal must learn to distinguish reward-promising cues from other stimulus patterns all may evoke increased activity at first, but the response to a stimulus pattern not followed by a reward gradually subsides. As learning progresses and the decision on whether to respond to a cue becomes more automatic the activity in ventral striatum reduces, leaving that in dorsal striatum looking more conspicuous. A slenderer, more efficient pathway becomes established as the animal works out exactly what to do. Dopamine flows freely while the experimentation is going on, and information about reinforcing inputs is valuable. Less is needed when the response to a sensory input becomes a habit, requiring minimal attention.

An important component of the machinery, regarded as a functional part of the striatum, is the nucleus accumbens, where a great deal of the reward information arrives. The dopamine innervation is particularly strong here, and when something attentionworthy happens dopamine is released there in bursts. These signals are important in most forms of learning.

One result of such bursts is that GABA neurons running from the nucleus accumbens to the basal nucleus of the forebrain (the structure described in the previous chapter, source of acetylcholine and GABA fibres which project to finely targeted areas of neocortex) are temporarily inhibited. That leaves the basal forebrain neurons on which they synapse free to fire - one of the ways cerebral attention is directed to sensory inputs linked with reward.

In addition to the dopamine bursts in the nucleus accumbens there's also a fairly steady, ongoing or tonic dopamine input, forming a background to these signals. Furthermore there's always a certain amount of dopamine floating around in the intercellular space, since a significant amount of the neuromodulator escapes the reuptake system. It reaches dopamine receptors scattered over the cell surfaces outside the synapses (thought to be more numerous than synaptic ones) producing a delayed effect. The background level of dopamine, reflecting the number of recent burst inputs, probably influences mood and alertness.

The nucleus accumbens first attracted particular notice when it was discovered that substances such as morphine, cocaine and alcohol produce a significant increase in the amount of dopamine released there, and a rather less dramatic effect on dopamine elsewhere. Moreover artificial brain stimulation that functions as a reward also has this effect, and it's only when activation of the electrodes produces this increase that the animal learns to stimulate itself. The circuits that record these unnatural rewards, however, may not be quite the same ones that are affected by such desirables as food and water.

Willingness to explore the environment, rather than merely responding to it, seems to depend on the level of dopamine in the nucleus accumbens. So does the approach to cues which have become associated with rewards. Lesions to the structure interfere with the learning of such cues, or with any learning which involves a delay between the response and the reinforcement. They also affect an animal's ability to wait for a large reward if that involves rejecting a small, immediate one. They don't have any effect on the consumption of a reward. It seems reasonable to suppose that the nucleus accumbens circuitry engaged by addictive drugs might be the same circuitry that is involved in the joys of exploration, and finding out how to obtain goodies, rather than in the consumption of them.

So far most of the experiments on addictive drugs have been done with species that are quite adventurous eaters, and fairly energetic explorers, such as rats and mice. Repeating some of the experiments with timider species that consume more readily available forms of food, such as rabbits, might be interesting. Dopamine is a ubiquitous neurotransmitter/modulator, but some species have more of it than others. Maybe higher levels will turn out to correlate both with more energetic lifestyles and with a greater vulnerability to addiction.

Apart from dopamine many other neuromodulators exercise an influence in the nucleus accumbens. Two types of serotonin neurons synapse there, and one at least seems to contribute to addiction. An opioid input influences the value put on whatever taste is currently being registered. A rat normally tires of eating the same flavour, but artificial activation of opioid receptors prompts continued consumption, regardless of satiety. Various neuropeptides that influence pair bonding and maternal behaviour also operate in the nucleus accumbens. Altogether the basal ganglia receive the richest variety of neuromodulator inputs of any brain region.

The amygdala

Another structure important to reinforcement is a cluster of nuclei called the amygdala. This has sites where the opportunity for self-stimulation is taken enthusiastically, and regions where self-stimulation, once tried, is firmly avoided thereafter. It contains, too, spots where artificial stimulation produces an increasingly angry and aggressive animal, whose ferocity fades only slowly after the stimulation ceases. Stimulation in another area induces all the signs of abject fear. In humans imaging experiments show an increase in amygdal activity in response to a conditioned stimulus associated with an aversive event.

Clearly the amygdala is deeply involved in the action-guiding influences we call emotion. Its central nucleus is heavily connected to the hindbrain, and influences many of the bodily signs of emotion, such as altered rates of heartbeat and respiration, and dilated arteries - all the adaptive changes which precede actions inspired by fear or anger. It also sends signals to the hypothalamus, the brain area which controls the release of hormones, and thereby produces other changes in physical state.

Some neurons in the basolateral nucleus of the amygdala respond to negative events and connect to the central nucleus. Rats with damage there may have difficulty in learning to avoid negative reinforcements, or having learnt one trick don't adjust to an alteration in the rules. Or they may fail to develop fear responses when appropriate. Other neurons are activated by rewarding inputs, and these send their axons to the nucleus accumbens. So lesions in the amygdala can lead to problems with learning how to obtain desirables too.

The amygdala also contributes to a special sort of learning, the acquired aversion to the taste of a food after consumption of it has been followed by an upset stomach. And monkeys with amygdal damage don't seem to get satiated with one type of food but will continue to work for the same thing, even though they have the opportunity to adapt their actions to obtain something different. There also seems to be a more general role in regulating appetite.

It's long been known that events linked to emotions are more likely to be remembered. Scans of human brains while pictures were studied showed that the amygdala was active when the pictures carried an emotional connotation, whether agreeable or aversive. The degree of activation correlated with the subjects' ability to recognise the same pictures, among new ones, four weeks later.

The amygdala has strong reciprocal links to other areas involved with reinforcement, such as the basal ganglia and frontal cortex, and a strong dopamine innervation, so it's not surprising that it plays a role in some aspects of addiction. It also transmits to brain areas involved in attention and arousal, providing a means by which stimuli connected with emotion or with various sorts of reinforcement can attract attention. Along with the nucleus accumbens connection to the nucleus basalis it's probably responsible for the fact that an emotion-tinged stimulus can sometimes reach awareness where a neutral one would have been masked by a subsequent input following fast upon it.

There's a good deal of input from sensory neocortex, but some also arrives direct from the thalamus, which means that a response to a simple stimulus pattern can be conditioned without the involvement of sensory neocortex. Lesions of auditory cortex didn't prevent rats from acquiring a fear response to a tone, but lesions of auditory thalamus, or of the midbrain auditory centre which supplies its input, did. Connections also lead to the amygdala from the parabigeminal nucleus, mentioned in the last chapter, with its neurons activated by looming visual stimuli, suggestive of approaching danger. These short pathways mean that instructions evoking visceral and autonomic preparations for action can be despatched while the neocortex is still analysing the sensory data.

In rats and mice the predominant input about the external world comes from the olfactory and vomeronasal systems, and the amygdala is especially sensitive to the pheromones which characterise conspecifics. The portion which receives this odour information is vital to maternal behaviour. We humans show an amygdal response to faces, one that is actually stronger if the photograph is blurred and the details are missing - an archetypal face. Faces with emotional expressions, which can be as easily read from blurred photographs as from clearly defined ones, produce a particularly strong effect. But the response is lower in people who have been classified as psychopathic criminals.

The amygdal activity occurs even when the face is not consciously perceived, and indeed may even be stronger then - as if the primeval response was partly suppressed as a result of conscious perception of the photograph. The eyes seem to be the really crucial feature.   Moreover, as Paul Whalen and colleagues showed, there's an especially strong sensitivity to eyes with a large amount of white showing, as in fear. People with amygdal damage are bad at reading faces, and the expressions which cause most difficulty are those conveying fear or anger.

However, various strands of evidence suggest that the actual identification of face-shape and expression is done elsewhere – perhaps in superior colliculus or pulvinar. What the amygdala may do is build on the report and ensure that attention is directed to the significant stimulus pattern. Investigators studying a woman with bilateral amygdal lesions, who had great difficulty recognising expressions of fear, found that she didn't look at the eyes. If she was constantly reminded to study the eye region she did as well as normals. And in a 14 year-old boy with bilateral damage caused by disease fearful faces attracted spatial attention with unusual rapidity, but it faded very fast.

In the monkey amygdala there areneurons which respond specifically to the stimulus pattern which is an eye when it falls on the fovea - in other words when eye contact is made. This too is something which demands attention, sometimes pleasurable, sometimes alarming, and no doubt we possess something similar.

The automatic attention we give to human voices, and especially to emotional voices, can also be attributed to the amygdala. When it's damaged or removed on one side the auditory cortex on the same side no longer gives the same priority to voices, and if it's the left amygdala that's affected emotional voices lose their power.

All this may explain why male monkeys with amygdal lesions may slip down the pecking order, losing interest in fighting for status. They may also go about their business in a rather detached sort of way, paying little attention to the more dominant individuals. Presumably they lack an essential part of the machinery for attention to conspecifics, as well as lacking the emotions that might motivate them to compete with their fellows.

It's likely that the original function of the amygdala - a very ancient structure, present in some form in all vertebrates - was to promote attention to conspecifics, and to trigger the appropriate autonomic and hormonal reactions if mating or a fight seemed to be on the cards. In early evolutionary history the signal indicating the presence of attentionworthy conspecifics is likely to have been olfactory, and to have been conveyed by the direct connections from olfactory apparatus to hypothalamus which still exist. The interposition of the amygdala between the two, it can be postulated, allowed the mechanisms for identifying conspecifics to grow more complicated. Anything that promoted efficiency in procreation would allow numbers to grow faster, and increasingly delicate methods of identifying breeding partners would become essential as the number of species expanded.

Elaboration of the amygdala made it possible to register expressive behaviour in conspecifics. The evolution, in mammals, of increasingly mobile and expressive faces will have increased the potential. And when the effects of visceral responses became available to consciousness the possibility of education by transferred emotion came into being.

Other extensions to the amygdala, meanwhile, can be supposed to have supplied other sorts of feedback arrangements, making it possible to learn, for example, about what causes a stomach upset. A reasonable guess is that the amygdala is involved in learning about anything that may need an autonomic, visceral or hormonal response.

Pain

News about damage to the organism, or the threat of damage, is registered by an assortment of high-threshold mechanoreceptors, thermoreceptors geared to extreme temperatures, and certain chemoreceptors. It's relayed to the arousal systems of the hindbrain, to central nuclei of the thalamus, to the hypothalamus and the amygdala, and finally to prefrontal cortex. Prefrontal evaluates the significance of the pain, in the light of current circumstances, and sends messages back to the spinal cord, via several routes, to enhance or suppress the input. In some areas there's increased activity not only when pain is experienced but also when it's anticipated. It would appear that analgesia by an opiate prevents this increase, however - and so does a successful dose of placebo.

When opiates or placebos suppress pain there is a coincident increase of activity in the very front of anterior cingulate cortex, an area at the bottom of the frontal lobe. Using real-time fMRI imaging as the feedback, people can be trained to control the activity here, and thereby increase or decrease their perception of pain. This worked not only with healthy subjects tested with a noxious thermal stimulus but also with patients suffering from chronic pain. The area must be part of the arrangements for tuning down the pathways through which pain reports are relayed, and for activating the centres which have descending connections to the sensory pathways in the spinal cord and can modify the pain message at that level.

Also activated by pain is a diencephalic structure, the habenula. It responds, too, when an anticipated reward fails to arrive, and provides a route by which undesirable effects are signalled to the dopamine and serotonin systems.

A pathway by which sensory information particularly relevant to reinforcement reaches prefrontal cortex goes through the insula - in primates a deeply buried bit of cortex which lies underneath the sulcus, or groove, which divides frontal and temporal lobes, and just above the striatum. This area seems to be particularly concerned with processing and transmitting information from the internal and boundary senses. It's activated by the taste and feel of food, and in hunger, thirst and disgust as well as in pain. Rats which have learned to avoid a certain taste forget to do so when the insula isn't functional; nor do they show the usual behavioural signs of gastrointestinal upset when dosed with lithium. However, damage to the area doesn't affect the desire to eat, or the pleasure of doing so.

It does seem to destroy the attraction of addictive drugs. Smokers with lesions in insular cortex lose the urge to smoke and find it quite easy to give up. Rats which had become used to receiving a dose of amphetamine in the brighter of two compartments spent a lot of time there, until the insula was temporarily inactivated, and then they resumed the normal preference for darker surroundings. The degree of activation in the insula, meanwhile, seems to correlate with a drug addict's own rating of current desire for the drug.

Cognitive and social emotions

The brain areas responsible for emotional and motivational experience contain what may be called emotional mirror neurons, activated by evidence of someone else's pain or pleasure. For instance insular cortex, dealing with taste and with information from visceral and interoceptive neurons, and seat of emotions related to food consumption, responds to the sight of another's disgust. Some prefrontal neurons respond to the signs of another's pain. Such neurons must have played a huge part in the evolution of infant education in mammals, since they clearly provide the means of learning by emotional infection.

They also provide a foundation for insight into the behaviour of others. An expansion of their number and variety must have been crucial in the development of social lifestyles in mammals, increasing the potential for co-operation. It can be deduced, furthermore, that a further expansion of emotional mirror neurons underpins the evolution of altruism. If the brain re-echoes the sufferings of a fellow group member there's a motive for alleviating them.

In human brains the areas concerned with material pains and pleasures are implicated in more abstract ones as well. The insula is activated by moral disgust, as well as by rancid or otherwise unacceptable food. Orbitofrontal cortex, the part of prefrontal cortex mainly involved in emotion, responds to social rewards and discouragements, and to monetary gains and losses. Amygdal activity reflects monetary loss in a gambling task, and indignation at a division of rewards perceived as unfair. The ventral striatum is active when someone learns the meaning of a new word. Brain mechanisms that evolved to deal with primary reinforcements and motivators have clearly expanded to deal in more sophisticated emotions.

The significant fact may be that the action prompted by a reward, whether it's physical, social or abstract, is in many cases much the same. The sight of a friend, the smell of good food, the sound of pleasant music are all likely to attract an approach, or at least the thought of it as a possibility. They may also evoke smiles. Rancid food, on the other hand, raucous noise, or an intensely disliked person tend to inspire a retreat. The function of emotion is to shape behaviour, so stimulus patterns and events that elicit similar forms of movement are also linked to similar emotions. Motion, emotion and motivation are as tightly linked in the brain as in the word forms.

This flexibility in the reward areas seems likely to be a significant factor in our ability to think with the help of metaphor and analogy – in effect by transferring a schema to a new subject.

Recapitulation

Several brain structures join forces to convert desirable and undesirable inputs into feedback that can shape future behaviour.

The amygdala is sensitive to a whole range of things which are worth approaching or which constitute threats. In response to one of these it despatches messages to the hypothalamus or the brainstem to elicit suitable hormonal, autonomic and visceral reactions.

In the basal ganglia circuitry the ventral striatum and especially the nucleus accumbens react strongly to an unexpected encounter with something desirable.

If a cue regularly predicts the appearance of a desirable an initially modest response to it is gradually extended to include a much larger number of neurons.

As the appropriate behavioural response to the stimulus pattern becomes habitual the activity becomes confined to narrower channels, and concentrated in the dorsal striatum.

The neuromodulator dopamine is important to learning about how to obtain a reward, and to establishing habits. When a cue is regularly followed by a reward the dopamine response to the goody is gradually transferred to the cue.

Serotonin may be more concerned with the actual experience of reward, though its influences are so numerous and contradictory it's difficult to draw definite conclusions.

Various other neuromodulators are also involved in reinforcement, including opioids, cannabinoids and oxytocin.

Much of the information about reward and about noxious inputs, is relayed to prefrontal cortex via the insula.

Abstract rewards activate much the same areas as material ones.

Artificial stimulation in some of the areas mentioned, among them parts of the hypothalamus and the bundle of dopaminergic axons which ascends to the neocortex, can function as a reward, and animals will work hard at a pedal that provides it, or show a strong preference for the place where it's experienced.

Artificial stimulation in certain other areas is thoroughly aversive.

Chapter 18

Learning

One of the big questions in neuroscience has been: how does learning happen? Santiago Ramon y Cajal and Llorente de No, pioneering neurologists, both hypothesised that it involved the modification of neuronal circuits. In the nineteen forties the idea was further developed by Donald Hebb, who suggested that synapses might be strengthened when transmitter released by the presynaptic neuron regularly resulted in activity in the postsynaptic neuron. It was some time before recording equipment of sufficient delicacy to test the hypotheses was developed, but it turned out that there are indeed many synapses in the brain which are modified as a result of the amount and nature of the use they get. Quite a number of different mechanisms contribute to this synaptic plasticity, and as Hebb predicted some of them come into play only when activity in the pre-synaptic cell leads to activity in the post-synaptic cell - something which, in the major information-processing neurons, only happens when there are a good many inputs to the post-synaptic cell. It's been confirmed, moreover, that the strengthening of synapses does coincide with learning, while interfering with the mechanisms of synaptic modification prevents it.

It was in the nineteen sixties that the first modifiable synapses were identified. Tim Bliss and Terje Lømo, working with rabbits, were investigating a structure called the hippocampus. This is an ancient, three-layered part of the telencephalon, hidden away on the inner curl of the temporal lobe. Bliss and Lømo discovered that when neurons in one of the main pathways through the hippocampus receive a strong burst of stimulation they become more sensitive, and begin to respond to modest inputs which previously wouldn't have been sufficient to produce an effect. The heightened sensitivity could last for some time - minutes, hours, sometimes even days. The effect was christened long-term potentiation.

Synapses that can be potentiated, it transpired, are also subject to the opposite effect, a reduction in sensitivity, or long-term depression. Both presynaptic and postsynaptic influences are at work, it turned out, and different combinations of mechanisms operate in different neurons. Glutamate synapses on to glutamate neurons received the most extensive study initially, but potentiation and depression are also found in various other types of neuron, including those deploying GABA, dopamine and norepinephrine.

At glutamate synapses there is a very immediate form of potentiation which arises from the different modes of performance among glutamate receptors. They come in three main types, distinguished by the substitute chemicals which can be used to fool them, AMPA, NMDA and kainate. (Readers who want to know what the initials stand for can find the answers in the glossary.) The AMPA receptor behaves in a straightforward manner, opening its channel when glutamate latches on to it and letting sodium ions flow through, and in some cases some calcium as well. The NMDA receptor is different. The structure of its ion channel is such that a magnesium ion usually gets lodged in the entrance. It takes a reversal of the membrane voltage to shift it, and only then can glutamate unlock the channel and let in a current - usually one that's composed mostly of calcium ions, with some sodium. The NMDA receptors don't come into operation, therefore, until there has been sufficient activation of nearby AMPA receptors to produce an excitatory post-synaptic potential. But once the AMPA receptors have got things going the NMDA receptors ensure that further post-synaptic excitations can be achieved more easily.

Here, then, is an arrangement which produces a rapid strengthening of a connection when an action potential in the presynaptic neuron produces a result in the postsynaptic dendrite. The recruitment of the NMDA receptors can lead to stronger postsynaptic potentials, and an increased chance of a current that reaches the soma and triggers an impulse in the axon. The direct effect of NMDA activation is only temporary - the magnesium ion takes up residence again when it gets the chance. But the way has been opened to other effects, and the importance of these receptors is confirmed by the fact that in many cases blocking them interferes with both potentiation and learning.

In AMPA receptors, meanwhile, a major burst of stimulation leads to an extra component being temporarily attached, which makes the ion channel open a little more easily, an effect which lasts for at least an hour. Inputs through metabotropic receptors further influence the receptivity of the receiving cell. There are also various means of increasing the amount of transmitter released when serious activity occurs.

In the late nineties another important contribution to long-term potentiation was discovered. In some of the major neurons of the brain an excitation that reaches the axon and triggers an action potential rebounds back into the dendrites. The backward-propagating impulse opens voltage-sensitive gates in the dendritic membrane again, and the calcium they let in further modifies the local chemical environment, making for increased sensitivity. But this influence is only exercised if the backward-propagating impulse follows very precisely on the forward-running excitatory potential, which means that the effect is confined to those synapses which have contributed to the action potential - or at least were active at precisely the same time.

The crucial point is that the effect is greatest at synapses which have been only weakly stimulated. The backward-propagating potentials have a particularly strong influence at synapses where the most significant changes are possible.

This must be a mechanism by which, when a cell gets seriously active, potential input synapses can be converted into actual contributors and recruited into an existing circuit. Let's imagine a very much simplified neuron which fires strongly as a result of excitatory input at synapses *a, b, c* and *d*. The nearby synapses *e* and *f* also receive input at exactly the same time, but the input is not initially strong enough to produce a wave of excitation that can travel beyond the immediate area, so it doesn't contribute to the action potential. However, after *e* and *f* have produced a succession of subthreshold post-synaptic excitations, and experienced a succession of backward-propagating potentials as a result of the simultaneous and more potent activity at *a, b, c* and *d*, they begin to produce a wave of depolarisation which reaches far enough to join the flow to the axon.

Thus these synapses are lifted out of impotence into effectiveness, and perhaps this correlates with the incorporation of a new bit of information into a neuronal representation. In future, inputs at *b, d, e* and *f*, say, may be sufficient to initiate a message to the next neurons in line, even if *a* and *c* are silent. Potentiations of this sort might mean that something known heretofore only by its most conspicuous features will in future also be recognised by a combination of less salient ones, when part of the whole is concealed.

Alternatively, the whole neuron may be brought into a circuit in which it hasn't been active before. Suppose that synapses *a, b, c* and *d* between them can produce enough excitation to send some impulses rolling down the axon. But when the impulses reach the far end inhibitory connections prevent them from passing on the message. Or else neurotransmitter is released and excitatory messages are passed on to the next neuron, but the balance of excitatory and inhibitory inputs produced in that cell by the current stimulus pattern is such that no action potentials result there. If synapses *e* and *f* become potentiated then the impulses in the first neuron may become more frequent. The more intensive bombardment of action potentials may be sufficient to overcome the inhibition at axon-end, or in the receiving neuron. If this is happening at a number of input synapses it may tip the balance in the receiving neuron so that it starts to fire action potentials in turn, and potentiation may develop at the activated synapses on this neuron. Conceivably this is what's going on when one stares, puzzled, at a confused scene, and then it turns into something recognisable; or when a new idea breaks through into consciousness. (One must imagine, of course, the potentiation being built up at many of a major cortical neuron's ten thousand or so synapses, and in many neurons.)

A significant point is that for this mode of potentiation timing is everything. Only synapses which are excited in unison get in on the act. This indicates the importance of the synchronised neuronal firing, extending over quite large areas, which was described a couple of chapters ago, and which seems to correlate with attention. If the excitations related to one stimulus pattern all arrive at a neuron in the processing pathway at the same time there is a much greater chance of potentiation being achieved.

Synapses can be expanded fairly rapidly by shifting additional AMPA receptors to the site. These receptors lead a mobile life. In addition to those clustered at the synapse there are many scattered individually about the membrane, which are regularly internalised and then inserted back into the membrane, sometimes at a different place. They can also slide through the membrane at quite a rate, and they are readily recruited to an active synapse, and anchored there. (NMDA receptors are moveable too, but seem to be less energetically peripatetic.) Further scope for potentiation is provided by synapses composed entirely of NMDA receptors, known as 'silent' synapses since they can only fire when an excitation at a nearby AMPA synapse shifts the magnesium ions. But when the silence is broken at sufficient length AMPA receptors are attracted to the site, and it's transformed into a fullscale, bipartite synapse.

Modifications are found in the axon too. For instance, the initial segment, where there is a dense array of voltage-gated sodium channels to trigger the action potentials, is flexible. If inputs to the neuron grow weaker the initial segment can be extended, making it possible to produce an action potential with a rather smaller input. But if the neuron is subjected to some heavy inputs the whole array of voltage-gated channels may shift a little way down the axon, which has the opposite effect.

There is, clearly, a whole range of ways for stepping up the efficacy of a synapse when it becomes busy. Not all of the mechanisms are found in every type of potentiatable cell.

Longer-lasting potentiation

Long-term potentiation was perhaps unfortunately named, for it turned out that its lifetime can vary considerably. What was initially known as the early form of long-term potentiation is brought about by a comparatively modest train of excitatory input, and depends on the sort of mechanisms just described. Stronger or longer stimulation produces longer-lasting potentiation, which was originally termed 'late'. It then transpired that the late form has two stages. Nowadays the short-term sort is sometimes designated LTP1, the medium-term LTP2 and the longest-lasting variety LTP3. The two longer-lasting varieties depend on the synthesis of new proteins for building up synapses.

Many of the excitatory synapses of neurons which support potentiation have a protein-manufacturing unit (a ribosome) situated conveniently close at hand in the local dendrite and already equipped with messenger RNA (despatched earlier from the cell's nucleus), so that it's ready to start turning out some of the materials required with a minimum of delay. LTP2 seems to depend on this local, rapidly initiated protein synthesis. LTP3 involves the manufacture of new proteins too, but these are proteins for which the messenger RNA is not prepared in advance. A message has to travel to the cell nucleus before manufacture can begin. Calcium plays an important role in labelling the synapses to which the new proteins must be delivered, and in some neurons calcium admitted through NMDA receptors is vital to longer-term potentiation.

The earlier stages of potentiation are readily reversible. NMDA receptors get blocked by a magnesium ion again; AMPA receptors are dephosphorylated (lose that extra attachment) or get transferred back into the cell and moved elsewhere. So a potentiated synapse can return to average strength if it doesn't get further use. Just how longlasting fullscale potentiation can be, once established, and whether it needs to be refreshed from time to time by further use of the synapse, pretty certainly varies among different brain areas.

Long-term depression

Synapses can not only be potentiated, they can also become less sensitive, so that a greater number of action potentials in the pre-synaptic axon is needed to trigger excitatory potentials in the post-synaptic dendrite. Long-term depression is produced by low-frequency stimulation of the synapse (but not by complete inaction). The precise levels of stimulation which cause potentiation or depression vary in different parts of the brain, and even in different layers of the neocortex. A fairly typical state of affairs is that a burst of excitation at about fifty pulses per second results in potentiation, while inputs that arrive at a rate of less than about three have a debilitating effect, and those that come at ten to a second or so leave the sensitivity of the synapse unchanged. But in some cells the rate of input that leads to depression is much higher.

Depression is not a mirror-image of potentiation. Some of the processes by which a synapse becomes depressed are the same as those by which potentiation is reversed, but some are subtly different. AMPA synapses are rendered less efficient by removal of an attachment, for instance, but the subunit which is dephosphorylated is not the one which is phosphorylated in potentiation. And depression, like potentiation, can involve the synthesis of new proteins.

The two processes often operate in tandem. As the active synapses in a busy neuron become potentiated those that are receiving only minimal inputs from their connections become depressed. There's also a form of depression attributable to those backward-propagating waves of excitation that rebound after an action potential to bestow potentiation on synapses which were active at the time of the input. Their influence causes depression at any synapses which were active over the same period but not at precisely synchronised moments. Thus weakly excited synapses with the correct timing can be elevated into major players, while those working to a different tune get pushed out of the team.

In short, potentiation and depression between them shape a neuron's performance to make it particularly sensitive to certain combinations of inputs, and much less sensitive to excitation from other connections. Indeed, totally unused connections may shrivel and die away. We can begin to see how it's possible for a cell in a cortical schema-area to become responsive to a certain set of visual features, or to a certain set of auditory qualities.

There are several forms of long-term depression, dependent on different mechanisms, just as there are several forms of long-term potentiation. Confusingly, the mechanisms are mostly very similar to those engaged by potentiation, and in some cases similarly involve calcium triggers. What makes the difference, it seems, is how much calcium arrives all at one go. Activation of NMDA receptors by a strong burst of input lets enough calcium into the synapse to produce at least the early form of potentiation. A similar amount of calcium admitted in small doses over a long period leads to depression.

How easily a synapse becomes potentiated or depressed often depends not only on these mechanisms but also on its recent history. One that has been inactive for some time is more readily potentiated than one which has been busy, so the neurons employed in the latest new circuit are those which have not been doing much of late. A synapse that has undergone an early form of potentiation is protected from depression for at least an hour, which leaves scope for making the potentiation more permanent if further relevant experience occurs.

The varied mechanisms for potentiation and depression confer considerable flexibility on the neurons which possess them. But there's a ceiling on how far potentiation can go. Further strengthening of the communications between two neurons is done by recruiting 'silent' synapses, or creating new ones.

In an under-used neuron, the potentiation of synapses may presumably increase the total amount of activity, but probably the more usual situation is that the potentiation of some synapses is balanced by the depression of others. Overall, therefore, activity is not increased, merely redirected. That's important, for neurons can be damaged by over-excitation. Various mechanisms serve to keep the overall level of excitability within limits.

The contribution of neuromodulators

In addition to the effects targeted to active synapses there are wider influences on the machinery of potentiation and depression. Neuromodulators such as dopamine, serotonin and norepinephrine have an important role, in some cases a very decisive one, as do the metabotropic receptors for glutamate. The effect may be to alter the rate at which one of the membrane's ion pumps works, or to modify one of the channels in the membrane which open to restore the ionic balance after an impulse, so that ions flow in faster, or more slowly. Such adjustments mean that the cell fires more readily, or less readily, or can maintain a faster rate of firing. Most importantly, some metabotropic messages cause the release of calcium from the cell's internal stores. All these factors can influence the chances of potentiation or depression occurring. Such modifications can be confined to one part of the cell, making it possible for just one branch of dendrites to become more receptive.

The effects of neuromodulators can be complicated. Activation of dopamine D_1 receptors is necessary for long-term potentiation at some striatal synapses, while both D_1 and D_2 receptors must be brought into play for long-term depression. Excitation at D_1 receptors in cerebral cortex is important to working memory, but too much dopamine also activates D_2 receptors and thereby blocks the effect.

Serotonin performs a variety of functions, operating very differently in different areas, in its usual confusing fashion. Norepinephrine and acetylcholine both promote the synaptic modifications which constitute learning, the latter operating in both transmitter and modulator modes. Its modulatory influence on cortical neurons has been shown to be very powerful in shaping their operations, and lesions to the nucleus basalis impair learning ability quite severely. The effect of neuromodulators, and the complex interactions between them, must account for the fact that attention, emotion and mood can all influence the speed and efficiency of learning.

The shape of flexibility

As we've seen, potentiation and depression are very precisely controlled. Just a few milliseconds separate the co-ordinated activity that brings potentiation from the unsynchronised activity that results in depression. This is possible because the influences that produce potentiation can be tightly confined, affecting just a few closely adjacent synapses. Tight timing and precise targetting are facilitated in major neurons of the forebrain by the fact that the excitatory synapses are situated on tiny projections on the dendritic membrane. These are known as spines, though modern microscopes have shown that they aren't in fact pointed but blobby, joined to the dendrite by a narrow stem. Calcium admitted to the cell through a synapse is largely confined within the blob, which explains how its effects can be deployed with such precision.

Large spines support well-established synapses. Smaller spines contain less developed ones, which offer plenty of scope for potentiation. They're also easier to potentiate, for they have narrower stems, so the calcium which enters through NMDA receptors has less chance to leak out before its influence can be felt.

One of the surprises of modern neurology is the dynamic behaviour of these spines, even in some parts of the adult brain. In brain slices viewed *in vitro* through the microscope the blobs may change shape even as they are observed, rising up on their stalks like mushrooms, or sinking back to become less defined. A fair proportion of them have only a temporary existence. In the course of a few hours some disappear, while new ones materialise on previously empty membrane. The sketch indicates the sort of changes that can take place over hours to days.

The conversion of small spines to large seems to correlate with learning. A drastic change in sensory input has been shown to bring about major alterations in the distribution of spines. With the aid of several of the tricks of modern neuroscience the somatosensory cortex of mice in which alternate whiskers had been cut off was repeatedly imaged for a month afterwards. Over this period new, tiny spines grew and became stabilised, while some of the previously large and persistent spines shrank and disappeared.

Sho Yagishita and colleagues discovered that in the striatum the enlargement of spines in the direct (inhibition-releasing) pathway requires dopamine. In their experimental animals, mice, this must arrive within two seconds of the glutamate input to achieve the effect. This seems to provide some insight into one of the most fascinating questions in neurology - how a stimulus pattern gets linked to a reinforcement that follows it rapidly but without quite overlapping in time.

New spines, meanwhile, tend to grow in close proximity to those containing active synapses, and can become functional very rapidly. Long-term potentiation of a synapse can be accompanied by an expansion of the number of synapses available for connection to the same network. The guess is that new spines can grow within the field of influence of the calcium flush created by activity at a nearby large spine. As the synapses on a spine become potentiated the neck of the spine becomes shorter and wider, making it easier for calcium ions to escape. A spillover of transmitters from the synaptic cleft on to adjacent membrane may exert an influence too, both on post- and presynaptic membranes. It's also possible that passing axons, when they are active, emit a signal which promotes the establishment of a receptive zone on nearby dendrites, and can lead to the creation of new synapses. Certainly long-term potentiation leads to an increase in the number of contacts an axon makes on a dendrite.

Axons too are turning out to retain some flexibility even into adult life. A study of primary visual cortex in macaque monkeys discovered that while the main axon branching patterns were stable over the weeks of observation the smaller branches were much less so. About 7% of them, along with their boutons for releasing transmitter, disappeared in the course of a week, and were replaced by new twigs elsewhere.

Aplysia - learning in an invertebrate

A good many neurologists have been attracted to the study of long-term potentiation and depression in the mammalian brain. Eric Kandel chose instead to spend the first decades of his career investigating neuronal plasticity in something much simpler. *Aplysia*, a marine mollusc also known as a sea-hare, has only 20,000 neurons, many of them conveniently large, and readily identifiable. Nevertheless it's capable of modifying its behaviour to suit the current environment, and even of some Pavlovian-style conditioned learning.

The ruffled skin-flaps which are shown here stretched out to the sides are usually folded up over the delicate gill, and the siphon protrudes only far enough to allow fresh seawater to be pumped over the gill. If the siphon is touched *Aplysia* withdraws it behind the mantle and retracts the gill. After several weak inputs of this sort, however, the defensive reflex habituates. On the other hand after a strongish prod to the tail it's triggered by a smaller tactile input than formerly - a development which is called sensitisation. And if habituation has already happened the strong stimulus to the tail revives the response. Kandel, in collaboration with a succession of co-workers, has over the years deconstructed the withdrawal reflex in detail.

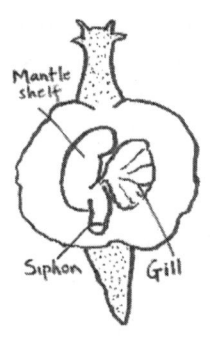

There are twenty four sensory neurons to record tactile input to the siphon, synapsing directly on to six motor neurons which control the movement of the gill. Habituation occurs because calcium is needed for the process by which vesicles of transmitter are fused to the membrane and emptied into the synaptic cleft, and the number of available calcium ions at the end of the sensory neuron's axon gradually declines. To reverse habituation their number must be replenished, while sensitisation is achieved by augmenting the supply beyond the usual level.

At the end of the sensory neuron's axon, close to the output synapse on to the motor neuron, is an input synapse from a serotonin neuron, one which brings information from a sensory neuron in the tail. A serotonin message prompts the opening of a channel that allows calcium ions to flow into the cell, and this triggers two chains of events. One re-energises the machinery for transmitter release. The other leads to a reduction in the inward flow of potassium ions that restores hyperpolarisation after an action potential, which means that further action potentials can be achieved a little more readily. As a result the rate at which the sensory neuron releases glutamate on to the gill motor neuron increases.

A single strong prod to the tail of *Aplysia* sensitises the gill withdrawal reflex for several minutes. Several such pokes produce a sensitisation that lasts rather longer. But if the prods are spaced out over several days the animal's defensive response to even gentle touch may last weeks. Moreover if modest touches on the tail are regularly paired with tactile input at the siphon the animal comes, in best Pavlovian manner, to withdraw its delicate parts behind the mantle shelf as soon as the tail is touched, without waiting for an insult to the siphon. Five paired inputs, spaced out at intervals, are sufficient for this conditioning.

The behavioural effects are parallelled when a single sensory neuron is studied *in vitro* and serotonin is applied by the experimenter. A single puff of serotonin at the serotonin synapse increases the release of transmitter briefly. Several puffs in a row create the same effect for a somewhat longer time, and modify the performance of potassium channels so that the cell is more readily excited. Five spaced-out puffs that are paired with action potentials produce a long-term potentiation of the sensory-motor synapse, involving enlargement by the insertion of new proteins.

The signalling mechanisms which organise this expansion are much the same as in vertebrates. And just as in vertebrates, another synapse on the same neuron can be potentiated too if it happens to be modestly busy at the same time as the very active one. Learning of this sort is clearly long-established.

And even in *Aplysia* the arrangements are turning out to be fairly complicated. There are several other routes between the tail and the gill-withdrawal motor neurons, and in some of them serotonin operates with an opposite effect, reducing excitability or causing depression at the synapse. The contradictions can be explained by the fact that there are at least four different types of serotonin receptor in this species. In addition, post-synaptic mechanisms are being discovered, to add to the pre-synaptic ones. Sensitisation or conditioning involves a serotonin input to the motor neuron as well as the sensory neuron, leading to a bigger response at AMPA receptors.

Studying the mechanisms of reward in *Aplysia* is also proving interesting. The animal has an oesophagal nerve in which high-frequency activity is observed when food is being ingested, and which presumably functions to register the presence of food in the oesophagus. When this nerve was artificially stimulated every time a spontaneous biting movement was made the movements became much more frequent than in a control group, even though no food was acquired. Following this behavioural experiment the neuron deduced to be responsible for the modified action was studied *in vitro*, and puffs of dopamine turned out to stand in very effectively for the reinforcing effect of oesophagal stimulation.

Dopamine, serotonin, or a neuromodulator not found in vertebrates, octopamine, seem to play a part in every form of invertebrate learning that has so far been analysed. But although the same neurotransmitters and modulators are found in many phyla they are not always employed in the same way. Some of the receptor types they operate on no doubt differ from vertebrate forms.

The fact that in *Aplysia* several spaced learning experiences have more effect than the same number occurring together parallels the way some sorts of vertebrate learning are consolidated by a later encounter with the conditioned stimulus. This means that a long-lasting synaptic modification isn't made until it's pretty certain that the neutral stimulus pattern is consistently associated with the noxious or rewarding one. Only a seriously noxious input produces indelible learning from a single experience. From the subjective viewpoint, meanwhile, the animal findings tie in with our own experience that facts encountered twice, on separate occasions, are often remembered better than those studied just once, even if the study has been diligent.

Where learning neurons are found

As might be expected, long-term potentiation and depression have been discovered in all the brain areas that have been implicated in connecting reinforcements to actions. Generally they are found in more than one form, with different assortments of mechanisms employed to achieve the synaptic modifications. Just how long a 'long-term' modification endures without subsequent reinforcement differs in different behavioural contexts.

In the amygdala potentiation can be induced at various synapses, including those between the lateral and the central nucleus (the one which despatches the messages that invoke autonomic reactions). In the lateral nucleus it occurs at both thalamic and cortical input synapses. When converging thalamic and cortical afferents are both active at the same time the amount of transmitter released at the connection from the cortex increases - regardless of whether a post-synaptic excitation is achieved or not. This looks like an arrangement whereby the thalamus, responding to a roughly defined stimulus, can tutor the amygdala to take notice of the more detailed information provided by the neocortex.

Joseph LeDoux and colleagues recorded the activity in amygdal neurons before and after animals learnt to take evasive action at the sound of a tone. They found that the response to the tone had increased dramatically after the conditioning. It reduced again when the sound regularly occurred without the unpleasant sequel, and the conditioned response had been abandoned. However, long-term potentiation in the amygdala, once established, may take a lot of destroying. The evidence suggests that if the danger ceases to materialise the amygdal response to the conditioned stimulus does not fade nearly as fast as the behavioural one. Instead it's suppressed by messages from prefrontal cortex. Animals with lesions in the medial prefrontal area continue to react to a warning signal for a long time after the nasty experience has ceased to follow it, and in normal animals the evasive action is resumed rapidly if it becomes useful again.

A permanent record of unwelcome contingencies seems sensible, especially when the cue has been associated with something particularly noxious. The long-term nature of amygdal records probably explains why stimuli that were once very threatening can still provoke a reaction decades after the threat has vanished - why, for instance, some of us oldsters still feel a twinge of disquiet at a sound like an air raid siren, or the drone of a heavy bomber.

The basal forebrain also plays a part in this sort of learning. An experiment with anaesthetised mice showed that a small electric shock to the foot prompted an acetylcholine input from the basal forebrain to GABA interneurons in layer 1 of sensory cortex, causing them to step up their firing rate. These cells connect to another set of inhibitory interneurons in layer 2, which are tonically active and connect in turn to the cell bodies of local pyramidal neurons. Inhibition from layer 1 suppresses the activity in the second set of interneurons, and leaves the pyramidal neurons free to use their full capacity on the signal that predicts the noxious input. When this disinhibition occurs the record of the signal is still being maintained, we must assume, by a sustained circulation of impulses around areas which include the auditory cortex, or in other words in echoic memory. This chain of connection from basal forebrain to layer 1 to layer 2 is another mechanism whereby reinforcements and discouragements can be linked to the cues that precede them, switching attention, if necessary, to the contents of echoic memory.

In the basal ganglia the synapses of cortical neurons on striatal ones are subject to both potentiation and depression - not only those on the GABA-using cells which project out of the striatum and account for 90% of the total, but also those on acetylcholine and dopamine neurons. The role of the basal ganglia circuits in the formation of habits is clearly dependent on this synaptic plasticity.

As we saw in the last chapter, as an animal learns how to gain a reward some neurons in the ventral striatum come to fire in a way that suggests anticipation of the reward, others in a way that may represent preparation for action. The number is substantial enough to suggest that much of the potentiation is fairly temporary. Once a learning task is mastered activity in the striatum grows sparser and more focussed, so many neurons must be depotentiated and liberated for further learning.

Dopamine levels are high when an animal is working out the rules that will enable it to gain a reward, and reduce when the challenge is mastered and a behaviour becomes habitual. It's no surprise, then, that dopamine is essential for the potentiation of both excitatory and inhibitory synapses in the striatum. A pulse of dopamine at exactly the right moment prompts enlargement of spines on the medium spiny neurons. Mice that were deficient in dopamine failed on a whole range of learning tasks, but object recognition, discrimination learning and spatial memory were all normal after the neuromodulator was restored to the dorsolateral striatum.

Different parts of the dopamine mechanism seem to be particularly important in different aspects of learning. Mice with a particular version of one dopamine-related gene are especially good at probabilistic reward learning, while a certain version of another gene makes for an above-average degree of efficiency in avoiding negative reinforcements. Yet another gene, associated with dopamine function in prefrontal cortex, contributes to an ability to adapt rapidly to new circumstances.

Potentiation and depression also occur in the thalamus and the superior colliculus, much of which presumably embodies learning about the implications of relatively simple sensory stimulus patterns. They are also found, predictably, in the neocortex, where much more detail can be discovered, and much finer discriminations made between the stimulus patterns that deserve a response and those that don't. And in motor cortex neurons modify their activity as a response to a cue is learnt, some firing more frequently, others less. The synaptic modifications are often accompanied by the growth of new synapses. The formation of new dendritic spines was observed in the motor cortex of mice as they learnt a forelimb reaching task. The spines grew within an hour, became stabilised after further training, and remained after training stopped. (Some pre-existing spines shrank away concurrently, leaving the total number little changed.)

But the picture is complex. Different forms of potentiation occur in different areas - that in motor cortex isn't like that in sensory cortex - and even in different layers. The sort of input that creates long-term potentiation in layers 2 and 3 of rodent sensory cortex results in long-term depression in layer 4.

A great deal of our learning is stored in the neocortex - the sensory schemata and the expectations connected with them, the knowledge of the self's potentials, the rules about how things work. Subjective experience indicates that this sort of learning can last a very long time, even if it isn't always readily accessible, so potentiation in what are known as the association areas - or at least in some of their layers - can presumably be seriously long-term.

In the cerebellum synaptic plasticity provides the means of engineering a conditioned response to a cue, but since this type of conditioning only happens when the two inputs overlap slightly the mechanism may be a little simpler than in some other areas. But the defensive response is produced at exactly the right moment, so the cerebellum obviously accomplishes some very subtle timing.

A most important site of synaptic plasticity, finally, is the hippocampus, the structure where long-term potentiation was first discovered. But there will be no more about it here, because it's about to get a chapter to itself.

As might be supposed, potentiation in different areas can proceed in a cooperative manner. That in the basal ganglia develops in tandem with that in the neocortex, with the former leading the way. Modifications to synapses in the locus coeruleus can be co-ordinated with those in the auditory cortex. Activity in the basolateral nucleus of the amygdala promotes long-term potentiation both in the cerebellum and in cortico-striatal synapses. In the latter case the influence operates even when the inputs are far from synchronised - which is probably why events that occur in an emotional context tend to be well remembered. Even within one structure, meanwhile, learning is likely to involve modifications at several different sets of synapses.

Addiction

A supernormal capacity for effecting potentiation and depression explains the power of addictive drugs. They mimic the effects of neurotransmitters and modulators, or distort the way they work. Nicotine, for instance, fools nicotinic acetylcholine receptors (hence the name). In the ventral tegmental area dopamine cells receive glutamate-deploying connections from the neocortex, and there are acetylcholine connections to the axons of these glutamatergic neurons, operating through nicotinic receptors. Acetylcholine input leads to an increase in the amount of glutamate released on to the dopamine neurons, so that NMDA receptors come into play, leading to potentiation of the glutamate-dopamine synapse. Nicotine consumed by way of cigarettes promotes this form of potentiation just as the natural transmitter does.

Experiments with rats have shown that a single dose of cocaine produces long-term potentiation at the same junction, which lasts for several days. Again, the strengthening effect is achieved indirectly, in this case by reducing the amount of GABA delivered to the glutamate neuron. Opiates do something similar. When a glutamate message is powerful enough to excite NMDA receptors on the dopamine neuron the latter normally releases a messenger which reaches a neighbouring GABA terminal, and prompts potentiation there. The strengthening of the excitatory input to the dopamine neuron is thus balanced by a strengthening of the inhibitory synapse. Opiates block the message to the GABA synapse, thus preventing its potentiation and making the dopamine influence more powerful. Benzodiazepines can similarly interfere with the inhibition of dopamine neurons.

Cocaine, amphetamine, nicotine and morphine can all trigger the earlier stage of gene expression. Addiction is established when full-scale protein synthesis occurs and synapses are more permanently altered. Twenty four hours after cocaine administration an augmented number of AMPA receptors is found at synapses in the ventral tegmental area. Repeated consumption of psychostimulants such as cocaine and amphetamine increases the dendritic complexity and spine density in the medium spiny (GABA) neurons of the ventral striatum, in dopamine neurons in the ventral tegmentum, and in some cortical pyramidal neurons. The soothing opiates have the opposite effect at most of these sites. In experiments with rats, interestingly, the alterations brought about in the ventral striatum by either morphine or cocaine are greater when the drug is acquired voluntarily rather than adminstered by an experimenter.

Once drug use has become habitual a whole series of structures - ventral tegmental area, insula, amygdala, thalamus and parts of frontal cortex - are activated by sensory inputs related to taking the drug. Any such input thus becomes a cue for craving. It's not yet clear, however, whether all the changes in neuron morphology and synaptic structure contribute to promoting addictive behaviours or whether some of them should be regarded as a way of compensating for distorted functioning in other parts of the circuitry.

Clearly, though, addictive drugs co-opt many of the mechanisms of learning. Moreover their effect can be harder to reverse than that which results from natural rewards. This limits the amount of change that can be wrought by subsequent experience, which is why addiction is hard to cure.

But the equation may also work in the opposite way. The fact that mice living in interesting surroundings are less susceptible to addiction than those housed in bare laboratory cages may be attributable to the greater number of synapses available for potentiation in bored animals. The same factor may help to explain why young people tend to be more vulnerable to addictive drugs than those whose brains are more tightly packed with experience.

Not just learning

Long-term potentiation and depression have been studied principally in the context of short-term and long-term memory. It's relatively straightforward to establish whether an animal has learnt something, and how long the memory lasts, and to correlate the learning with synaptic changes. But potentiation and depression can be deduced to achieve other ends as well.

The neuronal networks of the brain are inherently noisy, and the very short-term synaptic modifications provide a means of filtering out significant events more efficiently. Spontaneous firing among the neurons that feed into a cell may elicit a slow succession of excitatory potentials but only AMPA receptors will be involved. It takes a serious burst of input to activate the NMDA receptors and turn the cell on properly.

The earlier forms of potentiation and depression must clearly play a role in consciousness. If consciousness is created by a self-maintaining circulation of impulses around interlinked neuronal circuits, then the early form of potentiation should function to increase the sensitivity of the synaptic connections in the circuits, allowing the flow of excitation to proceed more freely. The minimum input required to produce such experience must be the minimum needed to set up an adequate degree of potentiation in the earlier links of the sensory pathway, so that the current grows strong enough to reach prefrontal cortex.

In the chapter on creating the picture in the brain experiments were described in which a stimulus pattern that wasn't consciously perceived at a brief first presentation became apparent at the second, third, or fourth. Presumably the repeated inputs gradually built up short-term potentiation in synapses along the early stages of the visual pathway, allowing impulses to reach a little further at each exposure until they extended as far as the schema-neurons, where synapses had already undergone long-term potentiation. Once the current reaches a schema-neuron return messages can be sent to visual cortex, and all the neocortical and subcortical components necessary to set up an adequately sustained flow of current are brought into the circuit, including prefrontal cortex.

With a normal amount of input familiar stimulus patterns can reach consciousness quickly. But if the pattern is novel and unidentifiable there is no established pathway, and no guidance from schema-neurons, so the initial conscious experience is a muddle. And ambiguous input may activate the wrong pathway. If an incorrect guess is made and the wrong schema activated, the guidance sent back to visual cortex may emphasise some of the less definitive parts of the input. Short-term potentiation established in an inappropriate circuit probably explains why it can be hard to sort out inadequately defined input correctly after making a wrong guess.

Meanwhile sustained excitations along a pathway must in turn promote a slightly longer-lasting potentiation at other synapses, including those leading to prefrontal cortex. These potentiations must underpin short-term memory and attention. The synchronised rhythms of excitation which are so important to attention must depend heavily on short-term potentiation of the synapses involved in the circuitry. The importance of dopamine, acetylcholine and norepinephrine in attention can be deduced to lie in their contribution here.

Inhibitory synapses are modifiable as well as excitatory ones, and it's becoming clear that the shaping of schema areas and other forms of neocortical learning owe as much to the potentiation of inhibitory synapses as to changes in excitatory ones. In some instances the inhibition no doubt serves to define the currently activated circuit more firmly. In others the synapses between inhibitory neurons are potentiated, so that activation of the first neuron interrupts the inhibition normally excercised by the second upon an excitatory cell. This looks as if it may turn out to be a widely employed mechanism for learning in the neocortex.

The more permanent varieties of long-term potentiation and depression, meanwhile, are involved in the very shaping of the brain. The growing neocortex, in embryo and newborn infant, produces many more neurons than it needs. They have to compete for synaptic connections, and those that fail to gain a foothold in the circuitry, and thus establish a job for themselves, fade away. In the embryo, where sensory input is obviously limited, spontaneous firing contributes significantly to the initial patterning of circuits. Early experience of the external world takes the process further. Long-term potentiation and depression are clearly at work here. Indeed a good deal of the study of the subject has been done with young, immature animals.

However, in a species where the period of immaturity is extended the modifications can take quite a while to become fully established, for potentiation works differently, in some ways, in the infant brain. Some receptors are constructed out of a slightly different selection of subunits, and the infant forms are replaced by adult ones as the animal matures. This arrangement probably ensures that learning doesn't become too firmly embedded while there's a chance that what's learnt in infancy or juvenility will prove misleading in adulthood.

In the older brain there's no longer such cut-throat competition for survival among the neurons, but there is still competition, and the neocortex remains fairly flexible. Areas of sensory cortex which get a lot of use expand, at the expense of less used or unused areas - an expansion which can be attributed to the space taken up by denser numbers of synapses and the growth of additional twigs on dendrites and axons. In the blind the cortical areas that are normally devoted to visual matters shrink, and auditory or other input invades what remains. Similarly, the deaf may put auditory cortex to visual use. The takeover is more thorough in those who are blind or deaf from birth, but it can still happen in the adult. If someone loses an arm, so that the tactile information once reported from it no longer arrives, adjacent areas of the somatosensory map in the neocortex expand and occupy the vacated space - which may mean that a touch on the cheek feels to the amputee like a touch on the missing limb.

Less drastic experiences also bring adjustments. Learning to make many fine discriminations leads to the relevant bit of cortex expanding its fief. Expert Braille readers devote an unusually large amount of somatosensory cortex to the input of their first two fingers - the ones used for reading. If monkeys are trained exhaustively on a task that demands delicate sensitivity in just one or two fingers the effect is similar. Neuronal plasticity brings a degree of plasticity to the whole neocortex, allowing it to adapt its architecture to the particular learning requirements it encounters as well as to the effects of injuries.

For the most part, however, the majority of the spines which support the synapses on the principal neurons of the neocortex seem to be fairly firmly established by adulthood. The essential wiring system has become fixed by then, as it were. A reduction in stability is one of the symptoms of Alzheimer's Disease, schizophrenia, and various other mental disabilities.

Recapitulation

A great deal of evidence indicates that learning is achieved by means of modifications to synapses.

Neuronal connections can be strengthened (potentiation) or weakened (depression), and both processes are involved in learning.

Glutamate, GABA, norepinephrine and dopamine synapses are among those which have been shown to be capable of such modification.

A whole variety of mechanisms contributes to these synaptic modifications, at the input synapse, along the dendrites, at the beginning of the axon and at its terminals.

Various neuromodulators play an important role in modifying the state of the neuron to support the learning.

The modifications begin with a temporary form, which may fade; or it may be strengthened, by the addition of new proteins to the synapse, into a medium-term and then a long-term version.

In the principal neurons of the neocortex and in the medium spiny neurons of the basal ganglia the synapses are situated on little blobby protrusions, known as spines, within which the various influences affecting potentiation and depression can be largely confined.

Potentiation and depression have been found in basal ganglia, amygdala, cerebellum, hippocampus, thalamus and superior colliculus, as well as neocortex.

The early form of potentiation must surely be involved in establishing the sustained circulation of impulses which creates conscious sensation.

The longer-term modifications do a good deal to shape the infant brain, and to adapt the adult one for the challenges it meets.

Chapter 19

Contextual and Episodic Memory

Back in the fifties, in Canada, there was an unfortunate young man, Henry Molaison, known to medical history as H. M., who had suffered since the age of ten from severe and debilitating epilepsy. To cure it the doctors decided to try removing the focal area in which the hyperactivity originated and from which it spread to engulf other parts of the brain. In H. M.'s case – as in many others - the problem area was in the medial temporal lobe (the bit that folds underneath the rest), and included the structure called the hippocampus. Most of this structure and a substantial amount of the surrounding areas were removed, which successfully cured the epilepsy. But a drastic additional result was that H. M. no longer remembered anything for more than a few minutes. His echoic memory still functioned, so he could keep something in mind a little longer by repeating it over and over. And he could still recall much of what had happened up to a year or two before the operation. But no new information, it seemed, went into either medium-term or long-term memory. He didn't consciously recognise the doctors he saw regularly, nor other new faces, nor could he learn to find his way around the hospital, or other places he hadn't known before the operation. He lived only, as it were, in the present moment, apart from the long-term memories that dated from before the operation.

The obvious implication was that the hippocampal area, while not necessary to consciousness or to echoic and very short-term memory, is essential to the medium-term memory that provides the ongoing context of our lives - the memory that holds a fairly good account of what you did an hour ago, a less detailed recall of yesterday, and a very incomplete recollection of last week. The further implication seemed to be that this medium-term memory provides the route to long-term memory, or at least to the form of long-term memory which can be consciously consulted. The story of H.M. focussed interest on this area of the brain, so when Bliss and Lømo discovered the phenomenon of long-term potentiation in the hippocampus some years later there was every reason to be excited. The two strands of evidence added up to the first major insight into the mechanisms of memory.

H. M.'s problems have been investigated in great detail over many years, and a good many other patients with various degrees of damage in the hippocampal area have also been studied. The results are not clear-cut enough to make the picture simple, since lesions inevitably cover varying amounts of the hippocampus and adjacent areas. (In H.M.'s case the amygdala was removed as well.) But they confirm that this stretch of medial temporal cortex does indeed function as a medium-term memory, and that it plays a very important role in the business of creating a cognitive map, and in acquiring other patterns of information that are spread out over more than a very brief stretch of time. It's also pretty much essential for getting many sorts of information into long-term memory in a form which can later be re-accessed by consciousness. In the other species which have been studied the story is the same. Rats still navigate efficiently in an environment that was thoroughly familiar before hippocampal damage, but are poor at mapping new places.

The hippocampus proper is a folded bit of ancient, three-layered cortex. (This micrograph is reproduced from The Hippocampus Book, edited by Per Anderson et al.) The name is derived from the fact that its curves were thought to resemble those of a sea-horse. It has also been known as *Cornu Ammon*, because it reminded someone else of the horns of the ram-headed god, which is why three of its sectors are known as CA1, CA2 and CA3. Nowadays the dentate gyrus is usually included. It's part of a closely knit formation which includes subiculum and pre- and para-subiculum, and links to neocortex through two long-established bits of cortex known as entorhinal and perirhinal, the former essentially concerned with 'where' information, the latter with 'what'.

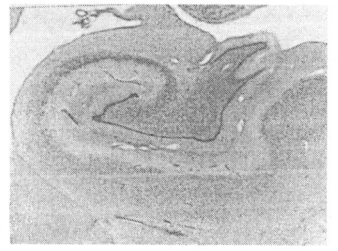

The hippocampal formation is not essential for all forms of learning. People with lesions in this area have shown a normal capacity for acquiring new motor skills, although they don't remember either the learning experience or the fact of their new ability. They may learn to use a new tool, for instance, but not recognise the tool when they see it again. It's as if the hand remembers what to do with it, although consciousness does not. Similarly, the eyes seem to remember where to look, even if there is no conscious idea of what was seen. Two patients who could not recall what had been shown on a computer screen, accompanied by verbal information, nevertheless directed their gaze to the appropriate spot on the screen when cued by the words. In another experiment the diameter of the pupil changed in the normal way when a previously seen stimulus pattern was re-encountered, though the hippocampally damaged subjects declared confidently that they hadn't seen it before. Motor memory, it seems, can still recognise cues that the conscious mind doesn't. Emotional associations are remembered too.

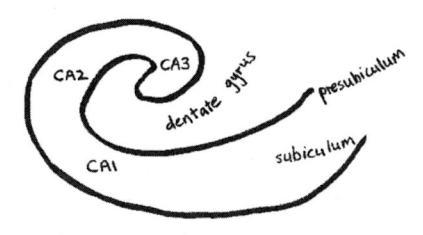

In other words learning that depends on the basal ganglia, cerebellum or amygdala is still possible. The hippocampally damaged can learn how to gain a reward or avoid a shock, though the learning episode is forgotten. They can also register patterns in a brief flow of sensory input, so that, for instance, they can absorb the grammatical rules of an artificial language, and apply them correctly. In one experiment subjects were presented with pairs of cues, out of a total set of four, and asked to make weather forecasts. Feedback was provided, so that guesswork about the implications of various cue combinations could, in principal, be gradually replaced by an appreciation of the recurring patterns. Subjects with hippocampal damage registered the patterns, and did quite well at forecasting, but when questioned later had no recollection of taking part in the experiment. In contrast, subjects with damage to the basal ganglia were insensitive to the patterns of cues and results and didn't learn the rules on which forecasts could be based, but they could describe what they had been asked to do.

Other sorts of information can also be retained, even though there is no conscious memory. Someone with hippocampal damage generally has little success if they are asked to memorise a list of words. But if the list contains a word like *reed* or *pane*, and if, a little later, they are asked to spell that word, then they are more likely to spell the less widely used word that was recently seen, rather than the commoner homophone, *read* or *pain*. So a memory trace must have been established somewhere, even if it's no longer available to consciousness.

Finally, information may get stored in recollectable form if it is enountered regularly enough. A few years after H. M.'s operation he and his parents moved from Montreal to a modest bungalow in Connecticut, where he lived for sixteen years. For a long time he had great difficulty in finding his way around the new home. But three years after another move he drew an accurate floor-plan of the Connecticut bungalow, and could still give the address. Frequent repetition of input, it seems, can make up for the absence of the long-term potentiation usually provided by the hippocampus, but the synapse-strengthening which creates long-term memory proceeds very slowly when it has to be achieved by this means.

This slow learning may depend on the basal ganglia, for imaging experiments show that even in undamaged brains navigation in familiar places comes gradually to involve less activity in the hippocampus and more in the basal ganglia. This tallies well with subjective experience. After a while we tend to respond to a familiar series of landmarks, rather than referring to our mental model of the way to the goal; turning left here, right there, becomes a matter of habit - and one sometimes finds oneself continuing on the more-practised route when the intention was to turn off to a different destination.

Experiments with animals have confirmed the picture suggested by damage to human brains. Rats fail on many learning tasks when their hippocampus has been destroyed, and anything which temporarily disrupts its normal functioning similarly interferes with the same sorts of learning. Distorting the natural pattern of electrical impulses in the hippocampus with artificial stimulation has this effect, as does an injection of a drug which alters the chemical balance in the neurons.

But again, not all learning is affected by the absence of a hippocampus. Rats in such condition are bad at learning their way around a new environment effectively, but they can master the task of locating a concealed objective if they can do it by simply heading towards a landmark. Hippocampally disadvantaged rats can also learn to associate stimuli with reinforcements, and perform straightforward tasks to obtain rewards or avoid discomforts. But they don't learn to associate rewards or discomforts with the place in which they were experienced if the place has to be defined by spatial relationships.

Furthermore they have problems with temporal contexts as well as spatial. They don't do well at alternating their responses in a repeated choice between two stimuli. Nor are they good at choosing from a pair of stimulus patterns the new one, the one that didn't feature in the previous pair a moment ago. A normal rat can master this 'delayed non-matching to sample' test with delays of up to a minute or two between the stimuli that must be compared. One with a disabled hippocampus copes with a delay of a few seconds, but its performance tails off much more rapidly than that of the undamaged animal.

Some of the distinctions between what can and can't be done without a hippocampus are quite subtle, and experimenters are devising ever more ingenious ways of discovering where the line should be drawn. Frédérique Mingaud and colleagues have suggested that normal animals probably use the hippocampus for tasks which are not absolutely dependent on it, and are therefore particularly disadvantaged when they have to perform the same sort of task after hippocampal damage. Animals in which the lesions occur earlier become adept at using alternative strategies. The results of some experiments, these authors showed, can therefore depend on when the lesion occurs.

The theory gains support from reports on three young people who suffered damage to the hippocampus (but not to any of the adjacent structures) in infancy or childhood. None of the three has a normal memory for events. None of them can reliably find their way around in what should be familiar surroundings, or remembers where they have put things. They are equally bad at orientation in time, and have to be reminded of schedules and appointments. And they don't remember stories, television programmes, conversations or messages any better than they remember the events of their lives.

However, all three have a good grasp of language, and did reasonably well at normal, not special schools. Their reading, writing and spelling, and their performance on verbal intelligence tests are all within the normal range, and some general knowledge has been acquired. So it seems that some sorts of information, at least, can be recorded, and produced when wanted. But the learning is more difficult, a study of one of these subjects suggests, and the context in which the new facts were acquired is not remembered unless there has been considerable repetition.

In most cases the hippocampus and the basal ganglia work co-operatively. For instance, in an experiment in which either the basal ganglia or the hippocampal system was rendered less efficient (by blocking dopamine input in the one case and acetylcholine in the other) monkeys could still discover which of a pair of visual stimulus patterns to touch for a reward, when the interval between the encounters with any particular pattern was up to four minutes. They learnt rather faster, however, when both systems were fully functional. Contrariwise, however, it seems that either system may sometimes work rather better on its own. Mice in which the functioning of the dorsal striatum had been disrupted mastered a spatial learning task more efficiently than unmanipulated control animals. And mice with lesions of the dorsal hippocampus showed enhanced learning of cues. The abilities and disabilities of the three people with very early hippocampal damage perhaps indicate just what the hippocampus is essential for, and where it plays a role that is generally useful but not critical.

It's clear that the hippocampus is not essential for linking consequences to simple behaviours, for learning cues that signal the likelihood of rewards, for motor learning, or for the detection of recurring patterns in sensory input over the short term. The human evidence indicates that the ultra-short-term memory which is the moment's conscious sensation is also unaffected by hippocampal damage. But the amount of the immediate past which remains available to consciousness is limited to a very few minutes. The context of the present moment, an awareness of what has led up to it, is lacking.

When the hippocampus is damaged, moreover, the sufferer not only fails to establish new long-term records but also forgets what happened immediately before the damage occurred. The period covered by this retrograde amnesia varies, and can be related to the extent of the lesion. Larry Squire and Peter Bayley found that in six cases where the damage was limited to hippocampus proper and subiculum the amnesia covered only a few years. Larger lesions to the area can cause whole decades of memory to be lost. The most distant memories usually remain intact, so far as it's possible to probe them. The greatest loss of long-term records is found when the damage extends into adjacent areas of temporal neocortex - the area where artificial stimulation in neurosurgery patients has been shown sometimes to evoke early memories.

When people with hippocampal damage recall the distant past, however, it seems that they may recollect less detail than age-matched controls. A likely explanation is that the occasional conscious access to an ancient record normally brings it back into the domain of the hippocampus. Without the hippocampus memory ceases to be dynamic, presumably, and there is no possibility of reinforcement for the long-term records. On the other hand this must also mean that there is no possibility of retrospective elaboration, so a possible corollary is that the sparse autobiographical records of a hippocampal patient might be more accurate than those of an undamaged brain - a difficult hypothesis to substantiate.

The hippocampus, then, is necessary for getting some kinds of information into long-term memory, and a certain amount of time is necessary for the transfer. In an imaging experiment subjects were asked to study three hundred and twenty pictures. They were reshown a quarter of them later the same day, another eighty the next day, a third batch after a month and the last eighty after three months. The activity produced in the hippocampus by these re-encounters was substantially reduced after one day, and declined further across the three months.

Experiments with rats point to a similar conclusion. A rat that has learnt to run a maze, or been trained on any task that involves the hippocampus, generally forgets everything if the hippocampus is disabled within a few days afterwards. If damage occurs after about four weeks there is no effect on the performance.

What the hippocampus does, it seems, is preserve information, in a gradually fading form, over a matter of days or weeks or months. The record must take the form of potentiated connections in a circuit running between hippocampus and schema-areas of neocortex. Just how long the potentiation in these connections lasts will depend on the significance attached to the input. While it persists an edited version of what has been experienced can be built up in the neocortex, by way of new connections among schema neurons. These new synapses take some time to achieve seriously long-term potentiation, so if the hippocampus is disabled before the process is complete the information is lost, or severely curtailed. But once the new neocortical circuit is firmly established the record can be maintained without hippocampal support.

Only a small part of actual experience is preserved in long-term memory, of course, and much of it only in the form of minor modifications to existing schemata. The role of the hippocampal formation was demonstrated by an imaging experiment in which people were asked whether they distinctly remembered seeing a stimulus pattern in a session a week earlier. Reports of actual recollection of the event correlated with activity in the area. Stimulus patterns that were merely felt to be familiar evoked no hippocampal activity.

Place cells

The third major landmark in hippocampal studies was the discovery by O'Keefe and Dostrovsky of hippocampal neurons which consistently became active when a rat was traversing a certain part of an experimental box, and not when it was in any other part. As the rat moved around a succession of neurons became excited, firing at anything from ten to a hundred times their normal rate. In a square box, for instance, some cells fired very rapidly when the rat was in one corner, and faded out as it moved along the edge of the square, where other neurons took over. Different cells were activated in other corners, along other sides, and in the centre of the box. The 'place fields' were found to overlap, so it's the particular selection of cells that fires at any one time that codes for the animal's precise position. If the animal was removed from the box and put back in it later the same set of cells came into operation again in the same way, showing the same place fields.

As well as relating to the overall shape of the environment the firing of place cells was influenced by salient landmarks, both near and far. A patch of a lighter or darker shade distinguishing one wall of the box could intensify the firing of the cells with local place fields. If the patch was then switched to another wall, the geographical relationships of the place cells remained the same, but the firing rates of some of them changed.

Whatever was visible of the laboratory beyond the experimental box also played a role in shaping place fields, enabling the rat to detect when the box had been rotated. Rats can be trained to pay more attention to distant or to nearer cues. And it's been shown that the CA3 portion of the rat's hippocampus is capable of establishing a different set of place cells for at least eleven different but minimally distinguished environments.

If the animal is lifted from one box to an identical one the same set of place cells may continue to operate, but if it moves between them under its own steam, through a door, a new, quite different, set of place cells comes into operation in the second setting. So it seems that locomotory effort is included in the information that's used to create place cells. (Maybe a human who found himself disembarking from an aeroplane at an airport which looked much like the one he'd left might wonder, if there were no signs to reassure him, whether he had really gone anywhere.)

Some place cells are maximally excited when the animal is actually in their field, others when the spot they code for lies ahead. And some of them, interleaved among the others, are sensitive to what's happening in the place they represent. They fire with especial gusto when something new and unexpected occupies the place field, or when something expected is no longer there. Some cells are excited by a particular sort of reward in their field, or something of equally crucial interest. Quite a substantial number of cells in rat hippocampus are sensitive to interesting smells. In female mice hippocampal neurons have been found which respond to the sight of a nest, or nesting material. David Dupret and colleagues found that the number of CA1 cells excited when a rat was in a certain location increased after it had learnt that food would be hidden there, and the number activated correlated with the efficiency with which the rat subsequently headed towards that location. The firing patterns changed when the food was hidden in new places.

In short, the hippocampus functions to link a variety of significant events to the location in which they occur. It's well equipped to do this, for it has extensive connections. Via perirhinal cortex it receives considerable input from the sensory pathways of the neocortex, all of it from their higher reaches and therefore consisting of fully processed information about objects and events. Spatial information calculated in parietal cortex is delivered through entorhinal cortex. The amygdala, the hypothalamus and the basal ganglia are notable contributors, as are the basal forebrain and the thalamus. Pretty well all these connections are reciprocal. There's also a dopamine input, which must help to highlight the presence of desirables and undesirables, and contributes to the learning process.

In some hippocampus cells the response is tailored to the current significance of a place, or the action required. In one experiment rats were trained to run around a figure-of-eight maze, earning a reward if they turned alternately right and left at the junction point. Some of the neurons with place fields along the straight central path fired differently according to whether a left or right turn was needed at the end. They served, in effect, to mark the stage of the journey, and where the animal planned to go from the current location, rather than the location itself.

In another experiment with rats, carried out by Robert Hampson and colleagues, the animal had to press a lever, move to the other side of the box and put its nose in a nosepoke until a cue sounded, and then turn back to where two levers now protruded from the wall. If it pressed the new one it got a reward. The experimenters found hippocampal neurons which fired for a left lever press, others which fired for a right, and also cells which fired at the first stage of the experiment and others which fired at the choice stage. Neurons with similar concerns tended to occur in clusters, and the neurons concerned with position were segregated, in stripes along the length of the hippocampus, from those concerned with the more temporal aspect of the task.

In addition to place cells the hippocampus contains other neurons which are sensitive to the direction in which the head is pointing, in relation to the current environment. Each individual cell covers quite a wide angle of view, but once again population coding is at work and it's the particular combination of activated cells which indicates head direction. These neurons must depend fairly heavily on information from the vestibular system, for when vestibular input has been interrupted rats are slower at spatial navigation even in the light, and severely disadvantaged in darkness. Cells sensitive to head direction are also found in the subiculum, the entorhinal cortex and in a thalamic nucleus closely linked with the hippocampus.

Place cells have been found in monkeys too. In addition there are neurons which fire when the monkey looks at a specific bit of the environment, regardless of the direction it's looking from. Many of these 'view cells' are also unfussy about the distance between the monkey and the scenery viewed. Like place cells, they can remain active for several minutes after the lights go out, suggesting they provide, along with head direction cells, the means of staying oriented for a while in the dark. In one part of the hippocampus the view cells are still activated even when the usual view has been obscured by a curtain. Studying monkeys which roamed freely around the laboratory, Edmund Rolls and his colleagues found that the view field of such a cell could cover as much as one whole wall of the room, or as little as a sixteenth of that area. Once again there seems to be a system of overlapping receptive fields of assorted sizes.

The entorhinal cortex, the three-layered bit of 'old' cortex that channels spatial information to the hippocampus, also contains place cells (discovered by Edvard and May-Britt Moser). These get much of their information from the hippocampus, and like the hippocampal place cells they fire whenever the animal moves into or through a particular spot. These cells, however, are excited in more than one place. When the locations at which the excitations occur are marked on a map of the environment they form the nodes of a lattice. The effect is to spread a set of overlapping grids of varying scales over the ground - grids which can be extended indefinitely as the animal moves into new areas. Each cell's grid is slightly offset from all the others, and the precise combination of node neurons excited at any one moment encodes the animal's position.

In a simple, symmetrical environment the lattice is formed of neat triangles. When an animal finds itself in a box which changes shape, elongating, for instance, or narrowing, the entorhinal grid may stretch or compress to match it. And if the potential travel space is shaped by more irregular boundaries the lattice gets a little distorted. Perhaps it can be inferred that the grid covers all the space currently detectable as being available for locomotion. Consistent with this idea, the rat's entorhinal cortex also contains cells which fire when the animal is close to a barrier or boundary that impedes normal progress.

In the rat the amount of ground covered by the triangles reduces in discrete steps from the dorsal extreme of the entorhinal cortex to the ventral, so that there are, in effect, several networks representing the environment, at different scales. Layer 2 of entorhinal cortex harbours pure grid cells, while in deeper layers there are both grid cells and head-direction cells, and cells which combine the two sorts of information, thus indicating which way the animal is pointing on the grid. In an experiment in which neurosurgical patients played a video game involving navigation around a 'board' some neurons were discovered in which the activity reflected whether the player was taking a clockwise or counterclockwise route.

This abstract way of representing the surroundings has the advantage that it's not tied to any particular place, but can provide continuing orientation while locomotion occurs and the scenery changes.

For all species other than ourselves the only sort of map possible is the cognitive map, created by covering the ground and putting many perceptions together to form a unified pattern. So naturally locomotion is reported to the hippocampal formation, including information about the speed of travel. A few cells are sensitive specifically to the speed of locomotion, regardless of direction or location, and these too must play a role in measuring distances and establishing the hippocampal place cells. Without their input the rat's hippocampus may work less efficiently. When rats were allowed to run freely on a circular track one set of place cells became established. When they were transported around the same track in a little motorised cart there was a considerable change in firing patterns, and the different patterns of activation were maintained over subsequent sessions of locomoting and riding.

Locomotion is also reported to the entorhinal cortex, and here the normal functioning of the grid cells is messed up completely if the rat is carried about in a little cart rather than travelling on foot. Speed-sensitive cells are found in this part of the system too.

Various distinctive rhythms can be discerned in the hippocampus, which are related to coincident motor activity. Large and somewhat irregular waves of impulses can be detected when an animal is sitting quietly, or performing such relaxed behaviours as eating, drinking or grooming. Small irregular jiggles appear in the record during transitions from one sort of behaviour to another - when the animal is waking up, for instance, or when it stops running and is about to relax. But the rhythm particularly associated with the hippocampus is the one known as theta, a relatively slow oscillation at about 5-12 Hz. It's produced by a circuit that also includes the medial septum and the entorhinal cortex. Theta comes in two forms. One occurs when something arouses attention and activates the basal forebrain. It depends on an arousal signal from there, conveyed by acetylcholine - damage to basal forebrain or blockage of acetylcholine receptors prevents its occurrence. The other form of this rhythm, not acetylcholine-dependent, occurs during locomotion, and when the animal is surveying its surroundings and paying attention, but without excitement. In bats this seems to mean specifically when the animal is employing its echo-location system.

The complex bursts of action potentials which occur in place cells when their place field is occupied are related to the theta rhythm. When an animal first runs a new trackway the place cells tend to fire at the peak of the theta cycle. As the route becomes more familiar and place cells are more firmly established they come to fire earlier and earlier in the cycle. One plausible explanation is that the connections to the place cells become potentiated so they come to fire a little more readily.

The theta rhythm also helps to co-ordinate hippocampal activity with that of whatever other brain area is currently being communicated with. In the classical conditioning experiment in which an eye-blink comes to be evoked by a sound cue, for instance, contributions from both hippocampus and cerebellum are needed if the unconditioned stimulus, the puff of air, begins after the warning cue ends. Learning proceeds faster when the theta rhythm of the two structures is synchronised. A theta rhythm in occipital and temporal cortex, meanwhile, is dependent on the hippocampus. During spatial learning prefrontal cortex and ventral striatum also join in with the hippocampal beat.

Other rhythms may contribute to more local synaptic alterations. Fitted within the theta cycles are periods of high-frequency oscillation (60-140Hz) alternating with rhythms of around 25-50Hz. The former synchronises activity in CA1 with that in entorhinal and perhaps perirhinal cortex, the latter co-ordinates it with CA3. This effective separation of inputs to CA1 seems to facilitate learning. Both rates of oscillation are influenced by the speed at which a rat is running.

It seems there are several different forms of plasticity in the hippocampus. Many of the synaptic connections are remarkable for very rapid modification in response to new input, but the changes may not last long if an exploratory period is only brief. Potentiation and depression may be reversed as soon as the animal enters a new environment. Repeated, spaced experiences are best for fixing the changes in synaptic strength, and establishing a record which can be transferred to neocortex. It's the places that provide rewards that are revisited, others can be forgotten, so this is a practical arrangement.

The hippocampus and the cognitive map

In the nineteen seventies O'Keefe and Nadel suggested that the hippocampus constituted a cognitive map. Subsequent discoveries about the relative impermanence of its records indicated that it could hardly be *the* cognitive map, the long-lasting one – something which was likely, anyway, to be too complex to fit into a small area of three-layered cortex. However, the hippocampal formation and adjacent areas clearly do provide a map of the current environment, and perform some vital functions in relation to the more permanent map. They indicate where one is on it, and how oriented - two pieces of knowledge without which a map is useless. In addition they are important to recording the context in which things of interest are found. All this information is essential to the process of creating the long-term map, as is a third facility provided by the hippocampus.

A cognitive map isn't necessary for navigating around the space within reach of the distance senses, only for knowing how to get to what's out of sight.

In order to build a useful cognitive map, therefore, an animal has to be able to remember where it's come from, and how long the journey took. The hippocampal formation provides the means of preserving that information for long enough to weave it into the pattern that gets recorded. It's also important to remember the order in which the various views and landmarks were encountered, and it's been shown that the hippocampus is critical for recording the sequence of such events.

In other words it preserves some temporal patterns. It's also important in other temporal matters. It responds to sudden change in the environment, which may of course signify danger. In addition, it's sensitive to re-encounters with recently registered stimulus patterns. This ability must be vital to mapping, providing the means of discovering when one has been going round in circles.

As we've seen, the process of transferring geographical information from medium term memory to long-term memory takes some time. A long-drawn-out process is more efficient for this form of learning. An animal exploring a new area may make many meandering diversions from a straight path, and what it needs to remember is the essential layout of the area and the best way home, rather than the details of the path it took on one particular occasion. Another day it comes upon a path it has established before, but from a new direction, viewing it from a different perspective. The truly effective map is likely to be distilled out of the accumulated discoveries of several visits. It must be built up gradually, over days of exploration, and it can involve establishing links between sensory inputs registered at different times.

It's the rather special means required for establishing such associations, it may be deduced, that forms records which are accessible for recall. The hippocampus builds separately acquired bits of knowledge into a network, which is such that activation in one bit of the network can lead on to activation in another.

The medium-term memory is also valuable because knowledge of how one has arrived at a place can be crucial to recognising it. Places that are known intimately are generally recognised when they are encountered without a context - in the form of an unlabelled photograph, for example. But there are many useful landmarks that can only be reliably identified in association with their situation. When you choose which way to turn by that glass-walled cube of office block, so like many others, you have to know which stage of your journey you have reached, and which city you are in. In the natural landscape, too, keeping track of the context is important for making use of landmarks. If that distinctive tree loses a branch it may be necessary to relate it to knowledge of where one's coming from to discriminate it from a similar landmark elsewhere.

The hippocampus is necessary, then, for the creation of a cognitive map, but not sufficient - not surprisingly, since the endeavour presents quite a challenge. There's no way of forecasting just how big a map of objective space an animal will need during its lifetime, and no way of telling whether the site of the animal's birth will prove to be the centre of its adult territory, or which way it will set off when it grows to adulthood. How can a completely unpredictable layout be fitted on to a limited set of neurons?

The evidence suggests that evolution chanced on a neat answer. Landscape is recorded in terms of separate elements, which are dealt with in separate brain areas. Knowledge about the stimulus patterns that function as landmarks is stored in an area abutting the hippocampal formation, known, in primates, as the parahippocampal gyrus. Imaging experiments show that this area is strongly activated when buildings are studied, or pictures of buildings, or features of the landscape which can guide navigation. If it's damaged buildings and other major topographical features are no longer recognised.

In one experiment normal subjects were shown a film sequence representing a stroll through a virtual museum, with various objects displayed along the walls, and asked to imagine they were being trained as museum guides, and needed to remember both the objects and the route. Their brains were then scanned while they performed an object recognition task, in which museum exhibits and novel objects were both displayed against a blank background. Exhibits which had stood at junctions, where a decision had to be made about which turn to take, evoked increased activation in parahippocampal cortex, even when they were not consciously remembered, while no such increase was observed for objects which didn't function as landmarks.

In contrast, people who have suffered damage in a bit of cortex just behind the parahippocampal gyrus, tucked around the rear curve of the corpus callosum and known as retrosplenial cortex, have no difficulties with landscape features. But they may no longer remember the spatial relationships between them. They still recognise the more conspicuous and individual buildings of their home town, but they don't know how to get from one to the other, even though the routes may have been familiar for many years. Rats with damage in the equivalent bit of brain have difficulty with mazes, and other spatial tasks, such as taking a direct route back to where a wandering progress began. Temporary inactivation of the area interferes with new spatial learning, and disrupts the navigation of familiar areas in the dark.

Retrosplenial cortex receives a strong input from both parahippocampal place area and hippocampus, and also from the nearby parietal area which translates egocentric space into allocentric space. It has cells which respond to locomotion, and a particularly large ration of head-direction cells. Importantly, there are cells which register the position and orientation of the self. Activity in this area seems to reflect the succession of turns made, the position in the larger environment, and the degree of progress made on an intended journey.

One might say that retrosplenial cortex records the geometry of locomotion - the lines and angles and distances invisibly sketched out on the ground by movement through the environment. As part of the process, it can be supposed, the overall geometry of the environment is also recorded - the relationships between landmarks. For land animals the well-worn paths, and the walls, mountainsides, stretches of water or dense vegetation that limit the potential for locomotion must also be included. Information about such boundaries is contributed by yet another forebrain sector, the occipital place area. Retrosplenial cortex can be seen as providing an adaptable template that can be shaped to any scene, a framework on to which defining landmarks can be fitted, along with one's own position.

There are dense interconnections between all these areas - hippocampal formation, parahippocampal area, perirhinal cortex, retrosplenial cortex and posterior parietal cortex. A couple of nuclei at the front of the thalamus (anterior lateral and anterior dorsal) also form part of the circuitry - damage here interferes severely with some sorts of spatial tasks. Yet another contribution comes from the cerebellum, which uses vestibular information to measure the amount of locomotion performed. Mice in which long-term depression is prevented at one type of cerebellar synapse can no longer find their way around in the dark.

The cognitive map is created, it can be deduced, by marrying the contributions of all these areas, and the hippocampal formation provides a particularly essential link between them all - as well as marking the locations of desirables and undesirables, and any other objects of interest that may be only temporary parts of the scene.

Recording landmarks and their spatial relationships in different areas not only solves a tricky problem, it must also offer great economic advantages. Landmarks can occur in different relationships, as the same spot is approached from different angles. A distinctive landmark is often not all that distinctive, considered in isolation. But with separate sets of schemata for spatial geometry and for landmarks it's possible to create different combinations. Two similarly shaped landscapes can be distinguished by the different landmark features they contain, while two similar landmarks can be distinguished by the fact that they are encountered on different journeys. By this means a great deal more information can be stored in a given volume of neurons than could be achieved if every scene had to be mapped separately. The system parallels the way the 'what' and the 'where' of visual input are handled through separate pathways.

Introspection supports the idea that the overall geometry of a scene can be separated from the landmarks. On the one hand there are broad sweeps of light and dark, any large areas of salient colour or texture, any conspicuous barriers to view and movement, any very distinctive lines - whatever takes up a really significant amount of the visual field. Out-of-doors that means the shape of the terrain on the largest scale - mountains, big stretches of water, field or forest, or perhaps clusters of large buildings, and walls, hedges or roads that significantly bisect the scene. Indoors the all-embracing context is provided by walls and ceilings, doorways and windows, arches and stairs. They impose an overall composition of horizontals, diagonals and verticals, curves and angles.

The landmarks which fit within this geometry need to be fairly conspicuous, and they are likely to take up a certain amount of retinal space, but not too much. And keeping correctly orientated in relation to the distant landmark tends to require more attention than following the edge of the forest, or otherwise being guided by the enveloping geometry of the landscape. But these distinctions between the landmark and the overall geometry of the scene are mutable. As one approaches the distant mountain peak or the distinctive building it becomes big enough to become part of the spatial geometry. And finally it may embrace the traveller, so that the features which were distinctive at a distance are no longer visible, and it becomes the immediate environment.

In short, retrosplenial and entorhinal cortex seem to be responsible for registering overall context, largely in terms of colour and texture and the barriers to locomotion. Parahippocampal cortex deals essentially with the landmarks which guide travel across the landscape. The hippocampus is important for registering the position of the self in relation to both overall view and landmarks, and the position of anything else of interest.

Episodic memory and other complexities

The cognitive map is put together with the aid of the hippocampal medium-term memory that records where the animal has come from, so in a sense geography is a form of history. And the map becomes more useful if it's combined with some episodic memory. The record of a place may embody the essence of many visits, but it's more valuable if it can be combined with records of particular visits. A fructivore benefits from not revisiting today the tree that was stripped of all its ripe fruits yesterday. It's useful to remember that a certain spot is a good place to go in times of drought, a bad one in heavy rain.

Some sort of memory for events, moreover, seems to be an essential prerequisite for some of the more elaborate forms of offspring care. A fox who decides the current nursery den is no longer suitable and transfers her young to a new one must remember she has done so. Species like crocodiles, which bury their eggs in the sand and come back to protect the youngsters when they hatch, must not only remember the site where the deed was done but also need to measure how much time has elapsed since.

Geography is thus intimately linked with history. Spatial patterns must be explored over time, and the information acquired must be preserved while it's shaped into a cognitive map. Conversely, the events of life unroll and a temporal pattern can be perceived which assumes a sort of shape, and can be perceived more or less at one go, like a map. It's often only when the shape of the whole episode is perceived that it becomes clear whether any of the individual incidents are worth remembering. The hippocampus has cells which are sensitive to time rather than place, and some in which the activity reflects both time and place.

There's evidence that CA1 is particularly important to registering temporal patterns. Rats with lesions here had difficulties with a temporal task but not a spatial one. Lesions in the dentate gyrus produced the opposite effect. Furthermore in CA1 (but not CA3) the activity prompted by a repeated experience varies according to the number of hours or days that have passed since the first occurrence. The hippocampus thus provides a means of estimating recency - something which is of course important to mapmaking, since if one goes around in a circle it's helpful to know how long it has taken to do it. In humans, where the increasing complexities of the matters dealt with may have led to greater segregation of processing facilities, the right hippocampus seems to concentrate on spatial matters, the left on temporal patterns.

An emphasis on geography and history doesn't quite do justice to the subtleties of the hippocampus, however. A popular experiment for testing learning about locations entails giving a rat a mild electric shock when it's in an experimental box, and observing how long it freezes when it's put back in the box. Hippocampal lesions lead to reduced fear conditioning for both simple and complex contexts, but the effect is greater in complex contexts.

Hippocampally damaged humans seem to be similarly challenged by complexity. Several such individuals were asked to study the locations of between one and five images in a virtual environment, and then to recollect the locations either from the same viewpoint or from a different one. Changing the viewpoint proved not to be a problem when there were only two images, but the performance deteriorated as the number increased.

Another experiment involved five hippocampal subjects. Four had lesions limited to the hippocampus, one had more extensive damage, and all suffered problems with topographical memory but were not perceptibly affected in other matters. A picture of landscape was briefly shown, followed by four variations on the same picture, from which they were asked to pick the one that matched the sample. In some cases the same bit of country was shown from somewhat different viewpoints, in others the view was the same but the lighting angle, the colour of the vegetation, or some other superficial element was different. On the non-spatial tasks two of the patients equalled the controls, while the others scored only slightly below them. But all five were significantly impaired on the spatial task - although they were not handicapped in simpler tests of mental rotation.

The explanation probably lies in a recent experiment with monkeys which found the grid cells of entorhinal cortex at work when visual objects were explored with saccades. This implies that the hippocampal machinery contributes to assembling the fruits of a series of fixations into meaningful form. If so this may explain some of the difficulties with mapbuilding, since large saccades are necessary to take in a whole view. And when similar stimulus patterns must be discriminated there is likely to be a high level of saccading around them, which may be why the hippocampus is busier than when more distinctly differentiated objects are on view. Perhaps this might help to explain why people with considerable deterioration in the hippocampal formation, as in an advanced stage of Alzheimer's, are sometimes unable to recognise old friends, and eventually even their nearest and dearest.

In some parts of the hippocampal formation, meanwhile, it seems that several varieties of information can feed into one cell, suggesting an essentially conceptual function. Neurons have been found (in patients undergoing surgery for epilepsy) which fired for any view of a particular individual, whether the face alone or the full figure, and even for the name that belonged to the face. It looks as if the input to this part of medial temporal cortex comes from schema-areas. Any scene from the television cartoon *The Simpsons* excited one neuron, while another responded equally to the Sidney opera house and a similarly styled Ba'hai temple. One cell fired both for Jennifer Aniston and for Lisa Kudrow, the two female stars of the television series *Friends*. The hippocampal formation seems to have expanded its original function of assembling temporally separated sensory inputs into an overall pattern and come to play a part in the development of more general concepts.

Renewal and Expansion

The hippocampal formation keeps recording new information, the more significant bits of which are gradually consolidated in the neocortex, which in large-brained species offers much more extensive storage facilities. The record in the hippocampus fades, if not renewed by further activation. Is the continual addition of new information achieved solely by depotentiation of potentiated synapses and restoration of depressed ones? This must be part of the answer, for the turnover of dendritic spines is very fast here, compared to other brain regions, but it's not the whole story. Another factor is that the dentate gyrus is one of only two structures in the mammalian brain (the other is the olfactory bulb) which has unambiguously been shown to keep incorporating new neurons and replacing old ones. In the rest of the brain neurogenesis seems largely to cease by adulthood. But in the adult human it's estimated that an average of 700 new dentate cells are born every day, maturing to replace old ones that die away.

Animal experiments show, however, that external circumstances influence the rate of replacement. Individuals living in an interesting environment acquire more new neurons than those living in boring laboratory cages. More new cells appear in rats allowed to run in a wheel than in sedentary ones, and the rate of production can be doubled or tripled by exercise. Clearly new cells are born when they are needed, most notably when there is much locomotion, which in the wild is likely to be taken for exploratory or investigatory purposes. Stress, on the other hand, reduces the number of new neurons produced, as does sleep deprivation.

When new circuits are created or old records updated and reinforced, new neurons are more likely than old to be incorporated. Their membrane characteristics and their virgin synapses are particularly susceptible to long-term potentiation. Old neurons linked to too many disused circuits may die away, so the total number of dentate neurons doesn't necessarily increase.

The infant hippocampus

The hippocampus is immature at birth. In humans it starts functioning after about a year, and only reaches its final form somewhere between two and five years. This timing is appropriate. The vital learning in early days depends mainly on amygdala and basal ganglia. The hippocampus gets going at about the time the child becomes mobile and begins to need some modest capacity for navigation, and matures as longer wanderings become possible. The delayed maturation, it's been suggested, explains why we can't recall our infancy.

Nadel and Jacobs have hypothesised that the earlier functionality of the amygdala may provide an explanation of phobias and irrational panic attacks. They may result from learning experiences which take place when the amygdala is at work but not the hippocampus. The learning experience would be potent but it wouldn't be remembered consciously, nor fitted into any context, so it couldn't be subjected to rational analysis. It would qualify as pure conditioned learning, the sort by which many species' lives are largely directed. But we humans expect to have some idea of why we are frightened, and a fear beyond the reach of reason is therefore exceptionally distressing. The initial emotion is thus augmented by the cognitive variety of fear, which springs from not understanding the nature of the threat.

Recapitulation

The hippocampus and structures associated with it, making up the hippocampal formation, provide the memory which is used to put together information gathered over more than a brief period of time.

This is important for creating a cognitive map, or for remembering a sequence of events.

The hippocampal formation must also play a part in relating the visual inputs gathered over several fixations.

The hippocampus contains cells which fire maximally when an animal is in a particular place, and others which respond to a place that contains something significant.

In creating the cognitive map the hippocampal formation works with the entorhinal cortex, the parahippocampal place area, which records the sort of stimulus patterns that function as landmarks, the retrosplenial cortex, which records the spatial relationships between landmarks, the parietal area that turns egocentric information into allocentric, the cerebellum, which measures distance covered, and two thalamic nuclei.

Some things can be learnt without the aid of the hippocampus. New motor skills can be acquired, regular combinations of information registered over brief periods of time can direct behaviour, and painful experiences can be connected with predictive cues, though not with the places where they occurred.

However, without the hippocampal contribution, and in the absence of extensive repetition to compensate, this sort of acquired knowledge is not accessible to consciousness.

The patterns which must be registered with the aid of hippocampal memory are transferred slowly into neocortical memory, where the necessary fullscale long-term potentiation takes some time to set up.

This is an arrangement which makes it possible to edit the long-term record as it's constructed. The developing cognitive map can be modified as additional information is collected. The event that yesterday seemed as if it might be significant can be forgotten if the experience of succeeding days indicates that it wasn't.

Chapter 20

Sleep, Dreams and Memory

For those interested in consciousness dreams must have a special fascination. Dreams, insofar as we know about them, are consciousness divorced from sensory input. Like thought, they seem to be part of the independent life of the mind. A creaking door or a cramp in the foot may sometimes get incorporated into a dream in some transmogrified form, just as a train of thought may be triggered by something seen or heard; but both dreams and thoughts can proceed under their own steam, and weave their patterns uninfluenced by what is currently happening in the external world. Both, too, seem to be made out of the same raw materials - images and ideas recalled from the memory store, and emotions. Freud, indeed, defined dreaming as *nothing more than a particular form of thinking, made possible by the conditions of the state of sleep.*

Styles of dreaming seem to vary between individuals, just as styles of thinking do. For many of us the figures of the dream world seem to be more like ideas than images. The sensory experience is hazy, but the knowledge is certain. There's something horrible chasing me, and it's imperative to get away, running as fast as possible, leaping over obstacles, there's no time to turn and see what it looks like. The pursuer is just an abstraction of horror, while the emotion and the action are more clearly conceived. The figures and objects perceived in dreams seem more like a succession of brief glimpses than fully realised images.

Gordon Rattray Taylor's descriptions of his dreams sound very similar. His, he wrote, seldom involved lengthy narratives, but mostly consisted of what he called 'feelings' - the feeling of having had a satisfying day, the anticipation of a meeting with friends, awareness of having a second thing to do after having done a first. *I remember a dream which started with me walking along with a friend, and knowing that his wife and my wife were walking behind us. At no point did I look round and see them.* He also writes of experiencing a confident knowledge of identity, even if a dream percept didn't match reality. *I see a white cat and know it is my cat Isabella, although in real life Isabella is a black cat...... I talk to someone I know is my sister, although my sister does not actually look anything like the person in the dream.*

So introspection suggests that dreams are built from the semi-abstract schemata that we build from our perceptions. By semi-abstract I mean that sense of an unchanging identity that is extracted from many encounters, and many different views of an object or person or landscape, the quintessential concept that allows non-signifying variations or changes to be ignored, and which is defined above all by an emotional evaluation, or by potential use. (I suspect that the dream-characters who don't look like themselves are born out of those disconcerting childhood experiences when loved figures make radical changes to one of the salient features by which they are recognised - adopting a new hairstyle, covering their hair, or shaving off a beard.) That these semi-abstract schemata form the basic material of dreams for many people seems a reasonable deduction, since V1 is often not very active during dreaming sleep, although more advanced stages of the visual pathway are, along with some other sensory cortex, and language and motor areas. This may also explain why dream images tend to be blurry, and hover in not fully realised spatial relationships.

Some people, however, report much more vivid dreams, and these seem likely to be the people who also think in very visual terms. Giulio Tonini and Yuval Nir, for instance, describe highly visual dreams, in full colour, with clear sensory character, not mere thoughts or abstractions. Presumably in this sort of dreamer the feedback connections from more forward areas of the brain to V1 still operate strongly in sleep.

Serious neurological interest in dreams began back in the nineteen fifties, when some of the physical correlates of dreaming were revealed by the electroencephalograph, the machine which records the synchronised fluctuations in membrane potential or actual firing shown by the neurons beneath each electrode. The pattern of activity develops in a consistent fashion from waking to sleep, and through the night. Active waking coincides with short, erratic ups and downs. In quiet relaxation the zigzags grow larger. As sleep begins and the outside world ceases to impinge on the senses the 'brainwaves' slow; then gradually their amplitude increases further, and the rhythm grows still slower. Four distinct stages can be identified. Stage 1 is a little ambiguous, and is regarded as a transition phase. As the sleeper advances to stages 3 and 4 it becomes increasingly difficult to rouse them. But after Stage 4 the brainwaves change back to a pattern quite like that of the waking state, and at the same time the eyes begin to move about under the closed eye-lids as if there were something to look at.

Wake someone up during this period of sleep and they will usually say that they have been dreaming. Roused out of any other stage of sleep they may report that there was something on their mind, that they had been having vague thoughts about something, but usually nothing of a narrative nature. We can't conclude that rapid eye movements in sleep are always accompanied by a conscious dreaming experience, but it seems clear that there is a particular state of the sleeping brain that supports dreams. It's known as Rapid Eye Movement or REM sleep, or sometimes as paradoxical sleep. Stages 1 to 4 are referred to as quiet sleep, or nonREM, and stages 3 and 4 more specifically as Slow Wave Sleep.

Other physiological changes also mark sleep. Heart rate and respiration slow, temperature falls a little, muscles relax and the metabolic rate slumps, as quiet sleep takes hold. In REM sleep, however, some of the signs of inactivity are reversed while others are intensified. The flow of blood to the brain increases dramatically, the brain temperature rises, and breathing and heartbeat become irregular. At the same time a firm inhibitory message is sent to the channels which control the use of skeletal muscles, so that practically all bodily movement is suppressed. The most that is seen is a little twitching of fingers and toes. This suppression is important, for cats in which this neuronal connection was severed stood up during REM sleep, moved around, and generally gave the impression that they were acting out their dreams. Occasionally the same thing happens in us, when the suppressive mechanism malfunctions. But normally only eye movements and facial expressions occur during REM sleep.

As quiet sleep grows deeper, meanwhile, the waking rhythms of the thalamus have been replaced by regular prolonged bursts of firing, which means that messages from sensory systems are unlikely to get through to the neocortex. In REM sleep they are even more thoroughly blocked. The regular pulsing of the norepinephrine system becomes gradually weaker during quiet sleep, and fades away altogether during REM. Only a very significant sensory input can reactivate the system, and wake the dreamer. The dopamine pulse is slowed during sleep but continues. And the acetylcholine connections from the brainstem which also contribute to attention go into an idling mode in quiet sleep. Deprived of these influences cortical neurons undergo prolonged hyperpolarisations, followed by spontaneous firing, producing the large slow wave-forms on the electroencephalograph. For many neocortical cells the mean rate of firing is more or less equal to that of waking, although the pattern is so different. Energy consumption in the brain nevertheless decreases progressively as sleep deepens.

Just before a REM period begins, however, the acetylcholine neurons of the pedunculopontine and laterodorsal tegmental nuclei spring into activity again. Thalamic function returns to something more like its waking form, and much of the neocortex is activated. Waves of excitation flow from hindbrain to thalamus to visual cortex. Metabolic activity increases in all these areas, and also in the hippocampus and in areas associated in one way or another with reinforcement and emotion: the amygdala, the basal ganglia, and the hypothalamus. Prefrontal cortex, however, mostly remains quiet, particularly that part of it involved in attention and perhaps particularly associated with reason - which may account for the way dreams operate with a logic all their own.

A complete sleep cycle in humans, from falling asleep to the end of the first REM period, averages about 90 minutes, and there are usually three more cycles during an eight hour night. The first stage 4 period is a long one, the second rather shorter, and stage 4 sleep is generally absent in the later cycles. The first period of REM sleep, in contrast, is a brief one, the second longer, and the third and fourth longer yet. Short sleepers - people who get by on around six and a half hours - seem, for the most part, simply to do without the last cycle, which is normally composed entirely of stage 2 and REM.

When people miss out on sleep the most noticeable effect, when they get the chance to catch up again, is an increase in the amount of stage 4 sleep. REM sleep doesn't get such priority. But if someone is specifically deprived of REM, by waking them up every time their eye movements and the electroencephalograph shown they are slipping into it, there does seem to be a need to make up the deficit. REM periods begin to occur sooner and sooner in the cycle, perhaps almost as soon as the subject falls asleep, and more and more awakenings are required on successive nights.

The average adult human spends altogether one and a half to two hours in REM sleep, and five to six hours in non-REM. Infants in the womb pass the major part of their time in something which looks very like REM mode, though it may be a mistake to assume it's quite the same thing. (At this stage the developing neurons don't work in quite the same way as they will later.) Full-term newborns mostly spend eight or nine hours in REM, about the same in nonREM, and between six and eight hours awake. The proportion of REM to nonREM falls off as the child grows and various brain mechanisms mature. By about ten years old the pattern is looking fairly adultlike. A significant feature of these figures may be the way the time spent in non-REM sleep remains relatively constant, while REM shrinks as waking time grows.

REM sleep has been identified in many species of mammal, and can be assumed to be common to nearly all of them. But the amount of time spent in REM varies considerably between species, and correlates with the risks and energy budgets of their lifestyles. Some large carnivores expend a great deal of energy on catching their meals, but then have enough food to last them two or three days. They spend a good deal of time asleep, with a large proportion of REM. Herbivores, on the other hand, apart from needing to devote a great deal of time to grazing and chewing, are very vulnerable to predators when sleeping. They're particularly vulnerable in REM sleep, for which, because of the accompanying muscle paralysis, they have to lie down. Cows, sheep and goats take only about half an hour of REM per night, with around two and a half hours of quiet sleep (which they can do standing up). But smaller herbivores sleep much longer - perhaps because they can hide in relatively safe places while they do it, perhaps because body mass is inversely correlated with metabolic rate, and the faster metabolisms of smaller animals seem to go with more energetic pursuit of food and longer periods of rest.

Some major anomalies among mammals are the marine species, notably dolphins and whales. They either have no REM sleep at all, or only a very tiny amount. Some of them don't go in for quiet sleep either, or only for the briefest of periods. Some dolphin species, and the odd whale, have been shown to have a particularly ingenious practice. One hemisphere of the brain sleeps while the other remains awake, and after a while the roles are switched. These unorthodox arrangements must be adaptations to the problems of sleeping in water, which even when calm requires constant minor readjustments of posture to accommodate its motion. The fact that marine mammals need to surface to breathe - even if they can make one inhalation last a very long time - must also be relevant. Even a modest level of anaesthesia causes dolphins to stop breathing, which suggests that being an air-breathing, deep-diving marine mammal has involved putting breathing entirely under voluntary control. That implies control that has to be learnt, so it may explain why newborns swim almost without rest, let alone sleep, in the early stages of life, and their mothers match this abstinence until the offspring ceases to need so much care. Fur seals can also sleep with half the brain, using only one flipper to maintain their position in the water, and one set of whiskers to monitor it. But when sleeping on land they tend to rest both hemispheres at once.

Ducks, like fur seals, seem to be able to sleep with half the brain at a time, and with one eye open, but are only likely to exercise this ability when in a vulnerable situation. Birds experience REM, but in the species so far studied, only in very small amounts. Some of the more advanced reptiles also have a suggestion of it, though not everybody agrees on how to read reptilian brainwaves. There are some grounds, therefore, for thinking that REM sleep might be a prerogative of highly developed forebrains. It certainly seems logical to suppose that it might be confined to species which can maintain their brains at a constant temperature.

Why Sleep, Why Dream?

Since we generally feel awful when we miss out on sleep the obvious assumption is that it enables an organism to restore itself, probably by allowing energies to be concentrated on rebuilding damaged cells, and similar important housekeeping functions. The idea is supported by growing evidence that a fair number of genes are expressed only at a particular stage of the twenty four hour cycle, and some of them only during sleep. One extensive study in the rat found that among 20,000 genes the transcription rate of about a hundred was influenced by the sleep/wake cycle, and another hundred were more directly regulated by the biological circadian clock.

Rats which were prevented from sleeping altogether died within three or four weeks. This was principally, it seemed, because the immune system works particularly hard in sleep, and the absence of sleep allows infections to take hold. (The sleepiness that we often experience in illness is no doubt part of the defence strategy.) Milder sleep deprivation results in a variety of symptoms. Energy expenditure almost doubles, which can be balanced to some degree by the consumption of an energy-rich diet, but still leads to signs of malnutrition. Metabolism is lowered in the hypothalamus and certain other regions, and the body temperature falls below normal.

Some of these effects are parallelled in humans. Sweet food can help one to keep awake, and prolonged lack of sleep makes for a greater susceptibility to infection. Too frequent misalignment of sleep with the circadian clock leads to adverse effects on the metabolic and cardiovascular systems. But, mysteriously, there are a few individuals who sleep for as little as half an hour a night - as confirmed by observation in the sleep laboratory - and who nevertheless seem to function efficiently.

Various other theories have also been advanced about the functions and advantages of sleep. One is that it is a way of conserving energy when there is nothing else that needs doing, which certainly must be true of slow-wave sleep. Another is that sleep serves to make animals inconspicuous during those parts of the light/dark cycle to which their visual systems are not adapted. A visual system used in inappropriate illumination must be in danger of being misled, so it's better to shut it down. Species which possess dark-proof sensory endowments such as whiskers or echo-location, meanwhile, can reap the fullest benefit by keeping out of the way of daytime predators. It must be a sensible policy to be as quiet as possible at times when other species' sensory systems have the advantage.

For species with sophisticated sensory systems it seems particularly sensible, since the capacity to build an elaborate sensory model of the world must lead to feelings of insecurity when an adequate model can't be produced. Just as we feel edgy in the dark, so must many other diurnal species. Rather than being alerted by every little rustle in the undergrowth, and peering anxiously about for visual information that the eyes can't provide, it makes more sense to find a safe retreat, close down the input systems, and save energy.

If the functions of sleep have inspired speculation it's nothing to the suggestions made about the functions of dreams. One of the most popular hypotheses has been that dreams serve to consolidate the neuronal connections that have been set up during the day, making them stronger by giving them further use. The events of the day may make their appearance in somewhat curious and distorted forms, but, it was supposed, the circuits that represent them are nevertheless reinforced by the dreaming process.

A directly opposite function was also proposed. Back in 1886 dreams were described as a process of excretion by which rubbish is cleared out of the brain. Dreams, W. Robert asserted, never arise out of fully completed processes of thought but only out of things which are in our minds in an uncompleted shape. *A man deprived of the capacity for dreaming would in course of time become mentally deranged, because a great mass of uncompleted, unworked-out thoughts and superficial impressions would accumulate in his brain and would be bound by their bulk to smother the thoughts which should be assimilated into his memory as completed wholes.*

A modern version of this hypothesis was propounded by Francis Crick and Graeme Mitchison in 1983. If memories and asociations are stored in the form of neuronal circuits, they argued, and thus embodied in the strength of synapses, the implication is that many memory circuits must be superimposed, and any given synapse may play a part in a large number of different circuits. Because of these superimpositions neuronal networks could become overloaded, which might lead to bizarre associations, or to a limitation of processing power. REM sleep, occurring when the brain is isolated from normal input and output channels, might be an 'unlearning' programme, which allowed inappropriate synaptic connections to be down-regulated.

Another possibility is that dreams function, as Freud assumed, to dredge up memories from the past. We know that a great deal more is stored in long-term memory than can readily be retrieved. Recognition is generally an easy matter, but recall is frequently hard. Do dreams have a part to play?

Freud is insistent that *in every dream it is possible to find a point of contact with the experience of the previous day*. He gives several examples which make it clear that the point of contact may be a slender one, merely a phrase or a relationship or some object of no apparent significance, perhaps even a metaphorical connection. But he is certain that this triggering event must have been in the experience of the preceding day - *whenever it has seemed at first that the source of a dream was an impression of two or three days earlier, closer enquiry has convinced me that the impression had been recalled on the previous day and thus that it was possible to show that a reproduction of the impression, occurring on the previous day, could be inserted between the day of the original event and the time of the dream.* Freud believed that this 'instigating agent' of the dream provided a means of plugging in to older memories. *Dreams can select their material from any part of the dreamer's life, provided only that there is a train of thought linking the experience of the dream-day (the 'recent' impression) with the earlier ones.*

Freud's interest, of course, was in the excavation of traumatically emotional memories which had been positively suppressed, rather than merely buried by time. But his theory could also be applied to the more mundane and more necessary acts of recall which many species need to make. Imagine an animal encountering conditions it hasn't met for some time. Winter has come round again, the snow is falling, and it needs to remember the best places to find shelter. Or there's a drought, and it's important to recollect where water was still to be had in the previous drought, a few years ago. Just as in us humans, the records are hard to reach, buried in the recesses of memory under more recent accumulations. It seems reasonable to suppose that the sight of browning vegetation and dried-up watercourses, and the sensation of thirst, should evoke memory traces of previous similar experience. In sleep, when there's no distraction, this point of contact with the past provides a link to records of places which provided comfort at the earlier time. As a result, images of those places come to mind in the next day's waking life. Similarly, in migratory animals the record of the route might be primed for use as the time for departure draws close; experienced mothers may start dreaming about last year's safe, comfortable den as pregnancy advances.

The idea that dreams provide access to long-term memory leads on to the theory that they function to supply the raw materials for creative thinking. As Stuart Dimond put it, *The dream provides a reservoir of internally generated images upon which the brain can draw in the solution of its mental problems ... is a powerhouse of images - a central generator for the processes of intellectual function.* Freud referred to *the collection of instances made by Chabeneix [which] seem to put it beyond dispute that dreams can carry on the intellectual work of daytime and bring it to conclusions which had not been reached during the day, and that they can resolve doubts and problems and be the source of new inspiration for poets and musical composers.* The most famous example of a problem solved by a dream is probably that of the chemist Friedrich August Kekulé, who, after working for a long time on the molecular structure of benzene, dreamt of writhing snakes, one of which took its own tail in its mouth - at which he awoke and realised that the carbon atoms of the benzene molecule must form a ring.

Nicholas Humphrey put forward a theory that seems at first rather different. He suggested that dreams act as a sort of rehearsal for waking life, allowing us to explore our own potential so that we are prepared for new experiences. The dreamer *can invent extraordinary stories about what is happening to his own person and so, responding to these happenings as if to the real thing, he discovers new realms of inner experience .... I have in my dreams placed myself in situations which have induced in my mind feelings of terror and grief and passion and pleasure of a kind and intensity which I have not known in real life. If I did now experience these feelings in real life I should recognise them as familiar; more important, if I were to come across someone else undergoing what I went through in my dreams I should be able to guess what he was feeling and so be able to model his behaviour.* Humphrey's proposal could be seen as a variation on the preceding theme. Dreams are again being identified as a source of creative thinking, but in the realms of social intelligence, rather than in art and science.

A final theory is that REM sleep, or something very like it, is important in the development of the brain during embryonic life and infancy. Roffwarg, Muzio and Dement pointed out that many types of neuron do not grow properly in the absence of stimulation, and that mature neurons generally atrophy if unused. The large amount of REM-like brainwave activity which occurs in the unborn infant, they suggested, acts to supplement the limited amount of external sensory stimulation available in the womb and promote the development of the cortical nerve-cells. This should be particularly useful in the early development of the visual system. The large amount of REM sleep in early childhood, meanwhile, coincides with the period when the growth and myelinisation of the central nervous system is still proceeding at a tremendous rate.

These varied hypotheses are not incompatible. Consolidating new records in longerm memory, retrieving old ones, discarding the less relevant-seeming stuff and finding new connections could all be part of one overall, co-ordinated process. And it's readily conceivable that a neuronal machine subtle enough to achieve all this needs activity in order to develop properly while it's still enclosed in the womb and protected from normal sensory input.

Learning and sleep

A good deal of evidence supports the idea that sleep plays a role in consolidating learning. Both rats and humans, having learnt a new skill or a new response, perform it better if they are tested after a sleep. Subjects who learnt a finger-tapping exercise showed a 20% increase in speed after a night's sleep, without loss of accuracy, but no significant improvement after a similar period without sleep. With a task that involved drawing a line on a computer screen while the computer mimicked the effect of pressure on the pen some improvement was seen after a four-hour break, but a much greater improvement after twenty four hours. Verbal learning has also been shown to benefit from sleep.

In some experiments a short nap during the day, though not as effective as a night's sleep, proved beneficial. And a sixty minute nap - which is enough to include a significant amount of stages 3 and 4 sleep and perhaps a little REM - was a great deal better than a 30 minute nap. Sleep deprivation, meanwhile, interferes with the retention of many sorts of learning.

This sort of evidence is not very conclusive in itself, since alternative explanations are possible. Sleep deprivation causes stress, which also interferes with learning. When tests were carried out after a few hours the subjects might have been beginning to grow tired, and functioning less efficiently than when the tests were performed the next day. A nap might simply serve to prevent other experience from piling up on top of the test task and distracting from it.

Happily there are other lines of evidence. Experiments with animals have shown that many sorts of learning are followed by an increase in either slow-wave or REM sleep, and often both. More difficult tasks tend to be followed by greater increases. Furthermore the amount of extra sleep correlates rather well with the efficiency later shown in performing the task.

Confirmation has also come from experiments in which the learning was spread out over several days. Rats which learnt to run a simple maze, with one trial a day, showed a sharp increase in REM sleep just before they started to perform really efficiently. The same thing happened, but faster, when the maze was modified and a new path had to be found. In another experiment rats learnt to avoid a shock by moving from one compartment of a cage to the other when a tone sounded, rather than making the instinctive response of freezing. They showed a gradual improvement over the first seven days, and the increase in REM sleep came towards the top of the learning curve. In both cases REM dropped back to normal levels once the trick was mastered.

If sleep helps to consolidate learning it could be assumed to do so by strengthening the synapses that have been newly brought into use. It was a logical deduction that the patterns of neuronal firing which occur during the learning process might be repeated during sleep, and the hippocampus was a logical place to look for such recapitulation. Pavlides and Winson confirmed the hypothesis, showing that hippocampal place cells which fired together during exploration of a new environment showed an increased tendency to fire together during subsequent sleep, as compared with previous sleep. Place cells which weren't active at the same time (in other words cells which didn't have overlapping place fields) did not. Co-ordinated firing occurred during both REM and quiet sleep, and was more noticeable during the latter. The co-ordination gradually unravelled over subsequent sleep periods.

As it became possible to record from larger numbers of neurons simultaneously it emerged that a whole sequence of such co-ordinated place-cell activity might be repeated during sleep, so that a complete trajectory of movement was replayed. These episodes of recapitulation could last for tens of seconds or more. The sequence repetition that takes place in the hippocampus during slow-wave sleep tends to be speeded up, however, or occasionally slowed down.

Then it turned out that the recapitulation goes on in waking life as well, sometimes during the course of exploration, sometimes later. Such replays occur during what are termed *sharp wave ripple events*, brief periods of very high-frequency oscillation (up to 200Hz) which are thought to indicate high levels of communication from hippocampus to cortex. When a rat is confined to a small track a whole journey can be reproduced in a single ripple event. On a larger track hippocampal firing sequences corresponding to quite lengthy runs are replayed in instalments. The beginning of the sequence may relate to a location quite distant from the rat's current position. These results, opined the investigators, suggest that extended replay is composed of chains of shorter sub-sequences, providing a flexible mode of storage for information which may on another occasion need to be consulted in a different way.

A couple of experiments indicated that hippocampal replay is indeed involved in planning journeys. Rats were allowed to run from one end of a linear track to the other, and back again, finding food, on each visit, in the food wells at either end. After consuming the food they generally paused for a little grooming, or whisker-checking of the surroundings, or just a brief quiet rest, before setting out again for the other end of the track. During this time many of the hippocampal place cells which had fired during the locomotion fired again, but in the reverse order, as if in contemplation of the return journey.

More recently some refined technical wizardry has allowed recordings from hippocampal neurons to be made while rats moved around a two-meter square arena. They were trained to search for a reward among 36 clearly marked locations, and then to return to a 'home' location for a further reward. The site of the first reward was constantly varied, but the home location was only changed between one day's testing and the next. Pfeiffer and Foster found that a sharp wave ripple event usually took place just before the rat set off to return to the place where it had learnt that a reward was regularly available, and that the sequence of firing encoded the route it would take.

Another experiment which recreated natural circumstances to some extent was carried out by a team led by Sidarta Ribeiro. Five rats, each with up to 159 electrodes in various forebrain areas, were allowed to become accustomed to a square box. Then four novel objects were introduced, one in each corner of the box, all maximally differentiated for shape, texture and behavioural importance to rats, and the rats investigated them thoroughly for an hour. Meanwhile extensive recordings were made from the electrodes in neocortex, hippocampus, thalamus and basal ganglia. The firing patterns associated with exploration of the novel objects reappeared significantly more often than patterns which had been recorded 24 or 48 hours earlier. They repeated strongly for over an hour and decayed only slowly after that, being still perceptible after 48 hours. During slow-wave sleep the number of repetitions in cortex, hippocampus and basal ganglia was higher than in waking, but the correlations during REM were more variable.

Many examples of replay activity have now been found during both waking and sleep. Continuing activity immediately after the learning can last for several minutes, and it can coincide with the beginning of new learning, provided it's of a different type, using a different area of neocortex. Evidence for this has been found in both rodents and monkeys, and fMRI studies with humans support the idea. The immediate recapitulation, it's been suggested, sets up the new circuitry for further strengthening during sleep. This would explain why sleep doesn't produce any improvement in the performance of a motor task if the training on the task is rapidly followed by training on a slightly different version of it.

In rodents the patterns of place-cell activity seen during exploration and repeated during sleep may also be found occurring in moments of waking rest, when nothing else is going on. Just as we have quiet moments of reflection on the facts just learnt, the paths recently travelled, the book just finished, so, it seems, do other animals replay immediately past experience. And such contemplation helps to establish a record which may later be more firmly consolidated in sleep.

Replay activity also occurs in the neocortex. Moreover a learning experience during the day, especially one shortly before going to sleep, can increase the amount of slow-wave sleep in the neo-cortical area most involved in the learning. Electroencephalography and PET scans have shown that in humans the parts of the brain activated by a task learnt during the day are specifically reactivated during sleep - during REM for some tasks, and slow-wave sleep for others.

All this evidence indicate that sleep and dreams do indeed play an important part in preserving knowledge acquired during the day. Circuitry which was set up then can be reactivated in sleep, consolidating the new connections. Gradually the long-term potentiation in the hippocampus fades, unless it is renewed by further experience. But meanwhile a schema has been established in the neocortex, where long-term potentiation may take longer to build up, but, once firmly established, can last for a long time. The transfer is effected quite rapidly, it's been shown, where the new knowledge extends an existing schema. For many sorts of learning both slow-wave sleep and REM seem to play a part, though one sleep-stage may contribute more than the other.

Not all the day's experience is preserved, of course. Only a selection of the synapses potentiated during the waking period are further strengthened in sleep. The events that have made a strong enough impression to be occupying the mind as sleep begins must be particularly likely to be conserved. But a good deal of depotentiation also occurs during slow-wave sleep, and the weakened synapses outnumber the potentiated ones - in contrast to the pattern during waking.

If learning can be consolidated during sleep can the whole process take place from scratch? In an impressive experiment by Elizabeth Hennevin and her team the stimulus, the reinforcement and the response were all intracerebral. The initially neutral stimulus was artificial stimulation in the medial geniculate nucleus of the thalamus - one of the structures along the auditory pathway. The unconditioned stimulus was a similar input to the pain pathway, and the conditioned response looked for was a increase in hippocampal activity after the 'warning' stimulus. Rats which underwent this procedure during REM sleep acquired the hippocampal response almost as effectively as those which received the treatment during waking. Stimulation during slow-wave sleep, however, had no effect.

In another experiment eighteen human volunteers attempted to memorise the positions of fifteen card pairs on a computer screen before they went to sleep, in an atmosphere perfumed with rose. If the same odour was present during the first two periods of slow-wave sleep recall of the card pairs next morning was much better than when only an odourless breeze was wafted towards the sleepers. MRI showed that the cue prompted significant hippocampal activation. In this case the cue had no effect when applied during REM sleep.

The learning of a motor task, which doesn't require the participation of the hippocampus, wasn't improved by a sleep cue. Motor learning does seem to benefit from an intervening sleep, though, which suggests that the synapses between motor cortex and basal ganglia are modified in much the same way as hippocampal connections.

Editing the records

For complex patterns of information, and for records that are accessible to consciousness - 'declarative' memory - it seems reasonably certain that REM sleep is necessary. REM sleep, perhaps, allows the record of the heterogeneous experiences of the day to be edited, the bits that relate meaningfully joined up, and new learning experiences to be more firmly linked to their contexts. For the sort of learning which requires a hippocampus - the sort which involves discovering larger-scale patterns in the sensory input, such as can only be perceived over longer periods than that encompassed by short-term memory - REM sleep seems to provide a crucial contribution.

What's meaningful or interesting depends on what's already known, and information that relates in some way to an existing schema is generally better remembered than that which doesn't. It seems probable that newly-potentiated neocortical circuits are more likely to remain potentiated if they can be incorporated into an already existing circuit - a situation which represents the augmentation of an existing concept by a new bit of knowledge. Some of the older neocortical circuits to which the day's new circuits are already linked become activated, no doubt, during REM, and by this means the day's new information is incorporated into older knowledge, confirming or extending it. The past day's explorations are linked to the previous day's, and more thoroughly fused into the whole cognitive map. The latest interactions with a conspecific are related to their earlier behaviour, to create a more complete idea of their character.

But very novel information does get remembered too, on occasion, and it seems possible that it's the degree of novelty that determines how long it takes for the record to become firmly established and assimilated. Judging by some of the animal experiments this is something that can vary quite widely. The odd associations, transformations and plot developments of dreams suggest that during REM sleep new information can be linked to old circuits in what might be called an experimental fashion. Connections can be tried that wouldn't have been made in waking life, and this must often be helpful when the information is particularly novel and doesn't fit well into existing schemata.

The idea that looser, more flexible associations can be made in REM has been supported by an experiment making use of the classical idea of semantic priming. Subjects see one word, the prime, and then must say whether the letter-string that follows is a real word or not. Normally a real word that is closely related to the prime elicits a faster judgement than one which is more distantly associated with it. When people are woken out of REM, however, primes that usually have only a weak effect tend to work better than more closely related words. This is not the case with awakenings from nonREM sleep.

All this processing of the records during sleep can be held responsible for the way episodic memory is gradually converted, over days or weeks or months, into semantic memory. One thinks of X as a nice person, although there is limited recollection of the many encounters that have led to this judgement. As newly explored routes become more familiar, recall of the various instances on which they have been traversed is lost in more generalised recollection. Only events with strong emotional associations are likely to be individually remembered.

Emotion, of course, can serve to make an experience indelibly memorable, even if it's a unique experience that doesn't link up to any established schematic framework. But what happens more frequently is that the records are edited so as to preserve the useful elements and lose the rest. New extensions are added to old schemata, or old schemata are modified. In the terms Piaget employed to describe the intellectual development of the child, there is both assimilation and accommodation. Or as Stickgold and Walker put it, continuing adjustments can *enhance both the efficiency and the utility of stored memories over time and in response to changing needs of the organism.* Memory is a dynamic business.

In other words this complex mode of learning, through hippocampal-neocortical dialogues, is extremely flexible. What's been learnt can be modified or expanded. New information can be fitted on to old, extending or limiting its application. Conditioned learning, dependent on comparatively simple neuronal connections, can only be reversed. If the cue that was followed by a certain event yesterday proves not to be predictive today or tomorrow the synaptic modifications that embody the learning will fade. Neocortical learning, in contrast, seems to be capable of suppressing the response to the cue, but retaining a record in case it should be worth taking notice of it again in the future. REM sleep, by providing a potential for an ongoing rearrangements of circuits, must be what makes this possible.

It also seems that sleep offers an opportunity of converting medium-term memory into long-term memory, and carrying out the business of serious long-term potentiation of synapses, without any interruption. Some of the genes involved in synaptic modification are activated immediately, but the major building work is probably done during sleep.

The other theories of sleep and dream function

It's reasonable to deduce that the process of weaving new information into a network of existing records is what creates the strings by which records can be pulled out of memory. Sleep and dreams, and probably the dreaming process in particular, function to store information in a way that makes it accessible to the conscious mind, and therefore available for use in predictions and plans.

The same processes may also contribute to actual recall, especially in the case of information that has not been accessed for some time. This could explain why it is that the name which goes with that vaguely familiar face you encountered yesterday doesn't come back to you until today. It's a likely hypothesis, since dreams often seem to combine odd fragments of recent experience with evocations of the distant past. We may guess that in many cases most of the synapses involved in a new neocortical circuit will already be involved in other circuits, making for potential links. Under the conditions of REM sleep some of these other circuits are prompted into action. Since the associations that are made in dreams are sometimes bizarre, we can deduce that some of these connections would in waking life be suppressed, but in dreams emotion and perhaps sheer chance can wield a less inhibited influence. Consequently some unpredictable links are evoked, as well as obvious ones. Schema-circuits which are activated in REM by means of these connections become linked to the hippocampus again, and can be accessed in waking life the next day.

The reactivation of stored information and the connections brought into play in a perhaps somewhat random manner during REM can also account for the way dreams sometimes provide inspiration. Both William Dement and Morton Schatzman tried setting people problems to solve and asking them to think about them as they went to sleep. In a significant number of cases their subjects reported having dreams which pointed to the correct answers. Ullrich Wagner and colleagues asked their subjects to perform transformations on strings of eight digits by applying two simple rules. They didn't tell them that there was a hidden abstract rule underlying all the transformations. Once a subject discovered this underlying principle their performance speeded up rapidly. Some achieved the insight on their first session, but more than twice as many got there after a night's sleep.

Henri Poincaré, the famous mathematician, observed that his discoveries usually involved periods of prolonged thought, but that the solution to the problem he'd been considering often popped up later, and often when he was in the middle of something else. He gave a delightful metaphorical description of how he thought the inspirational process worked: *Figure the future elements of our combinations as something like the hooked atoms of Epicurus. During the complete repose of the mind, these atoms are motionless, they are, so to speak, hooked to the wall .... during a period of apparent rest and unconscious work, certain of them are detached from the wall and put in motion. They flash in every direction .... as would, for example, a swarm of gnats or, if you prefer a more learned comparison, like the molecules of gas in the kinematic theory of gases. Then their mutual impacts may produce new combinations.*

What is the role of the preliminary conscious work? It is evidently to mobilize certain of these atoms, to unhook them from the wall and put them in swing. We think we have done no good, because we have moved these elements a thousand different ways in seeking to assemble them, and have found no satisfactory aggregate. But, after this shaking up imposed upon them by our will, these atoms do not return to their primitive rest. They freely continue their dance.

Now, our will did not choose them at random: it pursued a perfectly determined aim. The mobilised atoms are therefore not any atoms whatsoever; they are those from which we might reasonably expect the desired solution. Then the mobilized atoms undergo impacts which make them enter into combinations among themselves or with other atoms at rest which they struck against in their course ....

However it may be, the only combinations that have a chance of forming are those where at least one of the elements is one of those atoms freely chosen by our will. Now, it is evidently among these that is found what I call the good combination.

This sounds like a good description of what REM sleep may be supposed to accomplish - the hooking together of some of the items currently held in medium term memory (by means of hippocampal-neocortical connections) with records established in the longer-term store. In dreams, it would seem, current problems can be submitted to a process of chance encounters with the contents of long-term memory, thus sometimes making useful connections which have not been found in waking life. What's necessarily unclear is whether dreams contribute to creative thinking even when they're not remembered, but it seems more than possible. With all this in mind it's even possible see why dreams have often been regarded as prophetic. The way information is brought together in dreams may sometimes provide better guesses about the future than are made in waking life.

With its strong contributions from the emotional centres of the brain and its divorce from more rational influences, REM sleep probably also allows the emotional aspects of problems to be weighed up more thoroughly. That could be why it's sometimes easier to make a confident decision after a night's sleep, even if one doesn't know what has swayed the decision. *The fact that dreams concern themselves with attempts at solving the problems by which our mental life is faced is no more strange than that our conscious life should do so*, said Freud.

If the conditions of REM sleep are important for the synaptic modifications that constitute learning they are also likely to be important for the synaptic strengthening which is a vital part of brain development, and which employs pretty much the same mechanisms as learning. That theory has found some experimental underpinning too. If kittens, when their eyes first open, are only allowed to use one of them, the primary visual cortex doesn't develop normally. The neurons which should receive input from both eyes, and which are important to depth perception, become entirely dominated by input from the functioning eye and the usual paired columns of ocular dominance don't develop. Moreover this situation rapidly becomes permanent - uncovering the deprived eye after the initial sensitive period, which lasts just a few days, doesn't mend the situation. However, sleep has been shown to contribute considerably to the maturation of the synapses involved in establishing eye dominance. Sleep deprivation can slow it down, especially deprivation of REM sleep.

Infant rodents, meanwhile, spend around 60% of their sleeping time in active sleep (their version of REM), and depriving them of it results in brains that are significantly smaller than normal at 35 days old. Activity, it seems, is necessary for a healthy brain. Perhaps it's significant that, while some species seem to be able to do without sleep, no mammals have been found which spend long periods in quiet sleep without interrupting it with REM. And perhaps it's no coincidence that people undergoing total sensory deprivation tend to experience hallucinations. Total sensory deprivation might be said to mimic the the suppression of sensory input that occurs in sleep and most powerfully in REM. In both cases the resulting lacuna is filled by dreamwork.

Recapitulation

Both quiet sleep and REM sleep contribute, in different ways, to the process of transferring the more significant parts of the information acquired during the day into a long-term neocortical memory that gradually ceases to require support from the hippocampus.

This is a flexible way of preserving information, enabling the fruits of several days' experience to be knitted together when appropriate, and further modifications to be made when new evidence appears.

Quiet sleep probably serves to strengthen the neocortical synapses directly employed in these records, while REM sleep can be supposed to integrate the new information into existing circuits, and to establish the connections that will make it accessible to conscious recollection.

REM sleep may also help to bring long-buried information to the surface again.

Much synaptic potentiation is also undone during quiet sleep, so that the less significant of the day's events are sooner or later forgotten.

Chapter 21

The Long-Term Memory Store

So, the more attention-getting of a day's events are recorded in a medium-term memory, embodied in a network of potentiated linkages between hippocampus and neocortex. Parts of the record are gradually consolidated, with the aid of sleep and dreams, and reflection in quiet moments, in neocortical networks, while fading from the hippocampus. In the process information is gradually filletted, and blended with other experience. The essential significances are preserved, the rest lost. New schemata are added or old ones extended, new knowledge is filed about what can be expected of a certain stimulus pattern, or what action needs to be taken when it is encountered, but the events that led to this knowledge are largely forgotten. Only a small proportion of a life's experience is remembered in the form of history, and most of the historical narrative is heavily edited or merely fragmentary. In short, most episodic records are gradually rendered down into what the psychologists call semantic memory.

All the evidence indicates that the stores of semantic knowledge, the schemata records, are neatly ordered, each major subject with its own filing area. This area is selectively activated when that sort of percept is dealt with, while damage to the area impairs the ability to discriminate between examples of the same subject, and transcranial magnetic stimulation causes a temporary disruption.

Language

The first discoveries concerning the filing system date back to the nineteenth century, and were deductions based on the effects of damage to the neocortex. Damage around the lower rear corner of the left frontal lobe, where it adjoins the temporal lobe, causes great difficulty in speaking, or the complete loss of speech. The area was named after the surgeon who collected most of the evidence, Pierre-Paul Broca. A few years later Karl Wernicke established that damage to an area further back on the left, covering an upper part of the temporal lobe and a margin of the parietal lobe, also causes difficulties with language.

With damage to Wernicke's area the sufferer can string words together to produce a fluent flow of sound, but many of the content words are likely to be inappropriate, or replaced by substitutions such as *thing, whatsit, thingamyjig*. Writing too may be fluent but not very meaningful, and the patient is likely to have as much trouble understanding what's said to him as in making himself understood. Broca's area patients, in contrast, speak in telegraphese, producing the essential content words and omitting the frills. Comprehension is relatively undisturbed, as long as the content words can only add up to a single, semantically probable meaning. Limited damage may produce difficulties with only some aspects of grammar. The assumption has been that grammatical rules are stored here, and an experiment using transcranial magnetic stimulation has supported this inference. Disrupting the functioning of Broca's area interfered with judgements of syntax, but not of semantic probability. The area is close to motor cortex, and is part of a larger region which is concerned with the ordering of all sorts of patterned actions, not just sentence construction.

Damage to the pathways which connect Wernicke's and Broca's areas also affects language badly, as might be expected. In addition fair number of other areas are activated during language production and comprehension, but it's a tricky business sorting out their precise contributions. To some extent levels of activation have been found to depend on whether the language places more emphasis on word order rules for its syntax, as English does, or on modifying the shape of its words, like Latin. Some of the active areas must be primarily concerned with the ideas being expressed rather than the means of expressing them. (But the really interesting question is how far these two processes are separable.)

There's evidence that a variety of other species deal with conspecific vocalisations on the left side of the brain, and other sorts of auditory input on the right, so the left-hemisphere specialisation for language looks like a logical development. Certainly it seems to be the communicatory function that is important to the siting of the language areas, and determines what types of sensory information are channelled to them. In native speakers of languages like Chinese, in which intonation is important in differentiating meanings, the left hemisphere is engaged by the inflections given to words to a much greater degree than it is in speakers of non-tonal languages. Deaf and dumb users of American Sign Language show activation of the same areas as vocal communicators, and their language competence is similarly susceptible to damage in those areas, although in them there is also a greater right-hemisphere activation which must reflect the strong spatial element of sign language.

Broca's and Wernicke's areas are larger than the corresponding areas on the right. However, the right hemisphere comprehends words with concrete meanings, though it may be less good at abstractions, and it's sensitive to semantic anomalies. It's thought to be involved in weaving sentence-length pronouncements into larger patterns (just as it seems to deal with the perception of larger scale spatial patterns). But the degree of language lateralisation has been found to be variable. What's certain is that there must be a high degree of co-operation between the hemispheres. Semantics and syntax have to be linked to semiotic signals such as intonation and gesture, processed on the right. Moreover the ideas expressed by language clearly come from both hemispheres.

The storage question gets more complicated when there's more than one language to be accommodated, and much seems to depend on the age at which a language is learnt, whether it's by formal or natural methods, whether the second language is of a similar type to the first one, and on just how many languages there are to be fitted in. Sometimes the right hemisphere may be used for a second language. Maybe, if seriously multilingual brains were to be investigated, it might turn out to be capable of providing a home for more than one.

Reading is also dealt with in a left-hemisphere region, regardless of whether the input comes through vision or, as in Braille, through touch. The precise pattern of areas activated does differ, however, in ways that reflect the nature of the writing system. A temporo-parietal area concerned with sounds is involved in reading languages in which an alphabet or a syllabary is used to convert spoken words into visual signs, but not for reading Chinese pictograms. The enormous number of word-signs to be mastered in Chinese means the use of wider swathes of both visual and frontal cortex. In Italians, who benefit from a language in which the spelling consistently reflects the pronunciation, there's a greater degree of activity in the cortical area concerned with phoneme processing than in English subjects. English readers show greater activation in areas associated with word retrieval. Learning to read leads to an intensification of the connecting links between various areas, including some between the two hemispheres.

The hatched area here is Broca's, the starred patch represents Wernicke's, and the small dotted bit is an area important for speech comprehension. Writing activates areas in parietal and motor cortex. Oddly, writing doesn't entirely depend on being able to read. A case has been recorded of a temporo-parietal lesion which seriously impaired the patient's reading, but left his writing ability relatively intact.

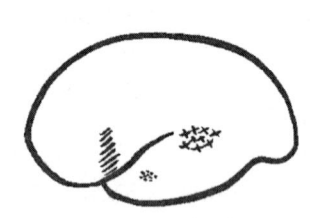

Inside the broadly defined areas the detailed records seem to be stored in a neat and logical manner. When a minor stroke knocks out only a small area of brain, and destroys only a minimal amount of information, the information is generally all of a closely related nature. One man could no longer name fruits and vegetables, but the rest of his vocabulary was unaffected. Other patients remembered concrete nouns but not abstract ones. A few subjects have been found who have lost their command of English prepositions as used in a spatial sense but can still use the words in a temporal sense, or *vice versa*.

Slips of the tongue provide another strand of evidence that fits well with what is implied by the effects of limited damage. When we come out with the wrong word it's usually one that's related in an obvious fashion to the one that should have been uttered. It's a word of the correct grammatical case, with the right number of syllables, and stressed in the right places, but one or two of the phonemes are wrong. Or someone is miscalled by their brother's name, or their mother's, but not by that of someone totally unconnected. Freudian slips happen too, but more commonly the error suggests that the selector mechanism has only just missed picking the right item.

A further inference from slips of the tongue is that a stimulus pattern can be filed several times over, in different areas. As David Rumelhart put it, the organisation of memory can be compared to that of a library, with each book carefully cross-referenced by subject, author and title. Meanings of words must be organised in one store, the sounds of them in another, spelling and grammatical functions in yet other places. The word is represented by the whole neuronal circuit. Just occasionally the neuronal impulses are misrouted at a late point, and the wrong word is produced. This hypothesis also feels right as an explanation for the tip-of-the-tongue phenomenon - that feeling that the word one wants is almost within reach, one just can't get the complete shape right. One knows the meaning one wants, often the initial letter, and there's an approximate idea of the size and shape, but the details are evasive. It's as if one has got to the right bit of shelf but can't quite make the final selection.

While significant sectors of the left frontal and temporal lobes busy themselves with semantic aspects of language an area in the right temporal lobe is concerned with the music of language, and the emotions expressed by it, in parallel with the words. Intonations can be vital, of course, to understanding the true import of words. They are also essential for the infant's first learning of words. This division of responsibilities between the two hemispheres seems like a good way of dealing with two aspects of the same input which must be processed simultaneously, but nevertheless be distinguished.

Close by the right temporal area which registers the emotion in voices is one which responds to emotional facial expressions, and between the two is a region where the response when both face and voice express the same emotion is greater than the sum of excitation levels evoked by either alone. Another nearby area deals with identifying individual voices.

Faces, bodies, animals, movement

There are other classes of percept for which the location of the schema-areas seems to be fore-ordained. Activated neurons are found in such an area when a stimulus pattern of the relevant category is being studied, and also when it's merely imagined. Furthermore, these neurons maintain their activity when the stimulus pattern they represent has to be held in mind for a short period.

Visual input, it may be recollected, is processed through various subdivisions of the occipital lobe to extract information about lines, edges, orientations, colours, distance and movement. From here one pathway leads on into the lower part of the temporal lobe, where the cells are less fussy about just which bit of retinal field a stimulus occupies, but sensitive to the way the features group together. Receptive fields increase in size along this pathway, and so does the complexity of the stimulus needed to impress the cells. First come neurons which are activated when particular combinations of features are registered, such as add up to simple objects, or to component parts of larger objects. Beyond them are cells which are seriously excited only by more complex stimulus patterns.

Back in the nineteen sixties it was discovered that some neurons in the inferotemporal cortex of macaque monkeys respond specifically to faces. This tied up suggestively with the fact that brain-damaged people sometimes lose the ability to recognise individual faces, a disability known as prosopagnosia. Prosopagnosics may fail to recognise the faces of their nearest and dearest, or even their own face in the mirror. They have to identify people by their voices, or depend on unreliable clues, such as the colour of the hair, a beard, or clothes.

The human equivalent of that monkey area forms part of one of the convex bulges of temporal cortex, the fusiform gyrus, with the right fusiform gyrus playing the stronger role. In imaging experiments this area is activated by the sight of a face, and when the face/vase illusion is viewed, activation in the fusiform face area coincides with the perception of faces, and not with seeing the vase. Other sectors of the gyrus deal with other complex stimulus patterns among which fine discriminations are made. It's part of the temporal cortex that wraps around under the rest of the brain, so the sketch shows the view from underneath. Nearby is the parahippocampal gyrus, where the parahippocampal place area is found, as described in the chapter on Contextual Memory.

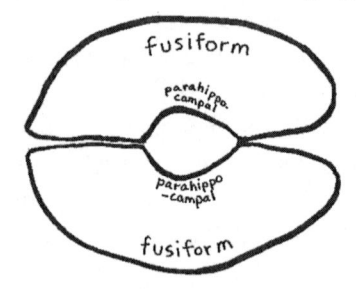

Neurons in the monkey face area can be prompted to mild but significant activity by roundish shapes, such as an apple or a clock, but only get really busy for faces. The normal spontaneous activity may be suppressed by other shapes, such as hands. And micro-stimulation here can bias the decisions of a monkey trained to discriminate between pictures of faces and of other things. Many of the face-selective cells respond to any face, including human and cartoon faces as well as monkey ones, indicating they are tuned to a very basic archetype. But some are fully excited only by one particular face, or are influenced by the viewpoint, or are sensitive to the facial expression.

Even the neurons which are activated by any face are more excited by some than others. In experiments with monkeys Doris Tsao investigated the preferences with cartoon faces, while David Leopold and colleagues used photographs of human faces which were gradually morphed from one extreme of a dimension to the other. The experiments showed that some cells go more for long, thin faces, others for short fat ones. Some like large eyes, some prefer smaller ones.

A good many are influenced by the spacing of the features. An individual cell is likely to be influenced by several such features, giving its maximum response to the perfect combination. The selectivity takes a little time to emerge, however. Many neurons fire briefly for a wide range of faces, but only continue firing when the shape of the face, or the current expression, fits their individual preferences. Presumably this reflects the build-up of inhibition among neighbouring neurons which are all responsive to faces but specialise in different examples.

Natasha Sigala and Nikos Logothetis proposed that the fusiform face area codes for individual faces by constructing a 'standard' face, based on experience, and measuring how far each new face departs from it, along various dimensions. The dimensions are the ones which are difficult to disguise, and not much affected by the passage of time, such as the overall shape of the face, the size and shape of individual features, and the spatial relationships between those features. For an fMRI experiment with humans Gunter Loffler and colleagues created a picture of a very average face and gradually morphed it along some of these dimensions during the brain imaging. They found that the fMRI signal from the fusiform gyrus grew stronger as the distance from the 'mean' face increased.

Goedele Van Belle and colleagues confirmed that it's the overall shape and spatial arrangement of a face that most people rely on, and that prosopagnosics don't register. When, in a face matching task, only one central facial feature could be viewed at a time the performance of the control subjects sank to the level of the prosopagnosic. When the whole face could be seen but only the feature on which the gaze was centred was masked the controls' performance was almost normal, that of the prosopagnosic subject seriously impaired.

It seems that we all, apart from the prosopagnosics, have a Platonic 'ideal' face, as it were, with which to compare the real ones. It's founded on a genetically determined outline, but heavily influenced, no doubt, by the faces encountered in early life, and subject to ongoing modification as more faces are studied.

More recent work indicates that the fusiform face area is part of an extended system devoted to faces. There's a site that prefers moving faces to static ones, and others where neurons are responsive to cues indicating gender and ethnicity. The area dealing with emotional expressions has already been mentioned.

Analysis of faces begins in the occipital lobe, in an area where the activity reflects the current manifestation of the face more closely than that later on. From here information is passed on to the more forward areas. And from them it looks as if the path leads to an area at the front of the temporal lobe, which discriminates yet more firmly between individual faces, and is insensitive to changes in viewpoint. This, it's thought, may be where faces are linked to voices, related emotions, and any other relevant information. Perhaps it can be regarded as the schema-area which holds the semi-abstract idea of the individual, which includes all their aspects and changing manifestations.

One part of the fusiform gyrus prefers faces that are visibly attached to a body. Activity can be evoked for a very blurred representation of a face if it's in the correct relationship to a body. On the border of temporal and occipital cortex, meanwhile, in the right hemisphere, is a small area where the neurons' strongest response, as tested by photographs, is to a human body, and the response may be maximal if the face attached to the body is occluded. Another bit of temporal cortex is especially sensitive to bodies in terms of their posture, and often to posture as shaped by an action. Yet another area is responsive to the sight of hands. Our records of animal shapes, meanwhile, seem to be filed close to those for humans, with the species more closely related to us generally in closer proximity and those built according to different plans further away.

The middle temporal regions responsive to visual movement similarly show signs of being subdivided into category-specific sections. In the broadest characterisation biological movement is distinguished from inanimate. Human styles of movement have their place as a subdivision of biological movement. Neurons in that area can be triggered by film of people with lights attached moving in the dark, as in Johansson's experiments, almost as strongly as by the movements of visible people. Photographs of people or objects that are clearly in motion also produce activation here, whereas pictures that don't imply movement do not. In a spot just in front of this area excitations occur when subjects are asked to produce action words.

Another part of middle temporal cortex deals in the patterns of visual movement created by locomotion, and other movements of the self. Cells here are sensitive to optic flow - coherent change over the whole visual field. Some are excited by the transition of features outwards towards the edges of the visual field, the effect of moving forwards. Others are activated by the opposite pattern, contraction towards the centre of the field, or by the sideways slide of the whole visual field which accompanies rotation of the head or whole body, and some by the erratic sorts of optic flow that occur when one takes a tumble.

Various other sectors of neocortex have been located which are necessary to other types of percept or activity - sites in which activity correlates with attention to a subject, and damage impairs the ability to deal with it. Music, geometry and arithmetic are among the subjects for which associations with certain areas have been firmly or tentatively suggested. Music activates the temporal cortex bilaterally, but especially on the right, in a region close to that concerned with the emotions expressed in human voices. This provides support for the idea that the capacity to make and appreciate music must have evolved through an elaboration of the ability to interpret the emotions expressed in vocal intonations. *All art aspires to the condition of music* said T. S. Eliot, which might be another way of saying that music is purified emotion.

Parahippocampal cortex, already mentioned, responds strongly to pictures of houses and other buildings, and to anything that can mark the goal of locomotion, or the point at which a change of direction must be made. It's activated by a view of a room, whether empty or furnished, but is less than half as impressed by a picture of the room's furniture shown against a blank background. Right retrosplenial cortex seems to be concerned with recording the spatial relationships and distances between such features. Left retrosplenial cortex seems to be more concerned with large-scale temporal relationships.

Mirror neurons have been described earlier too - neurons which fire both when an action is performed by the self and when another individual is seen performing the same action. In the monkey these are found in and around premotor cortex, in the part of parietal cortex that contributes to planning motor actions, and in what's regarded as corresponding to Broca's area. Some are excited by the sound of an action being performed, as well as by the sight of it. Others respond even when the goal of the action is hidden from view and has to be inferred. For instance a neuron which reacted when a human picked up a fruit was also excited when the human reached behind a screen which the monkey knew to conceal a fruit. The idea of purpose is important - a cell which responds to a purposeful action may not fire when the movement is mimed. Some fire only for a precisely defined action, others for any variation that achieves the same end.

When humans watch other people performing actions fMRI reveals activity in both premotor cortex and the strongly connected areas of parietal cortex. A study of ballet dancers and capoeira dancers showed that these regions were particularly busy when they watched their own specialty. Since this might be because the dancers were more accustomed to seeing those movements a further study of ballet dancers was carried out. This ruled out that possibility by revealing greater activation when the dancers they watched were of their own sex, performing movements they were accustomed to perform themselves.

Mirror neurons fit perfectly the idea of schema-neurons - cells which encode a concept, in this case the concept of a particular action, abstracted to the extent of being divorced from a performer. It's likely that the simplest versions appeared a long time ago, enabling infants to copy the where and when of their parents' actions. Increasingly flexible motor systems were probably accompanied by an increasing number and variety of mirror neurons, allowing closer imitation, applied to skilled actions.

Schema-areas as nodes in a network

Clearly the synapses of schema-neurons must be extensively shaped by experience. The resulting synaptic modifications record whatever proves to be relvant to the import of a stimulus pattern. In other words, schema-neurons indicate where it's most important to direct the attention.

But how exactly do the schema-neurons which record the essential elements of a certain stimulus pattern contribute to recognition? What, indeed, constitutes recognition? The answer has to be that the schema-neuron connects to other schema-storage areas holding other types of information relevant to that particular schema. Consequently, when we look at a familiar tool we know what we can use it for, and how, what it will feel like, and what sort of noise will accompany the use. When we see a familiar face we know what body goes with it, and how that body moves, and we can anticipate the sort of voice that will come from the face. We know what sort of facial expressions are likely, and how the owner of the face is likely to behave. We know the name that goes with the face. Most importantly, we know how we feel about the owner of the face, how we want to behave towards them, and what sort of activities we might be prepared to indulge in with them. Or perhaps we know, equally usefully, that it's just a face we see in passing from time to time.

A sensory schema-neuron, in short, forms an essential node in an extensive circuit which constitutes a more comprehensive schema, a cross-referencing between senses which allows an input in any one sensory channel to provide a forecast of what can be expected in others. The outstanding feature of the neocortex is that it embraces numerous complex webs of this type, and makes many more interconnections than can be managed at subcortical levels.

The resulting forecasts don't generally occupy the forefront of consciousness. They're most likely to draw attention to themselves when they prove false, when the wrong voice comes out of what seems a familiar mouth, or something that looks like a fruit proves to be an imitation. But the forecasts are available if needed, and can be used for decision-making and planning.

These deductions from subjective experience are supported by imaging experiments. The fusiform face area shows some activation when familiar voices are recognised. Tactile exploration of objects can excite visual areas. In professional pianists watching a film of piano-playing the auditory areas were at work even though the film was silent. In the temporal area which responds to the sight of facial, hand, or body movements some neurons also respond to the sound that goes with the action. People who have been trained to use novel objects as tools not only show a more selective response in the temporal visual pathway after training but also develop activity in premotor and parietal cortex, and in areas concerned with visual movement.

As well as the neocortical areas that specialise in one category of percept, deriving most of their input from one sensory channel, there are others which derive equal levels of excitation from two, three or more sensory channels. These can be designated as the stores for truly multisensory schemata, for the essential concepts that combine all the evidence. The neocortex provides space for massive interconnections between sensory channels, and thus for extended networks of significance. The larger the neocortex the more numerous and more capacious the networks can be. In human brains the interconnections that characterise a schema-area, especially one of the larger and more important ones, are obviously very complex. The ultimate schema-areas might be envisaged as the centre of one of many overlapping and interconnected spider-webs.

If a small part of a schema-area is damaged and some of the neurons are lost there is usually some recovery of function. New synaptic connections can be formed, which bypass the lost circuitry, creating new routes for linking the information derived from one sensory channel with that obtained from others, and with the emotional evaluation. After major damage, however, there is no recovery. This seems understandable - if the whole centre of a web is lost, it's difficult to see how all the connections could be re-established. It's generally thought that no new neurons are added to the neocortex after infancy, and although one or two investigators have found evidence suggesting this might not be entirely true there is certainly no large-scale neurogenesis.

People who have lost the use of a certain schema-area are obviously unable to relate input of the sort it dealt with to what happens in other senses. But what is their subjective experience like? What's it like to lose the ability to deal with a certain sort of percept? In *The Man Who Mistook His Wife for a Hat* Oliver Sacks gave a telling description of a patient who, after a stroke, suffered from severe prosopagnosia. *He faced me as he spoke, was oriented towards me, and yet there was something the matter - it was difficult to formulate. He faced me with his ears, I came to think, but not with his eyes. These, instead of looking, gazing, at me, 'taking me in' in the normal way, made sudden strange fixations - on my nose, on my right ear, down to my chin, up to my right eye - as if noting (even studying) these individual features, but not seeing my whole face, its changing expressions, 'me', as a whole. I am not sure that I fully realised this at the time - there was just a teasing strangeness, some failure in the normal interplay of gaze and expression. He saw me, he scanned me, and yet ....*

This suggests that the usual well-learnt saccade-pattern that should be triggered by a face doesn't happen in prosopagnosia. What must be lost, too, when a visual schema-area is destroyed, is the source of all those messages that normally travel back to primary sensory cortex to turn up some parts of the input and tune down others. There's no way of influencing the earlier stages of the visual pathway, of checking whether a bit of data can be clarified, whether some lines and edges are worth more emphasis, whether a darkness is only a shadow rather than a significant feature. The connection to prefrontal cortex, where attention is organised, must also be lost, with equally serious consequences.

As a result the potential for that self-maintaining circulation of neuronal impulses which is thought to be necessary to the normal conscious experience must be seriously depleted. It seems likely that under these circumstances the conscious visual experience is rather like the confused jumble seen when a complicated image is displayed very briefly in the tachistoscope, and there is no time for the usual dialogue with the schema-storage area. Only in this case the confused visual experience is continuous.

Moreover without the sustained circulation of neuronal impulses there is no ultra-short-term memory to maintain the record of each fixation's worth of input data, so that several fixations' worth can be put together to create the larger picture. Once the gaze moves on the data it produced is lost, and can't be related to the next bit of input, so there's no way of measuring the relative distances between features. This idea receives support from a careful study of a prosopagnosic with a smaller lesion. It was concluded that, unlike the normal subjects, she *encoded local facial information independently of the other features embedded in the whole facial context.*

I think the most comparable experience to that of any sort of agnosic must be what we feel when listening to a wholly strange language, a language for which we have no schemata, no expectations. The sounds which don't occur in one's own language aren't distinguished because one hasn't learnt to pay attention to them, and the distinctions which don't carry any weight in one's own language aren't noticed. Nor can we spot the cues which would help us to know where words begin or end. The stream of sounds tumbles past, and we distinguish only tiny portions of it, and have great difficulty in registering a long enough chunk for useful dissection. The agnosic must be supposed to have similar problems, even though the stimulus patterns don't rush past at speed.

For some types of pattern the most important association is emotional, and this is particularly true of face schemata. The most vital information incorporated in a face-schema must be the emotions which are linked with the face. The various feelings which have been evoked by the person behind the face blend into a whole which might be described as a complex emotional flavour. This, I think, is what essentially defines the individual in the memory stores. When in dreams appearance is divorced from reality, and a figure is confidently known to be someone familiar despite not looking at all like them, it must be because the emotional schema has been activated independently of its associated sensory schema, and paired instead with a different one.

There's a rare neurological syndrome known as Capgras Syndrome which produces the opposite experience. A Capgras sufferer recognises the appearance of their nearest and dearest, but insists that the person is not their husband, wife, parent or whatever, but a double carrying on a mysterious masquerade. V. S. Ramachandran suggested that Capgras Syndrome occurs as the result of damage to the channels that allow emotions about people to be related to sensory information about them. Feeling no emotion when seeing a familiar face, the Capgras patient concludes that it's not really the face it seems to be. The fact that the familiar face and figure are accompanied by the appropriate behaviour is regarded as just part of the charade.

How are schema-areas set up?

Why is it that for many types of schemata the filing system is fairly consistent across all brains, so that a neurologist looking at the scan of a damaged brain, or a pathologist looking at a dead one, can diagnose pretty confidently just what difficulties will be or have been experienced by its owner? How is it that acquired knowledge can be arranged so predictably?

The likely answer is that the layout of the essential areas is determined by connections reaching the neocortex from subcortical areas. This would explain the fact that stroke patients often recover the ability to deal with the lost class of percepts when only cortical cells are affected, but the loss is permanent when the damage extends to the axons bringing input from elsewhere in the brain.

The emotional responses evoked by the most significant stimulus patterns must be one of the forces at work. As we saw earlier, it's not only in various neocortical areas that a response to faces is found, there's also one in the amygdala. The preferences of the amygdala are rather different from those of the fusiform face area, however. The most energetic activation here is evoked by a blurred representation of a face, sufficient to indicate outline, features and any strong emotional expression, but revealing little detail. In contrast, the fusiform face area in the neocortex is only mildly impressed by such an indeterminate representation, and responds more strongly to a clearly defined face.

The amygdal response depends essentially on subcortical pathways, and it's very fast. Mostly it's been studied with faces showing emotional expressions, and it's been shown that they can be distinguished in the absence of a conscious experience, and even in the absence of primary visual cortex. Subjects exposed to a very brief image of an emotional face which is immediately masked by a neutral picture of a different nature, so that there is no conscious experience of the face, nevertheless produce the normal skin-conductance response to the emotional stimulus - a response that is triggered by the amygdala. A 'blindsight' subject with an extensive lesion of left striate cortex, and consequently no conscious vision on his right side, can guess well above chance at the emotional expressions of faces shown on the blind side.

The unconscious response seems to be fastest when the face shows a direct gaze towards the viewer, rather than an averted one, which suggests that the 'bull's-eye' pattern of pupil surrounded by iris surrounded by sclera is important.

This, of course, is the pattern which attracts the gaze of a human infant most powerfully. The perfect double circle - which indicates a gaze directed towards the self - must be a particularly important component of the archetypal face pattern.

Information adequate to define that pattern, along with that about expressions of fear or anger, reaches the amygdala from superior colliculus *via* thalamus, specifically a bit of thalamus just above the lateral geniculate nucleus, sometimes defined as part of the pulvinar, sometimes called suprageniculate thalamus. An implication is that the primate superior colliculus extracts more information from visual input than it has been given credit for – and indeed evidence to that effect is beginning to emerge. It's also possible that some ganglion cells in the retina register a perfect bull's-eye pattern potently enough to define an eye.

The relevant sectors of amygdala and thalamus must constitute the ancient, hardwired mechanism for identifying a conspecific by visual means. The short route by which the message reaches the amygdala means that a signal can be passed to the hypothalamus very rapidly, and the autonomic response evoked by the unexpected advent of a conspecific can be triggered even as the conscious experience is still developing. It's a response that can be suppressed as recognition of the individual conspecific develops, but in the meantime it no doubt contributes to the arousal of attention that any conspecific tends to elicit. The pathway from superior colliculus to thalamus to amygdala must be at least partly responsible for the way infant attention is attracted to faces, in preference to other visual stimuli. And it seems likely that projections from the amygdala and the supra-geniculate thalamus to the neocortex help to determine the siting of the various face areas.

The parahippocampal place area, meanwhile, is not only close to the hippocampus and the retrosplenial cortex but also conveniently near the hypothalamus, which provides the sort of feedback about internal bodily conditions that must be particularly relevant to places. With similarly neat economy of wiring, tools are dealt with in proximity to motor cortex.

A possible further hypothesis is that the siting of some major schema-areas may be influenced by the sort of actions they evoke in the newborn infant. Facial expressions tend to prompt imitations. Small nearby objects soon come to inspire grabbing actions. Some foundations for the schema-areas for faces and for graspable objects may be laid by these actions. Eye movements may play a role too. Faces and other stimulus patterns which are differentiated on the basis of fine detail require many saccades around a small area. Landscapes, for species with a retinal fovea, are studied with large saccades. Landmarks can be defined as stimulus patterns which influence locomotion.

The earliest responses of the newborn infant are genetically determined ones, triggered by genetically determined forms of sensory input, basic stimulus patterns which are registered in the midbrain – the superior colliculus/optic tectum, and adjacent areas. The news is reported to the basal ganglia system, which releases the genetically determined response from inhibition. This results, it can be posited, in a message to neocortex about the action. The axons bearing such messages may go to pre-ordained areas, just as those carrying information from the lateral geniculate nucleus of the thalamus run to visual cortex, those from the medial geniculate to auditory cortex.

Stimulus patterns which present some of the same features as the genetically defined ones are likely to be filed close to them, both because they have some of the same attention-catching qualities and because the initial steps in the sensory processing pathway will tend to overlap. The essential features of the monkey face are not too far from those of the human, for example. Sheep are rather more different, but not drastically so. Thus schema neurons for animals which share the essential structure of humans and move in a similar style tend to be located closer to those for humans than the records for things like fish, snakes, crabs, insects. Possibly there's an emotional factor too. It may be that the child who keeps pet spiders in matchboxes grows up with spider-schemata filed rather differently from those of the person who is frightened of them.

It's also likely that the degree to which a general schema is subdivided, in other words the number of subtle discriminations that can be made, is significant. The fusiform gyrus, as well as containing the area which specialises in face identity also has other specialist subdivisions. The recognition of letters depends on a part of it. Experts who can distinguish similar car models or closely related bird species show more activity in the fusiform gyrus than non-experts, when dealing with their speciality. Some prosopagnosics have difficulty in distinguishing other types of complex stimulus pattern, as well as faces. This suggests that the pathway dealing with a particular class of pattern extends as finer discriminations are learnt. The fusiform gyrus must represent the end of the pathway for analysis of visual input, performing the most complex sorts of operation.

The role of experience

The siting of Broca's and Wernicke's areas is more closely similar in identical twins than in other individuals, so it must be deduced that there's a strong genetic element to the organisation of the neocortex. But the filing system develops very slowly. The schema-areas mature later than earlier stages of the sensory pathways, and imaging studies indicate that in children the cortical activity evoked by a sensory input is significantly more extensive than in adults. It's likely that in a very young infant any sensory input causes widespread neocortical activation, which is slimmed down only slowly as schemata are developed, probably through the long-term potentiation of inhibitory connections.

Slow learning seems desirable, for many stimulus patterns are likely to change their significance as experience accumulates. To the young infant an adult's finger, or their spectacles, are most interesting as something to grab at. It takes time and experimentation to sort things into useful categories. There's plenty of time, for the brain matures very slowly, undergoing various changes over the first twenty or so years of life. Synaptic connections multiply lavishly in the first year, and are then considerably reduced in subsequent years. Excitatory connections are slowly pruned, and there's a gradual elaboration of inhibitory circuits. And axons slowly acquire the myelin coating which insulates them and makes for faster conduction, but at the same time reduces the potential for establishing new synapses. The myelination occurs later in the association or schema areas than in earlier parts of cortical sensory pathways.

It's a reasonable deduction that the logical layout of neo-cortical association areas is the result of a long period of competition, during which the areas which get the strongest feedback or have the closest connections to a particular motor area gradually come to specialise in the patterns of sensory input which evoke that particular sort of feedback, or require that sort of action. Thus the basis of the system is genetically determined, ensuring that essential matters are dealt with, but there's also room for experience to exert an influence.

What may make the process simpler is that infants begin by concentrating mainly on one subject at a time, in each sensory channel. Prompted by signals from pulvinar and amygdala, they take every opportunity to study faces, and soon learn enough to distinguish mother's from any others around, since hers is accompanied by the voice and smell that have already begun to be familiar in the womb. Faces show interesting, complex patterns of movement, and mother's comes attached to another moving pattern that is associated with warmth and comfort. Shortsighted and not yet capable of locomotion, the infant has limited opportunities for studying other things, so probably the visual world is initially divided into face-shapes and everything else - a wide range of colours, movements, and other things that provide some interest but aren't as rewarding as faces. Given this concentrated study, faces that are encountered frequently come to be distinguishable, though for some time there is a heavy dependence on salient features rather than essential structure. By the time the infant is mobile friends can generally be distinguished from strangers, though it won't be until the teenage years that face recognition skills are fully developed.

Auditory learning in the early months, meanwhile, is similarly focussed on conspecifics, and a collection of schemata for individual voices begins to be set up. Within weeks the infant has firm convictions about the voice that should come out of mother's mouth, and gets distressed if these expectations are upset.

Meanwhile they are grabbing at small objects that come within reach, but if they succeed in getting hold of one they immediately put it to the mouth and suck at it. Only at about seven months do they begin to study the objects captured. By now they must have learnt quite a lot about faces, and the bodies attached to them, so they can appreciate that small, graspable objects are something rather different, both as visual objects and in respect of what you can do with them. It's at this point, it seems reasonable to assume, that the arrangements for storing visual schemata start getting more complicated.

By this time a number of voice schemata have been acquired, and the infant is beginning to work out that voices convey other things as well as identity. First of all they register the tunes of the language they hear spoke, and learn how to reproduce them. Then come actual words, and the exciting discovery of their potential uses. And then they begin to acquire the rules for putting words together.

It's only when the infant gets seriously mobile that the need arises to pay attention to large visual stimulus patterns that can function as local landmarks. Control of the eye muscles is well practised by now, and distance vision much improved, so they can focus on these patterns, and it's at about this stage that the hippocampus begins to become functional. We can guess that another perceptual category is initiated at this point. It takes several years for the hippocampus to mature fully, however, and it's also several years before the child becomes competent at finding their way around on their own.

So it seems that the business of setting up the major schema-storage areas in each sensory pathway may be rendered rather more straightforward than it might otherwise be because the major subjects are tackled one at a time. The basis for one visual schema-area, or one auditory one, is established before the next one has to be initiated. The subjects that are learnt first are the most vital ones for a social mammal, and they are also the ones that require the finest distinctions to be made. Conspecific faces are all built to the same essential pattern, conspecific voices may exhibit only subtle variations, language is a serious perceptual challenge. It's just as well that we make an early start on these very difficult subjects.

It may be that the size of the challenge explains why it is that the schema-areas which specialise in the predeterminable and supremely important classes of percept apparently need to be established at the right time if they are to be established at all. There's a window of opportunity which closes after a certain period if the relevant sort of stimulus pattern has not been encountered. The first evidence of this concerned language. The rare recorded cases of 'wild' children, and those born deaf and not introduced at a reasonably early age to lip-reading or sign language, suggest that those who are not exposed to language by somewhere around seven to twelve years of age will never properly master it. The most thoroughly studied case is Genie, daughter of a mentally unbalanced father and a timid mother, who spent most of the first twelve years of her life locked in an attic. With careful coaching, she subsequently acquired quite a lot of vocabulary, but her grammar remained haphazard.

It's difficult to rule out the possibility that genetic factors are at work in such cases, but recent discoveries about prosopagnosia point the same way. Neurologists giving popular lectures that covered acquired prosopagnosia found that occasionally they were approached afterwards by a member of the audience who confessed that they had never been good at distinguishing faces, and had always depended heavily on other cues to identify people. Skill at facial recognition has a heritable element, and sometimes there may be a genetic cause for this disability, as has been shown for some other agnosias. But sometimes it seems to be coincident with a lack of pattern vision in very early life, usually caused by being born with severe cataracts. Various individuals who had cataracts removed within two to six months of birth have been shown to be impaired at distinguishing between faces that are only subtly differentiated, although they perform normally when geometrical patterns with only minor differences are to be discriminated. It seems as if foundations must be laid in the first few months of life for a skill that won't be perfected until several years later.

Why should it be necessary for certain perceptual subject matter to be encountered at the right time? One possibility is that the emotional feedback pathways have to be brought into use at the right time, or else they wither away. An infant who has to do without the comfort provided by the familiar visual stimulus pattern which is mother learns to get along without it. He obtains all his reassurance and encouragement from voices, odours and touch. When his eyes become functional there is no reward to be gained from the close study required to distinguish faces, and just as much attention is given to easier stimulus patterns.

Similarly, the growing infant learns language because it's an interesting way of interacting with adults, because it turns out to provide an efficient means of obtaining things they want, and because it supplies an excellent means of satisfying curiosity. If they have to manage without language at the time when developing curiosity needs it, then they either find other ways of satisfying their needs or never realise that they can be satisfied. The incentives to undertake the very demanding task of learning language, and especially grammar, therefore grow weak. Indeed, since language supplies the means of building and exploring more complex thoughts it's likely that an individual who spends the first few years without it doesn't produce the sort of ideas that need grammar for lucid expression, and never realises how ambiguous language is without grammar.

Perhaps, also, it's no coincidence that the infant's study of faces and language begins with instinctive imitation. Conceivably this genetically prompted exercise of the muscles somehow helps to develop the neocortical schema-area. The unthinking imitation of facial expressions fades within weeks, and its absence might explain why even a few months of visual deprivation can lead to prosopagnosia.

Another possible factor has been revealed by a study of mice, which, just as some humans have difficulty recognising faces, can suffer from an inability to recognise the odour of a fellow mouse. A male who encounters an unfamiliar female will sniff her very thoroughly. If he meets her again half an hour later his investigation will be much briefer, whereas a new female gets the full treatment. However, a mouse which has been genetically engineered to lack oxytocin in its amygdala treats the female encountered for the second time as if she were a stranger.

We met oxytocin in the chapter on the neurology of reinforcement. It's one of the neuromodulators that doubles as a hormone, and is much involved in social interactions and maternal behaviour. Among other effects it facilitates the release of norepinephrine, and a surge of this neuromodulator in the amygdala accompanies emotional arousal. Norepinephrine is important to the establishment of long-term potentiation at thalamo-amygdala synapses, so learning must be impaired without it. This suggests that oxytocin promotes the amygdala to a state in which it responds fully to signals indicating conspecifics. It's thanks to this bit of wiring, we can reasonably hypothesise, that infant primates focus with particular intensity on the faces and voices of conspecifics – though other neuromodulatory influences connected to emotion may also play a part.

Maybe this circuitry has to be brought into use at an early stage, so that really strong thalamo-amygdala synapses are established before the competition for them grows stronger. Otherwise it becomes difficult to direct sufficiently intense attention to very fine distinctions between stimulus patterns.

Finally, the mechanisms which make one cortical area more sensitive than another to a certain class of stimulus pattern probably come to be solidified, as experience accumulates and as the brain matures, by inhibitory connections. In the first few months of life inhibitory neurons continue to be born and to migrate into prefrontal cortex. There they may help to solidify ideas about what's important and what isn't.

Perhaps the GABA neurons that run to the neocortex from the basal forebrain also play a role. This structure contributes to selective attention, and the GABA neurons can be activated both by dopamine and by norepinephrine. Conceivably they can switch out non-relevant bits of circuitry - shutting off the fusiform face area when an object is being viewed, for instance. The point at which it suddenly becomes clear that the indistinct burbling in the background is actually a human voice may be the point at which GABA neurons block the way to other schema-areas and focus all the excitation on to speech areas. Maybe the basal forebrain is responsible for the difficulty of achieving a satisfactory perception of a tachistoscopically displayed picture if it's wrongly identified at first.

Clearly it would not be a good idea for such a mechanism to come into operation too early in life, curtailing the flexibility of the young brain. But once sufficient experience has been amassed it should make, under normal circumstances, for faster processing of input.

However organisation is normally imposed on the schema-storage areas, the infant brain is extremely adaptable. In someone learning to read a visual word-form area becomes established in left fusiform cortex. In the illiterate that bit of cortex remains sensitive to faces and various other things. Furthermore, if the area that should store one sort of information becomes non-functional in an infant brain, that sort of information can be stored at another site. In the most spectacular demonstration of this an entire cortical hemisphere can be removed in early life without conspicuous effects. This drastic operation is occasionally required because if one hemisphere is seriously malfunctional it interferes catastrophically with the development and functioning of the other, which consequently can perform more successfully on its own. The essence of language is still mastered, whichever hemisphere is missing, though verbal intelligence and short-term verbal memory are usually below average, and the trickier forms of grammar are likely to cause confusion. The inference must be that in the very early years the connections that determine the siting of the schema areas can be re-routed. But as sensory stimulus patterns come to be more firmly classified, flexibility is gradually reduced.

Recapitulation

The knowledge of the sensory world that is acquired during infancy and juvenility gradually comes to be stored very neatly in the schema-areas of the neocortex.

The records take the form of complex networks, and the schema-areas represent crucial nodes.

If such a node is seriously damaged, and especially if the connections to subcortical areas are destroyed, a meaningful perception of the relevant type of stimulus pattern is no longer possible. The conscious experience of such stimulus patterns becomes muddled and confused.

Part of the reason, probably, is that the means of paying attention to that type of pattern is lost.

And most importantly, the circuit through which sustained impulses should circulate to create a very short-term record of sensory input, allowing it to be related to the immediately ensuing input, is interrupted.

The more crucial schema-areas are sited with remarkable consistency.

The connections from subcortical areas probably determine their location.

The nature of the action likely to be evoked by the percept-class may be relevant, and perhaps, in the case of visual percepts, the size of the saccades needed to study it.

Neocortical responses to any input are very widespread in the infant, however, and the segregation of schema-areas proceeds slowly, allowing for the contribution of experience, and for adaptation to early brain damage.

Chapter 22

Where the Thinking's Done

To summarise the plot so far, the circuits of occipital, temporal and parietal cortex analyse the sensory information reported by external or boundary senses and discover the patterns in it. At the same time parietal pathways work out where the stimulus pattern is in body-centred space, allowing any action directed at it to be accurately aimed. Subcortical structures such as amygdala and basal ganglia, meanwhile, respond to the sort of sensory inputs that can function as rewards or disincentives, and turn them into feedback which indicates whether any action taken has been profitable. When a stimulus pattern proves to be significant some of the synapses in the circuitry it excites are strengthened, which means the stimulus pattern is recognised if it recurs: a schema-circuit is established. Places can be linked to this sort of feedback too, and also to the internal situations that are monitored in the hypothalamus. The schema-areas of the temporal cortex constitute vital nodes in circuits which allow complex stimulus patterns to be classified according to their significance and the type of action they may require.

All the pathways run on to terminate in the frontal lobe, in the part known as prefrontal cortex. Here the information provided by the rest of the brain is subjected to the final rounds of processing. For actions which are not dictated by habit this is where the choices between different possibilities are made - decisions to be passed on to the adjacent motor cortex. Similarly, it's where attention is organised, when it's not 'bottom-up', or automatic.

Particularly exciting was the discovery of the crucial role prefrontal cortex plays in what's generally termed 'working memory'. Many neurons here keep firing to maintain a record of a stimulus that's no longer present, when there's reason to keep it in mind. In addition, there's activity when recall of items from a more distant past is required. In short, neuronal activity in prefrontal cortex correlates nicely with several forms of the subjective experience we call thinking.

Evaluating sensory information

The whole of prefrontal cortex is thoroughly interconnected, but there are different specialities in different areas. Orbitofrontal cortex (above and behind the orbits of the eyes) is where the sensory information that can function as reward or discouragement arrives. There are major inputs from the hypothalamus, and strong connections, mostly reciprocal, with the amygdala, the nucleus accumbens, and the rest of the ventral striatum. The visual, auditory and somatosensory pathways, meanwhile, deliver the fruits of their analyses. The area is therefore thoroughly equipped to relate information about the external world both to the internal situation and to information from the boundary senses. In addition, it's closely linked to the hippocampal formation, and to entorhinal and retrosplenial cortex.

The importance of orbitofrontal cortex in linking reinforcements to actions and to the stimuli that are worth reacting to is shown by the fact that when it's damaged discrimination tasks are not readily learnt. Once the task is mastered, moreover, it takes even longer to learn to switch a response to a previously unrewarded stimulus. This is because reversing the learnt behaviour entails more than just learning the new stimulus-reward association. As a behaviour becomes automatic basal ganglia and cerebellum take over its management. Their contribution has to be suppressed before a new rule can be applied, and that's part of orbitofrontal's job.

While a situation still requires attention many orbitofrontal neurons fire when a cue appears which indicates the chance to earn a goody. Some fire with increasing intensity until the arrival of the reward, and then subside. Other cells fire in anticipation but more strongly when the reward appears, as if registering that the anticipation was justified. The activity of such neurons is unrelated to the type of action needed to obtain the reward (other cells in the area deal with that). Some are activated by any kind of reward, but most are more selective. There are also neurons which fire only when the cue indicates no reward. Yet others respond to a noxious input, or to the cue associated with it. In monkeys the anticipatory activity prompted by cues predicting rewards or undesirables can last for a matter of seconds.

A notable feature of the area is that many of the cells appear to be making judgements, or registering associations. A neuron which responds to fruit juice when it's obtained as a reward may not respond to the same substance when it reaches the mouth in a different context. Clearly that cell signals not the taste of juice but the successful outcome of the monkey's choice of action. The anticipatory firing of another neuron may reflect the value currently being put upon an expected reward, rather than its particular nature. For instance, when a monkey was working to obtain a reward which was sometimes apple, sometimes lettuce, and the advance cue indicated which to expect, some neurons fired more strongly in anticipation of a piece of apple than for lettuce. When the situation changed slightly and the rewards on offer were apple or raisin the higher rate of firing was for the raisin, and apple qualified only for the lower rate. Neurons which responded only to one particular food were found, but not very many.

Even something normally regarded as a treat can lose its attraction after a surfeit, and another feature of orbitofrontal cortex is that neurons which are initially excited by a particular food go silent as the animal becomes satiated with it. This change no doubt correlates with the gradual transformation of a flavour from rewarding to repulsive when one consumes too much of it, thereby discouraging overconsumption of some of the substances which are good in modest amounts, harmful in excess.

As well as relating the potentials of the external world to internal needs, the area can balance the attractions of quantity against quality. This capability was explored by Camillo Padoa-Schioppa and John Assad, by means of an experiment in which monkeys could choose to receive one of two different drinks, in varying amounts. When the menu offered Kool-Aid or water the monkeys clearly preferred the former, but six or more drops of water outweighed one drop of the flavoured stuff. When the ratio was four-to-one the choice was variable - presumably it depended on how thirsty the monkey was feeling. Preferences between other drinks were less predictable, and might vary from day to day. Recording the activity of 931 neurons in total, the experimenters found that activity in just over half of them varied in line with what was on offer. A good many, active as the monkey surveyed the symbols on the computer screen, reflected the values the animal revealed by its choices. Many encoded the actual choice, and these remained active until the signal at which the monkey had to make the saccade which indicated its decision. Some, active immediately before and just after the delivery of the drink, were only excited by one type of drink, so clearly encoded the expected taste.

A not dissimilar problem is balancing the value of a potential reward against the odds on obtaining it. O'Neill and Schultz taught monkeys visual cues which indicated the probability of achieving a juice reward, and identified neurons in which the activity correlated with this information. They were distinct from the cells which coded for the value put on a particular type of juice.

Another important function of orbitofrontal cortex is judging the significance of pain. In cases of severe, intractable pain which fails to respond to any other treatment it's been found helpful to cut the connections between the pain pathway and orbitofrontal. Patients report that the pain doesn't feel any less as a result - but somehow it doesn't seem as important any more. The evaluation of the pain changes, rather than the sensory report itself.

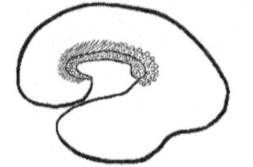

Orbitofrontal cortex can be summed up as weighing together the many factors, both internal and external, which affect the current value of a desirable or the significance of a noxious input. In this it's supported by the front end of cingulate cortex (the central, innermost core of neocortex, wrapped around the corpus callosum). Anterior cingulate is involved in balancing the desirability of a reward against the cost - just how much effort is it worth? In a T maze which offered the chance of obtaining a small reward in one arm or one twice as big at the cost of climbing over a 30cm. barrier, rats consistently chose the higher reward. Lesions of orbitofrontal cortex did not change this behaviour, but lesions in anterior cingulate cortex left the animals less inclined to make the effort required for the large reward.

The same area also contributes to working out the odds when rewards are erratic, favouring sometimes one action, sometimes another, with the less frequent rewards tending to prove larger than the more regular ones, so that the probability of potential rewards has to be balanced against their size. Its function could be described as devising strategies in these conditions, monitoring the effects of chosen and planned actions, and identifying errors. It's notably activated by unsuccessful responses.   Mark Walton and colleagues, imaging human subjects, found that when the rules of a task were switched and the subject had to guess what to do and monitor the result, activation in anterior cingulate was greater than that in orbitofrontal. The balance was the other way around when the subject was told what the new procedure should be, suggesting that anterior cingulate is particularly involved in discovering how to obtain rewards, orbitofrontal in storing the results. This conclusion is supported by another imaging study, which showed higher activation in anterior cingulate cortex in the early stages of learning how to make money at a gambling game, and greater activation in orbitofrontal later on, when decisions were grounded in experience.

Working out which actions are most profitable when rewards are variable obviously requires a memory for the recent pattern of rewards. An early interpretation of the effects of lesions in the anterior cingulate was that they undermined the ability to switch responses when circumstances change. More recent work suggests that they reduce the ability to remember which action is currently profitable. Monkeys with such lesions proved quite good at switching their response after an unrewarded trial, but not so good at maintaining the same response after a rewarded one, which implies that negative results are being remembered more successfully than positive ones. The same lesions make monkeys bad at learning to play the odds when some stimuli produce rewards more frequently than others. A similar effect was found in a foraging task, in which food was to be found in various places on a probabilistic basis, imitating the uncertainties of the natural world. But there was no difficulty in choosing a large reward over a small one. The frontward portion of anterior cingulate cortex functions, it would seem, to record the pattern of past rewards, and to keep track of which stage of the sequence has been reached. Other areas of anterior cingulate, where activation correlates with the experience of pain or of negative emotions, no doubt perform similar operations in managing avoidance behaviour.

The evaluations made in these innermost parts of prefrontal cortex naturally have a considerable influence on what happens elsewhere in the brain. Connections from here, for instance, influence the amygdala neurons activated in fear and anxiety, some to augment their activity, some with suppressive effect.

Social emotions and economic calculations

The evaluative division of prefrontal cortex is proving to be involved not only with simple, immediate, material reinforcements, but also with the more distant and abstract rewards that humans tend to pursue, and with more abstract discouragements. When the subjects in imaging experiments play a gambling game orbitofrontal cortex is activated during the feeling of regret that follows what proves to be a losing decision, and people with lesions in the area don't feel this regret. Another part of it is activated when people look at pictures they judge beautiful or listen to music they rate highly, and when mathematicians think about theories they find particularly pleasing.

The anterior cingulate is particularly involved in interactions with conspecifics. It contains neurons which respond to the cries of offspring, or to the soothing sounds made by mother. When it was ablated in female rats maternal behaviour was so severely disrupted that only 12% of the young survived. Imaging experiments show that it's activated in humans when the subject believes he or she is interacting with another human being, not when they make the same responses believing they are dealing with a computer. Close analysis of the activity during an "economic exchange" game indicated that different portions of the area are engaged, during such trading of benefits, by actions of the self and actions of the other participant, so it may be involved in keeping track of who does what.

The area also seems to track social situations. Male monkeys can be distracted from a small item of food by film of a female monkey with sexual swellings, or of a large male with a threatening stare, or of a medium-sized monkey making friendly, lip-smacking gestures. Macaques with anterior cingulate lesions fail to show any interest in such displays, suggesting they have lost the ability to care about important social cues. Another finding is that there are emotional mirror neurons here, which respond to the sight of someone else in pain.

Orbitofrontal cortex too is involved in social emotions, and in the judgements based on them. When people were asked to decide which of two unfamiliar faces they liked better the activity there increased at about the time they might be supposed to be reaching a decision. Lesions in either orbitofrontal or anterior cingulate, meanwhile, often reduce or abolish the capacity for social emotions. There may be difficulty with recognising emotional facial expressions, or emotional intonations. In the most severe cases the ability to perceive emotion in others, or even to conceive of it, may be completely lost. Failure to allow for the emotions of other people leads to great difficulty with social relationships. Friends may be treated in a very cavalier fashion, and unintentional wounds inflicted. At worst such damage may produce thoroughly anti-social behaviour, and an apparent lack of any concept of morality. The case of Phineas Gage, a quiet, conscientious construction foreman on a New England railroad, who amazingly survived an accident which shot a large tamping iron right through his forebrain but was transformed into a totally different character, is a famous example.

The area which has been most particularly implicated in experiencing emotion and understanding emotion in others lies in subgenual prefrontal - the bit under the bump at the front of the corpus callosum. This is also the site from which the autonomic and neuroendocrine responses that result from emotional judgements are triggered. The skin conductance response - caused by a tiny increase in the secretion of sweat - provides a handy way of measuring autonomic reactions. People with subgenual lesions produce the normal skin conductance response when they are startled, so they are obviously still capable of autonomic adaptation to current circumstance. But they show no response to pictures which would normally arouse emotion, whereas with most people the wiggles on the chart clearly differentiate emotional pictures from neutral ones. In a sample group of depressives whose family history suggested an inherited disorder, PET and MRI scans showed abnormalities of function and structure in the same area. Damage in this region, meanwhile, can sometimes produce a permanent effect on mood, making the sufferer constantly gloomy or irrationally cheerful.

An early indication of the significance of the evaluative and emotional part of prefrontal was a study by Antonio Damasio of someone who had been a successful businessman until, in his mid thirties, he suffered a brain tumour, the removal of which left him with damage to orbitofrontal cortex. Elliott's intellectual abilities and his memory seemed unaffected, and tests indicated that he could still analyse situations effectively. But he had great difficulty making choices, pondering at length over every factor even for minor decisions, apparently unable to weigh them all together. The result was many obviously bad choices. He also became inefficient at carrying out tasks and applying his time sensibly, and could no longer hold down a job. Yet he remained curiously undisturbed by his reversal of fortune, and showed no emotion.

Damasio studied more patients with similar prefrontal damage and found they all showed similar effects. A great many plans are enthusiastically made, but they are seldom put into action. Moreover there's no sign of regret after poor decisions - which may be why these patients have difficulty in relating actions to goals, and don't learn from their mistakes. The severity of the rudderless state seems to correlate with the degree of alteration in emotional experience.

Damasio suggested that *the cold-bloodedness of Elliott's reasoning prevented him from assigning different values to different options.* In other words, without emotion there was no real evaluation. Introspection indicates that 'feelings' contribute a great deal to decisions, and often provide an economical way of reaching them, especially where lengthy analysis of the situation isn't appropriate. The choice between going for a steak or a curry, a pizza or a paella, whether to try a new restaurant or travel to an old favourite, is generally based on feelings, which tend neatly to sum up the current level of appetite, the degree of effort needed to get to the restaurant, current energy levels, and other considerations. And of course the degree of pleasure associated with a prospect is the result of past experience, and reflects it in a very economical fashion. Similarly one's feelings towards another individual tend to provide a convenient shorthand way of remembering how they have behaved in the past, without having to recall many individual incidents. Orbitofrontal cortex, we might suppose, functions to decide and store the significance for the self of all sorts of stimulus patterns. Conscious emotions and feelings are orbitofrontal cortex putting this information to work in influencing behaviour. Pain no longer seems important when it can't be reported to orbitofrontal cortex and when orbitofrontal cortex itself is disabled the means of judging importance is lost altogether.

We might also guess that orbitofrontal cortex functions to power the maintained pursuit of a goal. What keeps one at work is an underlying sense of the joys of fame or fortune, or of the satisfactions of achievement, or of the distinct displeasures of not earning and not being able to eat. If one couldn't imagine what it's like to go seriously hungry how much consistent effort would one put into making sure that it didn't happen? And perhaps a goal is less likely even to be remembered when there is no emotion attached to it.

These hypotheses can be linked to the monkey experiments. The orbitofrontal neurons which keep firing as the monkey prepares the response which will earn it a reward may constitute components of both the short-term memory which allows the animal to remember what it's doing and (as a result of long-term potentiation) the long-term memory which stores the fact that the thing is worth doing. If this memory is missing it must indeed be hard to make intelligent decisions, and harder still to pursue a goal consistently. Someone with the sort of damage Elliott suffered presumably remembers the concept of goals and plans, but lacks the means to apply them.

Emotions and awareness of internal conditions, meanwhile, clearly provide the material for the moment-to-moment model, the combination of sensations which reflects the current state. On the succession of such models is founded the concept of the self as an individual, an ongoing entity, experiencing many states but preserving a certain consistency, an 'I'. And from that develops the concept of other individuals as parallel entities, essentially similar - a concept which allows self-knowledge to be used to predict the behaviour of others.

The inner part of lower prefrontal has been particularly implicated in the development and use of such knowledge, or in what is commonly termed a 'theory of mind.' Damage to lower prefrontal cortex presumably has to be very severe to destroy a concept of 'I' that has already been constructed. But a lesser degree of impairment may, the evidence indicates, seriously reduce the ability to use the model to understand and forecast the behaviour of others.

The increasingly fine resolution provided by modern imaging techniques is enabling some delicate distinctions to be made. The area activated by thinking about the self seems to be very close to, but distinguishable from, the spot that is most exercised when family and friends are contemplated, while that which is at work in thinking about public figures or strangers is a little further off. Quite distinct areas are active, meanwhile, when subjects watch a video of someone performing an act in a way that seems calculated to deceive an observer and when they observe a scene in which the performer is the victim of deceit.

The importance of this whole brain region, meanwhile, is indicated by the fact that really extensive damage here can so completely destroy emotion and motivation as to produce a state of complete apathy.

Attention

The judgements and evaluations reached in the lower part of prefrontal do a great deal to influence attention, and that's essentially the concern of higher reaches. Imaging studies show strong activation in dorsolateral prefrontal when subjects are required to exercise attention, particularly when it's selective attention and there are many distractors to contend with. This area is also busy when a perceptual task is difficult, or when an automatic response must be suppressed in favour of a considered one. In short, it's involved when we're doing the sort of thing that requires concentration. Damage to the area leads to problems with this sort of task. It's also important for maintaining the record of vanished stimuli that still require attention. It was here that neurons were first discovered which can keep firing over a period of delay.

This part of prefrontal is concerned with the signal that predicts a reward, and the information that must be remembered in order to obtain it. Some dorsolateral neurons keep firing when it's necessary to remember the place from which a stimulus has disappeared. Monkeys were trained to maintain fixation on a central point while a spot of light appeared briefly to one side, and then, when a cue was given, to saccade to where the light had been. Neurons were found which maintained their activation continuously during the delay between the offset of the light and the appearance of the cue. Others slackened their firing rate to begin with, but then increased it as the time for the cue to appear drew closer. Sustained activity correlated with successful performance of the required saccade, while abnormal firing patterns predicted failure.

Another experiment was concerned with a 'what' rather than a 'where'. A stimulus pattern appeared while the monkey fixated, and a saccade had to be made to the matching stimulus pattern when it appeared as one of a pair. Again, there were memory cells which kept firing through the delay, and activity in these cells correlated with successful performance. In humans a brief burst of transcranial magnetic stimulation to this area as the stimulus is registered causes a simple memory task to be performed less successfully. It can also make the subjects more likely to choose an immediate reward rather than waiting for a larger one, presumably because it's difficult to keep the prospect of the larger reward in mind.

Place-memory cells seem to be somewhat more concentrated in the higher reaches of dorsolateral prefrontal, getting their major input from the parietal areas that are concerned with the 'where' of sensory information, as well as some from early stages of the visual pathway, where the spatial coverage of receptive fields is smallest and therefore most precise. Neurons that can hold a record of where a sound came from have been found in the same part of prefrontal. Cells that preserve a record of the nature of a visual stimulus pattern receive their information from the final stages of the ventral pathway, and are mostly found a little lower down. But to a considerable extent the two sorts of cell are intermixed. And some cells have been found to respond to both place and stimulus, though more strongly to one than the other. These presumably provide a means by which the 'what' and the 'where', after being treated separately at earlier neocortical stages, are reunited.

Prefrontal memory cells are particularly important when more than one item has to be remembered. Their contribution is crucial to success, for example, when an animal has to investigate several containers to find food and there's no point in revisiting the ones it has already emptied or discovered to be empty. They also have a vital role when information must be remembered during encounters with irrelevant stimuli. Imaging experiments showed that higher levels of activity in lateral prefrontal, sustained during the appearance of the distractors, accompanied successful performances. Furthermore, the activity of prefrontal cortex and more posterior areas was more firmly correlated in successful trials.

Left dorsolateral prefrontal is particularly important to encoding new information, and damage on the left seems to have a greater effect on working memory. The retrieval of information seems to be more impaired by damage on the right. Right dorsolateral prefrontal plays the larger role in spatial attention, meanwhile, just as right parietal does.

In complementarity to its memory function, dorsolateral prefrontal is also heavily implicated in registering the advent of something new in the environment. One of the symptoms of damage in this area can be a failure to notice the sort of sudden change that normally attracts attention pretty reliably. (In this it presumably works particularly closely with the hippocampus.)

Sustained activity in prefrontal cells clearly plays an important role in the short-term memory that allows an animal to remember what it's doing, or to relate a cue to a reward that follows after the cue has terminated. How is it achieved? It's been shown that the overall rhythm of firing in the prefrontal memory cells becomes synchronised with that in the sensory area from which the information-to-be-remembered comes, which implies that the two areas become temporarily linked in a self-maintaining circuit. When it's a place that must be remembered the link is to parietal cells, and when it's a visual object that has to be held in mind the synchronised firing is in inferotemporal cells.

If prefrontal cortex is incapacitated temporal or parietal cells can still function as short-term memory stores, as long as there is only one thing to remember. If there is further distracting input, however, these areas need the feedback link from prefrontal to boost the memory activity. And a temporary inactivation of dorsolateral prefrontal cortex causes a much higher level of distractibility than a similar inactivation in parietal cortex.

The circuits linking the neocortex with the basal ganglia and thalamus also make a major contribution to sustaining prefrontal memory, as might be expected. The hippocampus would also seem to be involved, since synchronisation between hippocampus and prefrontal cortex increases during the successful performance of working memory tasks. Synchronisation of activity across different prefrontal areas plays a role too.

As already noted, prefrontal cortex is essentially concerned with significance, rather than with sensory detail. In a particularly jolly experiment, two monkeys were taught to distinguish two sets of stimulus patterns, cats, represented by three feline species, and canines, represented by three breeds of dog. They then performed a delayed 'matching-to-category' task, in which they were required to decide whether the second stimulus came from the same category as the first. Once the monkeys had mastered the idea the experimenters made the task more difficult by creating morphed images which were combinations of any two of the six originals, with varying proportions of each. This meant that a stimulus-pattern which was 60% dog and 40% cat could be more similar to one that employed the opposite proportions than were two unmorphed exemplars from the same category. Nevertheless, the monkeys learnt to classify even the very ambiguous images successfully for about 90% of the time.

Recordings were made from 395 neurons in the lateral prefrontal cortex of the two monkeys, most of which were active either when a sample picture was shown or during the delay. Nearly one third of them were more responsive either to cats or to dogs - they were divided about equally. And the degree of activation was the same for a stimulus that was only 60% dog as for 100% dog.

The monkeys had never seen real cats or dogs, so it could be assumed that the categorisation system was learnt. Consequently it was possible to retrain one monkey to divide the same set of exemplars into three categories, each containing one cat and one dog, so that the boundaries between classes were entirely new. The monkey became nearly as proficient at this task as at the earlier one, though it took longer over its decisions. And the neurons which preserved the record of the first stimulus pattern by firing strongly during the delay period now operated according to the new categories.

Anterior cingulate cortex, meanwhile, also contributes to attention and decisions about action. While the forward part is connected to orbitofrontal cortex and to all the structures implicated in emotion and evaluation the rearward sector is strongly linked to lateral prefrontal, to premotor and supplementary motor areas, and to parietal cortex. This part is activated by tasks that involve divided attention, the suppression of automatic responses, or care in selecting the appropriate response. In macaque monkeys, cells here were active when an animal decided to change its response as a result of receiving diminishing rewards, not when it was instructed to do so by a signal.

An interesting imaging experiment with humans studied lapses in attention. Subjects were briefly shown large letters composed of small letters, as here, and asked to identify either the large one or the multiple component. Inappropriate answers occurred when anticipatory activity in anterior cingulate and right prefrontal regions was below the usual level, and activity in other prefrontal areas was higher.

```
HH      HH
HH      HH
HH      HH
HH      HH
 HH HH
  HH
```

Anterior cingulate also contributes to spatial attention by means of long-range connections to inhibitory neurons in primary visual cortex. Its inputs there enhance the response to features at the attended location, and help to reduce that in neighbouring areas.

Organising action

The business of organising movements seems to be concentrated in lateral prefrontal (which includes Broca's area), and in the innermost or medial part. Here action plans are drawn up, before responsibility for their execution is passed on to the motor areas of the frontal lobe.

Lateral prefrontal is involved when decisions are made about unpredicted inputs. It's also busy when the precise order of appearance of a number of visual stimuli must be memorised. Once a sequence of cues is learnt the area is less activated in carrying out the required series of actions than when the sequential task must be performed as dictated by unpredictable cues. When a task is being performed in which the correct procedure is altered from time to time there is transient activity here when behaviour must be adapted to the new rule.

Keisetsu Shima and colleagues have shown that lateral prefrontal neurons can also encode somewhat abstract rules about actions. They trained monkeys to perform a series of four movements (constructed out of the possibilities *push, pull* and *turn*) in eleven different sequences. The sequences could be four repeats of the same movement, an alternation of two movements, or two pairs. They found 376 lateral prefrontal neurons that fired in some sort of relation to the task, of which 165 modulated their activity during the period between the appearance of the cue and the initiation of movement. The behaviour of more than half of these cells related not to the specific sequence of movements that was required but to the general category - repeats, alternations or pairs. The activity that had built up by the last half second of the preparatory period served as a good indicator of the type of movement sequence the monkey would perform.

Sensitivity to regular patterns of input has also been found in this area. Humans were imaged while viewing random sequences of circles and squares. A response was seen whenever a chance pattern was violated, in other words whenever either a run of repeats or a run of alternations came to an end.

Finally, lateral prefrontal cortex is important to suppressing the automatic or easy response. In humans asked to name the colour in which a word is printed it's busy when the word is the name of a different colour, though less so when a second example of incongruence follows. It's at work when immediate rewards are rejected in favour of waiting for something better. And when it was temporarily disrupted by transcranial magnetic stimulation during a game designed to test social behaviour the subjects seemed unable to suppress the selfish action which gained them an immediate reward, though still perfectly capable of recognising that obtaining a reputation for fairness and altruism might pay off better.

Medial prefrontal, in its lower part, seems to connect current emotional and motivational circumstance with relevant records of past experience, in the interests of choosing the most profitable action. There's an abrupt reorganisation of activity here when circumstances change and an animal has to work out new rules for obtaining a reward.

Kenji Matsumoto and colleagues trained monkeys on a GO/NO-GO task, with two visual cues, one of which indicated that the monkey should, after a delay, pull on a joystick, the other that the joystick should be held without moving it. After another delay there was a liquid reward either for successful GO responses, or for appropriate abstentions. Within a single block of trials the significance of the cues and the reward contingency remained the same. Between blocks the combinations were changed, so that either the 'pull' signal became the 'don't pull' signal and *vice versa*, or the previously unrewarded response became the rewarded one. All eight combinations of cue, response and reward were used, making for a demanding task, but after a few months of training the monkeys were able to cope with the switches fairly efficiently, usually with fewer than eight errors.

In medial prefrontal cortex cells were found which were activated after the cue, but only when a particular combination of response and reward was in prospect. Some, for instance, showed a burst of excitation when no response was required but a reward could be expected, others when the opposite conditions applied, and so on. The four possible combinations of action and reward were about equally represented. Some of the cells were activated only briefly, others maintained a fair level of excitation over the second that elapsed before the dimming of the fixation point indicated that it was time to pull, or not pull, the lever.

Prefrontal cortex has extensive inputs about emotional matters, as we've seen, which can alert it to what needs attention. In its role as the ultimate arbiter of input from the external world it also influences emotion. When an ambiguous pattern of light and shade amidst the undergrowth is classified as a concealed danger prefrontal sends messages to excitatory neurons in the amygdala which in turn trigger the autonomic responses that prepare the animal for appropriate action. If the sensory input is reinterpreted as harmless, messages go to inhibitory neurons which cancel the instructions to the lower brain areas, allowing the effects of the preparation for urgent action to fade away. In humans this pathway can be strengthened by practice and counteract the effects of stress.

A similar pathway is active when dieters manage to resist the temptation of attractive but fattening food. The extension of pathways which suppress emotional and instinctive reactions was no doubt an important part of the evolutionary process which enabled us to live successfully in large groups, linked together in super-groups.

Prefrontal control is weak in drug addicts, which is no doubt why sacrificing an immediate satisfaction for the sake of long-term health and happiness is difficult. Addictive drugs seem to wreak their havoc partly by damaging this subdivision of the frontal lobes, or by corrupting the inputs to it. Unfortunately they are particularly likely to effect such damage in adolescent drug-users, since prefrontal cortex matures quite late, and is still very susceptible in youth.

Prefrontal as 'workspace'

A remarkable feature of prefrontal cortex is the number of neurons activated by any one task. Whatever a neurologist trains a monkey to do, a significant proportion of the cells encountered by the electrodes turn out to be involved. This implies that prefrontal neurons can be tuned to many different employments. In other words their inputs can change radically when a new subject occupies the mind. A clearly relevant factor is that the pyramidal neurons here have many more spines on their dendrites than those in other neocortical areas, which suggests a potential for receiving an exceptionally large variety of information. Moreover there are many long and slender spines, in which potentiation and depression can be induced very rapidly.

When a monkey must discriminate between two stimulus patterns many cells respond to the rewarded pattern, only a few to the unrewarded one. When the situation is reversed a substantial proportion of the cells switch their response to the new reward-predictor.   Other cells which at first responded only to the rewarded stimulus cease to be selective, and come to respond to both. Presumably their activity reflects the fact that both stimulus patterns have reward potential.   If the monkey learns a different set of discriminations on another day, the same neurons may come to respond to the new stimulus patterns.

In one experiment monkeys learnt to perform a 'delayed-matching-to-sample' task in which they had to indicate whether the target object in an array of three appeared in the same location after the delay as before.   Each block of 80 trials began with 10 in which the target object appeared alone, then threesome and singleton trials were intermixed.   For the next block of trials one of the other two objects became the target.   In many of the 97 prefrontal neurons which were observed the activity depended on which object was currently the target.   Once its object had been established as the current target such a neuron might be busy before a trial began, as well as when the array was shown and during the delay period, suggesting that it was preserving a record of which object was relevant to success in this particular block of trials.   The response of many other neurons related to the position of the current target, while only a few cells showed activity influenced by the location of non-targets.

Prefrontal cortex could be described as painting a picture of the current situation not in sensory terms but in functional ones.   Neuronal activity reflects whether a stimulus pattern should elicit action or no, what action is required, and what result is expected.   The neurons that respond in large numbers to so many different sensory inputs clearly provide the means of making these judgements. The essential business of prefrontal cortex is working out what to do in novel situations - the time when fully focussed consciousness is required. As John Duncan put it: *prefrontal cortex acts as a global workspace .... on to which can be written those facts that are needed in a current mental program.*

The facts - the sensory inputs - are encoded in prefrontal by means of temporary links to the sensory pathways, synchronising the circuits involved into a common rhythm.   Semantic knowledge is added into the picture by the same means, so that working memory is to some extent (as Daniel Ruchkin and colleagues have it) a state of activated long-term memory. The messages which are sent back from prefrontal to subcortical areas, and to the pathways from which the sensory input originated, no doubt play a role in imposing this synchronisation between prefrontal and the source area. In the sensory processing pathways the synchronisation of synaptic input at each stage means that greater detail can be extracted from the data. This is how attended input is turned into a clearcut experience, leaving the rest as fuzzy background.

The implication is that prefrontal involvement is necessary to conscious experience - and as noted earlier, a conscious sensory experience only occurs when the activations caused by the sensory input reach prefrontal cortex. Other evidence points to the same conclusion. Victor Lafuente and Ranulfo Romo taught monkeys to respond to a tactile stimulus, and then gave them tactile stimuli around threshold level. The activity of prefrontal neurons correlated with the monkeys' reported perceptions, but that in somatosensory cortex didn't.

Hakwan Lau and Richard Passingham extended the finding to humans. They produced in sighted subjects something akin to the 'blindsight' found in people with lesions in primary visual cortex. As we saw earlier, when a briefly exposed visual stimulus is rapidly followed by another stimulus it's often 'masked', and doesn't reach consciousness. However, if experimental subjects are obliged to guess what the masked stimulus was, choosing between two possibilities, they choose correctly at a rate well above chance. Lau and Passingham established for each subject the conditions under which a correct choice didn't coincide with conscious experience. Then they scanned their subjects while they performed the task with these intervals. They established that conscious perception correlated with a significant increase in activity in left dorsolateral prefrontal cortex.

The mechanism in dorsolateral prefrontal which contributes to the larger, synchronised neocortical circuitry presumably takes a little time to set up. Further input which arrives too soon after the first prevents the mechanism getting into full operation, but once it has done its job the conscious experience can survive. Similarly, when the sensory input is very small and brief it may not reach dorsolateral prefrontal, or not with sufficient strength to trigger return messages.

The major neuromodulators play important roles in prefrontal cortex, and in underpinning fullscale consciousness. Norepinephrine, acetylcholine, dopamine and serotonin are all heavily present. Acetylcholine is essential to prefrontal memory, no doubt because of its role in arousal, and because of the widespread influence of the basal forebrain's acetylcholine-deploying neurons. The correct level of dopamine is also crucial, as Patricia Goldman Rakic and her colleagues discovered. Either too much or too little impairs working memory, an effect that has been observed in several species. The D1 type of dopamine receptor, which has an essentially inhibitory effect, must be the major culprit, since it's more numerous in prefrontal cortex, and more extensively distributed across the layers than other types. But why should too little inhibition be bad for the memory cells?

One reason has been elucidated by Susheel Vijayaraghavan and colleagues, experimenting with monkeys performing a saccading task. They found that when dopamine levels were low not only were the memory cells which coded for a saccade in the required direction excited, there was also a highish level of activation in those coding for adjacent directions. With a normal level of dopamine the latter excitation was suppressed, leaving the correct direction more clearly indicated. With too much dopamine all the memory cells were inhibited, and the spatial information was lost altogether.

Prefrontal cortex and intelligence

There's disagreement about whether human prefrontal cortex is proportionally larger than that of great apes, as well as absolutely. A key fact, though, is that a great deal of space is taken up by white matter - the myelinated axons which provide long-distance communication - and by the extra-dense distribution of spines on prefrontal dendrites. Bigger brains are fed with larger amounts of sensory data than small ones, and extract a greater amount and variety of information from it. The ratio between the areas that distil out the information and the one which juggles with it may perhaps remain fairly constant.

It's reasonable to suppose that the larger the prefrontal cortex, the more extensive the circuitry that can be kept in operation at one time, effecting more wide-reaching correlations and comparisons. It must be the number of subsidiary circuits that can be brought into play simultaneously which determines just how widely attention can be spread, and how much information can be held in working memory. That in turn must determine just how complex a pattern can be extracted from one intake of sensory information, and how elaborate a super-pattern can be constructed out of several such intakes.

It seems likely that the size of prefrontal cortex is relevant to all those feats which involve focussing attention on two distinct stimulus patterns at once. Realising that the actions of the image in the mirror correspond with what the self is doing is one example. Correlating the pattern which is a word with the concept for which it stands is another. Putting words together to build a sentence must require a complex assembly of prefrontal attention-circuits, and an extra-large ration no doubt explains why we construct much longer sentences than apes do.

However, human attention, and working memory, is nevertheless limited. George Miller's observation on how many human cognitive efforts seem to have a capacity confined to around seven units was noted earlier. Give someone a brief look at a number of assorted objects on a tray and then ask them to name as many as possible and they seldom remember more than seven or eight. Remembering long telephone numbers is difficult - though clumping them in threes or fours is helpful. It seems likely that it's the functional capacity of prefrontal cortex that determines this limitation.

The amount of information that can be correlated at any one moment may be restricted, but the problem can be minimised by nesting concepts within larger concepts. The immense flexibility of prefrontal cortex means that it can be employed to develop many categories. Then it can group those categories into classes and super-classes. The level of organisation might be expected to be hierarchical, becoming increasingly more abstract from rearward to forward areas, and this seems to be the case.

It's thought that individual differences in intelligence may correlate with differences in the capacity of working memory; or perhaps the efficiency with which it's deployed is equally important. An experiment in which subjects performed a challenging working memory task tested these ideas. Presented with a succession of stimuli (letters in some trials, faces in others) they were asked to identify any that matched the third one back - with repeats of the two-back or four-back item thrown in from time to time to confuse the issue. Those who had scored higher on an intelligence test also did somewhat better on this task. Activity in lateral prefrontal, an area associated with attention and with inhibiting irrelevant information, correlated with above-average performance.

Another experiment indicated that the individuals who performed most successfully in a visual working memory task were those who were best at concentrating on relevant information. Electroencephalography suggested that the less successful subjects were actually retaining more information, which presumably made the judgement required more difficult. This implies that the proper management of attention may be the key to a good working memory.

It's reasonable to suppose that an individual who can perceive relationships among eight stimulus patterns rather than seven will make more interesting discoveries than someone who tends to work only six components into a pattern. Working memory can be applied to any subject, so its capacity may correlate with the measure of what psychologists call general intelligence. Perhaps this can be tied in with a recent finding that the degree of global connectivity found in prefrontal cortex correlated with performance in a working memory task believed to reflect fluid intelligence. Lateral prefrontal and parietal cortex provided particularly important components of the network, but it extended much further.

Some psychologists believe that intelligence is not an indivisible unity, but comes in several varieties, such as spatial, linguistic, social, mathematical; and that an individual blessed with one sort is not necessarily endowed with all the others. A likely hypothesis is that the nature of the intelligence provided by an above-average ration of pre-frontal memory circuits depends on which bits of the brain those circuits are most heavily connected to. The artist perhaps has a working memory that is applied most strongly to visual input, and can thus encompass more complex visual patterns than the rest of us, and remember their perceptions more accurately as they transfer them to canvas. Similarly the musician may have heavier connections from right auditory cortex to prefrontal than most people, while linguistic intelligence involves stronger connections from language areas.

How far the strength of these connections might be shaped by genetic inheritance, how far by experience is an open question, though both must be at work. The issue is confused by the fact that people tend to work hardest at the activities they enjoy, and enjoy the activities they are good at. But in both mice and men it's been shown that working memory can be improved by training, with some indication of beneficial effects on general intelligence in the case of the mice. A particularly suggestive discovery was that in a mixed bunch of mice (not, that is, a single purebred strain) rapid learning in five different tasks was found to correlate with the level of expression of certain dopamine-related genes.

A bit of prefrontal cortex that must be particularly important to the sort of intelligence involved in abstract reasoning is the subdivision right at the front of the brain, behind the forehead. The neocortex as a whole can be divided up into around fifty areas on the basis of the different patterns of cellular architecture, and the most commonly used system is the map that was published by Korbinian Brodmann in 1909. Anterior prefrontal, sometimes known as the frontal pole, coincides pretty much with Brodmann's Area 10. Whereas all other parts of prefrontal cortex are extensively interconnected with the other cortical lobes and their assorted pathways BA10 gets all its input from within prefrontal (although its neurons project back to a much wider field). We can deduce that it's the place where the final correlations are made. It must be what allows us humans to carry categorisation to extreme heights and develop ideas founded not in physical resemblances but in behaviours, effects, values or emotions.

Anterior prefrontal is activated when a main goal has to be kept in mind while a succession of subordinate goals is achieved, or distractions dealt with, or when two different plans are being executed simultaneously. It's also busy when a complex strategy must be worked out. It was found to be at work during a delay between the instruction telling experimental subjects which of four possible memorising tasks was to be performed, and the appearance of the display of items to be memorised - in other words, while they kept in mind the rule to be applied on the upcoming trial. And it's particularly needed when a task involves more than one cognitive operation. All these functions can be deduced to involve co-ordinating, or supervising, what goes on in other parts of prefrontal cortex on the way to producing action.

Anterior prefrontal is also active, along with several other prefrontal areas, during recollection - top-down accessing of stored knowledge, as opposed to the bottom-up process which is recognition. Subjects were shown words and famous faces and asked to classify the former as pleasant or unpleasant, the latter as being involved in politics or entertainment. They were then scanned with fMRI while recalling either which decision they had made for each item, or whether it had been presented on the left or right of the screen. BA10 was busy for both tasks.

Perhaps most intriguing is the fact that this area is often active when the rest of the brain isn't - during the boring bits of experiments, the periods when the baseline or resting condition is being assessed and there's no stimulus to consider. This suggests that activity here can correlate with thinking, or daydreaming. And of course the bit of brain at the very end of all the pathways might be expected to be thoroughly involved in the business of thinking. The evidence provided by lesions suggests that the area plays an important part in general intelligence.

The default mode network

After decades of studying what happens in the brain when sensory stimuli are processed or muscles activated, neuroscientists have begun to look at what goes on in idle moments. Using fMRI they've studied the brain activity of subjects who weren't working on any tasks. A network was discovered that tends to be distinctly busy at such times. Along with BA10 it includes a medial part of prefrontal cortex, some cingulate cortex, and some parietal. All are concentrated along the central line of the brain, close to where the two hemispheres adjoin each other. Interestingly, these are areas with especially high metabolic rates, consuming energy even more greedily than the rest of the neocortex. The system was christened the default mode network. Much of it is actively inhibited when attention is turned to the outside world, or to action.

The network is busy, along with other prefrontal and parietal areas, when recent memory is consulted. At such times retrosplenial cortex, the hippocampal formation and parahippocampal cortex may be linked into the circuit. Areas such as visual cortex and even bits of cerebellum seem to be included when appropriate. It looks as if there is a central core to this network and, as might be expected, different brain areas are associated with it in different forms of mental activity.

Damage to the posterior cingulate is rare, but Guillaume Herbet encountered a patient with a tumour in this region. During pre-surgical exploration of the area he and colleagues discovered that artificial stimulation of the subcortical connections to this area induced a few seconds' interruption in consciousness of the outside world, with a glazed expression and no response to verbal stimuli. When he recovered the patient recollected dreamlike experiences of vague landscapes. This evidence seems to fit well with the idea that this bit of posterior cingulate cortex might be particularly involved in the forms of mental activity which give precedence to the contents of memory rather than current sensory inputs.

It's suggestive, meanwhile, that prefrontal cortex seems to be almost continuously in communication with parts of parietal cortex, and various cognitive operations are performed more efficiently when the operations of the two areas are synchronised. Parietal co-operates closely with prefrontal in organising behavioural attention, and is heavily involved in preparations for action. It looks as if these functions, all essential for the conscious control of action, may have been expanded to cover more abstract applications as a part of the capacity for abstract thought.

Recapitulation

Prefrontal cortex is where all the sensory information is put together for the last and most vital correlations, and where actions and action sequences which have not become automatic are decided and organised.

Information from internal and contact senses is weighed up here, in the context of the external situation, while the external situation is evaluated in the context of internal needs.

Prefrontal is the seat of conscious attention - the place where messages from the lower brain about inherently attentionworthy sensory inputs end up, the place from which messages are despatched when attention is deployed by choice.

It's vital to working memory, which involves paying attention to entities no longer physically present.

It can reasonably be regarded as the bit of the brain that provides conscious thought - though of course it is entirely dependent on the information supplied from more rearward parts of neocortex and on the underpinning machinery in the subcortical brain.

The frontmost bit, known as BA10, may have a particularly significant role in purely human capabilities.

Chapter 23

More about the Basal Ganglia

It's clear that the basal ganglia pathways running via the thalamus to the neocortex must have had a central role in supporting the evolution of the neocortex since they are heavily implicated in both attention and reinforcement, essential factors without which complex analyses of sensory input wouldn't be a practical proposition. A neocortex couldn't have been a good investment without a means of ensuring that its machinery was employed on the things that really matter. Equally important, if more and more sensory evidence was to be collected and analysed, was the ability to use feedback to augment the inbuilt knowledge of which stimulus patterns were significant, and which could be ignored. The attentional system and the capacity for learning where to direct it must have expanded in line with the growth of sensory systems, since one would be useless without the other. Both must have been well developed by the time the ancestral telencephalon was rearranged and expanded to form the mammalian neocortex.

The basal ganglia are certainly long established, for at least some of the circuitry is found in the lamprey, that jawless fish quite distantly related to the rest of the vertebrates. Indeed a similar network, produced by a similar genetic programme and performing a similar function in suppressing and releasing behaviours, has been found in arthropods.

In non-mammalian vertebrates the conspicuous basal ganglia connections are those with the thalamus and the sensory centres of the midbrain, such as the optic tectum. It must be deduced that in non-mammals most of the learning about where to direct attention, and when to follow it up with a full motor response, is based on the sensory information provided by these areas. The evolution of the neocortex was accompanied by an increase in the number of pathways leading from basal ganglia to thalamus, and the extension of many of them into the new structure, along with return pathways from neocortex to basal ganglia. There was also a considerable expansion of the thalamus. All this meant that the new sensory processing facilities could be combined with the long-established learning mechanism.

In mammals the basal ganglia circuitry is also employed in a subtler way to underpin another talent, the ability to learn not only when and where it's profitable to act, but also how to shape the movement. Many mammalian infants are born with a motor system that produces only rough-hewn, inefficient movements, along with feedback arrangements that allow these basic gestures to be refined by extended practice, and converted into a repertoire of subtle and finely directed ones. It can be deduced that the process of learning to produce such movements is what creates movement schemata - conscious ideas of the movements of which the self is capable. This leads to the possibility of planning a movement in advance, and choosing the means of achieving a desired end.

What further makes the basal ganglia circuitry particularly intriguing is that it must make a major and essential contribution to the sustained circulation of impulses that creates consciousness.

To recapitulate, the principal cortical input arrives in the striatum, with the sort of information that acts as feedback or reinforcement particularly strongly focussed on the nucleus accumbens. The two output stations are the interior part of the globus pallidus and the substantia nigra pars reticulata. The messages routed through the former output station were at first thought to be principally concerned with body and limb movements, those through the latter with eye and head movements - the outward signs of attention - but this idea has been somewhat undermined by further studies. The output from both areas is inhibitory (conveyed by GABA), and the default condition of the output neurons is a continuous, steady activation. The target of this ongoing inhibition is the thalamus, where it probably serves to reduce, rather than completely suppress, messages from thalamus to cortex. (An illustration of the whole network is on page v of the colour section.)

In mammals the pathways to the two output stations are of three types, modifying the inhibition in different ways. The direct pathway from neocortex to thalamus goes through two sets of inhibition. The excitatory message from the neocortex (dashed line) triggers an inhibitory message to the internal globus pallidus or the substantia nigra pars reticulata (solid line), suppressing the inhibitory message to the thalamus and leaving the targeted neurons' communications to neocortex unencumbered.

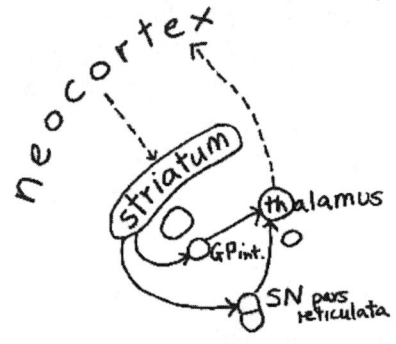

In the indirect pathway an excitatory message goes from the striatum to the external globus pallidus (the part nearer to the outside of the brain). This triggers an excitatory communication to the output stations, which are thereby prompted to intensify their inhibition on their thalamic targets. The message takes fractionally longer than that by the direct route.

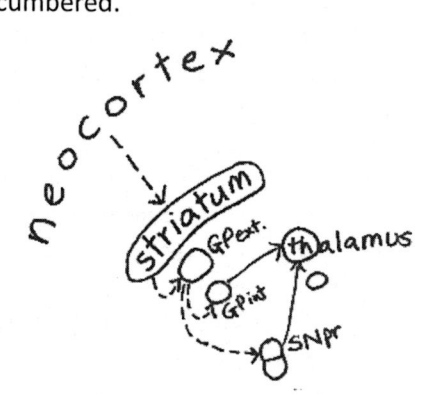

The third pathway is the hyperdirect, in which the excitatory message from the neocor-

385

tex, routed through the subthalamic nucleus instead of the striatum, similarly triggers an intensification of the inhibitory message, but more rapidly. This pathway operates when a planned or automatic action must suddenly be suppressed, or when it's necessary to consider alternative options carefully before committing to action. Malfunction here leads to impulsive behaviour. The pathway is probably also responsible for suppressing previous activity (both neuronal and behavioural) when attention must be diverted to salient and significant new inputs.

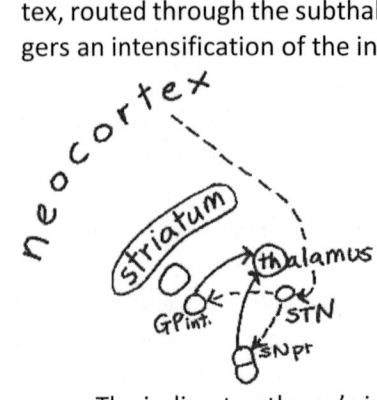

The indirect pathway's inhibition is widespread. The disinhibition effected through the direct pathway is, in contrast, quite focussed. It frees just a small group of cells to communicate with the thalamus, and thence the neocortex. A significant aspect of this machinery, vital to any interpretation of basal ganglia function, is the nature of the thalamic nuclei to which the pathways run – mostly some of the small, interlaminar nuclei tucked away among the larger ones.

The three pathways are not wholly constituted of separate neurons. Some of the fibres arriving from the cortex divide, and run to different goals. This means that in some cases the same originating impulse will be responsible both for a message that takes the direct route to the output station and for one that goes by the indirect way and arrives a fraction later. This implies that two pathways can be active at the same time. In guiding fine-tuned movements, presumably, a succession of rapidly alternating messages arrives at the output station. In addition, there are many interconnections between the various structures, further confusing the original concept of three straightforward pathways.

The ventral striatum receives input from the orbitofrontal cortex, and also from all the subcortical areas which have evaluative and motivational functions, such as hypothalamus and amygdala. Some of this information comes direct, some via the thalamus. Much of it is processed through the nucleus accumbens, that structure so heavily involved in addiction. Some neurologists also include a part of the olfactory tubercle. Odours must often play a significant role as evaluative stimuli, especially the ones with genetically determined effects which can no doubt lend significance to other odours which occur in association with them. In short, ventral striatum manages the feedback information which is essential to learning how to obtain straightforward classical rewards - something which also, of course, entails discovering where it is worthwhile to direct the attention.

Different sectors of the dorsal striatum get their cortical input from sensory areas of the neocortex, from motor cortex, and from prefrontal cortex. All these areas can be very busy, along with the ventral striatum, when a new situation is being investigated. The connections between prefrontal cortex and dorsal striatum are especially active if the task is complex. They become less so as the most profitable way to behave becomes clear. Gradually the reward-gaining responses become habitual, and may be continued long after the reward ceases to materialise. By then their management is concentrated in the lower part of the dorsal striatum.

It's the highest part of the dorsal sector, linked to the motor areas of the frontal lobe, that deals with motor learning. The feedback for motor learning comes mostly from distance senses - visual feedback indicates by how much one missed achieving one's purpose, auditory feedback reveals whether sounds came out in the intended form. So it's interesting that auditory mirror neurons have been found in this part of dorsal striatum, though only, so far, in a species of songbird. In parallel to the visual mirror neurons of monkeys, activated either by performing an action or seeing it performed, these fire both when a conspecific's song is heard and when the bird itself sings. A gene which in humans has been implicated in language - it's mutated in members of a human family with language problems apparently stemming from a difficulty with imitation - is heavily expressed in the basal ganglia of songbirds. This lends support to the idea that, despite the differences between bird brains and mammalian ones, avian song learning may in some ways resemble human language learning.

Dopamine

As we've seen, the dopamine influence is very important in learning. In the direct pathway D_1 dopamine receptors predominate, so a dopamine flux increases the chances of excitation and long-term potentiation, and of suppressing the inhibitory output from globus pallidus or substantia nigra pars reticulata. D_2 receptors rule in the indirect pathway, with opposite effect. Since the D_1 receptor needs a bigger dose of dopamine to activate it than the D_2 it's thought that the direct (inhibition-releasing) pathway is prompted into action by the bursts of dopamine firing evoked by an unexpected event. The dopamine centres are under ongoing inhibition from GABA neurons of the basal ganglia, which can be suppressed by an input from the hippocampal formation. This is probably what happens when an unexpected event occurs. Excitatory input from the pedunculopontine tegmental nucleus, that important component of the ascending activating system, can also elicit burst firing in the dopamine neurons, and no doubt contributes to their attentional response.

Another powerful influence on basal ganglia operations is wielded by the acetylcholine neurons which spread long axons through the area in a dense network. They tend to fire pretty much continuously, but pause briefly when a salient stimulus is registered, especially if it's a desirable. And they quickly come to produce this response to an input which regularly predicts the appearance of a desirable. Much of their sensory information comes from the superior colliculus, via some thalamic nuclei, but there are also inputs from the neocortex. They seem to be linked to the dopamine neurons in reciprocal inhibition.

Dopamine neurons, meanwhile, are turning out to be less standardised than originally thought, with variations that tend to reflect the order in which they appear in the developing brain. Substantia nigra and ventral tegmental area receive input from different cortical areas. There are also some distinctions between different sectors within these centres.

Despite the complexities of the system, however, an experiment with mice by Kravitz, Tye and Kreitzer showed that the circuitry can work in a straightforward way to support learning, as has been supposed. Artificial stimulation of neurons in the direct pathway functioned to reinforce the response to a stimulus, as if the effort had been rewarded. Stimulation of the indirect pathway discouraged the action, as if it had failed to produce a reward, or led to a noxious result. An interesting wrinkle was that the positive learning persisted for some time, while the negative learning faded fairly rapidly. That suggests that striatal learning can allow for the fact that what is unprofitable today may prove more rewarding tomorrow. Learning about serious dangers that should be avoided without further experimentation probably depends more on the amygdala and other structures.

The same laboratory also demonstrated that direct and indirect pathways have the expected effect on movement. Artificial stimulation of medium spiny neurons in the indirect pathway, this time in the motor sector of the system, caused increased freezing, reduced locomotion and slow initiation of movement. Activating the direct pathway had the opposite effect. And in mice bred to show the symptoms of Parkinson's disease stimulation of the latter sort temporarily cured the movement difficulties.

A filter for sensory input

In the mature, experienced brain how do the operations of neocortex and basal ganglia combine? If there's an alerting input from the hindbrain or if a genetically fore-ordained stimulus pattern is registered in the midbrain, there'll be a 'bottom-up' alert to activate the basal ganglia. If the significance of the input is already clear there'll also be a message to the appropriate cortical schema area, bringing that into the circuit. For more complex sensory data there must be a first pass through the analytical machinery of the neocortex before its significance is registered. The first tickle of excitation in a schema area, indicating a combination of features that has proved significant in the past, must be sufficient to trigger a message to the striatum, which will arrive at potentiated synapses, and evoke a returning message to the same neocortical site. This is conditioned attention.

A message must then be sent from basal ganglia to prefrontal cortex, where a decision can be taken as to whether the stimulus pattern is interesting in the current circumstances. Prefrontal no doubt has to find out what has caused the message by linking up with whatever parts of sensory cortex are dealing with the stimulus pattern, and the message from the basal ganglia must be important in prompting it to do this. The link seems to depend on a synchronisation of the larger-scale local oscillations in membrane potential in the areas involved - the synchronisation which has been shown to occur when distant brain areas need to work together. The basal ganglia messages to prefrontal and sensory cortex must contribute importantly to the beat.

If prefrontal cortex decides that attention is justified it despatches instructions to continue favouring this particular input. A stimulus pattern that has been registered several times without anything of interest ensuing, on the other hand, is likely to encounter depressed striatal synapses, so that no signal goes to prefrontal cortex. Furthermore, as something familiar classified as having no interesting connotations it no longer causes activity in dopamine neurons. This must be the explanation of learned inattention. When a stimulus pattern evokes neither a prefrontal response nor the dopamine flush that is important for the potentiation of synapses, it becomes hard to learn to take notice of it - unless of course it occurs in company with effects which elicit an amygdal response.

Clearly the basal ganglia pathways serve as a filter, promoting important sensory inputs to the foreground, leaving the rest as hazy background, or registering only at unconscious levels. Thanks to all those modifiable synapses in the striatum it's a filter that's shaped by past experience. The first instalment of data about a familiar stimulus pattern - a report of its most salient features - can be supposed to arrive from sensory cortex at striatal synapses that have been potentiated in the past. If the record indicates that the stimulus pattern has been associated with significant effects the inhibition on a certain route through a thalamic nucleus is lifted and a signal returns to the cortical site from which the sensory information was despatched. It encourages continued processing of that bit of input, and if it was a transient input, maintenance of the temporary record.

The medium spiny neurons of the striatum - the ones that convey the inhibitory message to the next basal ganglia station - each have ten to fifteen thousand spines on their dendrites, supporting an enormous number of input synapses. Inputs at a hundred or so of these synapses, or perhaps several hundred, are needed to fire the neuron, but there are only a few contacts from any one individual cortical cell, and in many cases the messages may originate in assorted areas of neocortex. It must be deduced that the combination of excitations that fires a medium spiny neuron can add up to a very solid dollop of information.

However, the medium spiny neurons in the striatum vastly outnumber the recipient cells in the output stations. And a single neuron in the latter may receive inputs from as many as a hundred medium spinys. This architecture is clearly appropriate for mediating between sensory inputs and action. Animals with elaborate sensory systems are capable of registering a potentially infinite number of stimulus patterns, but even with a very complex musculature the number of possible actions is limited. The basal ganglia is geared for applying a finite number of possible motor responses to a huge variety of possible stimulus patterns.

It seems probable that the flexibility of the striatum underlies the remarkable ability of primate prefrontal cortex to switch its neurons among many different sorts of sensory inputs, creating a sort of semi-abstract picture of the current situation, painted in terms of what's most desirable, and what action it's best to take. The striatum can be seen as a vital part of the mechanism which allows dorsolateral prefrontal cortex to function as a 'global workspace', dealing with whatever subject currently requires attention. It can arrange for any selection of current sensory input to be channelled to that workspace. It also seems that the same neuron can be fired by quite a variety of different combinations of input. Just as activity in a large proportion of prefrontal neurons reflects the task currently in hand, so it is in the striatum. A substantial number of striatal neurons fire as the choice between two actions is being made, and another goodly number are activated in anticipation of a reward, or when the reward arrives, or when an expected reward fails to arrive. A principal difference is that a larger proportion of cells seems to be influenced by the expectation of reward than in prefrontal.

How did it all begin?

A clue to the origins of the basal ganglia surely lies in the fact that they also communicate to the diencephalic and mesencephalic locomotory centres. Furthermore a deficiency of either dopamine or acetylcholine in the nucleus accumbens on one side of the brain causes a rat to turn constantly towards the side of the deficiency. Perhaps all the complex circuitry might have emerged from what began as a steering mechanism? This seems probable, since the basal ganglia are much involved in the control of movement, and in chordates as well as several other phyla locomotion was clearly the earliest form of centrally organised, muscle-powered movement. It's worth considering, therefore, just how locomotion began.

Some power of movement is present even in single-celled animals. They can change their shape quite notably, as they do when they multiply by dividing in two. They're also equipped with sensory receptors which enable them to withdraw from mechanical or chemical contacts. The cells of the very first multicellular animals will have retained these capabilities. Consequently those bag-like creatures I've postulated could probably shape themselves so as to fit against their fellows without losing access to the drifting food particles, and to avoid occupying unsuitable bits of rock.   Perhaps they also pulsed gently to increase the flow of water in and out of the bag.   Any that could move sufficiently to direct the 'mouth' towards the current that was bringing the food particles would have the best chance of multiplying.   It must have been the potential for doing one or all of these things better that made the evolution of a double-layered bag worthwhile, with the outer cells acquiring tactile and chemical information from the outside world and the inner ones providing most of the shape-changing ability, elaborating on the mechanism used in cell division.

The withdrawals from undesirable sensory inputs will have been managed in the first place by direct connections from the sensory cells to those which provided mobility.   It must have been these defensive movements that became organised to produce locomotion.   The shape-changing cells were elaborated, multiplied and bundled together to form muscles.   The insertion of interneurons among the motor neurons produced a locomotory pattern generator.

The essential chordate pattern generator is seen in the adult amphioxus, and in smoother versions in lampreys and hagfish.   A contraction of the muscles on one side at the front of the body produces a bend, which compresses the water behind the bend and reduces the pressure in front.   It's followed by a contraction on the other side further along, which in turn fades into another opposed contraction, and so on.   As the contractions ripple along each side of the body the animal slides through the water in S-shaped curves.

The control of steering in early chordates must have been straightforward. If a potentially harmful stimulus impinged on the body the old defence system would still work.   A direct message from sensory receptors to motor neurons would produce a contraction of the muscles on the side of the input.   If they were already contracting and producing a curve away from the threat the contraction could be intensified and the curve deepened.   If they were at the relaxation stage contraction would be initiated, overriding the locomotory pattern generator, so that the affected part of the body was withdrawn.   The rest of the body, following the re-shaped curve, would also, with luck, avoid contact with the threat.

Once there was a neuronal channel by which to send a message forwards this manouevre could be followed by a contraction at the front of the animal, on the side opposite to the threat, so that the head turned away, pulling the body after, and the danger was more thoroughly distanced.   Again, the message would be one that overrode the pattern generator.   It seems reasonable to suppose that it might have been conveyed by the same transmitter used between neuron and muscle, acetylcholine.   When vision improved sufficiently to provide advance warning of potential collisions the same mechanism could be used to produce a turn away from a visual threat by means of a rapid contraction of muscles on the opposite side of the body.

The approach to a cue suggesting something desirable may have worked differently. The odour which predicts a rich cloud of edible particles would be registered as the animal turned its head towards it, as would a visual cue. What's needed here is to continue the turn. This may have been achieved not by intensifying a contraction on the side of the promising odour but by suppressing the next contraction on the opposite side, either by direct inhibition of the motor neurons or by modifying the working of the locomotory pattern generator. Depending on the current state of play this would either prevent the head turning away from the desirable or prolong the turn towards it. The same mechanism could also function to slow the animal by applying the inhibition on both sides of the animal at once. It's the nature of neuronal pattern generators to be self-sustaining, working without prompting, so they must be inhibited when they're not needed, rather than excited into action when they are.

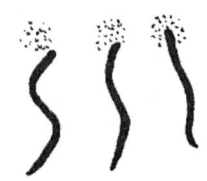

An attraction of this scenario is that it explains the rather puzzling fact that sensory inputs on the left side of the body are reported to the right side of the brain, and *vice versa*. (A slightly but not very different hypothesis has been put forward by Gerald Schneider.)

The steering and braking mechanism would have been derived from a means of reorientation that existed before there was locomotion (no point in locomotion without steering and braking). And it's likely that in the first place it worked through a neuromodulator. That would have meant turns and decelerations were effected rather more slowly than escape actions, but in a primitive filter-feeder with very limited sensory abilities that wouldn't matter. The obvious candidate for the neuromodulator is dopamine.

This speculation is inspired by the tiny nematode worm *Caenorhabditis elegans*. About one millimetre long, it lives in the soil, and wriggles between the soil particles by an alternation of opposed contractions not unlike those of *Amphioxus*, except that the contractions are dorsal and ventral rather than lateral. It eats bacteria, of a sort usually found in clumps, seeking them out by their odour. When there is no evidence of a meal in the vicinity it travels relatively fast, and makes occasional random turns. When chemical cues suggest that the right sort of bacteria are near the turns become more frequent and tighter. And when its tactile receptors brush against bacteria it slows down - tactile input at the rear end of the worm, indicating bacteria currently out of reach of the mouth, has a particularly strong effect.

Even a well-fed *C. elegans* does this, a tactic which helps to ensure it hasn't left the food source behind by the time it needs nourishment. The deceleration depends on dopamine, which doesn't operate through synapses for this purpose but is merely released in the general area of the motor neurons, where it encounters a D_3 type dopamine receptor. The motor neurons in *C. elegans* synapse on to muscle with acetylcholine (as in chordates) and the dopamine slows the rate at which acetylcholine is released.

If food is needed and the worm starts eating the deceleration is greater, and turns become yet more frequent, or there may be some wriggles backwards. This more complete slowing depends not on dopamine (it still happens if the dopamine cells have been destroyed), but partly on serotonin. And the serotonin signal only occurs if bacteria are actually reaching the gut; if the nematode is surrounded by bacteria which are too big to swallow, it keeps going.

The situation here seems to have some similarities to that which operates in the very much more complicated system of vertebrates. Dopamine release relates to the promise of something desirable, rather than its actual acquisition, and to a modification of locomotory activity. Serotonin plays a more notable part in closer encounters with the desirable, and its release is influenced by the animal's internal condition.

We can't read too much into this. Chordates and nematodes are only very distantly related, and while dopamine and serotonin are very ancient transmitters neither the receptors they engage nor the circuits they participate in are the same in all phyla. Even in *C. elegans,* moreover, there are at least two other types of dopamine receptor, which have augmentary rather than suppressive influences, and both dopamine and serotonin serve several other functions.

However, the evidence from the nematode prompts the thought that slowing or interrupting locomotion, thereby allowing more time to collect additional sensory evidence, must qualify as the most basic form of behavioural attention. With very limited sensory systems it won't have paid to pause for further evidence before escaping from a potential threat, but it will have been sensible to acquire all the available evidence before committing to action on a possible desirable. So this form of attention is likely to have been prompted by the possibility of obtaining the sort of goody that will come to function as a reinforcement or reward. It's reasonable to conceive, therefore, that many of the complexities of attention, reward and learning in modern vertebrates might have been elaborated from neuronal arrangements that began as a means of controlling locomotion.

As distance senses evolved and became increasingly effective dopamine may have come, with the aid of a new receptor carrying the opposite sign to the older one, to prompt acceleration towards a distant reward, or exploration in search of reward, as well as slowing and turning towards a nearby goody. Indeed several neurotransmitters which originally must have served only to convey a hardwired signal to motor neurons, or to modify locomotion to suit internal conditions, have become involved in learning. This must have come about through the evolution of novel receptors, some of which might carry the opposite sign to the older ones. In some cases the effects produced through the new receptor might become more conspicuous than those attributable to the original type.

Dopamine seems to be involved in the control of movement and/or feeding in quite a variety of phyla. For instance Thomas Riemensperger and colleagues accomplished a neat bit of genetic manipulation to produce fruit flies with nervous systems that couldn't synthesise this neuromodulator. These mutants show low levels of activity, can't be bothered to eat very much, and sleep a good deal, symptoms implying that neither the motor nor the attentional system is functioning properly - just as in rats depleted of dopamine. So dopamine may have influenced locomotion in the common ancestor of several phyla, including nematodes, arthropods and vertebrates, and developed its role in pretty similar ways in the latter two. As the locomotory system grew more complex and more efficient the influence would of course be operated through an increasingly complex array of neurons, many of which would use fast neurotransmitters.

Two sorts of locomotion?

In earliest days actions could probably be divided into two types. One was responses to local stimuli, things quite close to the body or actually making contact with it, including such visual objects as could be registered by lensless eyes. The sensory information involved will have been processed in the midbrain, and reported from there to the basal ganglia. Reorientation towards or away from these nearby stimuli and any ensuing accelerated locomotion was probably organised via a midbrain locomotor region.

I think a less finely targeted form of locomotion may have been guided by information from the thalamus, which will have registered the overall distribution of light and darkness, and when contrast-sensitive retinal cells evolved probably also registered something about visual textures. The resulting account of the surroundings, though vague, should have helped to guide the animal towards areas where food was likely to be found, though not necessarily to the food itself. The control for this more leisurely form of movement would be from the diencephalic locomotor centre, the message from which could be overruled or augmented when there was any sensory input that required a quick response. Both locomotor regions would be controlled partly by inhibition from the basal ganglia.

Part of the rationale for this hypothesis is that the diencephalic motor region gets input from the hypothalamus, presumably about matters such as the need for food, and from the olfactory system - the sort of information can supply reasons to keep travelling. The midbrain locomotor centre, on the other hand, is close to the nucleus which receives alerts from the spinal cord - the sort of thing that needs a more urgent response - and transmits them to the basal ganglia and elsewhere. It may be relevant, too, that dopamine from the substantia nigra pars compacta, which supplies the basal ganglia, seems to be important for both fast and slow locomotion, whereas dopamine from the ventral tegmental area, which goes to the forebrain, only seems to be necessary for slower, more exploratory types of movement.

As modifiable synapses emerged in the basal ganglia the second layer of inhibitory neurons must have evolved, and it became possible to use inputs to the contact senses as feedback on approach actions. This might have allowed an animal to learn, say, that turning or slowing down for the hardwired chemical cue wasn't profitable when it occurred in conjunction with a certain other chemical. A simple chemical signal would thus promote a reduction of locomotion or a turn, but a bipartite signal would lead to inhibition of that response.

I pointed out earlier that learning to discriminate among different versions of the essential hardwired stimulus patterns could only be possible where there was adequate sensory equipment, but that there was no advantage in additional sensory receptors before the learning ability evolved. I suggested that the answer to the conundrum might be that discrimination learning began when a means evolved of combining the inputs of existing sensory receptors in new ways. The basal ganglia look like the machinery for doing just this.

Motor Learning

The infants of larger-brained mammals and birds are born clumsy and have to learn to deploy their muscles in subtle ways. This adaptable control at the brain level has evolved in conjunction with increasingly fine control at the neuro-muscular connections. Muscle fibres are grouped into units which may be composed of just a few fibres or over a thousand, and a muscle can be made up of anything from twenty five to several hundred units. In most vertebrates one motor neuron controls many fibres - the number's high in the jawless fishes, and reduces in increasingly complex species – but in mammals most fibres are innervated by a single neuron. Individual control of fibres obviously makes for enormous flexibility. Clearly mammalian motor cortex needs to discover just what pattern of muscle contractions is needed to shape actions for maximum efficiency.

Happily, motor learning is easier to observe than perceptual learning. Just how it proceeds in humans has been beautifully described by Jerome Bruner, who analysed how an infant's reach-and-grasp gesture develops. The very young infant's grab at a nearby object, he noted, is a sort of flinging out of the hand, an all-or-nothing movement, with the fingers closing towards the end of the reach regardless of their position in relation to the reached-for object. At around four weeks this not very efficient but clearly delineated movement fades away. It's replaced by object-capturing attempts that are, if anything, rather less effective -

wild swipes, usually preceded by a sort of energetic waving of the arm that 'looks like a priming action'. If the palm of the hand should happen to make contact with the object a grasp is triggered, which is sometimes successful.

Then the action, Bruner recorded, is slowly resolved into a sequence of separable units - reach, grasp, draw in, performed in rather awkward succession. But by around eight months or so the performance has become quite skilful, with both reach and grasp well adapted to the position and nature of the object, and it's generally successful. Having been broken down into separate units the action grows smoother, as the separated units begin to flow into one another again. There's no longer a need for intense and total concentration, precluding the possibility of attention to anything else. The process of seizing objects becomes more or less automatic.

The original grab at any nearby, suitably sized and stereoscopically defined object must be a genetically programmed movement, common to all primates. The hardwired movement is taken apart, as it were, and reconstructed in a more flexible way. As part of the process the infant learns to make use of visual feedback, and to correlate it with proprioceptive feedback.

It's not difficult to link this to what's known about the workings of the basal ganglia. Breaking the movement down into separate components must mean learning to release just one instalment of muscle contractions from inhibition, then to re-inhibit some of those motor neurons while releasing others to contract other muscle fibres for the next stage of the gesture, and so on. To begin with the alternations of inhibition and release are neither smooth nor rapid, the segments into which the movement is divided are too large, and the infant has to learn how to use the visual feedback, concentrating hard. This is the stage at which there's a good deal of activity in the striatum, and also in motor and prefrontal cortex, as attention is focussed on the action and the feed-back. Developing a practised movement equates with potentiating and depressing synapses so as to establish a well-defined circuit through the dorsal striatum and its connections to supplementary, premotor and motor cortex.

The cerebellum, meanwhile, is learning how to apply reports from proprioceptive and vestibular systems to prevent anything going wrong. Thanks to all this synaptic adjustment the components of the gesture eventually come to be released in a rapid and fluent fashion. Just how fast the alternating inhibition and release from inhibition can flow, once they are well practised, is demonstrated by another sort of action, speech, and the rate at which phonemes can be uttered.

Once a hardwired movement has been dissected the parts become available to be put together with other components, in other combinations, opening up all sorts of possibilities. Motor learning, one might say, is a process of deconstruction and reconstruction. Converting those first clumsy responses released by the basal ganglia into smooth, efficient and varied movements means shaping the pattern of activations with well-directed and well-timed inhibitions, and conserving the record of the pattern with a great deal of long-term potentiation and depression.

The first disinhibitions of a motor response, we can suppose, are signalled from basal ganglia via the thalamus to motor cortex. As the infant begins to experiment with the action motor cortex begins to take over control, inefficiently at first. The signals sent from basal ganglia to muscles are copied back to the thalamus and thence to motor cortex, and the axons which carry instructions direct from motor cortex to the spinal cord have offshoots to the basal ganglia which implies that they too are fed into the circuit. This suggests a short-term record in both basal ganglia and motor cortex which can be correlated with the success or failure of the action. As the gesture becomes more practised a permanent record of the required sequence of muscle contractions must be built up in the synapses connecting dorsal striatum and motor cortex. Then, when a performance of the gesture is contemplated, an account of the whole sequence can be despatched from the basal ganglia to motor and supplementary motor cortex in one smooth flow, and a smooth flow of instructions can be transmitted down the spinal cord. They're the sort of instructions which go by the fast route, direct to spinal cord or to the motor neurons of the extremities, and they take into account how one set of muscle contractions may be subtly influenced by those which will follow.

Neurons have been found in the basal ganglia which, in an animal carrying out a habitual sequence of movements, fire in relation to the place a movement occupies in the sequence. A neuron fires for the first component of the ritual, or for the fourth, without regard to the nature of the movement. It seems probable that something similar happens on a smaller scale - that the muscle contraction messages which make up a single movement are labelled in the same way.

Once the learning is complete much less attention is needed to produce the action. Prefrontal cortex merely has to decide on the action and motor cortex, guided by basal ganglia, to despatch the appropriate instructions, so the activity in prefrontal cortex, premotor cortex and both ventral and dorsal striatum is much reduced. Meanwhile the concept of the movement and its possibilities has become firmly established in motor and supplementary motor cortex.

Perceptual learning

Complex perceptual learning, like motor learning, can be seen as a process of deconstruction and reconstruction. And this is just what the neocortex would seem to be designed for, with its pathways which first sift out minimal features, then lead on to areas which detect the larger patterns made out of those features.

In the first place it's a fairly large, undetailed pattern which commands an infant's attention, such as the overall shape which represents a face, with appropriately placed eyes. Then attention is directed around the most salient of the other features, and the most distinctive features of frequently seen faces are learnt. Finally, as the infant becomes capable of uniting the fruits of several fixations, they begin to learn about the subtly varied spacing which distinguishes different faces more efficiently than features alone. An innately programmed stimulus pattern leads to an elementary form of schema which can then, as the brain develops, be deconstructed, so that the bits can be put together again in new ways. The techniques learnt in these first perceptual exercises are subsequently applied to other types of pattern.

It's likely that schema areas and individual schemata are created by carving out increasingly narrow pathways through the neocortex. In the newborn babe, it may be hypothesised, the visual input provided by a face - any face - excites quite wide areas in the later stages of the ventral visual pathway. These later stages of the visual pathway are, in turn, indiscriminately linked to other sensory pathways, so that there are no preconceptions about the sounds, odours or textures which should be associated with a face. But the most salient features of a face which is seen repeatedly are learnt very fast, and they rapidly come to be associated with the voice and odour which have already begun to be familiar in the womb.

It's becoming apparent that the potentiation of inhibitory synapses plays an important role in learning, and we can imagine that perceptual learning proceeds through the establishment of new inhibitory connections in the schema areas of cortex. If hair or skin is the wrong colour it's not mother, so that especially comforting and familiar odour and voice can't be anticipated. If there are bushy eyebrows it's not mother. The pathway is slimmed down so that the links to other neocortical areas, which provide the expectations of odour, voice and behaviour, are only activated by the correct face. Moreover, only mother's face and a limited number of other faces come to prompt a special emotional response. As more and more faces are distinguished, by means of more and more attention to detail, the individual pathway dedicated to each becomes ever more restricted. Much of the inhibition that shapes the increasingly selective schema-neurons at these early stages must be attributable to the basal ganglia, and the emotion-sensitive neurons of the ventral striatum.

Recapitulation

The basal ganglia-thalamocortical circuitry, playing an important part in both attention and learning, was clearly a vital part of the underpinning that made the complexities of the neocortex worthwhile.

In the more sophisticated mammals it's important to shaping limited and clumsy infant movements into an assortment of well-honed ones.

It's hypothesised that the neuronal machinery evolved from early arrangements for braking when a cue suggesting a desirable was registered, and for turning towards it.

The emergence of modifiable synapses made it possible to inhibit turns when the genetically defined stimulus representing a possible reward was accompanied by some other sensory input that had proved to negate the promise.

At this early stage the sensory data must have been supplied to the basal ganglia by midbrain areas and thalamus.

The elaborate sensory processing effected in the neocortex was worthwhile because it could build on the foundation provided by this system.

Motor learning looks like a process of deconstruction and reconstruction, and this must be how complex perceptual learning begins too. With the aid of the basal ganglia the neocortex breaks down a genetically determined stimulus pattern into different variations with varied significances.

Chapter 24

Some Guesswork about the Thalamus

It has long been clear that the thalamus is a central hub in the vertebrate brain. Most of the sensory information reaching the neocortex goes through it, and it receives a great deal of input back, which modifies the way the sensory information is processed and forwarded. Other important communications from subcortical areas also takes this route.

In particular it carries signals from the basal ganglia to the neocortex, signals which take the form not of excitations but of releasing thalamic neurons from inhibition. Just what neurons are being disinhibited? Answering that question would make the thalamus much less mysterious.

As well as relaying information up to the neocortex the thalamus sends many connections downwards to lower areas of the brain. In addition some of its smaller nuclei function as a gate to the telencephalon, under the control of one branch of the ascending alerting system. When the latter is turned off, in sleep, input from the lower brain ceases to impact on the thalamus, and the telecephalon is left to its own resources, free to dream. A more complete interruption of this pathway causes coma.

It's thought that the gating system can also operate in a half-open state. When sensory information doesn't have any particular importance in the current situation, or when the action it's guiding is being performed more or less automatically, the reports sent to the neocortex may be fairly modest. If new sensory input is registered, or unforeseen feedback about the result of an action, a burst of firing in thalamic neurons serves to alert the neocortex.

The two thalami lie close to each other on the central line of the brain a position which hints at their antiquity. It has also made them somewhat inaccessible, but what goes on there is becoming clearer.

In mammals the thalamus is made up of a complex collection of nuclei. We've already encountered the lateral geniculate nucleus, a small bump towards the rear of the thalamus, which receives visual information from the retina, refines it a little further, and sends the results to the visual cortex. Close to it is the medial geniculate nucleus, another modest bump which does something similar for another distance sense, hearing. In this case the sensory data comes from the hindbrain and midbrain, where auditory centres have already wrung a certain amount of information out of it, as the retina does for visual input. Somatosensory inputs are similarly processed in more rearward parts of the brain, and the results are passed on to a thalamic nucleus near the lateral geniculate – one that's distinctly bigger than the geniculate nuclei. (Tactile inputs and messages from the interior of the body must get more processing in the thalamus than do visual and auditory inputs.)

In all three nuclei the sensory messages, from auditory or somatosensory centres or from retina, arrive at large ionotropic glutamate synapses and can evoke significant excitatory post-synaptic potentials, which, if there are enough of them, will cause the recipient neuron to fire. They're therefore termed driver synapses. They're considerably out-numbered by those delivering input from other subcortical areas, and by even more numerous synapses from layer 6 of the neocortex. However, these are small synapses composed of metabotropic receptors, so their effect is only modulatory.

The essential function of these three nuclei clearly lies in relaying sensory information to the neocortex. Neuroscientists call them first order nuclei. The information they transmit can be influenced by attention, emotion, and the general behavioural state, thanks to inputs of neuromodulators such as norepinephrine and acetylcholine from the lower brain, delivered by axons which spread their terminals widely.

In contrast, the input from layer 6 of the neocortex is conveyed by glutamate, to metabotropic glutamate receptors, and is much more focussed. In the case of the lateral geniculate nucleus the input is derived from visual cortex cells which cover a limited area of retinal field, though a larger one than that from which the retinal input comes. It must serve to adjust the firing of the thalamic relay cells in the light of the interpretation of the thalamic data that is being developed in the neocortex. Indeed this has been shown to happen in the case of the feedback to lateral geniculate from MT, the cortical area which puts together the information from the tiny receptive fields of movement-sensitive neurons in V1 to reveal the direction in which the whole stimulus pattern is actually moving.

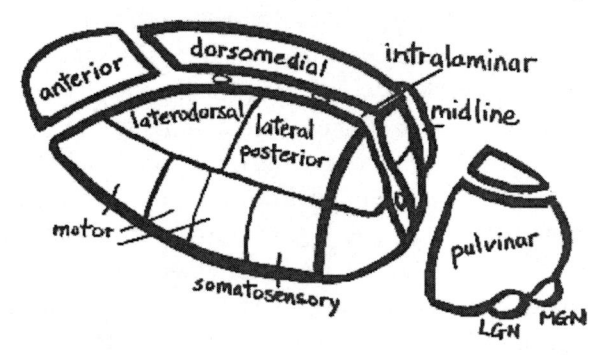

In all the other nuclei where the situation has been thoroughly elucidated the driver input comes from the lower parts of the brain and from layer 5 of the neocortex. These have been dubbed higher order nuclei. One such is the pulvinar, which featured in the chapter on the neuroanatomy of attention. It takes up much of the rear of the thalamus, shading into the lateral posterior nucleus. As we saw, this area plays an important role in selective visual attention. There's a small input from the retina, but most of the visual information comes from the superior colliculus in the midbrain, so it must concern whatever is currently attracting attention and may be about to receive a saccade. The output goes to all the cortical areas involved in vision - the occipital lobe, the temporal 'what' pathway, the area that deals with visual movement, the parietal areas which establish where a stimulus is and prepare the way for action on it.

The pulvinar complex is next to another higher-order nucleus which deals similarly with somatosensory information. (For many species this of course includes input from the whiskers.)   Perhaps the juxtaposition reflects the importance of combining visual and tactile evidence, especially for species which are active in dim light, as early mammals were.

Along the ventral part of the thalamus are three substantial nuclei, all variously connected to motor cortices and to the basal ganglia. One also gets input from the cerebellum and the red nucleus. Somatosensory news, including visceral and nociceptive sorts, is relayed to one of them

The anterior nucleus is connected with the hippocampal formation, to which it transmits head direction data. It's one of the two included in the circuitry which links the hippocampal formation, parahippocampal cortex and retrosplenial cortex in the production of the cognitive map, as described in the chapter on contextual memory. It's also linked with prefrontal cortex, especially the parts concerned with emotion and motivation, and it plays an important role in the interactions of emotion and attention.

This frontal area and the dorsomedial thalamus are divided from the lateral nuclei by layers of insulated axons. Tucked in amongst these fibres are some relatively small bodies, the intralaminar nuclei. Some more smallish nuclei are strung out on the inner side of each thalamus, and are grouped together as the midline nuclei. All these nuclei receive notable inputs of acetylcholine and norepinephrine from the brainstem, and some are heavily supplied with histamine and dopamine, neuromodulators which also contribute to alertness. They have been regarded as the terminus of one arm of the ascending alerting system - the system which is turned down during sleep. The biggest share of acetylcholine and norepinephrine goes to two of the intralaminar nuclei which also have an especially strong input from the basal ganglia, and project back very strongly to them. Damage to these two nuclei in particular can result in the ongoing loss of consciousness which is coma. Two other intralaminar nuclei contribute to a different aspect of attention, the control of eye movements.

The connections of the large dorsomedial nucleus implicate it in similar matters. It receives input from the superior colliculus and communicates to the frontal eye field, and it gets a pretty strong boost from the ascending activating system. It's strongly interconnected, moreover, with prefrontal cortex, the ultimate director of attention. The pattern of inputs and outputs for the dorsomedial and anterior nuclei suggest that they are involved in the overall direction of attention. Damage to either impairs the capacity for establishing the records which make recognition and recall possible.

All these nuclei are elaborated from the same embryonic origins and are sometimes lumped together as dorsal thalamus. The thalamic reticular nucleus, which wraps around them like a blanket (not shown in the diagram), is usually classified as part of the ventral thalamus, sometimes called the subthalamus. It may be remembered that the GABA-deploying neurons of the reticular nucleus penetrate the main body of the thalamus to effect a widespread inhibitory influence, which is selectively relieved by input from the neocortex. It thus forms one of the several pathways for the cortical, top-down control of attention.

An obvious question about the 'higher order' nuclei is what sort of information do the 'driver' connections from the cortex transmit? Some receive inputs from the neocortex made via collateral branches of axons carrying instructions to subcortical motor centres. That means the thalamus is kept informed about movements that are about to happen. The technical term for the message is an efference copy. An important function of efference copies lies in allowing changes in sensory input that are caused by movement of the self to be distinguished from those that aren't. If the scenery whizzes past the retina it's important to know whether it's the eye, the body or the scenery that's actually moving. It's suspected that one function of the thalamus may be to transfer this sort of information between different cortical areas. But clearly there is a lot more to be discovered here.

The thalamus as temporary map

The information which converges on the mammalian thalamus from the lower brain and from 'old' cortex would seem to constitute the essence of everything the animal most needs to know about the current situation. The data from the retina provides a rough idea of the visual surroundings, in terms of light distribution and texture. The news that arrives in the pulvinar indicates what's currently receiving behavioural attention and may require action. The sectors which connect with hippocampus, parahippocampal cortex and retrosplenial cortex can be supposed to place the self in the current geographical context. Input from tactile and internal neurons reports the current state of the self. And the whole picture can be coloured with shades of emotion and with alerting signals from the lower reaches of the brain. In short, the thalamus brings together all the really vital information of the moment. The result could be described as a sketch-map showing the elements of the current situation, and highlighting anything that might need a response. I think that knitting all the current information together in this way, from moment to moment, was probably the earliest function of the thalamus. Back then, of course, it will have been a much simpler structure.

Frogs and toads may offer some hints as to the nature of a fairly primitive thalamus. Theirs is divided into only two, minimally distinguished sectors. The wiring pattern suggests that two-way communications between thalamus and basal ganglia are undertaken by one sector, transmissions to the telencephalon by the other, including the forwarding of visual information. The very earliest thalamus we might guess to be something like the first of these sectors.

As recounted earlier, very little information goes to the thalamus from the five types of ganglion cell found in the frog's retina. What does is mainly about the general distribution of light and shade – which enables the animal to head for sunshine when it needs to warm up, to somewhere darker when it needs to cool down. In at least some species the thalamus also receives just enough information about texture to guide the animal towards the sort of environment that suits it – the vertical lines of grasses for some, more horizontal textures for others. The ganglion cells which register stimuli that need a rapid response - the small moving darkness which represents a potential meal, or the large approaching darkness which indicates a danger of being trodden on – report to the tectum. The hypothesis is that early vertebrate brains worked in much the same way.

This is not a very secure asssumption because we don't know whether the frog brain has grown simpler as the specialised lifestyle evolved. However the ancient, ancestral vertebrates must have been very similar to frogs in significant ways. The repertoire of movements will have been modest, and hardwired, and there will have been limited learning about when to suppress them. And in earliest days the eyes didn't rotate, as the frog's don't (except insofar as is necessary to maintain a steady view when the support on which the animal is sitting is unsteady) so there was no need to adjust a map of the surroundings for the changing views produced by a swivelling eye.

The guess is that the primitive thalamus similarly received information from the retina about areas of light and shade, and perhaps something about textures. Information about stimuli which were moving across the retinal field and might require action would go to the midbrain – to what in modern vertebrates would become the tectum, or in mammals the colliculi – from where it might be relayed to the thalamus. No doubt there would also be input from the tactile system about threats of potential harm. Early thalamus would thus combine something resembling elementary versions of parvocellular and koniocellular pathways of the lateral geniculate nucleus with some of the elements of the modern pulvinar.

This thalamic account of the current scene would be constantly changing. But the hippocampus was evolving, we can suppose, with varied forms of long-term potentiation and depression. Co-operation between thalamus and hippocampus could create the beginnings of a medium-term memory. That made it possible to turn the thalamic account of the current moment into a slightly longer-lasting map of the surroundings.

Its possessor would be able to direct their locomotion with knowledge of what was behind them, where they had just come from, where they had just depleted the available food, where a sign of danger had been spotted. This could be energy-saving and sometimes life-saving knowledge, which would provide a useful edge in the competition for survival. Another benefit would be a greatly improved capacity for pursuing odours to their source. It would become possible to retain a record of where an odour had registered most powerfully, and to continue searching for further traces if the input ceased, even if there was no convenient plume of scent molecules to be criss-crossed in an up-current direction.

This is the sort of cognitive map that must have existed before it became possible to distil a more complex, far-reaching and permanent map out of many temporary ones. Even today there may be species, roaming featureless areas of ocean in search of unpredictable supplies of food, which would have little use for a full-scale permanent map but can benefit from a short-term one. The hagfish which scrape the flesh from carcasses on the ocean floor, locating these sources of food by smell, come to mind as species that might answer to this description. Dead animals are likely to be distributed randomly, so there's probably little need for a long-term map. But it would be useful to a scavenger to remember how it had squeezed into the bony structure that was providing the current meal, and which bits had already been scraped clean.

The assumption is that some of the retinal cells sensitive to light/dark contrasts began sending output to thalamus as well as midbrain at a fairly early stage of vertebrate evolution. In primates and other species whose visual systems have been most extensively explored the visual processing done in the thalamus seems like a modest addition to that done by the retinal ganglion cells and a very minor prologue to the multiple levels of analysis carried out in the visual cortex. However, even the modest amount of information known to be extracted in the primate lateral geniculate nucleus could provide a useful if impressionistic sketch of the environment. It should certainly be adequate, in most cases, for distinguishing between rocks and seaweed and empty water.

Meanwhile visual inputs which required an immediate response, such as moving stimuli which implied a threat of collision, were relayed from the retina to the tectum of the midbrain. Other sensory inputs that required or might require action went to the same place. From there a message would be despatched to the basal ganglia, which released the appropriate response by suspending the normal ongoing inhibitory message to a motor centre. The guess is that simultaneously a copy went from midbrain to thalamus, pinpointing the position of the stimulus on the thalamic map. (The neurons carrying the message can be regarded as forerunners of those that run from superior colliculus to pulvinar.)

On this scenario it's possible to build an answer to the puzzle about the basal ganglia-thalamus-neocortex circuit, namely what is the function of the thalamic neurons that are under ongoing basal ganglia inhibition from which they are occasionally released. The hypothesis is that they're neurons which receive input about sensory matters from the midbrain. This input, it can be hypothesised, produces a modest degree of excitation, despite the inhibition by the basal ganglia. When an action is released in response to the input the inhibition, probably exercised by collateral axons of the same basal ganglia neurons that run to the relevant motor centre, is suspended. The firing rate of the thalamic neurons increases, and the stimulus that evokes a movement is thus more strongly registered in the thalamus than one that potentially might, but fails to.

In early days, when muscles were applied only to locomotion and the basal ganglia's main function probably lay in controlling steering, the signals could only indicate *something-to-be-approached* or *something-to-be-distanced*. When eye movements were added to the repertoire the thalamic/hippocampal map would register things that attracted a saccade, as well as things that prompted larger action. The significance of the stimulus pattern would be implied by the nature of the movement.

Further developments

Meanwhile the basal ganglia became increasingly well equipped with modifiable synapses and a second layer of inhibition evolved. The disinhibition of the response to a genetically determined stimulus pattern could be suppressed if there were additional features in the pattern which, in the light of past experience, suggested that a response would be unwise, or a waste of effort. The thalamus would provide an increasingly nuanced picture of the current priorities in the external world. As bodies became more complex and the repertoire of possible actions expanded there would be a yet greater variety of information to mark on the map.

At the same time a more permanent and elaborate map was evolving in the telencephalon. The information about actions released from basal ganglia inhibition would be forwarded to it and if such things occurred regularly in the same place they would be marked on the permanent map.

Then new sensory processing facilities evolved in the telencephalon, and basal ganglia-thalamus pathways were extended into the new telencephalic areas. We can deduce that an initial communication from basal ganglia reporting a hard-wired response prompts sensory cortex to apply itself to the stimulus-pattern that evoked it, and prefrontal cortex to focus attention on it. (The basal ganglia message no doubt arrives fast enough to overlap with the neocortical processing.) This ensures that the advanced facilities are directed where they are most needed.

Return pathways evolved from telencephalic sensory areas to basal ganglia, supporting a dialogue which allows fine distinctions to be made between the stimulus-patterns that repay a response and those which don't.

At some point, meanwhile, acetylcholine and norepinephrine pathways from hindbrain to thalamus evolved, upward offshoots from the pathways to the spinal cord, carrying alerting signals. Here too it can be supposed that the signal from the hindbrain to the thalamus produces a certain degree of excitation, which grows stronger if inhibition from the basal ganglia is lifted. Just how complex the brain had to become before these alerting signals were needed is anybody's guess. But once the basal ganglia's learning power was in operation there would be considerable benefits in allowing that system to modify the alerts, in either direction, in the light of past experience. Combining the norepinephrine and acetylcholine reports with basal ganglia learning means that the degree of emergency and the need for a response can be evaluated. This must be what happens in the two intralaminar nuclei that receive particularly strong inputs both from the alerting systems of the hindbrain and from the basal ganglia. These are the nuclei which constitute a gateway to the neocortex, open in waking consciousness, closed in sleep, and where damage can result in coma.

Most of the messages from basal ganglia to telencephalon will have been essentially concerned with action, released or suppressed. It's not difficult, therefore, to see how the system could evolve into one that would dissect movements into finer components, making it possible to learn not only when to act but also how – a development that accompanied the evolution of finer control over muscles, accomplished by targetting single neurons to ever fewer muscle fibres.

The messages which first went to thalamus and the evolving neocortex are efference copies, we can suppose, carried by offshoots from the axons that carry the disinhibitory message to the subcortical motor centres. As motor cortex developed the efference copies would go there too, carrying reports of the instructions to muscles sent out from the basal ganglia as an infant animal performed a hardwired response. The clumsy, genetically determined movements must then be modified either by directions from motor cortex to basal ganglia or by direct intervention from motor cortex via the corticospinal pathway – perhaps by both.

In the ancestral brain, then, co-operation between thalamus and hippocampus is envisaged as creating a temporary map of the current surroundings, which recorded something about the directions of recent locomotion and the changing spatial relationship between the self and any significant stimuli. Connections from the basal ganglia would pinpoint the stimuli that were attracting any sort of action or attention. These arrangements would provide a suitable foundation for a neocortex – and no doubt for the various forms of brain extension that occurred in non-mammalian vertebrates.

Educating the infant neocortex

The scenario about the basal ganglia-thalamus pathway can be extended further. I propose that this pathway makes a vital contribution to the priming of the infant neocortex. A significant player here is the reticular nucleus which wraps around the thalamus with a blanket of inhibition and functions as a filter between it and the neocortex, editing the information forwarded from the former to the latter. Enough must be transmitted, however, to allow the neocortex to decide what's worth its consideration and to inhibit the inhibition over whatever bits of the thalamus seem to have interesting news to impart. But how does an infant neocortex know what's interesting?

I think the answer may lie in the messages from lower brain and from midbrain sensory areas to thalamus to neocortex which, if they evoke an innately determined response, are augmented by release from basal ganglia inhibition. This leads to an alerting signal to neocortex via pulvinar and intralaminar thalamic nuclei. The signal to neocortex must trigger the message to the reticular nucleus which interrupts the inhibition on the region below and allows that bit of thalamus to communicate more extensively with the neocortex, thus setting up a sustained circulation of impulses that allows the sensory input to be fully processed. It's by this means, I suggest, that the infant's sensory education is focussed on the really important stimulus patterns – which are of course the ones which elicit hardwired responses.

These important stimulus patterns tend to produce emotional responses as well as action, or to be followed by feedback from a contact sense. The emotional and contact information which reaches the basal ganglia can be supposed to reinforce the disinhibiton in thalamic neurons, and further strengthen the message to neocortex.

As the neocortex learns what deserves attention, and how to apply the reticular filter, neocortex and basal ganglia can be supposed to communicate more directly. In other words long-term potentiation is established in the connections from neocortex to basal ganglia. Then the guidance provided by the report of a hardwired response is no longer needed.

The arrangement which educates the infant by directing attention may also be important in establishing the layout of the cortical schema-areas. I suggested in chapter 21 that percepts are classified not only by any emotions they arouse but also according to the sort of actions they evoke, including the size of the saccades by which they are studied – the information that I propose is conveyed from midbrain to neocortex by the thalamic route that's fully activated only when basal ganglia inhibition is suspended.

The hypothesis can explain those cases of prosopagnosia which are caused by cataracts in the newborn. The very young infant's gaze is drawn to eyes in particular, and to clusters of features, and to movement, and a face provides all of these. This means she's likely to establish a pattern of saccading around the essential features of a face – thereby setting up the template of action which will later allow her to measure the relative distances between features. It's clear, moreover, that infants derive pleasure from studying faces, so there's feedback which reinforces the learning. Add to this that the newborn infant is short-sighted, so there are usually few other equally interesting stimulus patterns to compete with a mobile face.

In addition, infants in their first days of life, when fully awake, often reproduce a parent's facial expression, especially if it's a smile – something that no doubt helps to strengthen the parent's bonding to the child. Infant monkeys also tend to copy a facial expression. This should probably be classified as an innate behaviour which is released by the sight of someone else performing it. It's one that it soon becomes possible to suppress, however.

Several of the usual influences must be distorted or completely removed by late access to sight, and by the cataract operation itself. The new, artificial lenses provide a more extensive field of view than that of the short-sighted newborn, so a larger range of salient stimulus patterns competes for attention with the demanding, subtly varying patterns which are faces. Moreover if the operation is performed after a certain stage the attraction exercised by the direct gaze of a pair of eyes may be beginning to be diminished by the maturation of a similarly hardwired distaste for being stared at.

Meanwhile the innate tendency to reproduce a seen smile has faded. And since the artificial lenses don't attach to the eye muscles the infant can't adjust the focus. Consequently two of the hardwired responses that are normally elicited by faces are missing. This means, we can posit, that the input from midbrain to thalamus to neocortex isn't relieved of basal ganglia inhibition, and is insufficient to set up the circulation of impulses which would lead to making a gap in the reticular blanket and enabling full communication between thalamus and cortex. The stimulus pattern which is a face thus fails to attract the sort of attention that is needed to make sense of it.

As a result no cortical face area is established, and other, less demanding types of stimulus pattern fill up the cortical space. If the theory is correct the inability to exercise the innate responses during the crucial weeks would explain why a lack of pattern vision for quite a short period of early life can result in prosopagnosia. It would also account for the fact that the whole business of learning to see is very difficult when cataracts are removed in later life from someone who has had them since infancy. But moving objects and stereoscopically defined ones can no doubt attract saccades at any age, so setting up the circuitry for recognising less demanding stimulus patterns after delayed exposure to patterned visual input isn't quite so difficult as dealing with faces.

A similar argument can explain why there's a crucial period for learning language. Infants soon begin to shape their burblings to the patterns of the language they hear around them, and this instinctive imitation clearly operates for several years. It's what allows pronunciation to grow gradually closer to the models as the child's vocal apparatus matures, and grammatical rules to be easily acquired. It must be why a second language is learnt more easily in early childhood than later.   But when the automatic tendency to imitate has faded reproducing the sounds heard requires conscious concentration and becomes much harder.

Perhaps some amelioration of prosopagnosia might be obtained through training in where to direct attention. As we know from trying to hear new foreign languages properly in adulthood, it wouldn't be easy. Nevertheless a few people do meet that challenge successfully enough to codify the sounds and explain them to the rest of us, and some actors can produce very passable imiitations of foreign accents, so it must be possible.

The thalamus can be seen, then, as evolving from a map registering the most vital aspects of the current situation into a filter which ensures that the most vital inputs are the ones conveyed most powerfully to the telencephalon – in our case the neocortex. In neonates the filter is operated by the basal ganglia through the hardwired responses it releases. It's gradually modified by conditioned learning, and the neocortex acquires an increasing influence.

An implication of the theory is that the workings of the 'second order' thalamic nuclei are directly influenced by inputs from midbrain and basal ganglia, while those of the 'first order' nuclei are more indirectly shaped, via neocortical influence on the thalamic reticular nucleus.

Recapitulation

The thalamus channels all sensory information other than olfactory to the neocortex.

In primates the thalamus is a complicated structure composed of many nuclei, having grown in complexity as the neocortex expanded.

The frog's thalamus is much simpler, divisible into two barely distinct sectors. The original, ancestral thalamus was no doubt simpler still.

It's hypothesised that it served to bring together information about any stimuli currently evoking a response, and to relate it to a rough account of the environment in terms of light and shade and a certain amount of texture.

In association with the hippocampus it came to provide a temporary map of the current situation, which would allow an animal to remember where it had just come from, and where danger had recently been registered.

It was on this foundation that a longer-lasting map evolved in the telencephalon.

Initially sensory information about stimuli that required immediate action went only to the midbrain.   From there a message to the primitive basal ganglia system released the response.

A simultaneous message, it's hypothesised, went from midbrain to the thalamus, evoking by itself only a modest excitation because of ongoing inhibitory input from the basal ganglia.

However, when the basal ganglia inhibition was lifted to release the motor response it was also lifted from the thalamic neurons activated by the stimulus that had prompted the action.

The stimulus was thus registered more firmly on the thalamic map, and in a way that indicated the nature of the response.

When the telencephalic map developed the information would be forwarded to that.

As the learning powers of basal ganglia evolved this arrangement became ever more valuable.

And as the telencephalon developed the system served to indicate where its increasingly sophisticated analytical powers needed to be concentrated.

It's proposed that the information about hardwired responses to genetically determined stimuli, forwarded from thalamus to telencephalon, helps to educate and shape the infant neocortex.

The message is necessary, it's theorised, to prompt a neocortical message to the reticular nucleus of the thalamus, lifting its inhibition on the underlying bit of thalamus and setting up a circulation of impulses in a brain that has not yet established much potentiation at its synapses.

As synaptic connections between basal ganglia and neocortex are potentiated a direct exchange of messages becomes possible.   Then the indirect prompt to prefrontal cortex is no longer a necessity.

If indeed the message the basal ganglia sends to the neocortex has always been essentially about action, then it's not too hard to see how the system could be elaborated into something that worked on a smaller scale, releasing muscle contractions individually and making motor learning possible.

Chapter 25

Overview

It's possible to detect a simple theme underlying the complexities of brain evolution. What brains do is discover correlations. Neurons which record different sorts of event, or different aspects of the same event, put their information together, usually by converging on another neuron. The more layers of neurons there are through which to sift the sensory input, the greater the amount of information extracted. And the larger the number of processing channels the greater the potential number of interconnections, and thus of cross-correlations. Bigger brains make more correlations. As a result the actions they direct can be more effectively fitted to the details of the environment. The larger the brain the more efficient its possessor can be about acquiring energy and the choosier about how it's spent. The neocortex, as it expanded, offered more and more potential for this. And in humans - thanks largely, no doubt, to that area at the very front of the frontal lobe - the capacity for making correlations has been taken to an extreme, and has created the potential for abstract reasoning.

Detecting coincidences and making correlations is not by any means unique to brains. It can be achieved elsewhere, as when *Aplysia*, after being buffeted a few times on the tail and then on the siphon, learns to withdraw its more vulnerable parts after the first prod, without waiting for further insult, thanks to a connection from tactile neurons in the tail to neurons in the siphon and gill. An arrangement like this offers limited scope for expansion, however. Brains, where information from many sources is reported within a limited compass, provide vastly more scope both for making connections between sensory inputs and for applying the results.

The most extravagant example so far discovered of multiple correlations within a single sensory channel is provided by the primate visual system. Preliminary rounds of processing in the retina and the lateral geniculate nucleus of the thalamus reveal small areas of contrast. This data is transmitted to the neocortex where inputs from adjacent retinal areas are compared to reveal lines, edges and corners. Inputs from receptors sensitive to different wavelengths are processed through several episodes of correlation to create umpteen shades of colour. Inputs from the two eyes are compared to establish distances. The conclusions about edges, textures, their relative distances and the colours attached to them are then put together to reach a decision about the nature of the visual object or scene. This decision is a prediction about the potential implications of the object or scene for other sensory channels, and about whether there is a need for action - a prediction that can be made because of the intersensory correlations that have been recorded in the past.

At the same time neurons in the parietal cortex are correlating what the visual receptors register with the direction of gaze and the position of the head, and establishing just where the stimulus pattern is in relation to the body, in case a motor response is necessary.

Higher cognitive powers also depend on neurons which make correlations. Mirror neurons, responding both to actions performed by the self and those performed by others, are particularly notable examples. Combining the activity of the mirror neurons with the evidence of the distance senses and with interoceptive feedback - the feel of performing a movement - allows the actions of the self to be distinguished from the actions of others. The ability to relate the actions of others to those of the self while nevertheless distinguishing between the two contributes a vital element to the development of a concept of self. It incidentally provides the means of working out that the figure gesticulating in the mirror is a reflection of the self - something that only larger brains achieve.

There are also emotional mirror neurons. Some cells in somatosensory cortex fire not only in response to painful input but also at the sight of someone else in similar pain. Neurons in the insula, through which so much information flows to prefrontal cortex, respond to pain, disgust and various other emotions when they are exhibited by others, as well as when they are experienced by the self. The correlations effected by these emotional mirror neurons can be supposed to be what makes emotions infectious, and creates the possibility of learning by sharing the reactions of others. They have contributed enormously to the benefits reaped from prolonged maternal care, and from living in social groups.

Much imitational and emotional learning, meanwhile, is facilitated by what might be called attentional mirror neurons, found in the parietal cortex, which fire both when an animal's gaze is pointed in a certain direction, and when another animal is observed to be looking towards that point in space.

The various mirror neurons supply a foundation for the idea of others as entities similar to the self, with similar reactions and motivations, a talent which provides the means of predicting the likely behaviour of a conspecific. A well-developed concept of others as similar entities to the self, acting and reacting in more or less similar ways, opens the way towards the invention of language. This clearly depends on being able to compare the sounds produced by the self with those made by others. It also requires relating the action of producing a word not only to the sound produced, but also to its reference. Full language becomes possible when the words can be put together in meaningful relationships. Abstract thought becomes possible when there are sufficient layers of neuronal processing to put several material concepts together and find an underlying theme.

However, the correlations most vital for survival are of course the most basic ones - those between events in the outside world and effects on the organism; in other words, between reports from distance or boundary senses and reports from boundary or internal senses. It's only when connections exist that can turn data about the self into feedback about the results of acting on sensory patterns in the external world, and use that feedback to modify future behaviour, that it becomes worthwhile to collect more data about the external world, in addition to that which drives hardwired responses. On the other hand, where there is a potential for 'reinforcement' of a behaviour almost any addition to the capacity for collecting sensory data about relevant events, and for winnowing information out of it, becomes worthwhile.

Machinery for turning inputs to the boundary and internal senses into feedback was well established long before the neocortex appeared. It was clearly accompanied by some very effective sensory processing, even if there was nothing like the elaborate neocortical correlations. All this provided a solid foundation, on which the neocortex could be built.

In the vertebrate brain four particularly important areas have been identified where links are forged which turn one sort of sensory input into feedback about the significance of another. They are areas where synapses can be strengthened by use, or weakened by sparse employment, a process that constitutes learning. All four are found in almost all vertebrates. Together these four pieces of neuronal machinery clearly contributed enormously to the survival of their owners in a competitive world. Probably they constitute important reasons why a vast number of less complex vertebrates and non-vertebrate chordates failed to survive, leaving an enormous gap today between the few remaining non-vertebrate chordates and the least complicated of the vertebrates.

The basal ganglia

If the basal ganglia have their origins in arrangements for steering, slowing and stopping, as seems likely, they must date back to the very beginnings of locomotion, or indeed earlier, since the means of reorientation probably evolved before locomotion did. The first elements of the circuitry must have been part of the conglomeration of neurons which grew at the front of ancestral bilateral animals long before there was anything that could qualify as a brain. The original components functioned, I've suggested, to inhibit contractions on one or both sides of the body, thus producing turns or decelerations. Possibly the transmitter in the very early stages was dopamine, operating on inhibitory, metabotropic receptors to fairly gentle effect. This would be adequate in a primitive, slow-moving creature which needed only to immerse itself in the richest distribution of food particles, rather than to aim for a specific target.

As both muscles and sensory systems grew more efficient more delicate controls would be beneficial, with faster signals, conveyed by neurotransmitters rather than a modulator. And a feedback connection evolved, along with some synaptic plasticity, creating an ability to correlate variations on the innately determined stimulus patterns that attracted an approach with what was gained, and to find that some versions didn't repay the effort. This development seems likely to have involved the evolution of the neurons that inhibit the inhibition, and perhaps a new application for dopamine.

Initially the stimuli amongst which discriminations were learnt were probably defined by new methods of combining the sort of sensory data already being acquired. However, the ability for this sort of learning meant that it became useful to collect additional sensory data if it helped to identify stimulus patterns worth approaching, so all sorts of improvements to distance senses would become worthwhile.

Meanwhile the chain of neurons controlling locomotion grew longer and more complex, as bilateral fins, jaws and other new accessories appeared, and the contractions of all the various muscles had to be co-ordinated. The need for well-timed inhibitions became even greater, and new feedback connections must also have emerged.

Muscles to rotate the eyeball probably appeared even earlier than fins, judging by the fact that lampreys are equipped with them. The direction of the gaze must of course be correlated with the orientation of the body if there is to be action on a visual object, so the machinery that organised bodily orientation and the direction of travel was extended to cover eye movements as well.

The deployments of skeletal muscle which constitute genetically determined responses to genetically determined stimulus patterns must be managed by the basal ganglia, with the stimulus patterns identified in the midbrain or the thalamus. Much of the straighforward sort of learning which defines the genetically determined stimulus pattern rather more closely - the sort that can be defined as conditioned learning - also depends on the basal ganglia. These hard-wired and conditioned responses must guide the development of sensory and association cortex, ensuring that its sophisticated processing abilities are applied where they're most needed. Then, as the neocortex matures, the modifiable synapses of the basal ganglia provide a vital component of the machinery that supports neocortical learning, both perceptual and motor. Potentiation at the lower end of the circuit supports potentiation at the cortical end.

The hippocampal formation

My guess is that the hippocampal machinery grew out of a mechanism which could correlate longlasting internal conditions such as are registered in the nearby hypothalamus with repeated or ongoing inputs from visual receptors (as processed in the thalamus), or from olfactory receptors. That would mean that the achievement of a satisfactory internal state could come to be associated with the location where it was achieved. The extended input on both sides of the equation would allow plenty of time for synaptic potentiation to develop - which would be the most practical arrangement when external senses were still primitive and only limited data about the external world was being collected. Furthermore a lengthy sampling of the external input would be necessary to ensure that what was registered was a permanent feature of the environment, not just a passing phenomenon.

Learning about the contexts in which desirables such as food or an optimal temperature were usually found would enable the animal to slow down when it reached an environment which provided the cue. The emerging hippocampus can thus be envisaged as complementing the function of the early basal ganglia. Working with widely distributed cues it promoted situations where more proximate cues could be registered.

Communication between hippocampus, basal ganglia and thalamus would bring new potential. The primitive hippocampus expanded into a means of keeping track of the immediate surroundings, and orienting the self in relation to them, using basic visual and olfactory information. The basal ganglia contributed information about cues inviting approach or prompting avoidance, and made it possible to mark places where noxious input had recently threatened, or where all available food had just been consumed. All this, I hypothesise, would be done in co-operation with the thalamus, to which reports of turns and reorientations would also be forwarded. The result would be an elementary map of the current surroundings.

The evolution of an effective permanent cognitive map, providing a long-term record of the sites that regularly supplied a fair amount of nourishment or an optimal thermal condition, would require a longer-term memory, and a means of measuring longish intervals of time. The navigating animal needs to know how long it has taken to get from A to B, and how much effort, and it's helpful, when encountering a landmark for the second time, to have some idea of how long ago the first encounter was. The hippocampus has the means of measuring such intervals. But the long-term record of all this information would have to be stored elsewhere, leaving the hippocampal machinery available for mapping new contexts in a way that enabled the more permanent map to be extended in any direction.

So far as I know no brain areas which seem to contribute to a permanent cognitive map have yet been identified in a non-mammal, but the obvious place to look is in the telencephalon, close to the area taken to be the equivalent of the hippocampus. In the simpler-brained extant vertebrates olfactory cortex takes up another big chunk of the telencephalon. We can safely assume that some of the remaining area is concerned with weighing up current needs and balancing them against other factors, calculations of the sort that in mammals take place in orbitofrontal cortex. But some portion of it, I believe, will prove to be taken up by a cognitive map, with equivalents of parahippocampal and retrosplenial neocortex, operating in close association with the thalamus.

From such beginnings we can imagine the hippocampal formation expanding to undertake other functions in an evolving brain. Building a map essentially involves weaving together information about geographical context with information about some of the things occupying the context, effectively creating a narrative, and thus creating links which make recall possible. With increasing capacity the narratives came to cover more obviously temporal patterns as well as spatial ones. As visual systems grew more complex the hippocampus would help in dealing with the more elaborate stimulus patterns, ones which had to be acquired over a series of fixations, or viewed from several view-points for full appreciation.

The hippocampus thus functions as a hub through which information extracted in different cortical areas can be linked. The links it provides can be reactivated, so that input to one cortical area prompts reference to related information recorded in another area. This makes the hippocampus essential to conscious recall, or explicit memory.

The amygdala

The definitive fact about the amygdala, I think, is that it's activated by the essential chemical or visual features that define a conspecific, and despatches messages to hypothalamus and hindbrain which evoke hormonal and/or autonomic responses. Some of the communications to the hypothalamus back up signals which go direct from the olfactory and vomeronasal systems. It may be significant that a small proportion of amygdal neurons are born in the hypothalamus, and migrate to their final position.

A reasonable guess is that, as the number of species expanded and an increasing amount of information was needed to distinguish them, the amygdala evolved to deal with the complexities, duplicating the function previously performed wholly by the hypothalamus, but using more complex information for the purpose. In many species genetically determined visual and auditory stimulus patterns would come to contribute to the identification of conspecifics, making it easier to identify potential breeding partners at a distance. These signals would be reported to the amygdala, and their import forwarded to the hypothalamus by much the same channels as the chemical ones.

We can suppose that the amygdala then expanded its role as an intermediary between sensory systems and hypothalamus by evolving sensitivities to other sorts of stimulus patterns for which hormonal responses became useful. For many it would also be profitable to adjust heartbeat, blood pressure, breathing rate and so forth. As predators grew more powerful and efficient, genetically determined attentional, hormonal and autonomic responses to threat will have become increasingly valuable to prey species. For the predators similar responses to potential prey will have been useful. The amygdal response to cues suggesting food presumably evolved where digestive systems became capable of dealing with more complexly structured food, and could usefully be primed in advance of its arrival. The amygdala and a closely connected structure also became involved in regulating appetite. Meanwhile more complicated interactions among conspecifics involved the development of amygdal responses to mates, offspring and rivals.

It's likely that in the first place the stimuli that prompted the amygdal danger response in prey species were odours. As an alerting signal an odour can be very effective, but it's less helpful at providing a rapid indication of just where the danger is that should be distanced. The evolution of modifiable synapses in the amygdala would therefore provide a great asset, since it would become possible to learn about stimulus patterns in other senses that were regularly associated with an escape-prompting odour.

The autonomic and hypothalamic responses which, when the results are elevated to consciousness, represent fear have been particularly associated with the amygdala (although at least one other brain area is proving to be significantly involved too). But the amygdala clearly contributes to learning about several other subjects. And when mammalian social lifestyles evolved it became worthwhile for faces or voices to become expressive and for the amygdala, already sensitive to conspecific faces, to evolve responses to some of the expressions. It must be largely due to the amygdala that mammals can learn from the facial and bodily expressions of mother or other conspecifics, in what I've called learning by emotional infection.

The cerebellum

On the evidence of comparative neurology an identifiable cerebellum evolved later than amygdala, hippocampus or basal ganglia. The part which receives input from the vestibular system and contributes to managing the relationship with gravity is thought to be the oldest bit, so that might have been the cerebellum's first function. An alternative hypothesis is that it grew out of an arrangement for co-ordinating the defensive responses prompted by local tactile inputs.

Modifiable synapses that make it possible to withdraw from a light contact before a heavier and perhaps harmful one follows are found in modern invertebrates with quite modest nervous systems, so something similar was pretty certainly present in the chordate line from an early stage. As animals became larger and more mobile the limited reflexive withdrawals effected through direct contacts between sensory neurons and motor neurons can no longer have been adequate for avoiding harmful collisions, so a central device must have become worthwhile. The mechanism could then have been expanded to integrate these withdrawals with locomotion that distanced the whole animal from the threat. The advent of a lateral line system would probably give additional value to the machinery, providing the means of retreating from potential tactile inputs before they could happen.

The hagfishes, those distant cousins of all other vertebrates, are the only ones without a detectable cerebellum, and their eyes lack exterior muscles for rotating them, as well as the lenses which would make rotation worthwhile. The other surviving jawless fish, the lampreys, possess both a detectable cerebellum and rotatable eyes. This suggests that the expansion of the earliest cerebellar machinery into a definable structure might have coincided with a need to co-ordinate eye movements with body movements. When bilateral fins appeared the arrangements could be extended further to co-ordinate their deployment with the locomotory ripples along the length of the body. Directing locomotion will no longer have been merely a business of imposing modifications on the locomotory pattern generator at the right point in the cycle, but will have needed additional proprioceptive feedback, and appropriately timed use of the fins.

A question that suggests itself is whether the existence of a modest cerebellum might actually have facilitated the evolution of lateral fins, and subsequent new body parts. Was it the learning ability of the cerebellar machinery that made it practicable to add new extensions to the body, the use of which had to be co-ordinated, like escape actions, with the workings of the basic locomotory pattern generator? Many infant reptiles and amphibians emerge from the egg with clumsy-looking thrashes and wriggles, which are quickly but not immediately resolved into smoother action. Do those early twitches indicate experimentation as to how the various muscular potentials of the parts can be organised into efficient, co-ordinated movement of the whole? In short, could it be deduced that body extensions were possible because there was a cerebellum which could quickly learn how to manage them, and that the lack of such a facility explains why the ancestors of the hagfishes didn't give rise to a more diverse range of species?

At any rate it may be suspected that the cerebellum was a remarkably fruitful invention. Hagfish, the only species without one, actually have a somewhat larger brain than that of the other extant finless fish, the lampreys, and it has a distinctive, rather squashed-up look. This suggests that in this respect they may have followed a somewhat different evolutionary route from all other surviving vertebrates, one which, though successful in itself, offered much less scope for further developments. The cerebellum certainly proved very versatile, as demonstrated by the fact that it's responsible for the electro-sense of some fish species.

In us and in the great apes - species which have elaborate and adaptable muscle systems and must educate the motor cortex on how to use them effectively – the cerebellum is exceptionally large. It's vital for achieving complex movements smoothly and efficiently, by means of short and therefore fast connections to motor neurons. It's also important in the business of maintaining an upright stance, making use of vestibular information. In addition the cerebellum plays a major role in motor memory, which, as I've stressed, has to be very precise. It's particularly important in the delicate business of uttering speech, and in controlling fine finger movements.

In humans its connections with the neocortex, and especially frontal cortex, are more extensive than would seem to be required for motor control alone, and damage to the cerebellum can produce cognitive problems. Some of these are no doubt attributable to a reduced fluency in language - even internal speech seems to be affected. Perhaps the ability to imagine actions may also be undermined, and with it some of the ability to deal with percepts and ideas that are closely related to action. It also seems possible that where the cerebellum's part in eye movements is compromised visual perception may be affected. Perhaps the cerebellum has a role in other forms of active perceptual exploration too, such as tactile investigation and sniffing. And conceivably it has a role in facial expressions, once they become voluntary. Perhaps it's relevant to remember that cognition has grown out of the means of directing movement, so that the two things, which subjectively seem quite different, are in fact closely interwoven.

Alerting systems

Between them the systems provided a foundation on which it was worth building increased sensory powers - a neocortex in the case of mammals, expanded midbrain structures in reptiles, birds and many fishes. Another major part of the underpinning must have been the evolution of norepinephrine and acetylcholine pathways leading upward from the lower part of the brain. Acetylcholine, the transmitter used by chordates for communication between neuron and muscle, must have been employed from the beginning in the contractions which produced a withdrawal from a potentially noxious input. The same transmitter presumably came to carry a message up to the proto-brain reporting such assaults and preparing the recipient neurons for any further suggestions of danger, and the possible need of larger responses. Norepinephrine perhaps began as a transmitter that influenced the operation of the locomotory pattern generators when such alarm messages were received.

The earliest upward connections were no doubt the ones that go to the hypothalamus and various other nuclei, from which hormonal responses are triggered and the pattern generators of the internal organs controlled. Expansion of these routes, targeted at the developing sensory-processing facilities in midbrain or forebrain, served to ensure that the new machinery was employed where it was most needed.

Visual processing before the neocortex

An implication of all these arguments is that well before the neocortex appeared a fair amount of information was being distilled out of sensory input by the tectum and thalamus. In non-mammalian vertebrates with well-developed visual systems the optic tectum is responsible for a good deal of the data processing, presumably doing much of the job that in us depends on the visual cortex. Even in some other primates many forms of action seem to be more handicapped by damage to the equivalent of the optic tectum, the superior colliculus, than by lesions in visual cortex – though this might be because the superior colliculus is important to guiding attention.

As yet the number of visual operations shown to be performed in primate superior colliculus is limited. The main sensitivity so far demonstrated is to movement, and probably - thanks to input from the neighbouring parabigeminal nucleus - to singularities. But in mice the superior colliculus contributes more to discriminating among different speeds of movement than does primary visual cortex. Mice also have orientation-sensitive neurons in this structure. Something along such lines may yet be uncovered in the primate. And then there is that tantalising idea that the superior colliculus may have the means of registering the archetypal conspecific face.

Similarly, studies of the lateral geniculate nucleus of the thalamus have suggested that in primates it doesn't extract a great deal of information from the retinal input. Some koniocellular neurons record the general distribution of blue light. The magnocellular layers concentrate on moving stimuli. And both magnocellular and parvocellular layers reveal contrasts, but apparently over receptive fields which reveal little about shape. But it's recently turned out that there is a small population of geniculate cells in the marmoset which respond both to orientation (especially vertical orientation) and to movement. Unexpectedly, they are among the tiny cells of the koniocellular layer. They communicate, in turn, to a higher layer of visual cortex than that targeted by the long-known pathways. It seems likely that there is still much to learn about the operations of both lateral geniculate and superior colliculus in primates.

However, it seems possible to sketch out a plausible scenario about how the vertebrate visual system might have begun. The very first photoreceptors employed in vision, simply registering the presence of light, must have served merely to guide an animal towards the light or away from it. They can be supposed to have reported to what would become the thalamus, close to the site of the internal clock which was regulated by the same data.

One possible speculation about the next step is that, perhaps after a gene duplication and a minor modification to one of the duplicates, another sort of photoreceptor appeared, with a maximum sensitivity to a different wavelength. A photoreceptor sensitive to the wavelength reflected by local rocks and stones might have found a function in helping its owner to avoid collisions with hard surfaces, thereby avoiding quite a lot of damage. The response needed was a rapid twist away from the danger, so the new receptor would be selectionworthy if its information went to the midbrain site where such actions were organised when they were prompted by tactile input. This site, in what would become the tectum, would expand to receive an increasing amount of visual information, and later auditory information too, as well as the original somatosensory sort.

Further types of receptor, selective for other wavelengths, might find a use in providing a visual cue to the presence of conspecifics, if the species sported bright colours. Each type of photoreceptor may have guided a different sort of behaviour, with a short pathway from receptors to motor neurons. Indeed, the visual guidance of many behaviours may perhaps still be organised in a basically similar fashion in some modern fish and invertebrates, for some species seem to have a large enough range of colour-selective receptors to serve several purposes, without any need for further processing of the colour information.

An alternative hypothesis is that the earliest development in the retina was something akin to an OFF-centre ganglion, capable of combining the data provided by several photoreceptors and detecting an edge between dark and light. That too might provide a cue to the presence of something hard, to be avoided. An interesting wrinkle in the visual system is that OFF neurons operate more effectively than ON ones, so that dark stimuli can be seen with somewhat greater spatial resolution than light ones. The OFF neurons also relay their information more rapidly. If, in early days, dark stimuli represented nearby dangers to be avoided as quickly as possible and light merely functioned to guide the slower business of locomotion, then efficiency in registering the former would have been more important. ON and OFF pathways do indeed serve different ends in zebrafish larvae, which are hardwired to swim towards light. The OFF pathway runs to the tectum, where it prompts turns away from dark areas. The ON pathway contacts serotonin neurons, causing an acceleration towards the goal.

One thing that seems certain about the first retinal ganglion cells is that they will have responded only to local contrasts, for there will have been no means of focussing a distant image on the retina - obviously there was no mileage in evolving such a mechanism before contrasts could be detected. It's likely, moreover, that the new device was sensitive only to moving contrasts, since the visual stimuli an animal really needs to know about are the ones that are moving in relation to the retina, whether the movement is attributable to the object or produced by the animal's own motion. An ability to detect moving contrasts would have vastly improved the collision avoidance system. It might have been one of the advances that made it practicable for somewhat larger animals to evolve.

As the variety of wavelength selective receptors and contrast-detecting neurons increased a potential will have opened up for guiding other forms of behaviour by vision, such as the search for food. Then yet further additions to photoreceptors or contrast-detectors would be useful. By this time, of course, any development that helped to focus the incoming light rays, a lens or a cornea, would also have brought benefits.

A little data from contrast-detecting ganglion cells, and possibly a little about colour, had perhaps been going to the thalamus as well as the tectum. There it would provide information about visual textures that could be used by the primitive hippocampus, which might be correlated with information about internal states from the hypothalamus. Only when a focussable lens evolved would it be possible to collect data about distant contrasts, and to detect distant landmarks, which could be used for guiding locomotion. At this point it would become worthwhile to report rather more visual data to the thalamus, and eventually, as vertebrate brains grew more complex, to the telencephalon.

In other words I think that there are essentially two visual systems. One evolved to manage local reorientations and actions on nearby objects, the other to direct locomotion to more distant targets. In many species the optic tectum is still the important structure for organising actions on nearby visual stimuli. But I think it's a thalamic pathway that guides travel. The hypothesis fits neatly with the fact that there are two centres for releasing locomotion, and the idea that one deals in locomotion to distant areas, the other in approaching nearby stimuli.

Many lifestyles might be managed quite satisfactorily with a limited degree of co-operation between two such visual systems and two locomotory centres. A much tighter knitting together of the thalamic and midbrain functions may be one of the things achieved by a neocortex.

Chapter 26

Some Pivotal Innovations

The structures covered in the previous chapter, with their modifiable synapses that underpin learning, play leading roles in the vertebrate brain and must have had a major influence on the way it evolved. (No doubt other vital structures will be uncovered.)

There have been other particularly seminal innovations which were central to the evolution of the vertebrate brain. The foundational one is inhibition, which shapes and controls the activity of the excitatory neurons. Inhibitory neurons must have appeared at a very early stage of multicellular animaldom. In the primitive bag-animal, we can guess, potentially noxious input caused an excitation, leading to a contraction that distanced the threat slightly. And perhaps cells with leaky gates and a fluctuating electric charge made it possible to pulsate gently and increase the flow of nutrient-bearing water across the internal, food-absorbing cells. Clearly the potential for extending movement to more elaborate purposes couldn't be developed until there was a means of halting or modifying it. The earliest inhibitory influence may have been modulatory, which would be sufficient at that stage. The evolution of a faster acting transmitter would add greatly to the possibilities.

Inhibition came to serve numerous purposes. It plays a part in most pattern generators. As muscle systems became more complex, and as the potential for learning when to apply them increased, locomotion and other actions came to be controlled by chains of GABA neurons. In sensory pathways reciprocal inhibition improves definition at the early stages. At later ones it sculpts schemata by making neuronal responses more finely tuned. Neocortical learning can involve potentiating an inhibitory synapse on an inhibitory neuron, so that the excitatory neuron to which the latter is connected is freed to fire strongly when the stimulus pattern which repays a response is registered. (This sort of mechanism is probably responsible for the tip-of-the-tongue phenomenon, when one just can't quite find the word one wants. It looks as if the failure at the last stage of selection lies with the inhibitory neuron which suspends an ongoing inhibition.) Potentiation of inhibitory synapses may play the biggest role in the education of the brain.

Moreover any sort of extended neuronal circuitry is impossible without inhibition, since interconnected chains composed solely of excitatory cells would rapidly spiral out of control. In the complex networks of the neocortex inhibitory control is more than ever vital. What can happen when inhibition is inadequate is demonstrated by epilepsy, in which perception and motor control both fail.

Inhibition is what makes elaborate computing possible, and the evolution of increasingly large brains has been accompanied by the appearance of an ever richer array of inhibitory neurons, as well as new applications for them. The monkey striatum, for instance, appears to contain a type of GABA cell not found in the rat. In the primate neocortex about twenty varieties of GABA neuron have been identified, weaving among the pyramidal and stellate cells. They are distinguishable by their shape, by the methods employed for storing calcium ions not currently in use, and by the inputs they receive from neuromodulators. They also differ in where they make their synapses on the pyramidal and spiny neurons. Some target individual spines, so that they wield a very focussed and selective effect on neuron's input, others influence a somewhat bigger spread of dendrites. Some contact the cell body, the main stems of dendrites close to it, the axon initial segment or the axon end, producing a wider-reaching effect. Different types are found, for the most part, in different cortical layers.

Functions for the half dozen most numerous types of neocortical GABA interneuron have been established. Some, heavily interconnected by fast electrical synapses, contribute to establishing the rapid, locally co-ordinated oscillations which enable data to be processed efficiently. Others provide the sort of interconnection that enables a powerfully firing excitatory cell to suppress a less strongly excited neighbour. The evolution of an increasing variety of inhibitory neurons has no doubt supported an increasing degree of precision in the way the brain works.

The importance of inhibition to sensory processing is emphasised by the finding (in mouse visual cortex) that inhibitory neurons are a good deal more active when an animal is awake than when it's anaesthetised. As some neuroscientists postulated more than half a century ago, inhibition really does work, quite powerfully, to slim down the active neocortical pathways, reducing the spread and the duration of responses and thus increasing both spatial and temporal precision. The reduction of inhibition in slow-wave sleep no doubt explains the way the rhythms of the brain, as recorded by surface electrodes, slow and become widely co-ordinated. Just what makes the inhibitory neurons of the neocortex work so much harder during waking isn't yet known, but an obvious possibility is that they are very much under the influence of the neuromodulators which are suppressed during sleep and anaesthesia.

The rotatable eye

Since the hagfish is the only extant vertebrate which lacks external eye muscles they must have originated at some point deep in vertebrate history. Furthermore some pretty ancient fish fossils show signs of them. It seems likely that the moving eye became an increasingly worthwhile investment as the retina recessed deeper into the head and the aperture through which the light rays reached it became smaller, thus limiting the field of view. The development of a lens (something the hagfish, an inhabitant of deep, dark ocean, also lacks) would restrict yet further the area over which vision was really effective. The moving eye could cancel out this disadvantage.

But new controls were needed to make use of the new facility. With a fixed eye an animal which is going to act on a visual stimulus - to seize some food, say - needs only to turn its body so that the stimulus is appropriately placed on the retina. A moveable eye requires not only a system to control the eye muscles but also a means of co-ordinating eye movements with body movements. Once the essentials were established new possibilities opened up. Eventually the mechanisms of eye control would provide major components of the foundation of cerebral attention. External eye muscles can thus be regarded as playing a significant role in the evolution of consciousness.

Moreover eye movements offer clues as to where visual consciousness must be present. Packing photoreceptors densely in a central fovea and increasingly sparsely towards the periphery of the retina can hardly be profitable unless there is a short-term memory which can put the products of several fixations together, so that the current view can be interpreted with the aid of information gathered from previous fixations. The short-term memory which can do this seems to coincide with consciousness. The same memory is needed to distil a decision about the nature of a stationary stimulus by comparing the slightly changing inputs which result from microsaccades. We can therefore deduce that visual consciousness had developed before either foveas or microsaccading evolved.

The system for managing eye movements was naturally built on to the existing machinery for turning the body towards a visual stimulus. The messages to the eye control centre which emerged in the hindbrain came from the area which organised whole body orientations to visual stimuli - the tectum, the area that was receiving the major part of the information about visual contrasts.

In the lamprey a brief artificial stimulation of the optic tectum produces eye movements, while more prolonged stimulation can evoke eye movements combined with lateral bending that reorientates the body, and a yet longer stimulation prompts locomotion. The precise nature of the action depends on just which bit of tectum receives the current. With the electrodes in one area there are co-ordinated horizontal eye movements combined with lateral bending, the amplitude of both depending on the precise site of stimulation; if the stimulation persists locomotion follows. In another area downward eye movements are elicited. Stimulation in a third area produces large body undulations that look like struggling behaviour, with large eye movements in contra-phase. Feeding current to a fourth area simply prompts locomotion. Things seem to work in much the same way in finned fish. In goldfish artificial stimulation in one part of the optic tectum causes the sort of saccade and turn appropriate when there is a goody to be approached, in another it elicits the movements appropriate to distancing danger.

In the ancestral chordates the first use of muscles must have been for locomotion, directed from the basal ganglia. The earliest function of feature vision must have been for directing the locomotory movement towards desirables or away from danger, with tectum and basal ganglia working in close co-operation. When eyes became moveable the basal ganglia circuitry took a part in managing their movement. The roles of the optic tectum and basal ganglia in orienting the head and body to visual stimuli were extended to cover eye movements.

We can deduce that by this time the hippocampus and thalamus were providing at least a simple map of the current surroundings, one that took into account which way the animal was pointing, and recorded the more important stimuli that had been in view before it turned that way, or just before it arrived at its current location. Now eye direction would also have to be taken into account, and there would be additional instalments of visual information to be plotted on to the map in the correct relationship. (A lack of adequate information about saccades may explain some of the difficulties with spatial matters experienced when there is hippocampal damage.)

The cerebellum too participates in the control of the external eye muscles. It manages the vergence movements that point the gaze towards nearer or more distant objects, and it contributes to the management of saccades. It's especially important in the business of keeping the gaze centred on a moving object, or on a still object when it's the self that's moving. The part of the cerebellum that makes use of this information is in the most ancient sector of the structure.

All the major brain areas discussed in the previous chapter thus had to co-operate in eye control, and increasing demands were made upon them. Similar demands would be made when lateral fins evolved, and some of the neuronal controls needed would parallel those which evolved for managing eye movements. It's likely that the two developments are connected. An animal with fixed eyes, approaching something desirable that needs to be fairly precisely located, must get the stimulus registered on mirror-image locations in the two retinas. When rotating eyes evolved this mechanism could be elaborated to measure how far away the stimulus was on the basis of how far forward the eyes must swivel for the mirror-image locations to be central. Lateral fins, which no doubt facilitated a more delicate aim towards the stimulus, would then become worthwhile. The two developments between them must have contributed substantially to a potential for grabbing at food, rather than just nuzzling up to it.

In modern vertebrates there are many ways of exploiting the capacity for eye movement. The style of movement is related to the nature of the visual system, to whether the animal has a neck and can turn its head without moving its body, and to whether it's a predator or a prey species. Some fish use their eye muscles mainly to keep their view aligned with gravity, regardless of the orientation of their bodies, so they can keep a wide, horizontal watch for potential danger, whatever else they are doing. Other species seem to benefit only in adjusting their gaze when they turn towards a target. These strategies are obviously appropriate where the retina is primarily geared for detecting moving stimuli, as it often is in species which consume non-moving food and themselves constitute food for predators. For the most part these are the same species which have their eyes at the side of the head, giving them a wide field of view without much need for eye movements. Rabbits, for instance, make very few saccades. The strategy can also be useful for some forms of predation, as with the frog's renunciation of all forms of eye movement other than the ability to keep the gaze steady when the support beneath it moves. This obviously helps frogs to react rapidly to fast-moving insects equipped with equally hypersensitive responses.

Many insect-eating birds have a relatively limited degree of eye movement. Typically their main use for the eye muscles lies in making very rapid adjustments to position when the head moves, thus allowing long fixations and presumably interrupting movement sensitivity as little as possible. Some other vertebrates scan the scene continuously, with muscle contractions co-ordinated by a central pattern generator. In some large fish there is a clearly defined triangular pattern of slow scanning movements which nicely reveals the operation of this system.

It can only be in species where frequent scanning of the environment is well established that any real unevenness in the distribution of photoreceptors is practicable. The area of most acute vision must be regularly applied across the whole field of view if it's to deliver its potential, rather than being a disadvantage. This is all the more crucial in that the ability to squeeze a great deal of information out of the pattern of light hitting the central part of the retina is paid for, to some extent, by a reduced amount of data from the periphery. Where ganglion cells are densely packed in the centre of the retina, receiving inputs from only a few photoreceptors, any masking of the incoming light would be unfortunate, so blood-vessels are kept away from the foveal area, and there are fewer nerve fibres intervening between the lens and the photoreceptors than in less central regions. That means that the light must penetrate a rather denser veil in the more peripheral areas. It seems likely that some species without central conglomerations of photoreceptors know rather more about what's going on around the edges of their field of vision than we do. If we didn't explore the scene pretty continuously we should miss a good deal, and the fovea would be a very poor bargain.

Only in mammals is pointing both eyes in the same direction obligatory - and even some mammals tend to keep one eye tipped slightly upward, one more firmly pointed at the business in hand. Other vertebrates move the two eyes independently, and survey the scene with unco-ordinated eye movements, bringing both views together only when accurate targeting is needed. A chameleon locates an insect with one eye, then turns towards it and rolls the other eye around to foveate it with both eyes before unrolling its sticky tongue at lightning speed. Some birds, head tipped on one side, use one eye for searching out food, while the other is directed upward to detect possible threats.

When mammals sacrificed the ability to look in two directions at once the initial advantage must have lain in being able to see better in poor light. With both eyes focussed in the same direction there was a double chance of registering the few photons available, and collecting all possible data. Co-ordinated eye movements also opened up the possibility of a new method of uncovering information about distance - the cortical system described in Chapter 6, which detects whether an elementary feature is registered in exactly equivalent sites on both retinas, and must therefore lie where lines drawn through the centre of both eyes would intersect; or how much it's displaced to one side or the other, which indicates how much nearer or further it lies than the distance on which the eyes are focussed. This range-finding by stereopsis is something else that will have been helpful in dim light. It will have been especially useful for capturing insects camouflaged against bark or among vegetation, and insects were probably a major source of food for some early mammals, as for many extant ones.

Yet a third characteristic also looks like an adaptation to low levels of illumination - those microsaccades which move the retinal image over several photoreceptors. The advantage must be that input from the retinal ganglion cells which are responsive only to stimuli entering or leaving their receptive field, conveyed by the magnocellular pathway from eye to lateral geniculate nucleus to visual cortex, can be used to harvest data about non-moving stimuli as well as moving ones. In primary visual cortex their input can be added to that from the ganglion cells which are continually excited by a stationary stimulus.

In us this seems like no big deal. It helps only when the contrast defining the stimulus is very low, perhaps in poor illumination. And the price is probably a slightly reduced sensitivity to actual visual movement. But for early mammals, no doubt equipped with a higher proportion of movement-sensitive ganglions than we have, it should have been very useful, making it a little easier to move around in early dawn and late dusk.

It seems likely that in the first place a single eye muscle evolved which could pull the eye forward when something attractive was spotted, and release the tension gradually as the body was reoriented towards the food. Swivelling the eyes forward to study potential food more closely or locate it more exactly would be useful. When danger threatened, on the other hand, the priority would be to move away quickly, and finely directed eye movements wouldn't be needed.

The evolution of more eye muscles, to produce saccades, would help to underpin the development of more complex lifestyles. A pattern generator, organising a constant survey of the environment, ensured that the muscles were employed effectively. When a fixation was required the pattern generator would be inhibited and the original gaze-directing mechanism would take over. In us and other species where saccades are largely directed by the frontal cortex and have become a matter of choice the pattern generator is pretty thoroughly suppressed, but it still forms a detectable part of the machinery.

The more intense the concentration of retinal photoreceptors in a fovea the greater the need to saccade around a scene. To assemble the information gathered into a coherent whole it must be linked to data from the motor system about the direction and length of the eye movements. The short-term memory that could hold a reasonably adequate record of what vanished from view when the head was turned would have to be expanded as foveas evolved and saccades became more frequent, since there was more information to be preserved, and it needed to be retained over rather longer periods.

Or perhaps this should be phrased in the opposite way. The more capacious the memory which can put together an extended series of movements and inputs, the more complex the pattern that can be captured, and the more profitable active visual exploration becomes. The memory seems to involve both basal ganglia and hippocampus, and we can assume that all the steps in the evolution of a moving eye are closely linked to expansion in both.

In mammals quite a chunk of neocortex is also involved in directing eye movements. What happens when the instruction to saccade goes out tells us quite a lot about the nature of the neocortex. The command issued to the superior colliculus from the frontal eye field is reported at the same time to many of the sensory areas that will be affected by the move, usually via a collateral branch of the axon bearing it to its main destination. This efference copy begins influencing the way the recipient neurons work even before the new inputs from the eye reach them. Receptive fields in V4 may shrink or extend, for instance, according to the nature of the upcoming saccade. Activity in the part of the parietal cortex that is concerned with relating the position of the self and its components to the outside world is particularly strongly influenced.

The thalamus is an important part of the circuitry. Two of the more forward intralaminar nuclei receive inputs both from the deeper layers of the superior colliculus and from the part of the substantia nigra which exercises inhibition on the superior colliculus. There is cerebellar input too, and various other structures of the lower brainstem contribute.

Thalamic areas close to those nuclei also get inputs from either superior colliculus, basal ganglia or cerebellum, and project to frontal eye field, supplementary eye field or prefrontal cortex. Lesions in these areas may affect particular sorts of eye movement. Alternatively, they may lead to spatial neglect, or problems with orienting to a visual stimulus. Inactivation of a particular bit of mediodorsal thalamus doesn't interfere with single saccades, but impairs the performance of a well-practised sequence. As we saw in the chapter on the neurology of attention the pulvinar is important in managing visual attention, and receives a good deal of input from the superior colliculus. Lesions there can affect eye movements, or cause spatial neglect. We can deduce from the thalamus' heavy involvement in eye movements in mammals that it was already making a major contribution to the business before the evolution of the neocortex.

Other forms of active sensory exploration

A variety of mammals have extended the principle of active sampling of the environment to other senses. Again, this means that the sensory input must be correlated with the action. We humans run our fingers over an object, and learn quite a lot about its shape. The succession of inputs wouldn't be very informative if it wasn't automatically correlated with the direction of finger movement.

Another form of tactile exploration is found in whiskers. Rodents hold their whiskers forward while running, but when they change direction those on the side to which they are turning are angled a little more rearwards. Rats can locate and distinguish objects by whisking, perceive subtle distinctions between textures, and even determine whether a gap is jumpable. This involves co-operation between motor cortex and the somatosensory area which receives the impulses resulting from whisker deflection. (It's known as barrel cortex because of the shape of the fat columns of neurons which each receive axons from a single whisker.) Input from motor cortex to the apical dendrites of barrel cortex neurons makes the latter fire more strongly when deflections occur during exploratory movements than they do when an experimenter bends the whiskers.

An intriguing observation is that infant rats, three to six days old, twitch their whiskers during periods of active sleep (the equivalent of our REM sleep). Are these twitches the counterpart of our rapid eye movements? Do infant rats have whisker dreams, as we have largely visual ones?

Olfaction has also become an active sense. All tetrapods acquire an input of odour molecules as they breathe in. Mammals can accelerate the flow by sniffing, which seems to make for more efficient distinguishing of odours. Odorants must engage with receptors spread out along the olfactory epithelium. Drawing the air across the receptors more speedily means that the sequence of activations accumulates more rapidly, and perhaps, since some odorants are absorbed more easily than others, the pattern of activation is slightly different too. The sniff also affects the way the receptor input is processed in the olfactory bulb, modifying patterns of inhibition, increasing the response to new odours as against those which have already been registered, and influencing neuronal oscillations.

For at least some mammals the situation is yet more complicated because the data about odour molecules goes to different brain areas according to which way the air is flowing. The inward (orthonasal) flow carries information about the outside world, while thanks to an interconnection between air and food channels the outward flow carries odorants from any food in the mouth. Both patterns of odour molecules are analysed. The report on the odours in the inflow goes to olfactory cortex, while the report on the outflowing or retronasal current goes to the bit of cortex that is concerned with taste, to be melded in with the information from the gustatory receptors. This must be part of the neuronal arrangement that makes it possible to correlate odour with taste, and to learn what odours go with stuff that seems at first like potential food but turns out to taste bad.

Whisking and sniffing are forms of active sensing that must be particularly valuable in the dark, and have probably been around for a long while. Sonar systems, evolved independently in bats and dolphins, must be of more recent origin. Many mammals no doubt make use of returning echos to judge distances, as we humans do, often without conscious thought. Only a few emit a specific signal and register the echo with such delicate precision that they can locate fast-moving prey. Again, correlating the sensory input with the motor output is crucial.

Moving eyes are not confined to vertebrates - they are found among some arthropods too. And other sorts of active sensing are not altogether exclusive to mammals. At least one fish and various crustaceans use behaviours that have the same effect as sniffing. Octopuses wrap their arms around objects in a way that suggests tactile exploration. And perhaps those fish which locate their prey by creating an electrical field in the surrounding water and detecting any disturbance in it should also qualify as active sensory explorers. But in the mammalian line of descent the ability to link movement in search of sensory input with the input that results can be attributed to the neocortex. And I think this is a significant aspect of the neocortex. Indeed knitting together sensory and motor information into a whole that is much greater than the sum of the parts - as opposed to simply using sensory information to prompt movement - may be the definitive function of the neocortex. It's increasingly being suspected that the thing we call cognition actually corresponds to this productive marriage between sense and action. Referring to bits of cortex as 'sensorimotor' is becoming increasingly popular, but I'm not sure the word couldn't reasonably be applied to the whole thing.

The fact that our eyes move when we are dreaming seems emblematic of the close connnection. There's evidence that a pattern of eye movements is part of the record of a visual stimulus pattern. Appropriate eye movements may be produced when a previously seen visual stimulus pattern is encountered, or when a spatial position is to be recalled, even though the visual input which originally prompted them is not consciously remembered. The movements clearly constitute a conditioned response. And the response serves to record which features of the pattern have seemed most informative.

Mirror Neurons

Another particularly seminal event was the appearance of the first mirror neurons. Those first discovered, which inspired the name, were cortical neurons activated both by the sight of an action being performed and when the self performs the same action. They were almost certainly preceded by something simpler, probably in the basal ganglia. The guess is that the earliest form served to release a genetically determined action when that action was observed in a conspecific. Such an arrangement would be of no value to an animal in which the action was elicited by a stimulus pattern that was fully predetermined. However, in a species which could learn to inhibit its response to some variations on a genetically determined prototype, and perhaps enhance its response to others, this elementary form of imitation might be useful, especially in refining innate knowledge about what to eat. It would point youngsters towards beginning their learning in areas where their elders were eating, areas where satisfactory food was available, and where their learning ability could be applied most efficiently.

I think this basic sort of mirror neuron was probably established before parental care of mobile offspring developed, and that new and more elaborate versions evolved in conjunction with that development. When a parent was regularly available for imitation the value of mirror neurons in guiding the attention of young learners would be greatly increased.

There may be a special benefit in mammals and in altricial birds - the sort that are fed by their parents when young. There's an awkward transition to be accomplished when the easy way of obtaining nourishment has to be replaced by the more demanding responses needed to feed the self. It may be suspected that a timely maturation of mirror neurons helps, by releasing the appropriate new behaviour. Young birds which look full-grown often seem to hang around a parent, squawking from time to time, in the manner which has previously prompted the parent to supply food. Then the youngster tries pecking the lawn for itself, with a tentative air, still approaching the adult from time to time to beg again. Eventually, after being consistently ignored, it starts pecking more energetically. The action may be an essentially hardwired response to appropriate stimuli, but perhaps the example of the parent provides a trigger to activate it.

Emotional mirror neurons - active both when an emotion is experienced and when it's observed - no doubt also evolved as aids to infant education. They enable the young to learn very quickly what inspires fear or other emotions in their mother, and they're clearly extremely useful where adults and youngsters keep company for at least a little while. The earliest communication of emotion was perhaps through smell, but the emergence of a more sophisticated type of mirror neuron must have been what made increasingly expressive faces, bodies and vocalisations worthwhile. The combination of expressiveness and emotional mirror neurons means that the experience of the ancestors, most valuably the learning about dangers, can easily be passed from generation to generation.

The imitative drive and the infectiousness of emotions must become even more valuable when animals live in groups and there are more models to copy. Facial, bodily and vocal expressiveness tend to be well developed in social species, and emotions can spread through a group rapidly, so there's a quick response to danger, and youngsters have plenty of opportunity to learn what causes it.

It may be relevant that adopting a facial expression tends to promote the emotion that goes with it, so conceivably when young infants imitate an adult's expression it contributes something to their emotional education.

As chronicled earlier, mirror neurons have been found in the basal ganglia of the song-birds which learn their song by imitating father's. It's likely, therefore, that the ancestor of the mammals possessed something similar, and that the infant primate's imitation of facial expressions can be attributed to a parallel mechanism. As facial muscles come under cortical control the innate response is suppressed and imitation becomes something the growing animal can choose to do. But just occasionally damage to human prefrontal cortex leads to a difficulty with inhibiting motor imitations. Patients directed not to reproduce the experimenter's action find their hands moving against their will. The inbuilt urge to imitate must remain.

Even in the undamaged brain it tends to retain some force – how else can we account for the power of fashion? The effect is also present in other species. When two individuals in a group of captive chimpanzees were taught two equally efficient ways of obtaining food from a trick container both had their imitators at first, but gradually the members of one set converted to the practice of the other. A similar experiment with wild birds produced similar results. In each of five subpopulations of great tits two males were taught one of two possible ways of opening a puzzle box to obtain live mealworms, a trick that was soon copied by many of their fellows. But ten of the fourteen individuals that migrated into a subpopulation with a different technique adopted the local one, and only three rugged individualists retained the one they had learnt first.

Orangutans seem to be even readier than other apes to imitate humans. Do they have an extra ration of mirror neurons? A strong taste for imitation might be especially useful in a species which is only intermittently social, so that models are not regularly available. Orangutans at rescue centres have been filmed in all sorts of unlikely activities clearly copied from humans - sweeping paths (with a modest degree of efficiency), washing textiles, sawing wood, and paddling a flat-bottomed boat on a quiet backwater among them. They tend to look as if they are playing at being human, rather than pursuing a useful goal – but who knows?

I hypothesise that in the infant the basal ganglia message which releases an innate imitative response is copied to motor cortex. There it coincides with input from proprioceptive neurons, and from the visual system. Something similar must happen as a result of early vocal imitations. The confluence of inputs leads to the establishment of mirror neurons in neocortex.

These must be necessary for the more complex forms of imitation found in larger-brained mammals, and their number no doubt grew as the control of muscles became ever more delicate. Apes copy actions in a way that shows they appreciate their purpose, and ape mothers clearly understand the value of providing a model for their offspring to imitate. Apes don't, however, reproduce the precise details of a performance. Among extant primates it seems that only humans do that, which suggests an unusually large supply of mirror neurons. It may also imply that the mirror neurons have wider cortical connections. Both numbers and connections must have grown as fingers became increasingly versatile and capable of delicate manipulations, so that tool-making became ever more sophisticated. An increasing number of vocal mirror neurons must similarly have contributed to the development of language.

The cortical mirror neurons also function, it can be supposed, to enshrine action concepts or schemata, making it possible to think about possible actions in advance, and to think about the actions and potential actions of others.

The evolution of highly developed forms of emotional and motor mirror neurons provided a means of modelling other individuals as beings essentially similar to the self, and predicting their behaviour on this basis. Conspecifics became stimulus patterns that could be understood to some extent, not just responded to, so that deceit and other means of influencing the behaviour of others became possible. A higher level of consciousness emerged, and the possibility of a 'theory of mind'. Then the stage was set for the invention of words, and ultimately language, which would take these developments even further.

Mirror neurons must have played an important role in the evolution of primate-style social living. Groups are likely to thrive better if they act in a coherent fashion, which means that the behaviour of fellow members is largely predictable. An occasional innovator brings benefits, but the advantages of group living would be lost in a group composed entirely of individualists. It's on the whole helpful, then, that the ability to imitate brings with it an inclination towards conformity.

Emotional mirror neurons underwrite empathy, making it possible to share a fellow animal's good or bad feelings and providing an incentive to care about their welfare. It's painful to see anyone suffer, especially relatives or friends, because one suffers to some degree in sympathy. There also seems to be an innate urge to inflict revenge on those who cause suffering. These factors help to bind the members of a group together.

As language developed the potentials of both types of mirror neurons were extended further. Verbal communication could make motor imitations more precise, empathy more informed. Mirror neurons made a vital contribution to the increasingly large-scale groupings which became possible with the development of language elaborate enough to establish structure and organisation.

A society which functions in a coherent fashion is more likely to flourish than one in which there is no agreement on the proper goals for both individuals and the group. The inborn inclination towards imitation and the infectiousness of emotions promotes the idea of norms of behaviour, and leads to distrust of deviations. Formalised laws were able to build on that foundation, and find acceptance because of it. And emotional mirror neurons made a foundation for a sense of justice, and for the behavioural rules that were established by the creators of religions and subsequently codified in a society's laws.

A shared code of morality, furthermore, helps to promote a sense of group identity, as does the religion which develops to support it. As societies grew larger, and benefitted from greater size, any trait that helped to knit them together must have been selectionworthy. This no doubt explains why humans have acquired a strong taste for morality, and a readiness to heap shame on those who flout it - despite varying opinions on just what constitutes moral behaviour.

The precise nature of a successful group's rules was influenced by the nature of the environment. They dictated the sort of behaviour that promoted the survival of the group's genes in the context of the territory the group inhabited. For most of history the group as a whole hasn't necessarily suffered if some of its customs seemed wicked, disgusting or ludicrous to another group.

Unfortunately this ancient foundation of morality, based on the tendency to conform, must also account for the way irrational moralities can be imposed on whole nations, to ultimately destructive effect, with little opposition. Mirror neurons can operate powerfully. Not many people question the accepted way of things, and few are prepared to stand out against it.

Familiarity and unfamiliarity

I pointed out in an earlier chapter that species which function solely by responding to genetically determined stimulus patterns, or to ones that are essentially innate but have been defined more closely through conditioned learning, have no need to take notice of sensory inputs which don't add up to something that may require action. Animals whose behaviour is governed entirely by innately determined responses have to respond only to the foreordained stimulus-patterns, and those which are capable of conditioned learning only to those variations which have proved to be associated with reinforcements or with harmful effects. Anything which doesn't evoke either an alert or an actual response must pass by unnoticed. In contrast, species with more advanced sensory systems get a great deal more information from them than is strictly necessary. That means they must be able to distinguish between stimulus patterns that have been classified as significant or insignificant and those which are newly met, and might be either. They need a means of recognising what's familiar and what's unfamiliar.

The implication is that familiarity is the sensation evoked when a stimulus pattern prompts a ready flow of excitation through well-defined neuronal channels, and triggers activity in neurons which encode a schema. These are linked to neurons in other schema-storage areas, which indicate what sort of inputs in other sensory channels can be expected to coincide, and what sort of behaviour the stimulus pattern should evoke, if any. An unfamiliar stimulus pattern, on the other hand, must initially produce less focussed neocortical activation. But how exactly does a neatly channelled flow of impulses through the later stages of a neocortical sensory pathway trigger a sense of familiarity, or the absence of a previously established circuit evoke a sense of unfamiliarity?

Part of the answer must lie in the hippocampus and its links with the neocortex. When a stimulus pattern is encountered regularly a connection must be maintained between the neocortical schema-neurons that represent it and the hippocampus, a connection that will be activated when the pattern reappears. And indeed there do seem to be hippocampal cells which respond to familiarity, and to unfamiliarity, presumably thanks to this link. They pretty certainly depend on the connections that run to hippocampus from the 'what' pathway through perirhinal cortex, which parallels the 'where' pathway through entorhinal cortex.

Unfortunately many experiments haven't distinguished between familiarity and recency, between novelty in the sense of something completely unknown and novelty in the sense of a new appearance on the scene. There are neurons in hippocampus, entorhinal and perirhinal that respond strongly to this last, as might be expected - the stimulus pattern that has just appeared may be the one that most urgently needs a response, regardless of whether it's a known or unknown quantity. There are also hippocampal neurons that respond when a recently encountered pattern materialises again, and in many cases it seems to be recency which is being registered. This is another very important consideration when one's exploring, so naturally it's noted in the hippocampus.

However, I think the real answer to the question of familiarity and unfamiliarity lies elsewhere. The important consideration, surely, is that a familiar stimulus pattern is something one knows what to do about - or whether, indeed, anything ought to be done. As part of knowing what, if anything, to do, one also knows how one feels about it - what one might be motivated to do. Most importantly, one knows whether the familiar requires attention or not. If it's a visual stimulus pattern one also knows how to give it attention, in other words, it evokes a practised sequence of saccades, though if it's very familiar it will be recognised at the first or second fixation, and it won't actually be necessary to execute the full scanpath. An unfamiliar object, on the other hand, is one for which a scanpath must be developed. When a stimulus pattern is unfamiliar we don't know which parts of it are definitive and repay attention.

Inputs to other senses, meanwhile, can prompt visual investigation, and experience indicates what sort of sound patterns or tactile inputs require such investigation. A familiar auditory stimulus or tactile input is likely to be one that's known either to be worth visual study, or to be safely ignorable. All this leads to the conclusion that identifying familiarity must be another basal ganglia function.

Maybe the initial response evoked in neocortex by a thoroughly familiar visual stimulus pattern (or a part of it) is sufficient to trigger a message to dorsal striatum and to superior colliculus, sent even as the conscious visual experience of the stimulus pattern is still developing. This first message can be imagined as too weak to prompt a saccade to another bit of the stimulus pattern, functioning merely to prepare the well-established saccade sequence for possible use. While not producing an action, however, this communication may perhaps trigger a return message from basal ganglia and superior colliculus to the neocortex, indicating the habitual saccade sequence, and thus the important features of the stimulus pattern. This may serve to prime the neurons which will receive excitation if the pattern of saccades is carried out.

Familiarity and recollection activate some areas in the rear part of parietal cortex, as might be expected if reference to eye movements and other movements is involved. There's also a notable response in the dorsomedial nucleus of the thalamus, one of those especially well connected with the basal ganglia. From parietal cortex and dorsomedial thalamus, it can be supposed, a message goes to prefrontal cortex, indicating that the pattern of input implies a certain sort of action. Prefrontal draws all the relevant brain areas into a synchronised circuit. It also decides whether to execute the full saccade sequence or not.

An unfamiliar sensory input engages no schema-circuit in temporal or parietal cortex, so there's no message to the basal ganglia and no returning prompt about possible action, no communication to prefrontal cortex about significance, and no synchronisation among different brain areas. Impulses that travel from prefrontal cortex to more rearward parts of the cortex encounter no circuits tuned to their rhythm and primed to be connected into the circuitry. And all is muddle for the moment, until some short-term potentiation sets up some new circuits.

A familiar stimulus pattern, in short, is one for which an appropriate motor response of some sort is available, or for which a response is known not to be necessary. If a response is available prefrontal has to decide whether it's needed in the current circumstances. If the stimulus pattern is unfamiliar, on the other hand, prefrontal has to direct the rest of the brain in learning more about it, most importantly how or whether to pay attention to it in the future.

For humans most familiar stimulus patterns evoke the possibility of a word, and a word is an action - something to be pronounced, or conveyed by signs, or written. Abstract ideas are pinned down by being attached to words, and this may explain how it is that ideas, as well as objects and actions, can be familiar.

Recapitulation

Several processes or factors can be highlighted which must have played an especially notable part in the evolution of the vertebrate brain.

Inhibitory neurons were vital to the evolution of any even faintly sophisicated type of neuronal system.

An increasing repertoire of inhibitory neurons has played an essential part in the evolution of increasingly complex brains.

The ability to rotate the eye in its socket involved all sorts of extensions to the arrangements by which an animal oriented itself towards a goal, and led to the evolution of some of the neuronal machinery which would ultimately support consciousness.

The benefit of the first mirror neurons probably lay in making it possible to apply conditioned learning in the places where it could be most profitable.

More elaborate versions contributed to the advantages of parental care of infants, and of social lifestyles.

The ability of more complex brains to learn about a wide range of stimulus patterns in addition to those relevant to immediate survival made it necessary to distinguish those with implications already discovered from novel ones with unknown import.

The essential machinery of familiarity and unfamiliarity must surely be in the basal ganglia.

Chapter 27

A Tentative History of Consciousness

Consciousness, then, is the effect of neuronal mechanisms that bring a great deal of information together, making it worthwhile to collect a great deal of data from which to extract information. How these arrangements produce the subjective effect remains a mystery. (But this is not the sort of problem brains have evolved to cope with.)

Towards the beginning of this book I surmised that consciousness probably evolved in relation to one behaviour at a time, with each additional behaviour promoted to consciousness by the evolution of new neuronal connections. Specifically, I proposed that some of the machinery which underpins consciousness might first have emerged to cope with the challenge of directing locomotion, since this is where it could be most valuable. Genetically ordained rules can provide some guidance, and conditioned learning can extend it. But the geography of the environment into which an animal is born is essentially unpredictable, so something more is needed for knowing just where the best chance of food or shelter is, and the quickest way to get there from any starting point. And that's extremely valuable knowledge, the more so since locomotion, unlike other basic behaviours, is an extended activity which consumes a good deal of energy. I've assumed, therefore, that a cognitive map was the first sort of schema for which neuronal mechanisms evolved. For any species inhabiting a featured environment, equipped to travel, and living long enough to build and use a map, the ability to record the layout of the local environment and confine activity to the profitable areas must be a major benefit.

On the other hand stimulus patterns which represent food or conspecifics or signs of approaching danger can be adequately defined by hardwired mechanisms. Conspecifics are stimulus patterns that not only can but must be written in the genes, with sufficient detail that potential breeding partners can be reliably identified. The shape of danger is not so easily predicted, but experimentation is risky, so hardwired responses to broadly defined stimulus patterns provide the best way of avoiding harm. There's a great advantage, though, in learning to respond to cues that signal potential danger, however roughly such cues may be defined. Similarly, the stimulus pattern that represents food may be genetically determined, but the search for it is more efficient if the animal can learn to recognise signs that indicate its probable presence. All three behaviours can thus be managed pretty well by inbuilt stimulus-response programmes, but the efficiency of the latter two can be notably improved by some conditioned learning.

All three, moreover, simply involve responding appropriately to current sensory inputs. Learning where to look for food is another matter. If the territory covered is larger than that within immediate reach of the senses a medium-term memory is necessary to fit the different parts together, and to provide the means of establishing the overall pattern in long-term memory. There also has to be a means of accessing the records that are stored in long-term memory, so as to decide where to go.

Subjectively it's hard to imagine these essential requirements being fulfilled without at least a faint sense of place and a faint sense of motivation. Could the track that's now out of sight be correlated with the current surroundings, or a distant goal chosen and reached, by wholly unconscious means? Perhaps - there seems to be no way of finding a definite answer. But I think we must at least deduce that the necessary records are established by the sort of sustained and extended circulation of impulses that would be elaborated and expanded to produce consciousness. The business of guiding locomotion seems to be the likeliest place to look for the beginnings of the machinery that produces sensory consciousness, and for the earliest examples of such consciousness.

The hypothesis based on logic and subjective argument turns out to be capable of being fitted to the neurological evidence. Learning large-scale spatial patterns, getting any sort of information into the long-term store in such a way that it can be recalled, and remembering the goal of the current journey, all depend on the hippocampal formation. People who have suffered extensive hippocampal damage can't do these things. For those of us with fully functional hippocampi, on the other hand, a conscious knowledge of the origin, path and aim of a current journey over known territory is constantly available, though for much of the time it doesn't occupy the focus of attention. There's a hazy, background awareness, against which other activities or thoughts about other things take place.

An equivalent of the hippocampal area seems to be present in all extant vertebrates, so it must be a very ancient part of vertebrate neuronal machinery. In simpler, smaller vertebrate brains it takes up a large proportion of the telencephalon, which implies that its function relates to a major part of what the telencephalon does. It's reasonable to suppose that it began by providing the medium-term memory that's needed to assemble a series of successively experienced scenes into a coherent record, creating a temporary cognitive map in association with the thalamus. As the telencephalon expanded the temporary maps could be put together and a longer-lasting, more extensive one assembled in the newly evolved area.

If these arrangements began to produce some sort of consciousness it may be imagined as a very vague consciousness, something like the minimal awareness of place and time that in us forms the background to events that are receiving focussed attention. Conceivably it might only operate when the animal was moving towards some goal, or about to do so.

The telencephalic map began, I've suggested, with a fair amount of input from the adjacent olfactory system and some minimal input about visual texture from the thalamus, which would also forward information about current actions. In addition to the olfactory and visual consciousness I think there was some sense of the need the locomotion was designed to fulfil - hunger, or unsatisfactory temperature, or, in the breeding season, hormonal arousal. It might be called a motivational or emotional consciousness.

Quite a vague, undetailed account of the permanent features of terrain or seabed, along with olfactory guidance towards food or mating opportunities, should enable an animal to find its way around in a useful fashion more often than not. Once the goal area was reached the good old-fashioned hardwired forms of action guidance would take over. Food would be seized or breeding behaviour performed without any need for sensory or motivational consciousness to guide the action. Perhaps, though, a conscious sense of something resembling satisfaction might result, based on feedback about long-lasting internal states. Channelled to the hippocampus, it would reinforce the marking of the site on the map.

Insofar as current knowledge allows us to guess, a minimal, hippocampally based, travel-guiding consciousness might reasonably be attributed to frogs. The visual information which guides their prey-catching behaviours triggers extremely fast responses, leaving no time for the luxury of creating a conscious experience. Visual input which may represent a large descending foot or an approaching predator produces another rapid reaction. These mechanisms are clearly hardwired, and in neither case could a conscious sensation serve a useful purpose; for catching fast-moving insects or escaping from imminent danger speed is more crucial than seeing fine detail or planning the nature of the action.

The limited visual information which goes to the thalamus must be what guides locomotion. Internal factors, such as temperature or appetite, will prompt the decision to expend effort on moving about, and the urgency of different needs must be weighed. These judgements are no doubt made in the telencephalon, on the basis of messages from hypothalamus and elsewhere. The guess is that the correlation of internal needs with visual and perhaps olfactory inputs – all sampled over an extended period - involves a recurrent circulation of impulses sufficient to create a dim consciousness. Which seems appropriate for the sluggish, pause-punctuated form of travel generally practised by frogs.

Foreground consciousness

For us humans the bright, fast-changing consciousness which paints the foreground of our subjective experience clearly depends on the neocortex. If a patch of neocortex is destroyed some part of the sensory experience is either lost, or becomes very confused. Without primary visual cortex there is almost no conscious vision, while damage to one of the sectors a little further forward creates a gap in the patient's visual sensations. There may no longer be any perception of colour, or objects may appear to relocate suddenly from one place to another without any perception of the intervening movement.

Damage to a schema-area doesn't leave a blank where the percept ought to be, but renders the afflicted person incapable of making sense of a stimulus pattern that should be matched to that sort of schema, as in the case of Oliver Sacks' patient who 'mistook his wife for a hat'. When a bit of neocortex and its connections are knocked out an essential node in the circuit dealing with a certain sort of percept is removed, so the circulation of impulses which would allow a succession of inputs to be assembled into a meaningful perception is prevented. There would seem to be some sort of conscious experience but it's incomprehensible. The stimulus pattern cannot be linked to any associated information.

A conscious experience also depends on impulses reaching prefrontal cortex, as we saw in Chapter 8. This, we can guess, is necessary for the synchronisation of large-scale rhythms across activated pathways which enables the prolonged circulation of neuronal messages. Probably it's achieved by means of an interchange of messages between prefrontal and the basal ganglia, which between them record the known implications of the stimulus pattern.

This is relevant to several early experiments on conscious vision. If a visual stimulus pattern is undetectable when very briefly flashed by the tachistoscope, but becomes quite clearly visible after a series of such brief exposures, it must be because synapses along the pathway through sensory cortex have been sufficiently activated to become potentiated. As a result they become sensitive to even very brief inputs, and start firing strongly enough to forward a signal to prefrontal cortex. Prefrontal then initiates return messages, completes a circuit and establishes synchronisation. And George Sperling's experiment showing that one row of a very briefly displayed three-by-three array of letters can be read out from iconic memory after the exposure has ceased, but only one, implies that while signals from sensory cortex must reach prefrontal cortex to produce an awareness that there is something there, returning, top-down signals are necessary for the analysis that reveals what it is, and gives precision to the conscious experience.

We can also see how messages from sensory cortex can activate innate and conditioned responses before they reach prefrontal cortex, sometimes without getting that far, so that it's possible for attentional and emotional effects to occur even when sensory inputs don't reach consciousness. An innate or conditioned response to a sensory input makes it more likely that the message reporting that input will reach prefrontal cortex, but that result is by no means certain.

The situation must be somewhat different in the wholly inexperienced infant brain, where there are only genetically determined responses, no conditioning has yet occurred, and prefrontal cortex develops only slowly. If the hypothesis put forward earlier is correct a sensory input only causes an extended circulation of impulses in the neocortex if it evokes a genetically determined response, released by the basal ganglia. Any such action, even it it's only a saccade, serves to lift the inhibition exercised by the basal ganglia on the pathway from a midbrain sensory area to thalamus and thence to neocortex. That inhibition isn't complete, it's theorised, but without it the thalamic neurons can fire much more strongly, and a more distinct message goes to neocortex. This more powerful input produces a level of neocortical excitation sufficient to trigger a message back to the thalamic reticular nucleus, punching a hole over a 'driver' nucleus so that all available information about the stimulus that prompted the movement gets through.

The theory explains how behavioural attention is converted into cerebral attention – or at least one of the mechanisms. It incidentally supplies another reason why the evolution of rotatable eyes would prove so seminal. The hardwired part of the machinery that governs saccades provides an easy and inexpensive way for the midbrain to tutor the neocortex.

As the infant neocortex learns about a stimulus pattern and its significance a pathway back to the basal ganglia is established. Instructions can be sent to lift the basal ganglia inhibition, and thereby, indirectly, the inhibition exercised by the thalamic reticular nucleus, even if no hardwired response is performed. As this pathway is established the infant learns to direct attention more flexibly.

The supposition is that sensory or 'foreground' consciousness happens if the pathway to the thalamus which reports the performance of a hardwired response to a genetically determined stimulus pattern extends into a brain area that carries out a sophisticated analysis of the sensory input which prompted the action. In mammals this is of course the neocortex. The message indicates both the nature of the action and the position of the stimulus at which it's aimed. The neocortex takes over from there.

It's reasonable to suppose that the pathways were extended one by one, and that the first use to which the evolving sensory processing facilities of the telencephalon were put would be to add visual detail and precision to the cognitive map. A concomitant of this development, I propose, is that the cues which attracted approach or other responses were promoted into consciousness as actual sensory objects. No longer were they experienced merely in abstract, emotional fashion, as goals defined in terms of the need to be fulfilled, or threats to be retreated from. Now the actual shape or sound - to whatever extent it was revealed by the cortical sensory pathways - was consciously experienced. The account of the external world would continue, of course, to be pervaded by emotional and motivational significances. But it would be a more vivid and shapely world.

Other subcortical structures would also evolve links to the neocortex, or where links to the telencephalon already existed more value could be developed from them. Interconnections between cerebellum and neocortex, for example, would make it possible to learn in more detail than before which visual textures indicated solid barriers, which it was possible to push through, and what sort of input correlated with surfaces on which it was safe to put the feet. Pretty well all vertebrates must have some means of learning to distinguish impassable solidity from something flexible that can be pushed aside. Tetrapods also need to learn what sort of surface will bear their weight. The expanding sensory processing facilities of the neocortex must have provided mammals with a potential for more detailed learning about these subjects, once tactile inputs were reported to the neocortex – promoting another subject to consciousness.

This information would be marked on the map, but it's a slightly different thing from the map. Knowledge of landmarks and the overall geographical layout is acquired through vision and locomotion, with the gaze mostly focussed at some distance. Information about the ground and barriers is acquired through touch, and with a different convergence of the eyes. Moreover the feedbacks are mainly negative - the discovery that trying to push through dense bushes doesn't work, and that treading on the wrong sort of ground leads to alarming consequences. Such information must have come to be stored separately in the neocortex, available to attach to more than one location on the map. A neocortical area devoted to such matters has not, I think, been identified yet, but I suspect it was one of the earliest additions after the development of the new, elaborated cognitive map.

A similar exchange of messages with the amygdala would allow the stimulus patterns which prompted hormonal and visceral responses to be analysed in the neocortex. Later it would be a dialogue between amygdala and neocortex that made it possible for mothers and infants to recognise each other by visual and auditory means, as well as by olfaction. As maternal behaviour evolved the maternal amygdala must have become extra-sensitive to the olfactory signals given off by infants. The resulting excitations in the amygdala would be forwarded to neocortex, where complex visual and auditory discriminations could be made, enabling mother and infants to identify each other at a distance. And maternal hormones, as well as promoting suitable behaviour, came to have the sort of effect in the neocortex which is experienced as emotion.

It's sensory consciousness, I've suggested, that makes it possible for mothers and infants to maintain contact with each other while simultaneously pursuing other ends and dividing their attention. Mother's and infants' conscious perception of each other functions as a form of reward, though only sampled intermittently. Moreover, if the percept suddenly turns out to be unavailable its absence functions very powerfully as an alarm and a source of distress, because there is a conscious sense of what ought to be there. The implication is that a well-developed sensory consciousness lies at the root of mammaldom, and must have been established before this means of nourishing the young evolved.

In some ways the challenges entailed in keeping mother and infants or the members of a social group together constitute a variation on those involved in travelling to a distant goal. The goal and the route to it have to be kept in mind despite any distractions along the way. The stimulus that has to be kept in mind by both mother and infant is a moving one, and the aim is to make sure it remains within view although it's not constantly observed. The hippocampal map must play an important role in keeping track of the not-constantly-attended figure or figures. And measuring the distance between mother and infant probably depends on much the same mechanisms that measure the distance to a visible goal.

To some extent it's easy to deduce where sensory consciousness is likely to be applied in other species. Many mammals chomp away at a genetically determined diet and presumably don't need to devote much neocortical space to the question of what to eat, so perhaps a consciousness of taste is unnecessary. Neocortical analysis is probably widely applied, though, to the cues that indicate where to find food, especially where it's the sort of food that is inconspicuous or widely scattered.

Things must be more complicated for carnivores, at least if they hunt a variety of species, for there seems to be a need to learn how the different prey behave. Doubtless such hunters have a neocortical area which links the appearance of each prey species to its escape strategies, and the complementary strategies needed to capture such a meal. The particular taste which can be expected at the end of the hunt might also be expected to form part of the schema.

Something more is needed in adventurous omnivores, prepared to experiment cautiously with new possibilities, and equipped with some remarkable feedback connections that can derive learning from effects on the stomach which occur some time after the eating. The neocortical area which combines olfactory and gustatory information, turning it into the conscious experience of flavour, might be termed a schema area dedicated to the subject of food. It seems to be responsible for the remarkable conversion of a taste once found pleasant into something aversive when consumption has been followed by feelings of illness. In the gustatory cortex of rats the activity evoked by a sweet flavour began to look more like that evoked by a bitter one after it had become associated with 'a visceral malaise.' This neocortical sector no doubt evolved in conjunction with the development of an increasingly versatile digestive system.

Perhaps the most fascinating part of the neocortex is motor cortex, which provides the ability to shape actions with the aid of conscious feedback from vision, touch, and the senses reporting on balance, muscle stretch and joint position. In the lamprey there is a telencephalic area where artificial stimulation elicits movements. This ancient bit of forebrain has been greatly expanded in species which produce finely adapted movements shaped with the aid of experience - in other words those which go in for motor play in youth. The amount of such play indicates the amount of learning that must be done, and no doubt correlates with the size of the motor areas. We can deduce that where there is play there is a conscious experience of movement.

The expansion of motor cortex will have involved new links with the basal ganglia, and a considerable expansion of the latter. It will have coincided with the evolution of more complex musculatures that could be employed in more varied ways and of new neuronal pathways to carry instructions to them.

Such developments led to new uses for paws. No longer just things for walking on, forepaws variously became capable of patting in an investigatory way, clutching at food, clawing a way into it if necessary, or climbing trees - all activities which could benefit from learning just how movements should be adapted to fit different challenges. In some species jaws became more versatile, and different tactics could be adopted for tearing at different sorts of food.

The existence of motor cortex and the ability to learn how to use muscles most efficiently must have facilitated further evolutionary developments. Motor cortex and play were clearly well established long before the advent of hands, and extremities that could be described as fingers. We might guess that fingers were worth having because there was already a means of learning about the many varied possibilities for using them.

Perhaps, moreover, it was because primate fingers could readily acquire a variety of foods that were, in different ways, a little tricky to eat - fruit with inedible seeds or hairy skins, nuts that needed a lot of chewing - that some parts of the eating equipment also came under neocortical control. That meant that some of the muscles used in the consumption of food could be used as well for voluntary vocalisations, such as the lipsmacking and pant-hooting of chimpanzees. This laid one of the foundations for the development of speech and language.

As animals learnt to use their forepaws in new and varied ways, meanwhile, a neocortical area must have evolved that dealt in the objects to which such actions could be applied in a useful manner, and the sort of objects that should be left firmly alone. It probably began as a subdivision of the area that recorded which surfaces were safe to walk on. In primates, as forepaws turned into hands and objects began to be used as tools, the new schema-area would expand significantly, reaching its greatest extent, of course, in humans.

The interlocking webs

The neocortex is well fitted for contributing to elaborate, self-maintaining circuits, of the sort believed to create conscious experience. Its major excitatory neurons, the pyramidals and stellates, have enormous numbers of input synapses, and the inhibitory interneurons which play such an important role are well endowed too. The thalamic nuclei which forward sensory information to the neocortex receive numerous projections back from it, as do other thalamic areas, and it's a general rule that any other subcortical area which sends axons to the neocortex also receives connections from it.

The largest proportion of the inputs to cortical neurons, however, comes from other cortical neurons, some local, some distant. Every stage of the visual pathway, for example, communicates directly with almost every other stage, and also receives inputs from the forebrain. These contributions synapse mostly in upper and lower layers, whereas incoming sensory information arrives mainly in the middle layers. In visual cortex the upper and lower layers play an important part in distinguishing small patches of one texture against a wider extent of a different sort, which no doubt indicates the sort of use that the horizontal connections within the cortex can be put to.

As we've seen, some parts of the neocortex are primarily concerned with processing a certain sort of sensory input, others with recording significant stimulus patterns, some with evaluating the input, and yet others with directing movement. But the whole is constructed like a three-dimensional net of interlinked spiderwebs, so that a small impact on a bit of one web can send tremors reverberating through other quite distant webs. Activation in a node that represents a familiar visual stimulus pattern can cause a subliminal level of excitation in the auditory node representing the sound associated with it, and in nodes representing other aspects or associations. The rate and pattern of neuronal firing quite early in the sensory pathways is modulated by attention and emotion, and influenced yet further during the development of a decision to act on the stimulus pattern being registered. Motor areas and motor neurons can show a small flutter of excitation in response to sensory inputs that merely suggest the idea of action. As we listen to speech there is a faint activation of the motor neurons that would be at work if the self was uttering those words. Tongue muscles can be faintly primed in anticipation of an expected phoneme, and the priming process is suspended if the expectation proves wrong.

This supports the idea that the neocortex is organised essentially in relation to action, thanks to those inputs from midbrain and basal ganglia via thalamus - a reasonable proposition, since the essential function of neocortex is to decide where action is needed, and if so, what sort. The result is a very effective filing system. As I've emphasised, mammalian sensory systems have to deal with an enormous variety of sensory inputs, with the potential number of classifiable stimulus patterns limited only by the amount of detail that can be registered. Inputs must be organised into schemata flexible enough to fit many variations on a theme. In contrast, the range of possible movements is limited, even in the most agile of animals, and the movement required in response to a stimulus pattern often has to be very precise, honed by much practice. The records in motor memory are thus much more closely defined that those in perceptual memory. Using the former to help deal with the latter is a canny strategy.

The extensive connections of the neocortex provide an enormous potential for large and complex circuits, with many possibilities for recycling a current back to points it has passed before. I envisage foreground consciousness as a flow of activation snaking through the neocortex along many bifurcating pathways in a constantly evolving pattern, fuelled by inputs from lower brain areas. Many sectors of pathway are covered repeatedly as a perception evolves, but there may be modifications to the circuit at every pass. Maintenance of the current is facilitated by a temporary co-ordination of the larger-scale rhythms in all the areas involved. Well-synchronised and strongly activated pathways equate with the focus of attention, while less synchronised rhythms elsewhere provide the unattended and sometimes unconscious background to the main sensory experience.

The reiteration of the conversations between cortical sensory areas and schema-areas means that information originating from many different sources can be assembled into a coherent whole, even if the processing of input takes a little longer in some pathways than in others - as in the magnocellular, parvocellular and koniocellular pathways from retina to visual cortex. Consciousness, I suggested earlier, is the answer to the philosophers' 'binding problem'. It's what makes it practicable to have so many and such extended sensory processing pathways. Their contributions can be brought together over the considerable number of milliseconds during which the conscious moment is created. Thus quite complex correlations can be discovered.

On a slightly longer timescale, the short-term memory provided by the sustained circulation of a large proportion of the impulses is substantial enough to reveal the pattern in a succession of inputs, and to allow components of the pattern to be evaluated in the context of the whole. The bigger the circuit, the longer the circulation is likely to be kept going. And the larger the chunk of data that can be fitted into one perception, the greater the potential variety of percepts.

The reiterated circulation of impulses around the extended circuits also serves to reduce the danger that the elaborate processing which produces fine discriminations might sometimes lead to unreliable or confused results. In the case of moving or evolving stimulus patterns the interpretation of each component can be influenced by the context of the whole. In the case of static ones the information extracted from the data undergoes multiple rechecking, and the interpretation can be refined or altered before action is undertaken. There's time, too, for correlating the input of different senses.

For emerging mammals, operating in the dark, it was no doubt quite practical to delay action while triple-checking such information as was available, and very worthwhile. There were probably no nocturnal predators to threaten them, and when some did evolve they would have to depend heavily on olfaction and hearing to locate their prey. A prey animal could afford to freeze until it had sorted out where the threat was located, and which direction was best for escape.

In short, consciousness functions, as Giulio Tononi and others have proposed, to correlate all sorts of disparate information, information which may be acquired and processed over a period of time. The correlations have to be established through experience, of course, which is why we can make a reasonable guess as to just which sorts of information a species is equipped to correlate in this way by observing what sort of exploration and experimentation the young undertake. In humans, meanwhile, the expansion of the final stages in the cortical processing machinery, at the very front of the frontal lobe, has extended the reach of consciousness and made abstract thought possible.

The major divisions of the neocortex are determined by axons arriving from the thalamus and perhaps other subcortical areas. The structure built on these foundations is to a large extent self-shaping. Neurons are produced in exuberant quantity, and those that find no employment die away. And the nodes of the neocortical webs are heavily shaped by experience, thanks to the plasticity of the synaptic connections. The evidence suggests that in the newborn infant a great deal of the neocortex is activated by any sensory input, and the interconnections are gradually pared down over years.

This may be true even of early stages of the sensory pathways. Some people enjoy a condition known as synaesthesia, in which some sensory inputs produce, as well as the usual experience, an additional one. For instance words or numbers or tunes may regularly evoke sensations of particular colours. In some of these individuals this idiosyncratic form of perception is lost as they grow up, in others it persists. One theory is that synaesthesia may be the norm in infancy. Perhaps the excitations from sensory inputs initially spread widely, but then the excitatory synapses that multiply profusely in the first year of life are pruned and inhibitory synapses are potentiated. The infant brain somehow discovers which links in the network produce useful predictions and which don't, and the synaesthesic effects disappear. But maybe some individuals find such reward in a world with extra colour, or other associations, that some links escape the editing process.

An important footnote here is that such a self-shaping structure is only likely to evolve in species where infants are nourished and protected by their parents, and don't initially need to use their brains for the vital purposes of finding food and avoiding danger. That provides an opportunity for educating the neocortex through experimentation and play. A potential for extending the shaping process over a longer period arose where parental protection came to last beyond the feeding, and especially where a social lifestyle added to the security.

Olfaction as a probable pioneer among senses

The shape of some of the most ancient fossilised fish skulls suggests that olfactory bulbs took up a good deal of space, implying that there was already the means of registering a fair variety of odorants. It looks, therefore, as if the means of collecting a quite extensive range of information may have been acquired earlier in olfaction than in the other distance senses, which seems quite probable. Odours can be intrinsically informative, so olfactory powers could be expanded simply by the evolution of an increasingly large variety of receptors. In extant olfactory specialists such as rats the number of different types is around a thousand, in dogs it's greater, and even we humans possess four or five hundred - in strong contrast to the limited repertoire of photoreceptors or mechanoreceptors possessed by any one species. The size of the olfactory bulb probably reflects the number of different receptor types, so we can deduce that many early fishes had a decent selection. Early tetrapods are unlikely to have been worse off.

With such a variety of receptors there would be less need for the correlations and comparisons required to extract better data from the inputs of just a few types of visual receptors. The earliest odorant receptors are likely to have been tuned to the things a species most needed to know about - conspecifics and food. Receptors responsive to odorants regularly associated with food, rather than emitted by the food itself, were probably then added. All this input was worth mining for subtler information, however, and the three-layered olfactory cortex, at the front of the brain, close to the olfactory apparatus and the olfactory bulbs, was well developed long before the neocortex appeared. When that new bit of telencephalon was added it probably elaborated on techniques that had first been applied to odours.

An olfactory receptor is activated when an odorant molecule keys in to a docking site. For some receptors there's only one odorant that fits, for others there may be several bearing the right sort of projection. To complicate things, some odorants can activate more than one type of receptor, but with different degrees of efficiency. Consequently some odours may be securely differentiated only by the varying proportions in which the same set of receptors is activated. In these cases the input levels of different types of receptors must be compared.

To add to the challenge a significant aspect of olfaction in tetrapods is the way the odorants are drawn across the olfactory epithelium with the intake of air. Encounters between odorant molecules and the receptors they fit must be somewhat random, on the timescale relevant to neuronal excitations. Moreover some odorant molecules engage with their receptors more rapidly than others. In rodents the different types of receptors seem to be located along the length of the epithelium in such a way as to allow for the latter factor, but it seems likely that chance still plays a role in the timing of excitations. The solution is to collect up the input over a brief period. This must be the only way to identify an odour that activates several types of receptor to varying degrees, and to establish at the same time just how intense it is.

The axons of the receptor neurons run to the olfactory bulbs, terminating in structures called glomeruli, each of which receives inputs from only one type of receptor. The overall pattern of activation develops over the course of an inhalation, reflecting both the temporal pattern of receptor activation and the increasing amount of inter-glomerulus inhibition which is triggered as the levels of excitation grow stronger. The initial pattern of responses may be common to several different odorants but the final pattern usually seems to be unique to the odour currently being inhaled.

So in effect the inputs over the whole intake of breath are compared, and treated as a unit. Tests indicate that a rat can recognise a familiar and easy odour before a whole sniff's worth of input has had time to register, but dealing with a more complex odour or an unfamiliar one takes longer. For discriminating between similar odours it's necessary to wait for the later-registering components, and weigh the relative contributions. In short the olfactory system uses a chunking system, very much like that of the cortical part of the visual system, in which the inputs gathered over a short space of time are compared. There is in effect an ultra-short-term memory, which can winnow more reliable information out of the input than would otherwise be possible. Some learning contributes to the process, for there is a degree of plasticity in the synapses of the excitatory and inhibitory neurons which weave among the glomeruli.

The evolving pattern of responses in the olfactory bulb is conveyed to olfactory cortex, the largest sector of which is piriform cortex. Here the terminals of the glomerular axons may spread themselves out over wide areas, and a piriform pyramidal cell may receive inputs from glomeruli connected to a few or many different receptor types. Imaging experiments with humans show that cells grouped together in the piriform cortex tend to deal with odours that are subjectively rated as similar. Moreover a piriform cell which is excited by a combination of odorants may not respond at all to just one of the components on its own, which correlates with the fact that quite different conscious experiences are produced by such inputs. On the other hand the neuron may be quite tolerant if a minor component of a complex odour is missing - which echoes our ability to recognise a whole visual stimulus pattern when part of it is concealed.

Piriform responses are modified by learning, and fine discriminations can be acquired when odours are predictive of noxious inputs or potential rewards. When two odours of very similar composition were regularly presented to rats, one indicating a reward on the right, the other on the left, the pattern of responses in piriform cortex became more differentiated as the animals learnt the trick of gaining the reward. When circumstances changed and there was nothing to be gained by distinguishing between the odours the piriform responses to both became similar again.

Piriform cortex, it seems, is where odours are classified by significance, rather than by molecular structure, and where significance can be determined by experience. Although it only has three layers rather than six it thus operates on the same essential principle as the association or schema areas of neocortex. Combinations of inputs are classified according to the action that needs to be taken in response to them. Neocortex performs a similar job, with the difference that it's able to give the data more intensive processing before the decision on significance is reached.

There's no way of telling how much of this olfactory processing power was present in distant ancestral species, but it seems reasonable to hypothesise that olfaction was the sense in which an ultra-short-term comparing memory first evolved, and the first to acquire the means of making subtle distinctions between similar patterns of inputs.

It also looks likely to have been the first sense to benefit from a somewhat longer preservation of data. Odours can provide useful contextual information, and probably that was their first use - indicating that it was worth slowing down and turning frequently because food or potential breeding partners were around. The next development would be an ability to pursue an odour to its source by comparing its strength in different places and turning when the signal faded, much as male moths do when searching for the female emitting a sex pheromone. A simple mechanism which prompts a reversal of direction whenever the olfactory input ceases can produce a reasonably efficient search pattern. But clearly the pursuit can be much more efficient if a record of the olfactory input can be maintained for a while, and the search can be continued even if the scent has been temporarily blown away by wind or current.

It should be facilitated even more if the intermittent olfactory inputs can be related to a map of the current environment. Probably the hippocampus evolved close to the olfactory processing system because olfactory information made a major contribution to the earliest form of map. The first maps perhaps received only a limited amount of visual information from the thalamus, just sufficient to relate the animal's position to stationary surroundings. Such maps may have been painted quite richly, however, with odours. The colour would shift, of course, as currents of water or air deflected the stream of odour molecules, which is why some visual information would be needed as an anchor. Keeping a record of the sites of recent olfactory inputs, so that odours could still be pursued if temporarily lost, may have been one of the most valuable functions of the early hippocampus. As vertebrates acquired a hippocampal consciousness odours may have played a big part in it.

Consciousness in non-mammalian species

Most mammalian species must apply consciousness over a narrower range of subjects than we do. Where the neocortex is small we may guess that foreground consciousness is quite limited. Indeed the richness of sensory consciousness must correlate with the complexity of the data that is collected, and the length of the period over which the data can be assembled.

What of non-mammals? Conceivably a hippocampal, geographical consciousness is common to many if not all vertebrates. It may be a talent closely intertwined with the evolution of the backbone and more efficient locomotion, and perhaps a significant factor in the success of vertebrates at the expense of most non-vertebrate chordates.

If the theories about sensory consciousness are correct there's no reason why it shouldn't also happen in brains that are structured quite differently from ours. It might be expected wherever the circulation of impulses around the sensory processing arrangements can be maintained after the sensory input ceases for long enough to allow multiple inputs to be analysed and correlated, and information to be sifted from 'noise'.

In reptiles the telencephalon – rather than mushrooming as the mammalian one did – has been extended with a cluster of nuclei which bulge inwards, into a ventricle, forming what's known as the dorsal ventricular ridge (DVR). In line with this expansion the thalamus grew and increased in complexity, and so did the tectum. Whether that led to sufficiently elaborate circuitry to produce some degree of 'foreground' consciousness we can only guess, but some crocodilians seem pretty intelligent.

Birds have expanded the tectum further, and turned the DVR into a collection of nuclei that probably does a job roughly similar to that of the neocortex, and achieves some of the same ends. Clearly birds not only have well developed cognitive maps, but some at least do things for which sensory consciousness seems to be necessary, such as giving parental care to mobile youngsters. Quite a few of the larger-brained species - parrots and crows, for example - show signs of a considerable and flexible intelligence. It seems reasonable to deduce that there is foreground consciousness here, the amount varying from species to species, as it may be supposed to do in mammals.

The larger cartilaginous fish, meanwhile, have brain-body ratios high enough to put them in the same camp with smaller-brained mammals, and in some bony fish the telencephalon is quite large. It's certainly conceivable that skilled hunters, such as tuna and the larger sharks, might have evolved a form of intelligence similar in principle to that of some tetrapods, though employing differently arranged neuronal machinery. Looking outside our phylum, the octopus brain is composed of two interconnected lobes which could well provide sustainable circuitry - as J. Z. Young proposed. The behaviour of some of the larger species certainly suggests a high level of curiosity, which in turn implies consciousness.

Whatever sensory consciousness other species possess it must vary, of course, according to the sort of sensory information they are equipped to collect, as well as the amount of processing it gets. The olfactory world of a mouse, for instance, must be much richer than ours, the visual one much poorer, drawn with much less detail. How else might the sensory worlds constructed by other species diverge from ours?

Some aspects of visual experience seem to present only limited possibilities for variation. The perception of edges and points and angles is likely to differ only to the extent that the range of orientation-sensitive neurons in primary visual cortex might differ. In this respect many species might have a less detailed perception of the visual world than we do, but the essential ingredients might be expected to be the same. Distinctions between orientations may be less subtle, perhaps, and lines, edges and angles may collapse into textures more readily, especially for those species without foveas. Indeed it's very possible that for some species textures form the whole story.

Colour is another matter. Language works on the assumption of shared perceptions, but there's no way of being sure that we don't all experience colours differently - not just those of us who lack one or other wavelength-sensitive photoreceptor, or have alternative versions of the red receptor. (It's ironic that when philosophers discuss 'qualia' they so often instance the 'redness of red', given that the uniformity of that sensory experience is particularly in question.) No chance, then, of guessing what the colour qualia of another species might be like. If there are species with conscious vision which can detect ultra-violet light but not red, is their colour experience different from ours, or is the same rainbow palette shifted a notch, as it were, to be evoked by the lower range of wavelengths?

Then there are those species which have sensory powers we don't possess, such as sonar. This, like vision and whiskers, is a sense involving active exploration, so cetaceans and bats could be expected to be endowed with a sonar consciousness. Is it quite different from visual consciousness, or is it fused with it, so that these animals 'see' objects detected by sonar in much the same way as those detected by their eyes? The latter seems quite probable, for two blind people who have been using tactile devices as a substitute for vision over many years report that they have come to obtain some degree of visual sensation thereby. The essential function of both vision and sonar is locating things in space. It seems reasonable to suppose that the nature of the conscious sensory experience might be dictated by the use to which the sensory information is put, as much as by the sort of receptor which has provided it. Occipital cortex may perhaps be the cortical sector concerned with depicting space and spatial relationships, rather than simply the visual area.

Pursuing this line of thought, it's conceivable that colour might be a sensation applied to whichever sensory input undergoes the greatest amount of neocortical processing, with the subtlest results. Maybe olfactory specialists such as dogs and rats smell in colour, rather than seeing in colour. A fascinating idea, which perhaps philosophers will like to have fun with.

Could a computer experience consciousness?

This has become a popular question in some quarters. It may be a futile one, for if computers did develop consciousness how would we know? I assume that my fellow humans experience the world in much the same way I do because you seem to be entities constructed along the same principles as me, behaving with similar motivations. I am happy, too, to deduce that other species have some form of consciousness if their brains look something like ours and their behaviour suggests it. But these factors don't apply to computers. The only possibility for checking on whether a computer has consciousness would seem to be to ask it. And how would we know it had understood the question?

But we might ask why a computer would need consciousness. Electronic circucits seem to be somewhat more consistent than neuronal ones, and the input comes in decisive form, so there's no necessity for a comparing memory to clarify it. And nourishment is provided by an outside force, so emotions and motivations and a means of weighing up internal needs are all unnecessary.

The most promising line of research, still at an early stage, would seem to lie with the development of robots which have to learn how to interpret input, how to act, how to get their battery recharged, and how to communicate with each other. But it will still be difficult to determine whether they experience anything parallel to this curious effect produced by numerous interlinked neuronal circuits.

On the other hand it might be argued that in a sense computers already have consciousness. The monitor screen constitutes their form of it, revealing the results of the umpteen calculations going on out of sight as a result of the sensory input received through the keyboard. The difference, it might be said, is that computers use their consciousness to communicate with us, whereas we use ours first and foremost to communicate with ourselves.

Recapitulation

A hazy, hippocampally-based form of consciousness probably evolved first, supporting the construction and use of a cognitive map – first a temporary map relating the animal to its immediate surroundings, then a more permanent and extensive one.

Fullscale, vivid sensory consciousness as we know it must depend on the neocortex. (But something comparable may well be created by other brain areas in some non-mammals.)

Consciousness involves a maintained circulation of impulses through complex circuitry.

It allows sensory evidence to be assembled into meaningful wholes even when the inputs arrive at slightly different times, or when simultaneous inputs are processed at slightly different rates.

The first application of the sensory-processing machinery of the neocortex probably lay in adding detail to the cognitive map, and to cues for approach or avoidance.

As a result, we can hypothesise, such cues were promoted from a vague motivational consciousness to something like the sort of sensory experience that we know.

Subtle ways of combining the evidence from different senses became possible.

The potential for subtle distinctions opened the way to new behaviours, such as maternal care of mobile young.

The olfactory telencephalon was well developed long before neocortex appeared, and olfaction was perhaps the first sensory system to process inputs with a very short-term comparing memory.

Chapter 28

Language

The evolution of language ability was clearly a crucial step in the development of reason. Concrete thinking can be done without language, but abstract ideas are slippery. It's almost essential that the abstract concept acquire a handle, in the form of a shape registered by both a sensory system and a motor system - a shape which can be hooked out of memory much more easily than the formless idea. With the aid of such handles concepts may be woven into new relationships.

Darwin was fully aware of the important relationship between language and thought: *A complex train of thought can no more be carried on without the aid of words, whether spoken or silent, than a long calculation without the use of figures or algebra.*

What's equally important is that an abstract idea can only be reliably conveyed into another mind by means of language. Even pointing to a possible correlation between two events widely separated in time is a challenge without language. It follows that a capacity for complex thought offers no benefit to a species which doesn't have language. If brains with such talents should occasionally appear there will be no selective advantage in them - probably a disadvantage, in that they will be more expensive to maintain. Where ideas can be communicated, on the other hand, at least some of them are likely to bring a benefit to offspring and relatives. Furthermore, communicated ideas have a chance to mate with the ideas they encounter in the receiving mind and breed additional ideas. New thoughts build on old ones, so that ideas are elaborated over many generations.

Language is defined here in the classical and most demanding way - as a system of arbitrary signs, and rules for their combination, which can convey an infinite number of potential messages. It's something which has to be learnt. It requires a heavy investment in neuronal machinery, and in time spent learning, which will only reap a dividend if there are useful messages to be conveyed. So the corollary of the first argument is that there will only be a worthwhile return on such investment where members of a species have useful information or ideas to communicate to each other which can't be conveyed by cheaper methods.

It's a reasonable deduction, therefore, that language appeared only when the hominid brain had reached a certain level of sophistication. And once it had begun to be invented it opened up new potential for further elaboration of that brain. With this new tool available the sort of neocortical expansions that provided scope for more complex forms of thinking became much more likely to prove profitable to the species. The evolution of language ability and of increasingly complex reasoning powers must have been inseparably intertwined.

George Miller made this point firmly: *Some scientists believe language is what it is because it reflects the way the brain works; others believe our brain works the way it does because language has shaped it so. I believe they are both right. Human brains and human language have shaped each other.* Or as Terrence Deacon later put it, language *is so pervasive and inseparable from human intelligence in general that it is difficult to distinguish what aspects of the human intellect have not been shaped by it.*

It doesn't necessarily follow that humans are unique in possessing a capacity for language. In recent decades this assumption has been somewhat undermined. The increasing sophistication of recording equipment and sound analysis techniques has made it possible to study the vocalisations of other species in unprecedented detail. It transpires that their patterns are often more elaborate and more consistent than was realised, modulated in ways that may be inconspicuous to us but must be readily perceptible to conspecifics. Furthermore, distinguishable dialects are found in different areas, indicating that learning is involved. The long and complex songs of dolphins and whales have received particularly extensive study. It isn't clear whether the message they carry is a detailed one. There's no proof so far that other species practise anything which satisfies the strict definition of language. But we can't be sure we're not missing something.

There have also been various attempts to teach language to other species. A grey parrot proved capable of making intelligent use of words, and dolphins turn out to be able to follow quite lengthy instructions. Of the most direct relevance here, however, are the various experiments in teaching human language to apes. These have provoked a fair amount of controversy, but the results indicate that our closest extant cousins have, as might be expected, at least some of the talents required for mastering language. They are capable of making use of words, and of what can be defined as a proto-language. A healthy side-effect is that attention has come to be focussed rather more sharply on just what's involved in language.

Since language must be learnt one essential requirement for developing it is an ability to communicate effectively without it. This is well established in mammals, thanks to those emotional mirror neurons. Infants of many species are extremely responsive to the emotions of their parents. They can also follow the direction in which a parent's head is pointing and discover what has evoked an emotion. But humans are particularly well equipped for this sort of communication in having a white sclera surrounding the pupil, which makes the direction of gaze more conspicuous, a feature that may well have been selectionworthy because of its informational value. A pointing gesture, meanwhile, seems to come instinctively to human infants.

Thanks to these sensitivities, when an adult points and says "Look, a cow" the infant generally attaches the word to the correct object. And when the adult screws up her face into a disgusted expression and says "nasty" the infant readily deduces what nasty means. It helps, of course, that both adult and child are endowed with the same attentional system, so that what one finds salient is also likely to engage the attention of the other. Adults vary, however, in their talents for conveying the meaning of new words, and children whose parents are good at it have larger vocabularies by the time they reach school age.

Another obvious necessity for language is a means of expressing it. A vocal means is generally the most practical, since sound travels to a distance, and around obstacles, and functions in the dark. The receiver of the message, meanwhile, doesn't have to be looking towards the sender, and in most cases neither need they stop what they're doing while the message is communicated. A further advantage is that we've been able to evolve specialised vocal equipment which not only produces a sufficiently wide variety of sounds to make an infinite number of combinations possible, but also produces strings of these units at great speed, and without much expenditure of energy. These modifications, furthermore, have been made without greatly compromising other functions. It seems likely that a gestural language, for instance, would have proved impracticable, because the sort of muscular adaptation necessary to make it as fast and efficient as speech (if indeed it ever could be) would have interfered with the other uses to which hands and arms are put. It can be no accident that the other emitters of prolonged and complex signals, cetaceans and songbirds, also use sound. Humans who need to communicate by sign language, meanwhile, tend to express themselves fairly concisely.

A spoken language demands not only the means to make a goodly range of sounds, but also the ability to distinguish them, as a hearer. Experiments suggest that even some non-primates can learn to recognise some of the fine distinctions that segregate many of our speech sounds. The inference is that this sort of perceptual ability was established long before new productive ability.

Producing the sounds that are perceived depends on a talent for vocal imitation, an ability to compare the sounds the self makes with the models being imitated, and to modify them if they aren't coming out right. In other words it's necessary to be able to use auditory input as feedback about the results of vocal efforts, just as visual input is used to evaluate the results of other forms of muscle use. The auditory equivalent of mirror neurons must have been established for some time, since monkeys which move to a new group readily acquire the group accent. It's been demonstrated, moreover, that young marmosets gradually adapt the sounds they utter to match those produced by their parents. Similar machinery for imitating vocalisations has evolved in songbirds, and bottlenose dolphins too are extremely good at imitating sounds.

Another vital talent that must be present before language can be developed is the ability to conceive of a recipient who will be influenced in some way by a message. Social primates which have grown up playing with other youngsters, and discovered in the process that their actions influence the actions of their fellows, certainly possess this talent. Requests for grooming, for instance, are plainly intentional communications, made by choice, and with the idea of achieving an effect on the addressee.

A fully effective use of language requires something more, however - the ability to conceive of the listener as a conscious being more or less like oneself. To use the current jargon, the speaker must possess a Theory of Mind. This is the social intelligence that also allows the practice of deceit and social manipulation. As we've seen, there's ample evidence of the practice of deceit among apes, and there are indications that it's also present in some monkeys.

In the young child language obviously helps the theory of mind to develop much further than it otherwise could, offering as it does the means of communicating feelings and sharing percepts. And with age and experience children work out that it pays to imagine themselves in the circumstances of the listener and consider just how much information they need to communicate. This is clearly one of those abilities that require neuronal circuitry sufficiently extended to hold two things in mind simultaneously, in this case both the message and the receiver at whom it is aimed.

There are, then, some fundamentals that must be in place before language is possible. Now we can turn to some of the more centre-stage essentials - the ability to handle words, and to put them together in a meaningful and rule-ordered way.

Words

A proper appreciation of words entails recognising that they are purely arbitrary signifiers, with no inherent connection (other than the occasional example of onomatopoiea) to what they signify. The sound which is a noun must therefore be associated with the object it represents in a different way from the sound the object can be expected to produce. In fact the relationship of a word to its referent is rather like that of a pretend fight to the real thing, and it's equally important to know the difference. The word *lion*, for instance, is linked both to the animal's appearance and to its roar, but not in the same way that appearance and roar are linked to each other. One suggests a strong possibility of seeing the animal, but this expectation is not necessarily aroused by hearing the word. Admittedly infants sometimes seem to attribute magic qualities to words, perhaps because uttering a desire often leads to its fulfilment. The fact that magic spells usually involve words suggests that the idea is not always wholly lost. But once we've learnt to appreciate words properly we know that one of their defining characteristics is the potential they provide for referring to things which are not present.

Another important aspect of words is that they function to categorise experience, rather than representing it precisely. Just as some face neurons in inferotemporal cortex fire for any face, so does the noun *face* cover all manifestations of physiognomies. A great many nouns are more concerned with use or purpose or effect than with sensory similarities, providing a way of classifying percepts according to what can be expected of them. Verbs, adverbs and adjectives are to an even greater extent generalisations of experience, categorising actions and qualities that are demonstrated or exhibited by many different entities.

It's often been observed that the categories that our mother tongue happens to emphasise tend to influence our perceptions. Benjamin Whorf argued that the implications of our vocabulary may on occasion function like blinkers, giving us a dangerously limited view. But at the same time, as various philosophers and psychologists have emphasised, it's one of the great strengths of language that it imposes classifications, reducing the infinite variety of the perceptual world to manageable proportions. It can do this, of course, because our perceptual systems already categorise experience into perceptual schemata. So the word can both represent the schema and have a strong influence on its shaping. Even infants as young as nine months seem to learn to discriminate between two stimulus patterns more readily if they're associated with different verbal labels.

The reason words function differently from cues, and the reason they work so well to organise perceptual experience, must be that they are not only percepts but also actions. We must learn not only to recognise them but also to produce them. Motor memory is a different sort of thing from perceptual memory, as I've already observed. For the most part we don't need to learn too precisely what things look or sound like, as long as we know what sort of response is needed. It's more valuable to have perceptual schemata that allow us to record the essential patterns in sensory input, so that we're not confused by lighting effects, or acoustic effects, or other accidentals that have nothing to do with identity. It's important to be able to recognise, but seldom necessary to recall with absolute precision. Movements, on the other hand, do generally need to be reproduced pretty exactly, and this is particularly true of the very delicate and elaborate muscular deployments that produce speech. So learning a word links a concept that is usually, and appropriately, a bit fuzzy to a finite and reproducible action.

All four species of great apes seem to be able to cope with the concept of words. At least the more intelligent individuals among them do - just how variable the potential for such learning may be in these species, where there has been no selective pressure for it, is not yet clear. The picture was initially confused by the fact that it took time to establish that apes learn best when they begin young, and when they are maintained in a supportive environment, with company, and have a friendly, ongoing relationship with their instructors, so that communicating with humans is rewarding. In other words they need pretty much the same circumstances that produce results in human youngsters.

An assortment of chimpanzees, bonobos, gorillas and orangutans have been taught American Sign Language, as used by the deaf. Washoe, a chimpanzee, was the first ape to learn ASL, under the direction of Beatrice and Allen Gardner, and subsequently in a project headed by Roger Fouts. Two gorillas, a female called Koko and a male, have also acquired ASL, in a study led by Francine Patterson. Under the aegis of Sue Savage-Rumbaugh, meanwhile, chimpanzees and bonobos have learnt to use a language of 'lexigrams'. These word-signs can be assembled into a message on a computer screen, by pressing lexigram-labelled keys; or an ape can communicate by pointing to the signs displayed on a portable board, which has become the preferred method. The apes often combine the use of these lexigrams with gestures. Both systems have their advantages. Ape hands don't shape ASL signs as dextrously as human ones, so lexigrams provide the clearer evidence of appropriate use of words, and when the computer system is used the utterances can be automatically recorded. But using the lexigram display is more awkward than signing.

The most talented apes have proved capable of amassing a vocabulary of several hundred signs. In the case of ASL some signs have elements of mime, but they are far from transparent, while many are necessarily arbitrary. The lexigrams have been carefully designed to carry no clue to their meaning. The apes must surely appreciate the arbitrary nature of the relationship between sign and meaning, moreover, since the lexigram users have sometimes been allowed to choose the signifier for a newly introduced object, and the signers have on occasion invented their own sign combinations.

Apes, moreover, seem to take readily to words such as *fruit* or *tools*, words which organise percepts into rather more general classes, with only limited sensory similarities. So it seems that their words really do function as categorisers. Moreover when they make mistakes the incorrect word is likely to belong to the correct category, or to have a similar shape - just as when humans make slips-of-the-tongue. Words are clearly appreciated as a means of communication, for the apes use signs among themselves, and often initiate conversation with humans. Some apes, meanwhile, have been seen using words as an *aide-memoire*. The gorilla Koko, for instance, signed *quiet, quiet,* when she thought she wasn't being watched and was doing something that wasn't allowed.

Infants born or adopted into the language-using groups acquire signs or lexigrams by imitation, and also get occasional instruction from mother or fostermother. The experimenters carefully refrained from signing in front of Washoe's son Loulis when he was young, apart from seven signs which were felt to be indispensable. Over the first eighteen months Loulis learnt twenty four signs by imitating other apes, and never imitated those the humans used.

Some complementary evidence, meanwhile, comes from a chimpanzee called Sarah, who was not part of the mainstream of language experiments. David and Ann Premack aimed not so much to investigate apish language potential as to teach language to chimpanzees by means of positive reinforcements, and use it to discover what sort of concepts the chimpanzee mind could encompass. The words were coloured plastic shapes, arranged in an order roughly equivalent to English grammar. Sarah first learnt to make requests, which were fulfilled if properly phrased. Once she could do this easily she was taught more words, and learnt to replace *?* in a sentence by the correct sign, so as to create statements such as *red colour of apple* and *round shape of apple*. She appropriately applied colour signs (all of a different colour from that they signified) to many object words, and adjectives like large and small, round, square, few and many. She also mastered quite a variety of sentence forms, including an if-then construction. Her introduction to this was the powerful proposition *If Sarah take apple then Mary give chocolate Sarah, if Sarah take banana then Mary no give chocolate Sarah.* In time she graduated to proposals such as *If red on green then Sarah take apple, if green on red then Sarah take banana.* The Premacks thus augmented the results of the other ape-language projects by demonstrating the conceptual abilities of chimpanzees more definitively.

Kanzi and other bonobos and chimpanzees who have been exposed to human speech from an early age have acquired a fair degree of comprehension of spoken language. Kanzi was still quite small when Sue Savage-Rumbaugh and his other human companions discovered they had to spell out words they didn't want him to understand. Furthermore, some of the bonobos have begun to produce untypical sounds which may be attempts at speech - sounds which acoustic analysis shows to be structurally distinct, which can be distinguished by humans who have been extensively exposed to them, and which appear to be used in a meaningful fashion.

This observation seems less startling after a careful study of the vocalisations of two captive populations of chimpanzees which were brought together in a zoo. Chimpanzees have a selection of grunts, uttered at the sight of different foods. Initially the vocabularies of the two groups differed, but over the course of three years, as social relationships developed, the immigrants gradually modified their food grunts to resemble those of the established inhabitants.

Observation in the field, meanwhile, established that monkeys too use signals that come close to functioning like words. Dorothy Cheyney and Robert Seyfarth observed that vervet monkeys have different danger calls. One is made when an eagle is sighted: it prompts hearers on the exposed outer branches of trees to retreat downwards. Another is made for a leopard, and causes monkeys on the ground to climb into the trees. A third type of call is made when a snake is spotted, and leads the whole group to pay close attention to the threat until it moves away. Young juveniles sometimes make the wrong call, or make the eagle call to the wrong sort of bird, which suggests that these vocalisations must be learnt, rather than instinctive. (Chickens have varied alarm calls too, but they make them appropriately even if reared in isolation.)

Similarly varied sets of alarm calls have been observed in several other monkey species. Adult male Campbell's monkeys have at least six different calls, which they string together in various ways to augment the amount of information conveyed. The presence of a predator, for instance, may be indicated by three or four different sequences, indicating not only what sort of predator but also how the caller has learnt about it.

Rhesus monkeys in Puerto Rico have five acoustically distinctive calls which are made when they find food. Three of them are produced by individuals who find high-quality, rare treats, the other two indicate more standard food items. The latter sometimes occur in other contexts as well. Experiments indicate that monkeys habituate to these meaningful calls on a semantic rather than an acoustic basis. A rhesus monkey which has already reacted to one sort of *super food* call is unlikely to react again if it's followed by one of the other vocalisations that has a similar import, and even less likely to pay any attention to a report of ordinary, less interesting food.

Similarly, a diana monkey which has taken the appropriate action in response to an alarm call doesn't taken any notice of a similar alarm five minutes later uttered by a different voice, but does react to a different type of alarm. Where several species of monkey share the same habitat they may respond to each others' alarm calls.

Another sort of call found in many monkey species is a request for help when attacked by another group member. And it's been realised that what were at first thought to be standardised vocalisations can convey slightly different information by means of subtle variations of intonation. A young monkey's call may vary according to the social standing of the attacker, and presumably monkey hearers can appreciate the implication.

There could be straightforward explanations behind all this. Perhaps the variation in calls for support reflects the degree of fear, since individuals of higher rank are likely to represent a bigger threat. In the case of alarm calls about predators it might be that the visual percept simply functions as a cue for the call, and the call as a cue for action. Hearing and imitating varied alarm calls could merely focus the attention of infants on potential dangers, and contribute to teaching them the appropriate reaction. Even seen in this light, however, these vocalisations constitute a step on the way to learnt utterances which can be divorced from action, and which do signify categorisations. Monkey alarm calls may have been established by accident, as it were, the call becoming attached to the particular threat because it happened to be applied to it several times in succession, or because calls made while looking up tend to come out differently from calls made while looking down. Real words must be defined as words which are consciously invented, and which are recognised, at least by adults, as having been invented. But the revolutionary innovation which was the purposeful invention of the first word must have occurred in a mind prepared for it by the use of learnt signals. Equally important, comprehension of this radical new idea is only likely in an audience already used to using learnt signals.

Syntax

Words become much more useful if they can be strung together. A single-word utterance can ensure that everyone's attention is directed at the same thing, or that the whole group knows what the current goal is. Yelling *snake* is more helpful than simply screaming. But for the most part, creating a message that couldn't be fairly well conveyed by a less elaborate vocalisation and a gesture tends to require two or three words. For this sort of communication grammatical rules aren't really necessary. If the ideas to be conveyed are simple and there are relatively few units of vocabulary to play with, the chances of ambiguity are limited, especially when there is context, facial expression and body language to provide extra clues. It's when the messages begin to get more varied and more complicated that it's necessary to have conventions about how the words are strung together, if reliable transmission is to be achieved.

Assessing our ape cousins' capacity for grammar is a little tricky. In spontaneous conversation the typical utterance tends to be only two to four words long. But this may be because sentences of two or three words are sufficient to say what the apes want to say. With a vocabulary of only a few hundred words there's a limit to the number of things that can be said, and the vocabularies that have been taught to the animals seem to be skewed rather heavily toward content words, especially nouns, which limits the possibilities for complex sentence construction. It's probably also relevant that both sign language and lexigrams are clumsy means of communication compared with speech, and encourage parsimony. Humans using the lexigram board construct the briefest possible messages. However, apes that have become thoroughly practised in the use of words do occasionally put together strings of up to about seven different signs.

Not all apes have caught on to the idea of imitating the word order rules used by their human tutors, but of course the need for fairly firm rules of grammar only becomes obvious when the creator of a message realises that the recipient may interpret it in the wrong sense. Apes are quite capable of taking the perspective of their conversational partners into account in the sense of allowing for a different state of knowledge - for what the partner hasn't seen or can't see, for instance. But they can hardly be expected to allow for the infinite number of utterances a human can produce, and the consequent scope for ambiguity without grammar. It must also be said that although ape sentences tend to sound clumsy when translated into English, they are often acceptable as American Sign Language, which has a different sort of grammar.

A chimpanzee called Ally grasped word order rules very easily, and Washoe generally put subject before verb. The difference between *me tickle you* and *you tickle me* seems to be readily mastered. But apes may develop their own grammatical habits. The ASL signers sometimes, for instance, form a verb sign against the body of the person they want to perform the action, often on the part of the body that is relevant (young human signers sometimes do this too). Chimpanzees using the ASL sign for *that* to express a question hold the sign for longer than when making a statement, and maintain eye contact for longer. This no doubt springs from the fact that they are looking for an answer, but it still serves as a convenient grammatical device.

Some individuals, meanwhile, tend to string signs or lexigrams together in a consistent pattern, even if the pattern is one of their own creation. The bonobo Kanzi, one of the most talented and enthusiastic language-users, usually puts action before object. He also combines lexigrams with gestures, conveying the action he would like performed by lexigram and the person he would like to perform it by gesture. When he does this he puts the lexigram first, even if he has to cross the room to get to the lexigram board.

Moreover at least some of the language-using apes have some sensitivity to the syntax of spoken English. Kanzi has been extensively tested on his ability to carry out spoken instructions, many of them containing subordinate clauses, and many of them bizarre enough to rule out the possibility of guessing what was required. His performances are successful enough to imply at least some comprehension of English grammar. However, the videotapes sometimes give the impression that he doesn't find the syntax all that easy to decipher. There's room for a suspicion that perhaps he understands the content words and has to guess at how they fit together. Sue Savage-Rumbaugh noticed that Kanzi responded readily to the instruction to *get the ball that's in the group room*, but hesitated when the phrasing was *go to the group room and get the ball*, and eyed the ball that was already in the same room with him. She wondered whether he might appreciate the clearer grammatical construction provided by the dependent clause. But there's an alternative possibility. It could be that Kanzi only registers *get ball group-room* and *go group-room get ball*. The former may be readily interpreted as a single instruction while the latter could conceivably intended as two separate instructions, of which the second could conveniently be carried out first.

The example of Sarah and her plastic signs indicates that chimpanzees are capable of handling relatively complex concatenations of concepts, even if they don't produce lengthy utterances spontaneously. It may be significant, however, that Sarah used a sign language which equates with reading rather than talking, one that remains available for study while the sentences are studied. It seems plausible that the capacity of short-term memory might be a significant factor in the construction of apish sentences. The working memory of the chimpanzee may well be smaller than that of the human, so far as vocalisations are concerned.

At any rate, the experiments with apes indicate that these cousins of ours can make good use of words, and can put a few together in a productive fashion. I think this was entirely predictable. The capacity for using language is something which has to evolve, but language itself has to be invented. Clearly words can only be invented by a brain that already has some capacity to use them. The talents which underlie language are likely to have been around for quite a while before inspiration dawned. The talents needed for language - the ability to communicate by means of gesture and facial expression, to observe where a conspecific's attention is directed, to imitate, to recognise a vocalisation as a warning rather than merely responding to a conditioned stimulus - are valuable in themselves. And an animal creative enough to use a tool and carry out an organised sequence of planned actions is likely to be capable of putting at least two or three words together. It seems perfectly reasonable to suppose that an animal with a brain not significantly bigger than an ape's might have happened to invent the idea of words, and a proto-language.

Only when the invention has occurred can natural selection operate on the brain's capacity for using it. Only then does the potential for thoughts that require language become selection-worthy, only then does it become worthwhile to be able to construct more elaborate grammars. Then the capacity for language and the capacity for thought begin their intertwined evolution.

Young children begin producing single word utterances at around twelve months, and expand their vocabulary fairly slowly until about eighteen months. Then they begin putting two words together, cautiously at first, but soon with an intonation that imples a unified construction. At this point vocabulary begins to grow dramatically. Then there's a rather sudden progression to sentences of several words, which are generally sequenced according to the conventions of the heard language. Between two and five years old many grammatical rules are mastered - the ones which are most important to making oneself understood. In English that means word-order habits come first, followed by the rules on modifying word forms. At a certain point these are applied with rather too much enthusiasm. Although the child may have been using words like *mice* and *brought* quite correctly they start saying *mouses* and *bringed*, and persist for a while, until they come to terms with the fact that the rules they have triumphantly worked out have exceptions.

The subtler rules of grammar are acquired more slowly, and some of the more ambiguous constructions are misunderstood for some time. Sentences in which the chronological order of actions is reversed may give trouble up to the age of ten or so. *He did X before he did Y* may be understood correctly, but not *Before he did Y he did X*. The passive voice can also cause problems. Perhaps we might conclude that conveying essential meaning has to become easy before shifts of emphasis can be mastered.

While utterances are very short the business of arranging object-words, action-words and descriptive words in a certain way is probably a habit, managed like other habits. We put the subject of a sentence before the verb (or *vice versa*) just as we regularly put the milk in the cup before the tea (or *vice versa*). Human infants, equipped with plenty of vocal mirror neurons and excellent echoic memories, acquire the same habits as the models they are imitating, and do it pretty rapidly. Apes learning an artificial language are more likely to develop their own habits - perhaps because dealing with words takes up all of working memory, leaving little spare capacity for noticing the word order habitually employed by the human models, except when using the wrong word order brings the wrong results. Their tool use shows they can order a modest sequence of actions, but there's a logic about the sequence of actions in tool use that doesn't apply to the early stages of language.

Broca's area, the site strongly associated with grammar in the human brain, is part of a larger region in the left cortical hemisphere involved in other sorts of sequence organising. Highlighted in the human brain by being rather larger than the corresponding part of the right hemisphere it's also a little larger in apes, and to a lesser degree in at least some monkey species. In both apes and monkeys Broca's area, and Wernicke's too, are activated by the vocalisations of conspecifics. It seems, then, that even monkeys have some of the neuronal machinery necessary for the foundations of language. In apes it has been elaborated sufficiently to cope with a proto-language. But a considerable expansion of working memory, and a major elaboration of both the machinery for recording patterns of input and that for producing finely crafted patterns of vocal action must have been required for complex syntax, with extensively nested structures.

Once grammar was developed it would not only have been valuable in itself but would also have provided clues for guessing the meaning of new words, and the means of explaining them effectively. Then language could really take off.

Is there a Universal Grammar?

Whatever the neuronal machinery of language it has evolved to a luxuriant degree, for grammars are extraordinarily diverse. One very basic division is that some depend heavily on word-order rules, some on morphological rules, while others, like English, use a combination of the two. Edward Sapir distinguished four main types. Chinese is an example of what he termed the isolating type. It's an essentially monosyllabic language, and totally uninflected. Considerations such as tense or plurality are indicated only if necessary, and conveyed by separate words. In Chinese, as in English, a word commonly used as a noun can be put to work as a verb or an adjective, but whereas in English this often involves an alteration to the form, in Chinese it requires only a change of tone, if that. Consequently Chinese philologists have never felt the need to label parts of speech beyond distinguishing 'full' words from 'empty' words. Full words are the ones we call nouns, verbs, adverbs and adjectives, a class referred to as *contentives* by English-speaking philologists, while empty words are *function words,* which serve to link the contentives together.

Most modern European languages are of what Sapir called the weakly synthesising type. They modify content words with suffixes and prefixes and vowel changes, to indicate things like number and tense, aspects of the message which it therefore becomes obligatory to express. The strongly synthesising languages are those like Latin, which vary the content words much more extensively, and express many of the relationships between them in this way, a system which means that word order can be more variable, though not entirely free. But even in Chinese, where content words can function so flexibly, word order is never absolutely fixed. As in English, it can be inverted for the sake of emphasis, or poetic effect, wherever context is sufficient to make the meaning clear.

The fourth type of language Sapir called polysynthetic; this type is found in New Guinea, Australia, Siberia, India, and among native American peoples. Here the basic units of meaning can be half-syllables or even small collections of consonants that are hardly pronounceable on their own. Speakers of such languages and the philologists who studied them seemed to agree that the strings into which such elements are assembled are indivisible, although several words are required to translate them into English. In English some small function words can be abbreviated, if there is no emphasis on them, to a form that cannot stand on its own - forms such as *'s* for is and *n't* for not. But in these languages, it seems, even content words may start out very slender and in need of support. As an example, here's a sentence in Chinook, recorded by Frank Boas: *Ania'lot*, which was translated as *I give him to her.* Boas analyses this as follows: *a* (tense), *n* I, *i* him, *a* her, *l* to, *o* (direction away), *t* give.

Conceivably these architectures aren't quite so varied as appears at first. Polysynthetic languages mostly don't have written forms, and it seems possible that their speakers, with probably a limited number of things to say, don't feel the need to separate their utterances into distinct words. Perhaps they are more closely related to Chinese than might be supposed.

Back in the nineteen fifties the philosopher Noam Chomsky revived and expanded an idea first put forward by French grammarians in the eighteenth century. Not only, he proposed, is language underpinned by specific, genetically determined neuronal machinery of some sort (refuting the Behaviourist B. F. Skinner, who had asserted that language was learnt simply as a result of conventional reinforcements) but there must be an underlying Universal Grammar which governs the way language can be constructed. No actual language corresponds to this Grammar. Rather, as Ray Jackendoff put it, it's *the prespecification in the brain that permits the learning of language to take place.* But the ways in which the Universal Grammar can translate into actual grammar are limited, and the infant learning the structures of his language therefore only has to choose between limited possibilities. The idea aroused a lot of interest at the time, and inspired some interesting research. But the universal laws suggested at various times by cognitive scientists have generally fallen by the wayside as an increasing number of obscure and previously unknown tongues have been investigated.

One of Chomsky's arguments for the universal grammar was that language is a complex business, and yet children manage to learn it rapidly. Terrence Deacon put a somewhat different perspective on this idea. He pointed out that learning a first language may appear to be a difficult task, but in fact it must be easy. If a language is to remain in existence it must be learnt, and since there is a critical period for acquiring a first language it must be suited to being learnt by children. Languages are like species, in that they must be suited to their habitat, and in this case the habitat is the human brain. Also like species, languages evolve, and are subject to natural selection. New usages that don't fit in with the perspectives and assumptions of most users won't last, while those that meet a generally recognised need will spread. Meanwhile, since we benefit from possessing language, the brain has also evolved in ways that suit language. Some ingenious experiments involving artificial words passed on through several 'generations' of subjects have supported Deacon's ideas.

An investigation inspired by Chomsky's ideas showed that children develop viable grammars of their own accord. Over three or four centuries of colonial expansion various European powers took people from their native countries and shipped them off to fertile but sparsely occupied bits of the globe to work on plantations, often as slave labour. Many unfortunates thus found themselves in countries where their own language was unknown, obliged to try and communicate with the colonialists, with native populations who spoke a third language, and often with other immigrants from elsewhere. Adults exposed to new languages under these conditions acquired vocabularies, of a sort, but no grammar, and came to speak a very debased form of the colonialists' language, something known as a pidgin. The children of the migrants, however, growing up hearing a pidgin, regularly managed to impose their own grammar on it and turn it into a more expressive form of language. This second-generation development of a pidgin is termed a creole.

Derek Bickerton found some remarkable similarities among the grammatical structures of all the extant creoles that had been studied. Certain rules about what must be expressed and what can be omitted were common to all of them. For instance creoles don't require tense to be indicated in most cases, but they all require that if an action might be either finite or ongoing then it must be specified which. The distinctions that are always expressed are also those which children learning language pick up first, and seem to find particularly easy. They must be those most salient to the young mind.

Evidence that nicely parallels that of creoles comes from deaf children. In 1979 the new Sandinista govenment in Nicaragua introduced proper education for deaf children, who until then had generally acquired only a few essential signs for communicating with hearing relatives. Brought together in school, but taught by hearing teachers who had no sign language, the children rapidly developed an expanding sign vocabulary, but used it in pidgin fashion. Subsequent generations, joining the school and acquiring this vocabulary at younger ages, developed a grammatical system and turned the pidgin into an expressive language. The evolution of this language was studied in detail from 1985 onwards. An interesting feature is that information can be split up into small units which can be combined in different ways, just like spoken words with their potential for infinite recombinations, whereas the same information conveyed by someone who is gesturing as an accompaniment to speech is usually rolled up into one mimetic movement.

The Nicaraguan children didn't have to conceive of the possibilities of grammar for themselves, for they were being taught to read and write. The means of applying it to the sign language, on the other hand, was their own invention. A study of very young deaf children who were developing their own gestural languages also indicates that the idea of putting words together in a consistent fashion comes naturally. Four children of hearing, American-speaking parents in the USA and four children of Chinese-speaking parents in Taiwan were observed at length, twice, between the ages of 3 years 8 months and 4 years 11 months. Combining gestures was common, and the way in which verbs were combined with their subjects and objects were pretty consistent. The actor of a transitive verb was often omitted; its object, or the actor of an intransitive verb, much less frequently. And there was more similarity in the gestural communications of the children than there was between their signing and that of their mothers. All this shows that young children like to order their words in a regular fashion, and feel a need for some sorts of information to be expressed but are prepared to deduce other sorts for themselves. It scarcely adds up to a universal grammar.

Grammars must certainly reflect something about the way the human mind works, which is why Chomsky thought the subject interesting. *By studying lan-*

guage we may discover abstract principles that govern its structure and use, principles that are universal by biological necessity and not mere historical accident, that derive from mental characteristics of the species. This is a nice idea, but it might perhaps be easier to go direct to the universal semantic. We humans all have similar perceptual systems, similar ways of dividing up the perceptual world, similar capacities for deducing relationships, and similar essential interests. What one person says, therefore, another is quite likely to guess the meaning of, even if the syntactical expression is novel.

Certainly semantics can always overrule syntax, if there is a conflict. We expect utterances to make sense, and when we're faced with a statement which must be either semantic nonsense or syntactically incorrect we assume that the mistake lies in the syntax. Nonsense can make us smile - as long as it's presented as nonsense. But meaningless utterances presented in a serious manner make us very uncomfortable, so we look hard for meaning.

In spoken language, moreover, grammar often gets a pretty rough ride. We don't bother with constructing a sentence if a word or two will convey the required meaning. And for much of the time we depend heavily on context - both that provided by the conversation and that of the place where it's happening. Phrasing and intonation also do a great deal to clarify the message. Even when context and semiotics are lacking it's often possible to distort conventional syntax to a considerable degree without obscuring meaning. Consider, for instance, Hamlet's report of his shipboard actions: *Up from my bed, my sea-gown scarf'd about me, grop'd I to find out them* - a construction which might almost be Latin.

Furthermore, if there are, among the world's huge variety of tongues, consistencies which tell us something about the underlying psychological machinery, it's extremely difficult to sort them out, since several explanations may be available. For example, in some languages with word-order grammars a simple statement involving a transitive verb commonly starts with the actor, in others it begins with the action, but only rarely does the object acted upon come first. At least three possible reasons can be postulated. One is that an action begins with the subject, and we have an inbuilt preference for chronological order. Another is that the subject of a transitive verb is likely to be animate, and frequently human, whereas the object is often just an object, and we are more interested in humans and other animals than in objects. The third possible reason is that actions and their performers tend to be more informative than objects. Once one knows who or what is acting the potential range of actions that are likely or even possible becomes restricted. Similarly, if one knows what the action is the potential range of likely performers is limited. The range of things that can be done to an object, however, tends to be larger. It's still dictated by the nature of the object, but not by its capacities, and on occasion actions may be directed at quite inappropriate recipients. Putting actor or verb first therefore prepares the hearer more effectively for the rest of the sentence. So there's a potential explanation in information theory, as well as two psychological ones.

It's also worth asking what influence writing has had on grammars. Once it had developed into something more elaborate than a means of noting down the tax-gatherer's records and the merchant's bills of lading some substitute would have had to be found for gestures, intonations and facial expressions. Grammars must surely have become more complex.

Moreover there must have been some interaction between grammars and modes of writing. Only an isolating language like Chinese could readily be written with pictograms. Where words must be modified to express tense, number and so forth, or where the form of a verb changes to agree with the subject, a modifiable way of representing them is needed - an alphabet, or at least a syllabary. Did writing, once invented, intensify the grammatical tendencies of the language it represented? Could Chinese once have had modifiable words, and lost them after its pictographic writing was invented? Do polysynthetic languages offer some insight into what a purely spoken language can be like?

Learning Language

Chomsky suggested that the ease and speed with which children learn language indicated not only that there must be a universal grammar, but also that the infant comes equipped with a special 'Language Acquisition Device'. His assumption was that infants must deduce the rules of language by analysing a flood of complex adult sentences, many of them ill-formed. In fact he greatly exaggerated this problem. Video-recording experiments demonstrated that while academics struggling with advanced ideas frequently talk in incomplete or ill-formed sentences, everyday conversation tends to be less ambitious but better phrased. Moreover mothers and other adults, when they talk to infants, usually shape their discourse to suit the listener's comprehension. Sentences are short and simple, the most important words are heavily stressed, and there is a lot of repetition. *Look, here's your DOLL. Yes, it's your DOLL. NICE dolly.* Furthermore, they tend to use a limited range of basic and exaggerated intonation patterns. Infants pay more attention to such utterances than they do to the more complex patterns of conversation among adults.

Nevertheless, it's true that even those infants who do not receive very much help from their elders do, as long as there is some interaction, acquire language, though not such a masterful and confident grasp of it as those who are encouraged. And the job young infants do in analysing the language they hear and discovering first how the phonemes are combined, and then how the words are put together, is a remarkable one. It requires a lot of work and a lot of concentration. It helps, no doubt, that they start on the project long before they become mobile so that there are only a few other forms of entertainment to distract them, and they don't have many ideas about what else they might be doing. One must nevertheless ask what inspires the necessary intense study. I think there is indeed something that might be called a Language Acquisition Device; it has several components, all of which were already in place when language began; but some of them have since become considerably more powerful.

Mammals have voluntary control over their skeletal muscles, and spend part of their youth working out how to make good use of it - an enjoyable process, also known as play. In humans the vocal apparatus is under voluntary control, so the baby has fun experimenting with that too. Doing things with lips, tongue and vocal tract results in interesting noises, which is very rewarding, and prompts further experimentation.

Then there is the innate tendency to pay attention to conspecifics, and the pleasure obtained from a familiar voice. Add to this the basal ganglia mirror neurons which prompt auditory imitation. These instincts and rewards combine to point infants towards shaping their early gurgles into linguistic sounds.

Multiple and various reinforcements come into play as learning advances. When the baby makes speechlike noises in the presence of an adult the adult often makes noises back. This is rewarding, and not only because the infant is innately programmed to derive positive reinforcement from the sound of a friendly adult voice. She realises that she can make things happen - *wowee, I can influence what goes on in the world, let's try it again.*

Then come the words. They produce perhaps even stronger reactions from adults, and more encouragement. It also turns out that they can sometimes help one to obtain something one wants (some classical reinforcement here). Better yet, they are extremely useful in the business of categorising one's percepts and thus anticipating possibilities. If that big black thing counts as a dog then one can be prepared for it to bark and bound about, even if it doesn't look much like yesterday's dog. If this thing is a chair then it must be OK to climb on to it.

When the infant has acquired a certain amount of vocabulary and begins to put words together, that makes for more effective ways of getting what she wants, or checking up on hypotheses, or showing off what she knows. The scope for obtaining cognitive, emotional and material rewards increases further as vocabulary and grammar expand. And as the child matures language offers exciting possibilities for penetrating into the minds of other people, and extending the schemata about this all-important subject. It also facilitates social interactions.

Does all this add up to a specific drive to acquire language? It seems rather to be a happy coalition of pre-existing capacities for reinforcement which combine to make language likely to be passed on to succeeding generations once it has been invented. As language developed these influences grew more powerful.

The early experiments with apes worked on the assumption that the animals would only be motivated to learn by the chance to obtain material rewards. The ape who demolished this idea was Kanzi, the bonobo. It was Kanzi's adoptive mother, Matata, who was supposed to be learning to use lexigrams. The infant Kanzi hung around during lessons, experimented now and then with the computer keys, and sometimes tried to grab the symbols as they appeared on the display screen. He was also developing his own methods of communicating with the experimenters by means of gestures. Matata, after two years of training, had mastered only six lexigrams, and used them only to request things, failing to respond to requests from others. At this point the experimenters gave up on her. Kanzi, however, turned out to have acquired a reliable grasp of eight signs by this time. Savage-Rumbaugh and her team decided to let him continue learning at his own pace, choosing for himself what words he wanted to know, just as human infants do. Within four months his vocabulary increased to over twenty lexigrams, and by the time he was three it was around fifty. Eventually he developed a vocabulary of several hundred signs, which he uses in a sociable and conversational way as well as for obtaining things he wants. He often announces what he intends to do before he does it, for instance. (Humans who come new to the lexigrams may take up to a year to become fluent.)

Another bonobo, Panbanisha, raised from infancy in the same environment as Kanzi, and in his company, did even better. Exposed from the beginning to the substantial keyboard of lexigrams that Kanzi had mastered, she expanded her vocabulary faster than he had done, and started combining lexigrams at an earlier age. She also composed somewhat longer and more complex utterances than Kanzi's. A chimpanzee called Panzee also did pretty well. All this suggests that apes can obtain cognitive pleasure from mastering a protolanguage, as long as they're doing it on their own terms. Using lexigrams becomes a form of play. Social emotions may also be involved, since the existence of an inter-species form of communication facilitates interactions with the human experimenters, and sometimes, too, with other apes.

But apes don't acquire words nearly as fast as children do. What particularly characterises human children, I think, is their use of language to satisfy their curiosity about the world. Curiosity can be exercised more easily and more safely where there is language. Some at least of the answers demands can be obtained without the dangers of practical experiment. It can be deduced that language has underwritten the evolution of the extreme and continuing curiosity that characterises humans.

An important part of the infant's motivation in learning language was well expressed by Ernst Cassirer: *Eagerness and enthusiasm to talk do not originate in a mere desire for learning or for using names; they mark the desire for the detection and conquest of an objective world.* This thought was echoed by Derek Bickerton, who asserted that language is *not .... primarily a means of communication. Rather it is a system of representation, a means of sorting and manipulating the plethora of information that deluges us throughout our waking life.*

Recapitulation

A whole range of talents had to be in place before language could be invented, including a capacity to communicate by other means, an ability to learn by imitation, an ability to produce a fair range of sounds, and an ability to recognise them, when produced by others, in different combinations.

Once this new means of communication was available larger brains capable of more complex ideas became selectionworthy, since the ideas could now be passed on to offspring.

Many species of monkeys have a varied range of calls which look like the forerunners of words.

Great apes of all four species have shown that they are capable of making intelligent use of words, if they are introduced to them at an early age. Some have acquired vocabularies of several hundred words, and occasionally as many as six or seven words may be strung together.

Apes often develop habits in the way they combine words but are not much inclined to imitate the grammatical rules practised by humans.

This was perhaps predictable, since their vocabulary, their situation, and the ideas they want to express are not such that there is much danger of ambiguity.

The development of grammatical rules must have gone hand-in-hand with the expansion of the brain that brought more complex ideas.

Language must inevitably reflect the nature of the human brain, its perceptual capacities and its interests. We can also be sure that it's beautifully fitted to the human brain, shaped by a process resembling natural selection, since otherwise children wouldn't be able to learn it.

Several factors motivate children to learn language. The most important, probably, is that language provides a means of satisfying curiosity, and making sense of the world they observe.

Chapter 29

The Evolution of Language

Just how language originated is a fascinating question - one of those about which we can speculate endlessly, since the answer is never likely to be provable. The obvious problem is that the earliest use of language, like the first production of tools, depended on both nature and inspiration. Language can only exist where evolution has provided suitable brain structures to support it, but it doesn't come into being until somebody happens to conceive the idea. The ability is innate but the words must be invented. Only then can there be a selective pressure for the ability to use them more easily and more creatively.

As I pointed out in the previous chapter, several abilities had to be in place before words could be invented. Experiments show that they are present in apes, and some are demonstrable in monkeys, so they were certainly available in the common ancestor of apes and humans. Several monkey species have been shown to use different calls to give warning of different threats, in a way that suggests the calls are learnt. Possibly they are learnt simply because a certain call, a certain threat, and the response made by adults to that threat, are regularly encountered in association. Even that, however, requires focussing attention simultaneously on two disparate stimulus patterns, both the sound-pattern of the call and the predator to which it refers. In addition there must be an ability to imitate the call made by adult animals. Clearly such talents are valuable in themselves, and need only to be elaborated to provide for the lengthier sound-combinations of language.

What defines a fullblown word, as opposed to a signal? I can think of three factors that seem important, all in the mind of the user. The user of the word should be able to appreciate, at least by adulthood, that it is something invented, which has no inherent connection to its referent. They must have a firm idea that they are communicating with another mind, similar to their own, evoking not merely reaction but understanding. And a content word must function as a way of defining something, so that knowing it clarifies the understanding of the world. On the experimental evidence apes can qualify as real word-users on the first two counts at least, and possibly on the third, but it's not easy to see just how to test that one definitively. Monkeys may have some appreciation of the perspective of their hearers, but there is no evidence to suggest that the other two conditions are met. It seems unlikely that their communicative calls were consciously invented, more probable that they emerged serendipitously. The crucial step in the evolution of language occurred when someone realised they could invent a new signal to add to those already in use, and their fellows appreciated that this was what was happening.

The abilities required for language parallel in many ways those involved in tool-making. Two disparate entities must receive attention simultaneously if a twig is to be recognised as a potential tool for catching ants, just as both sign and object must occupy attention together. And both tool and word are, as it were, elected to their purpose by the choice of the user. There are inherent qualities in stone or twig which make it suitable for a particular purpose, but what makes it a tool is the idea in the mind that directs the hand wielding the object. Similarly, sign becomes representation by the will of the user. The mind which can cope with one is perhaps somewhat prepared for the other. It seems reasonable to assume that by the time our hominid and Australopithecine ancestors were making tools their brains were capable of making use of words too.

There's reason to believe that the development of human language was assisted by some evolutionary developments that occurred when our ancestors moved out of the forest into more open country. Apes may have brains capable of dealing with words, but their vocal apparatus is not well adapted for forming a variety of distinct phonemes, or for projecting such sounds over any great distance. Some of the first adaptations that facilitated speech resulted from selection pressures that had nothing to do with communication. Making a habit of walking tall led to some changes in posture. The head came to be balanced on top of the spinal cord instead of thrusting forward, and consequently a curve in the vocal tract was replaced by a sharper angle, and the relationship with the nasal cavities was altered. Out in the open savannah it also became advantageous to be able to run faster, and for longer distances, making powerful lungs advantageous, and subtler controls over the breathing process. This would incidentally lead to a capacity for longer utterances, and probably also louder ones.

A bigger contribution to increased volume came from the lowering of the larynx, which meant space for amplifying the voice. A similar arrangement is found in various other species. The initial advantage is probably that it allows an animal to sound as if it might be larger than it really is, and for a comparatively small species braving open spaces a loud voice may have been a helpful gimmick, so it's been suggested that the development dates to this epoch. Possibly there was an especial advantage to males in a strong voice, since it seems likely to have been useful in competing for dominance and in attracting females - if only thanks to its attention-getting qualities. Certainly the lowering of the larynx must have brought a decisive benefit, for it also makes it possible for food to go down the wrong way and choke us.

These changes provided further potential for evolution to play with. Finer muscle control would develop so that the shape of that resonant cavity would come to be more modifiable, and tongue and lips too would come to be more delicately managed, so that all sorts of new sounds could be produced.

Speech as a substitute for fur?

Another adaptation to life out on the savannah, the disappearance of body fur, would have had some side-effects. One was that infants would have to be held when they were carried, since there was no longer anything for them to cling to. As the upright stance had freed hands to perform such jobs this shouldn't have been a problem. An incidental result, however, must have been that infants spent more time in face-to-face contact with mother, able to observe both her facial expressions and the direction of her gaze very easily, with every opportunity for learning by transferred emotion. Perhaps it was at this period that hominids evolved the white sclera that makes the direction of gaze much more conspicuous. Since an infant being carried has to turn the head quite far to see both mother's reaction and what's causing it, the attention-getting effect of the contrasting sclera would have been especially useful. The face-to-face contact between infant and mother would also promote early language learning.

The other result of fur loss would have been that there was a great deal less potential for the mutual grooming that plays such a large role in the social lives of other primate species, and which our hominid ancestors must also have practised. Grooming, it's generally agreed, serves the important function of social bonding, and effective bonding between group members must have become even more vital in open country, where the only defence against large predators, or groups of smaller ones, would have been co-operative action. What happened when the amount of hair available became vastly reduced? One interesting hypothesis, argued by Robin Dunbar, is that increased vocalisation, and in due course language, came to take the place of grooming.

Our ancestors, like other primate species, must already have possessed a fair range of vocal expression, by which individuals indicated their own emotional state, influenced each others' emotional states, and drew attention to threatening dangers and other important matters. There would have been reflex responses to certain situations - assorted roars and screams of anger, fright and pain. And there must also have been, as in extant apes, other sorts of utterance that were more a matter of choice - a vocal means of expressing submission to a dominant individual, for instance, soothing noises made to distressed infants, friendly noises made while grooming, and contented noises made while being groomed. As the scope for grooming was gradually reduced, it seems likely that our ancestors, emotionally aware of a gap to be filled, took to using this latter type of vocalisation more frequently, and at greater length.

The new capacities of the vocal tract thus found a use. Longer utterances would become possible, which would allow the development of new variations on old calls. In these circumstances any increase in the number of sounds that could be produced would be likely to prove advantageous, providing scope for new ways of expressing friendship and co-operative intentions, of showing submission to superiors, of soothing the irate, and of interacting with the young. As fur thinned we can imagine that grooming was gradually replaced by a sort of vocal massage. There would have been advantage, too, in loud, impressive calls for frightening off predators, emitted in unison, which would at the same time support group morale in the face of danger and promote a sense of group identity.

Any new means of binding the group together would be likely to prove beneficial in the new habitat. Indeed vocalisations offer some advantages. Grooming only cements bonds between individuals, but vocalisations can be addressed more generally, or can be carried on by many individuals in unison, thus spreading the effect further. Moreover vocalisations can usually be made without interfering with other ongoing activity. If social bonds can be strengthened by vocal means the effect can be achieved at the moment when it's actually needed, whereas bonding-by-grooming needs to be established in advance. If a group unites to scream at a predator, binding together the subsidiary coalitions within the group in a common purpose, then unity is at a peak when co-operation is most vital.

As a smallish primate (around four feet tall), living in open country, the early hominids probably judged it wise to move around in fairly tight-knit groups. They may also have tended to form rather larger groups than their forest-living ancestors did. If they did modify their behaviour in this way, however, the intra-group competition for resources would have intensified, and the stresses of being social would have increased. Any group that possessed the means to limit the amount of damage done in internal conflicts by soothing the stresses and promoting co-operation would have had a considerable advantage. So it seems likely that the changing vocal tract and the improved control over breathing muscles would have provided a potential for increased vocalising just at the time when it could prove really valuable.

The hypothesis allows us to envisage that a vocal apparatus of increased flexibility - though by no means as elaborate as it would eventually be - could have begun to develop before anything that could be called language appeared. Some

---

Another vociferous emigré

Baboons, like hominids, left the shelter of the forest for more open country. Adapting very successfully, they split into several species with different diets. Some of them live in larger groups than other monkeys - groups made up of smaller social structures. The gelada baboons of Ethiopia, which form the largest groups, are particularly noisy, chattering away nineteen to the dozen as they munch on mountain grass. It's hypothesised that all this vocalisation serves to smooth over the difficulties of large-scale socialising. Baboons still have their fur, but eating enough grass to survive must leave less time for grooming than is enjoyed by monkeys which eat fruit, insects or roots.

of the elements of speech would thus have been in place, ready to produce utterances of more precise meaning - words - when the idea arose. We can suppose that before words came into existence our ancestors had both acquired a vocal apparatus capable of more varied articulations than that of other primates, and had come to use it more extensively as a means of intentional communication.

They must also, like modern apes, have been making use of gesture. Gestures are so close to mime that apes and humans have several in common, and one species can often deduce the other's meanings. There must be a genetic foundation for the practice, for even people who have been blind from birth make hand gestures, and may do it when speaking to another blind person. At the same time communicative hand movements can obviously be shaped by cultural influences. This also seems to be true in apes, for different populations of bonobos or chimpanzees employ somewhat different vocabularies of gesture. In some apes, moreover, gestures have been observed that are not immediately interpretable - ones that look as if they began in iconic, largescale movements, and gradually got whittled down and formalised as the recipients became used to them.

Some theorists think that language began with such gestures, and certainly communicating a message by mime activates many of the same areas that speaking it does. At any rate the regular practice of intentional communication, using gestures as well as vocalisations, should at least have created a situation in which the invention of words could be appreciated. Minds would have been prepared for that epoch-making development.

Once there were words the ability to produce a wider range of consonants, and more clearly defined pronunciation, would have become valuable, and further modifications to the vocal apparatus would have been favoured. In the first place the new forms of vocalisation were pretty certainly produced by muscles involved in activities like biting, chewing and swallowing, actions which were already under voluntary control in the common ancestor, judging by the neuronal connections to these muscles in modern non-human primates. The emotional cries that involve shaping the vocal tract, on the other hand, can be triggered by mid-brain centres. Even in us such vocalisations often escape involuntarily, before they can be inhibited. The development of speech has involved the evolution of direct pathways from motor cortex to the neurons which innervate the speech apparatus - parallelling the direct connections from motor cortex to hands which facilitate delicate control of fingers.

These direct connections must also have allowed speech to be speeded up, and as the complexity of ideas and utterances grew there will have been a major advantage in increasing the speed of enunciation. Philip Lieberman has made the interesting suggestion that evolving a speedy method of transmission was important for communicating complex messages. The capacity of our very short-term memory would probably be inadequate if lengthy utterances had to be produced quite slowly. We would no longer have an absolutely accurate record of the beginning of the sentence by the time the end arrived.

The first words

Just how words came into being we can only speculate, just as we can only speculate as to how the first tools came to be made. But we can have some fun making up reasonably likely scenarios.

Ray Jackendoff has suggested that the first words were probably the equivalent of some we still use today, words that refuse to fit into any grammatical structure and are largely ignored by grammarians, words like *hello, no, ouch, wow*. These utterances are all, significantly enough, socialising words, or can be. They function as greetings, as agreement or refusal to co-operate, or as expressions of an emotion that is not quite strong enough to elicit an involuntary vocal response. But they don't help to categorise experience, so it could be said that they don't fulfil all the conditions for defining a true, fullblown word, in the sense which we need for the current context.   Indeed they function much like the vocalisations observed in other primates.

Very possibly the birth of the first category-defining word was a barely noticeable event. An utterance that had had a merely emotional content came to function in two or more brains - probably juvenile ones - as a means of defining experience. Perhaps, for instance, some mothers were accustomed to make an encouraging noise when an infant looked doubtfully at some sort of food not tried before. And then one day a juvenile uttered the same noise with a more interrogatory inflection when he encountered something new, was answered with the reassuring vocalisation, and came to understand the sound as something with the specific meaning of *good food*. Mother might never understand it in quite the same way, but other juveniles might catch on, use it among themselves, and perhaps eventually generalise it to indicate just *good*, or just *food*. Eventually someone might use the *good food* utterance when there was no food present, to indicate that they wanted some, or that there might be food to be found in such-and-such a direction. And so the possibility of reference to absent subjects would come into being.

Or there might have been an assortment of danger signals, applied to different forms of danger, as in monkeys. Originally they would have functioned simply as cues to trigger appropriate action, until one day a juvenile conceived of the idea of using one as a means of obtaining information, and refining his definition of that particular danger. Once the technique had been successfully applied to one matter the same or another juvenile might think of trying it out for another, parallel purpose. And so some emotional vocalisations would have been well on the way to being converted into words, by a change of intonation.

Or perhaps we might imagine a youngster experimenting, in play, with the sounds he can produce. He likes one particular pairing of sounds, and repeats it several times. Then he picks up, say, an unwanted fruit, and throws it at another youngster to attract their attention, while yelling his combination of phonemes. His friend imitates the cry and throws the fruit back, and the game is repeated a few times. Then the fruit gets rather squidgy or goes astray, so they start throwing something else, and the friend thinks of a variation on the phoneme sequence and yells that instead. And suddenly it seems quite logical that one cry goes with the first object, the other with the second. Possibly the game is sufficiently entertaining to be repeated, and a few more phoneme sequences are dedicated to different objects. Other juveniles join in. The idea of phoneme sequences which can be attached to particular objects becomes established in their minds.

Probably it's not immediately obvious that the idea has a potentially profitable application, and the adults don't take much notice. But when the juveniles grow up the utterances they invented in play prove useful for co-ordinating group activities. A dominant male can point in an appropriate direction and produce the utterance associated with a certain sort of fruit, and at least some of the audience will know why he wants to set off in that direction. The new signals will be part of the education of the next generation, who will encounter them while the neocortex is still being moulded by experience and their attention is unprejudiced, and will for the most part catch on to the idea very readily - just as juvenile macaques caught on to the idea of washing grain and sweet potatoes.

How fast might vocabulary have expanded? New words may have come at a faster rate than new tools, since the potential for combining phonemes in different ways seems more obvious, once the original idea has dawned, than the potential for exploiting different sorts of raw material for use as tools. But whether new words were accumulated very slowly, one at a time, or sometimes came in a rush, with one invention inspiring another and several being introduced within a single generation, is something else we can only wonder about. Most probably both patterns occurred, at different times.

On the evidence provided by apes and young children it seems likely that an animal which has the neuronal wiring necessary for learning words also has the potential for stringing two or three words together, once they are well learnt. Learning a word implies the ability to hold two disparate percepts in attention at once, first to link the sound-pattern with the meaning, and then to compare the sound-pattern with the self's attempts at reproducing it. The attentional capacity that can do this should be sufficient, once a few words are thoroughly learned, to link the concepts indicated by two words.

If indeterminate, semi-emotional utterances can help to oil the wheels of social life, words can obviously do the job a great deal better. A single word can convey a greeting, indicating economically that the speaker has friendly intentions towards the individual addressed, or that they know their proper place in the hierarchy. An utterance of two or three words can elevate co-operation out of the realm of emotion and into that of formal bartering. The individual who is begged for a share of the food he is holding can propose a bargain, naming something not currently present - something troublesome to obtain, perhaps, or something the supplicant alone knows where to find. Or he can bargain for knowledge, suggesting *show where* something, or *show how*. Or he can find a way to indicate that he reckons the other is in his debt already.

    The example provided by Washoe and her fellow chimpanzees' communications among themselves can't be taken as a very strong indication of what early word use might have been like, since this was a highly untypical grouping. At the time the study was made the group consisted of only five individuals, of which Washoe was by far the oldest and the two males were the youngest. Insofar as there was a dominant animal Washoe assumed the role, by virtue of her age, and there were no competitors. Moreover living under restricted but protected conditions, with food provided, must be very different from surviving in the wild. This said, it can be noted that a great many of the utterances were social in character. Invitations to play and suggestions about sharing particularly prized foods figured largely in the signing of the two youngsters, and Washoe frequently offered *hug* to her son, and sometimes to other individuals if they were distressed. It seems reasonable to suppose that a large proportion of early human speech must similarly have been devoted to social purposes - as it still is today.

    For animals coping with the difficulties and dangers of a particularly demanding habitat the potential for communicating information will have been of significant value. So will the possibility of making clear just what project the group is to co-operate in, and the scope for suggesting alternative ideas. It also seems likely that a group could psych itself up for some particularly demanding venture, such as a hunt, by chanting a word or phrase in unison. From its very beginnings, probably, language would have been entangled with ritual, providing the means of imposing co-ordination on the group, and a wholesale method of imprinting things of communal importance in everybody's minds.

    We may guess, too, that the group which could wield words would have had a significant advantage over neighbouring groups who lacked them, on whom the mysterious sounds would no doubt have made a great impression. The sight of individuals co-ordinating their actions just by uttering noises may well have looked like magic. When groups came into conflict this new form of communication must have been as good as a new weapon - until the non-languaged group worked out that they could use it too.

    Meanwhile a skill in the use of words must also have increased the individual's chances of passing on their genes. They would have been valuable in the education of offspring, and in this way perhaps especially valuable to females - though Tecumseh Fitch has suggested language would be particularly likely to have taken off where males were co-operating in the care of offspring.

It has also been suggested that words must have come to function for males as a form of display, one which could impress both rivals and females, and which (along with the resonant voice) helped in obtaining both dominance and mates. This might explain why we have evolved the ability to produce notably more phonemes than we actually need for the purposes of efficient communication. Our phonetic versatility may perhaps be an equivalent of the peacock's tail, although in our case the trait must have contributed to female success too.

Then again, when words were first invented they will no doubt have been very short collections of phonemes. Speech must have been quite slow, and utterances brief. Perhaps it was practicable and even profitable to make use of the complete set of possible phonemes. Maybe it was only as the speed of utterance increased that it was necessary for languages to trim the variety, and limit the number of fine distinctions that had to be made.

One of the more remarkable aspects of language is that we manage to make sense of phonemes that rush past at a speed at which any other sort of sound turns into a blur. This is because we recognise the larger units, the familiar schemata which are words, rather than the individual units which make them up, and we are guided by the context of the whole utterance. Just how much we depend on context and guesswork is demonstrated by people who are deaf enough to be barely capable of identifying a single letter or a single word when it lacks a context, but nevertheless mostly understand whole sentences if the subject matter is not unexpected. The complete alphabet of potential phonemes, some only subtly differentiated, must have become unwieldy as speech accelerated and the need for intelligent guesswork grew.

The development of syntax

Many relationships - ideas of condition and cause, for example - can be conveyed clearly without special forms of expression if speaker and hearer are both looking at the same thing. *I give pear you give apple* makes sense if the speaker is holding a pear the hearer an apple. *Stick hit apple, apple fall* is an adequate statement when the conversationalists are both gazing at a fruit which is out of reach. It seems likely that habits of word ordering tended to develop, thanks to the basal ganglia circuits which ensure that action sequences are regularly carried out in a certain way, and would generally be copied, thanks to mirror neurons. More complex linguistic architecture will only have become necessary when the potential range of utterances increased, more elaborate combinations of ideas began to be constructed, and ambiguity became a regular possibility. Some mathematical analyses have suggested syntax will only develop when the number of words in use passes a certain threshold. And the number of words seems likely to correlate to some extent with the number of schemata to be expressed.

The appearance of grammar can probably be correlated with the development of increasingly complicated tools, and other manufactures. George Miller observed: *both the use of tools and the use of language demand an ability to organise long, complex, hierarchically integrated sequences of behaviour and ... the acquisition of this general capacity underlies the appearance of both tools and language in the course of human evolution.* The proposition has subsequently received evidential support, for ape technology has proved to correlate fairly well with ape language constructions. Utterances usually comprise only a few words, and the typical bout of tool manufacture involves choosing the raw material carefully, modifying it in a fairly simple way, and perhaps taking the tool to where it's needed. Occasionally there's a somewhat longer utterance, and occasionally a sequence of tools may be used for some purpose or - especially where there has been a human model to copy - time may be spent on shaping a tool. But neither activity seems to be significantly more elaborate than the other.

On these grounds it's widely assumed that the archaeological record can provide some hints, if only slender ones, as to the development of language. By about 2.6 million years ago hominids were making a variety of tools, and it's likely that words were in use by then. By about 1.8 million years ago the tools were more finely crafted and seem to have been shaped for different tasks, and there's evidence that hominids were killing quite large animals, carrying them back to base in largish portions and then butchering them further. That suggests a degree of conceptual organisation that might have been expressed in word-strings of modest length. Words would have served, moreover, to facilitate the sort of co-operative planning that seems to be implied. PET scans have shown that making the sort of stone tools typical of this later period uses several brain areas also involved in language.

By the time archaic forms of *homo sapiens* had emerged, around 300,000 years ago, the tools come in standardised designs, and there are signs of hearths, huts and working areas. It's likely that some basic grammatical talent was in place by then, parallelling the greater complexity of the preserved handiwork, and the more elaborate co-operative efforts. And it seems safe to assume that some further syntactical ability had become established by 50,000 years ago, by which time there are traces of art and of burial rites, and quite sophisticated stone tools were being made, elegant and of standardised designs, and clearly intended for a variety of different purposes. A desire to impose standard forms on tools seems likely to co-habit with a desire to impose standard forms on language. And it might be supposed that as a capacity subsequently developed for building complex architecture, composed of interlinking structures and requiring long building times, so did the capacity for building complex sentences. But which came first? Did more elaborate language make longer and more complex projects possible, or did the projects create a need for more elaborate language? Or did the two developments go hand in hand?

The elaboration of syntactical complexity must be assumed to have coincided with expansion of Broca's area, that site on the lower border of the prefrontal lobe where damage can result in sentences being compressed to stark telegraphese, all inessential words left out. Close to the language area is one that organises sequences in other sorts of action. Next-door, meanwhile, is the bit of motor cortex that manages speech, and just above it that which controls hands.

A particularly interesting discovery is a gene which seems to be much involved with language, known as FOXP2. It came to attention because of a mutation found in one particular family. All the members who carry the mutation in one of their two copies of the gene have considerable difficulties with language, whilst the others have no problems, or relatively minor ones. The effects of this mutation are quite various. Not only are syntactical comprehension and production poor but control of the speech-articulating muscles is inadequate. There are also cognitive deficiencies, but it's hard to know whether this is a direct result of the mutation or attributable to the indirect effects of inadequate language.

The FOXP2 gene codes for what's known as a transcription factor, which is a protein that regulates the expression of other genes. This one governs the embryonic development of various structures involved in language. In carriers of the mutated version Broca's area is underdeveloped and there are abnormalities in basal ganglia and cingulate cortex.

It's been possible to produce mice with the same mutation. In those with both copies of the gene mutated it proves fatal after a few weeks, while those with only one mutated copy are poor at motor learning, lack the synaptic plasticity known as long-term depression in the motor pathways of the basal ganglia, and also show abnormalities in the cerebellum. A further clue comes from zebra finches, birds with elaborate and variable songs which are based on imitations of those they hear sung by the previous generation. The FOXP2 gene is normally brought more strongly into action during the period in which the young birds are learning their song, and also when, in adulthood, the song is modified as the seasons change. When expression of the gene was reduced in young finches they produced quite sloppy imitations, and a much more variable output in comparison with normal youngsters.

FOXP2 is an exceptionally well conserved gene. The protein it codes for is identical in chimpanzees, gorillas and rhesus monkeys, and differs by only one amino acid from that found in mice, implying that it performs a very important function and that alterations to it have nearly always reduced fitness. The human version of the gene, however, differs from that shared by chimpanzee, gorilla and rhesus monkey at two positions. So in us the gene has taken on a new form, after remaining almost unaltered since the lineage leading to rodents split from that leading to primates around 70 million years ago. It's estimated that the modifications to FOXP2 probably took place within the last 200,000 years - within the time window when many major changes to the genetic recipe for humans must have occurred. (Oddly enough, though, the orangutan version of the gene differs both from ours and from that of chimpanzees and gorillas.)

A talent for grammar would clearly have brought more of the same advantages that caused earlier linguistic developments to flourish, and brought them in larger measure. More detailed information could be exchanged. Experienced individuals could explain to the youngsters why it was better to do things one way rather than another, and point out potential dangers. Discussing procedures would help to fix them in the memory - which might have been a factor in the development of more complex tools. With efficient pooling of information more detailed and effective plans could be made.

Syntax seems pretty much essential, finally, for conveying the idea of the sort of technique that pays long-term rather than immediate dividends - for explaining, for instance, that if you hide these little hard bits in soft soil, whole new plants of the sort the hard things came from appear in due course. Noticing a connection between events that happen days, weeks or months apart has to be a lucky flash of inspiration. Drawing someone else's attention to it is greatly facilitated by a reasonably sophisticated language.

Equally important, there would have been new scope for promoting good social relationships within a group, and smoothing out conflicts. It would become much easier to convey ideas like *I didn't mean it, honestly*, or *we all love you, you know*, or *he's only a youngster*, or *you can't get away with that sort of behaviour, young man*. And quarrels with words could sometimes be substituted for physical blows. Eloquence would sometimes prove as good a means of impressing rivals and other group members as strength.

Observation of wild chimpanzees suggests that a good deal of time and energy can be spent on competing for leadership. Competing males may be badly wounded in the process, and occasionally females or infants who get in the way are wounded too. Quarrels may also break out betweeen subordinate individuals. No doubt our ancestors had similar problems. Any group which had the means to soothe some of the conflicts and reduce the amount of damage done would surely have a significant advantage over its neighbours.

A language with syntax would also have made it possible to draw up more clearly framed and detailed bargains. At the same time, it's been noted, language would make cheating easier - it offers the possibility of telling lies, and making promises that aren't kept. But it also provides a very good means of publicising the cheater's behaviour, so profiting from cheating requires more skill.

Anthropological studies of remote tribes have suggested that language probably found a major application in gossip about social matters. After analysing a great deal of public conversation Robin Dunbar concluded that things aren't much different in more advanced societies - a large proportion of it consists of gossip about the doings of other people. He proposed that discussion of our fellows serves a unifying function. Or as Max Gluckman put it, gossip is what defines the social group. Outsiders don't get to hear it, and anthropologists know that they have finally been accepted by the group they are studying when people start telling them the current tittle-tattle. Moreover the apparently idle chatter can provide a means of publicising the behavioural norms and standards of the group, so that members are regularly reminded of them, and the young are educated.

Once the capacity for lengthier utterances was established among several groups it would provide a means of reducing intergroup conflicts and establishing profitable relationships. Where neighbouring groups had somewhat different resources bartering would become possible. Trading surplus goods with neighbours would generally be less costly than raiding their territory and risking injury if noticed. As technologies grew, and different groups developed specialities, maintaining friendly relationships with neighbouring groups would be an increasingly attractive proposition. As with the improvement in intra-group behaviour, it should have meant a reduction in the number of injuries inflicted in fights. And migrating to new, uninhabited areas would be easier when there was a chance of making friends by barter along the way. So language should have brought an advantage that allowed numbers to expand rather faster.

As regular contact with neighbouring groups began there would be opportunities to develop perceptions of those other groups as something rather more than just tiresome nuisances who had to be kept off the home territory, and whose own territory was sometimes worth raiding. Applying the insight acquired in dealing with fellow members of one's own group could lead to successful dealings with members of other groups. That might well prompt a certain brooding on questions of group identity. Gossip might become more important, and more frequently practised, at this point. And yesterday's gossip would gradually mature to become history, or legend, the accounts of remembered and imagined pasts which provide that necessary sense of group identity.

Modern writers are not the first to stress the socialising functions of language. Edward Sapir noted its role in establishing *rapport between the members of a physical group, such as a house party.* He went on: *it is not what is said that matters so much as that something is said. Particularly where cultural understandings of an intimate sort are somewhat lacking among the members of a phys-*

ical group it is felt to be important that the lack be made good by a constant supply of small talk. This caressing or reassuring quality of speech in general, even where no-one has anything of moment to communicate, reminds us how much more language is than a mere technique of communication. Most people readily begin chatting when they have nothing else to do - and quite often, too, when there is something else they ought to be doing. It's often observed, justly, that sharing food plays an important part in social bonding; but conversation is a far more widely used glue.

But this only touches on the role language plays, for what's said often does matter. Consider just how much significance can be carried by the word that follows a greeting. *Good morning, your majesty* - or *your lordship, your reverence, sir, Mr Jones, Jones, Sam, mate, dear, darling, dearest.* Perhaps some of the more formal terms are used less regularly than they used to be, but they still get plenty of employment where organisational relationships need to be emphasised. The system is highly economical - much more streamlined than interrupting other activities to make practical demonstrations of submission, superiority or affection. It can also be very flexible, which must have been a significant factor in allowing the complex web of interlocking social organisations in modern societies to come into being. Someone can be *Sam* on informal occasions and *Professor* on formal ones, *Your Honour* at work and *Darling* off duty.

When conversation does involve the communication of real information precisely what is communicated is often influenced by the nature of the existing relationship, and may affect its further development. Fairly meaningless burbles will do to oil the wheels of casual acquaintanceship. But we don't just murmur *rhubarb, rhubarb* at each other, because casual conversation may uncover common interests and shared outlooks from which more profitable relationships may grow. With our friends we tend, when not just gossiping, to discuss subjects we care about - whether it be art or sport, Egyptian papyri or bee-keeping. Close friends may perhaps be told about embarrassing occasions, or theories we haven't quite finished working out yet. And the most delicate subjects, the secret hopes, fears and ambitions, are generally confided only to intimates. The degree of trust implied by the nature of the information revealed can serve to cement friendship more firmly.

The pleasure we get from social chat is important, for it ensures that we keep on practising our language. All the casual chatter means that when there is something important to communicate our speech faculty is in good working order. As R. A. Waldron pointed out, it enables us to maintain our vocabularies in line with the common standard too, and to keep up with new words and usages. Thus, *if we didn't talk to people when it wasn't strictly necessary for any practical purpose, we wouldn't be sure they would understand us when we really wanted them to.* Or to put it another way, the social value of gossiping means that language is kept in good shape for the transfer of important information. Indeed, the social reinforcement that is gained from verbal intercourse is probably important in ensuring that language survives.

Language and Memory

An important but somewhat overlooked aspect of language is the way it facilitates the use of memory. It supplies a new way of transferring information into memory, a new way of reinforcing a memory-trace, and a new means of access to the records. Some of its contributions are so obvious as hardly to need comment. Possessing language permits us to jog each other's memories both intentionally - *do you remember old so-and-so?* - and unintentionally, simply by bringing up in conversation some subject that has evocative connotations for the listener. Activating the record of an event by talking about it, meanwhile, strengthens the synapses of the memory-circuit. We also use language to help preserve intentions in short-term memory, muttering to ourselves *must remember to do so-and-so* - and it sometimes works.

Language can make things stick more firmly in long-term memory by virtue of encoding a complex and possibly unique sensory event into a relatively simple set of categorisations that doesn't take up so much memory power. Referring back to the verbal record can also be much less demanding than recalling an actual sensory experience. Just how helpful it can be has been demonstrated by asking people to match a colour sample when offered a pair of similar alternatives after a thirty second interval. The subjects performed more accurately when the colours had different names, rather than being classified under the same heading. Similarly, the Russian language distinguishes between lighter and darker blues, and Russian speakers distinguish between blues that can be differently labelled faster than English speakers who lack the verbal aid. Learning a name for something not only involves acquiring a word but also discovering just what percept it's applied to. Where subtle discriminations must be made to define the referent, learning the word means acquiring practice at making them.

Importantly, putting something into words can not only simplify the record of a subtle or complex percept, it also means bringing motor memory into play as well as perceptual memory. As I've already emphasised, the latter generally lacks precision, while the former must be pretty exact. (I suspect the crucial fact may be the involvement of the cerebellum.) The effect of using language to describe something, therefore, is to turn a fuzzy perceptual record into a well-defined motor record. The verbal account may perhaps distort the event somewhat. But the great advantage is that it is written into a memory that is designed for ready access. We take full advantage of motor memory when we learn the multiplication tables by reciting them aloud. Things of more essential interest than multiplication tables don't need to be recited so frequently to become fixed in memory. Just expressing them in words helps - even if the words are not spoken aloud. And of course for a lot of the time we think with words.

Perhaps the most important thing about language, however, is that it provides a new route into long-term memory. Getting things out of long-term memory is tricky. Even recognition can be rather slow, once an item is well buried under subsequent accumulations of records. Recall only happens when there is some sort of triggering experience to activate the thread of connections that leads to a long neglected shelf of the memory-store. Even when the thread has received an initial tug a night of dreams and sleep may be needed to develop the connections. Language provides an additional means of snagging on a useful thread and penetrating the labyrinth.

It can do this because it's constantly used, and because its parts are interlinked. Each word is recorded in several different ways. It's filed by sound, by sense, by grammatical function, and in the literate by spelling. That means that each word-circuit has connections to many other word-circuits, representing words of similar meaning, sound, or spelling. This is the obvious deduction to be made from slips-of-the-tongue. A further deduction is that when a word is used the schema-neurons representing all the closely connected words come near to being activated too, and are only switched out of the circuit by a last millisecond inhibition.

Each word-circuit is also linked to those for words with some accidental connection, such as the names of siblings, and to words which frequently occur in conjunction. It's these well-established neuronal connections that provide the expectations put to work when we listen to speech, or read, so that we anticipate an appropriate sort of verb to follow a noun, and in the wake of a preposition we are prepared for a noun or a noun-phrase. These links constitute the knowledge that allows a sentence to be interpreted even as it unfolds, and allow many of the ambiguities of heard speech to be sorted out. Other interconnections embody well-used catch-phrases, clichés, or well-known quotations.

Each word-circuit therefore has direct links to many other word-circuits, and indirect connections to many more. Those representing content words are linked to the circuits embodying the relevant perceptual schemata, and they in turn are linked to neurons which embody related perceptual schemata. So any use of language must cause reverberations through the language store, and on into the connected circuits that record the schemata represented by the words. Most of these reverberations will be promptly squashed as soon as the next word arrives. But when a content word gets a fair amount of use many of the circuits to which it is connected must be brought significantly closer to the threshold of activation. A schema-circuit thus primed must become a likely candidate for full activation in REM sleep. If the priming is strong enough it may even become activated during waking - a neuronal event which probably correlates with the subjective experience of an idea floating into the mind.

Thus everyday conversation (or reading), as well as frequently serving to keep open the route to currently used memory files, can also provide the key that opens up a route to dustier, neglected records. Just as a chance meeting with an old school-fellow, not seen for years, can awaken long-buried memories of the past, so can a name, or a word associated with some special context. The pathways opened up by a word may be more convoluted and unpredictable than those activated by a sensory experience, since the latter is more likely to be anchored to one particular context, while the whole point about a word is that it may appear in many. Consequently, although it's often easy to identify a perceptual event that precipitates the emergence of long-buried memory-files, a cue provided by a word may be much less noticeable.

This unconscious memory-jolting function of language doesn't supply unlimited access to the files. But it seems a fair hypothesis that the number of potential routes into memory is considerably expanded by our possession of language, and by our regular practice of it.

Advanced cognitive powers involve making use of the contents of memory - juggling with schemata to discover new potential relationships, and making plans based on acquired knowledge. The most advanced form of cognition, abstract thought, is entirely dependent on the schemata stored in memory. The evolution of a capacity for abstract thought is inextricably linked to the development of language, therefore, not only because there is no evolutionary advantage in abstract thoughts if they cannot be communicated, but also because language provides an improved means of access to the memory files. As John Morton put it, language *forces categorisation, makes possible a more efficient storage and retrieval system, and simplifies the computations upon which behaviour must be based.*

Recapitulation

We can only guess at how the first real words - consciously invented enunciations that could be used to classify a percept - came into being.

The first step may have been an ability to produce a greater variety of sounds, and to apply voluntary vocalisations to the promotion of group unity.

Language has continued to play a vital role in social matters of all kinds.

Rules for putting words together are likely to have evolved in line with the grammar of action. As the expanding brain came to organise a longer sequence of steps for the creation of a more complex tool, so utterances probably grew longer too. By the time enterprises that required co-operation and a fair amount of organisation were being undertaken, such as building shelters, a certain amount of linguistic syntax must have developed.

A gene called FOXP2 is important to language. This gene remained stable for a very long time, and then accumulated two substitutions after the hominid line split from apes.

Language obviously brings all sorts of advantages, some of which spring from the fact that it involves motor memory as well as perceptual memory.

One must be that it provides a new means of access to long-term memory. As we chatter or read and the relevant word-circuits in the brain are activated, other word-circuits connected to them receive a little preparatory priming - which is why we recognise a likely statement more easily than an unlikely one. Such priming adds to that provided by actual experience, and may lead to further activation in dreams, bringing subsequent recollection of some forgotten fact or event.

Chapter 30

Endword

    This has been a superficial and simplistic sketch of what has been elucidated so far about the mammalian brain, drawn in broad strokes. I've concentrated on those parts where enough is known to add up to a reasonably coherent story, ignoring a number of areas which are not yet sufficiently understood to embellish the plot satisfactorily. I've also skated over finer structure, rather than load the text with too many names of individual nuclei. Dozens of the numerous substances by which one neuron modulates what goes on in another have been omitted, as have the influences exercised by glia, the cells that support neurons in various ways, which are turning out to play a significant role in neuronal function.
    Even in the best-studied brain areas there is a great deal yet to be discovered. Similarly, the most conspicuous interconnections are known, but the immense tangle of smaller ones has yet to be sorted out. Excitingly, plans are afoot to map out the whole wiring diagram of the human brain, an ambitious project, but one which should prove just as profitable as the mapping of the human genome. As a first fruit of this effort a new map of the human cerebral cortex distinguishes 180 sectors, achieving much greater delicacy than Brodmann could.
    Meanwhile additional details of synaptic transmission and ways of modifying it are constantly being discovered. Broad classes of neurons are being subdivided as metabolisms, firing rates and axon distributions are more finely analysed. Continuing technical advances are constantly shedding new light. Exciting discoveries are made as more and more genomes are sequenced. Differences between the brains of even quite closely related species are being elucidated as more species are studied. And as facts accumulate many interpretations are being modified, and some will no doubt be redrawn altogether, as has happened in the past.
    The genetic variations that underlie personality differences are being unveiled. Different versions of both dopamine and serotonin-related genes, in particular, are proving relevant to many traits. But even inbred laboratory mice living in the same environment develop somewhat different patterns of behaviour, confirming that individual experience also influences personality. It's clear that the implementation of the genetic recipes is affected by emotional experience, especially in childhood; some of the ways in which early stress affects neuronal development have been elucidated. Neuromodulators have a considerable influence on the developing circuitry, doubtless through their contribution to long-term potentiation and depression, and individual 'connectomes' can vary distinctively.
    Much is also being discovered about the nature of the malfunctions which lead to neurological diseases such as Alzheimer's, Parkinson's and schizophrenia, which should make it possible to provide more effective treatments. And the

complex web of interactions between brain areas that leads to addiction is gradually being sorted out, along with the changes that occur as addiction takes hold.

The history of the vertebrate brain, meanwhile, will become clearer now that an increasing amount of study is being given to the brains of nonmammalian species such as lampreys, salamanders and toads, investigations which will no doubt be extended to other significant species. And thanks to the increasing amount of genetic information available it's becoming possible to estimate the evolutionary relationship between two species more accurately, especially where ontogenetic development can also be compared. However, there's still a large blank space to be filled in between the few surviving non-vertebrate chordates and the complexities of the vertebrates.

I've lashed the evidence together with a certain amount of guesswork, much of which will probably prove to be wrong. Both evolution and the neuronal structures it produces are all too capable of surprising our logic.

The important thing, I believe, is that it has become possible to conceive of the human brain as a biological machine, rather than an impenetrable mystery. For me the former is a more exciting idea than the latter, and not by any means one that leads to an underestimation of all our complex feelings. And it's possible (if still difficult) to stretch the imagination around the idea that even our conscious selves can be the products of evolution. I like to hope – though it's probably optimistic – that if we can learn to accept our position as creatures that have evolved by chance, with consequent definable limitations, our prospects for surviving the challenges we have created for ourselves will be improved.

I hope it's apparent, meanwhile, that the processes of reason are foreshadowed by the neuronal operations that lead to conscious perception and to the conscious experience of emotion, and can be supposed to have grown out of an elaboration of them. Perception and emotion constitute elementary forms of reasoning. By their means the various factors of a situation are weighed up, leading, for the most part, to a rather better chance of a survival-promoting decision than does a hardwired response or simple conditioned learning.

All the evidence indicates that conscious emotion and perception are the result of neuronal impulses whizzing repeatedly around some fairly extended circuitry. The longer and more complex the circuits the greater the amount of information that can be brought together, and the more synapses there are to be potentiated. And it seems likely that the lengthier the circuitry around which the synchronised impulses travel, the longer the activity can be maintained, and the brighter and more detailed the conscious experience.

Our capacity for abstract reasoning has evolved by the addition of further extensions to the neuronal machinery, thereby creating yet more extended and complex circuits. The large size of prefrontal cortex in *Homo sapiens*, and especially of its frontmost bit, provides the means of keeping several concepts in play simultaneously, and establishing complicated relationships between them.

Platonic ideals and Jungian archetypes

Not much has been added to the list of ideas that are shaped by the very nature of our brains – perhaps that has been accomplished elswhere in works I haven't encountered, or perhaps it will be a task for other brains than this ageing one. However, one or two things seem clear.

Platonic ideals can readily be equated with the schemata embodied in the more general schema-neurons, the ones that define stimulus patterns in terms of their more consistent sensory features, or their emotional import, or their potential use. An animal is identified by the way its components join together and the ways in which their spatial relationships can vary. The essence of a friend or a parent is how one feels about them, and if the emotion is missing the person seen must be an imposter. A chair is a chair because one can sit on it, and it appears to have been designed for that purpose – though other things can temporarily assume the function of a chair. The essential definitions are semi-abstract, which makes them hard to pin down, but in the light of modern neuroscience we don't need to imagine ideal exemplars in another world.

We can also see that the infant mammal is programmed to pay attention to the stimulus pattern that is its parent, and to look to that figure for guidance as it learns about the world, and for reassurance when it's frightened or upset. These early proclivities must provide the foundation for several concepts. The most obvious is 'mother', who generally provides the most guidance and comfort. Since human fathers usually also play some part in infant care and upbringing the same forces also support a concept of 'father'. The concepts built on these innate foundations are of course shaped by individual experience. But even a child entirely deprived of parental care can develop some sort of concept of it, and realise what he or she has missed.

A further idea that grows from these beginnings must be 'authority'. The infant becomes conditioned to accepting the greater wisdom of the adult, and though there will be rebellion in adolescence the general concept has been established. This makes it possible to practise a variation on the theme in adulthood. The mammalian style of social living, in groups with leaders and hierarchies, seems likely to develop only in species where there is extended maternal care and guidance of the young. Whether the individual becomes a leader or a follower depends, of course, on their temperament and capacities, and on chance.

This general concept of authority and superior wisdom can also lead on to the idea of a powerful and all-knowing god. Interestingly, this concept seems to have been advanced mainly in parts of the world where co-operative efforts were needed for survival, so that highly organised societies were desirable, a situation which demands a strong central authority with far-reaching powers. Polytheism tends to be found in places where the living is easier. Like the great variation in the relationships between parents and children this shows that our genetic heritage provides the foundation for our ideas, rather than shaping them inescapably.

How brains began

It's obvious, once one stops to think about it, that all the complex machinery of brains must have evolved from arrangements which functioned in the first place to control movement. Movement predates brains by many aeons. Even unicellular animals can move, and the first multicellular animals must have been doing so long before there was anything but the most basic of sensory abilities to guide them. It was the advantage gained from a more coherent form of movement that made muscle cells and neurons worthwhile. Patterned spontaneous movement developed to produce locomotion, along with arrangements for braking and steering. Distance senses were worth evolving because energy spent on locomotion could be applied more efficiently. The ability to learn about which distant stimulus patterns were worth approaching increased this efficiency further.

Increasingly complex sensory facilities made it profitable to expand the neuronal apparatus at the front end of the spinal cord, where the distance senses reported and where internal matters were managed, evolving something worth calling a brain. There different varieties of information could be correlated more easily, and the potential for learning expanded. But the sensory information and the learning were still applied solely to directing behaviour.

The inescapable conclusion is that the wonders of conscious perception and emotion and even our capacity for abstract reasoning must have grown out of neuronal arrangements that began as a means of directing movement. Where bodies have increased in complexity the nervous sytems and brains that control them have of course expanded too. Brains grew bigger as fins evolved, and as paws turned into hands. Further expansion accompanied active means of collecting sensory information. More occurred with the development of more complex control of muscles, with separate messages to individual muscle fibres.

The basis on which brains are founded is all about movement, then. And in the neocortex the machinery of perception and cognition is intimately entwined with that of movement. The more lavishly energy is expended the more important it is to ensure it's not wasted, so the more energetic species tend to have larger brains. Raichlen and Gordon found a correlation between brain size, relative to body mass, and maximum metabolic rate.

The locomotory control system must have been one of the earliest elements around which the brain grew. As bodies grew more complex additional rhythmic pattern generators would evolve to deal with the processes of turning food into animal - with pushing it through the gut, pumping the nourishment around the body, and later with such activities as breathing and chewing. The brain, too, evolved as a rhythmic instrument, not a passive one. Its neurons fire gently of their own accord when there is no excitatory input, and are only totally silent when they receive strong inhibitory input. Consequently brains pulse away quietly when nothing is happening, until sensory input impinges to cause a change in the pattern - just as some primitive creatures probably wriggled more or less steadily through the water until a sensory input caused a rapid deflection or a slower turn and a deceleration.   The rhythms of a sophisticated brain are many, and interact in complex ways, but the ancient grand-daddy rhythm from which they originally derived must surely have been the locomotory one.

The vexed question of free will

The study of the brain has brought a new perspective to the philosophical debate as to whether we really have free will in our choices and actions. Are our decisions truly rational, as they seem to our conscious minds, or are they wholly the result of genetic inheritance and acquired conditioning? Various experimenters have found that some sorts of decision can be predicted from the brain activity which precedes the manifestation of the decision. This has sometimes been interpreted in terms of 'it's not you that makes decisions, it's your subconscious.' But if it takes a measurable amount of time to create a conscious sensory experience it might be expected that reaching a tricky decision, and maybe double-checking it, takes at least as long, if not longer.

To me it seems that we have a concept of free will because we experience something that feels like it, so we might as well equate the idea with the subjective experience. Clearly some behaviours are both so rational and so firmly rooted in our genes that few people think to question them. Other actions and abstentions from action are habits, shaped by emotional or physical feedback in the past. The result will usually seem rational if we stop to think about it, but mostly we don't. The question of free will only really arises where a variety of relevant considerations have to be weighed, and the decision is not clearcut. Then we apply the resources of prefrontal cortex, and that's what feels like the exercise of free will. Quite probably an element of chance frequently contributes to the decision, by way of the precise state of the neurons when the choice becomes necessary, but I think the exercise of prefrontal cortex can reasonably define free will.

But prefrontal cortex is very immature at birth, and it takes time for free will to become established. Infants are clearly at the mercy of their emotions and their physical needs. Living in the moment, they don't know, until experience has accumulated, that such feelings don't last forever. The business of learning how to control emotions and exercise reason is necessarily spread out over several years.

Most children manage it fairly successfully, but some may need more help than others. It seems likely that a failure to provide a child with the degree of discipline suited to its character may in some cases lead to inadequate development of prefrontal control. Individuals who lack this control might be regarded as deficient in free will. (They are of course the ones who, as adults, will need help in learning to control anger.)

Cognitive reward

Sensory consciousness enables us to create flexible models of the world around us and deduce what the most profitable behaviours might be, rather than just applying rules acquired through conditioning. Making use of it involves the expenditure of energy. To reap all the potential benefit there must be active exploration of the environment, to the fullest extent permitted by a species' physical equipment. Eyes must focus, and saccade across the scene, the head must be turned towards salient inputs in other senses, whiskers must whisk, sonar capability must be used. The heavy investment in sensory and neuronal systems couldn't have come about without a means of ensuring that it was put to use. It had to be accompanied by a drive to expend energy on exploration, and by a capacity for obtaining reinforcement from the results. In short, cognitive capability must be accompanied by a suitable dose of curiosity.

One of the most interesting questions about brains is just what, in neurological terms, might curiousity be? It could perhaps be described as something that keeps the attentional system busy when nothing else is impinging upon it. This is also a description that could be applied to play. We might deduce that exercising attention is in itself rewarding. And as we've seen, the neurotransmitter/modulator dopamine plays a major role in attention, as well as influencing steering and locomotion, and the general level of activity. The dopamine system is especially active when an animal is sorting out a situation, working out what's significant, and what it has to do. Once the successful response has been learnt dopamine flows less freely. It seems likely that an elevated level of activity in dopamine neurons provides a major part of the pleasure of exploration and experimentation.

A modest level of norepinephrine activity is probably also involved. But the high norepinephrine response evoked by a very startling event, or something painful, or any excessive sensory input doesn't produce a feeling of pleasure, at least not in most people. The balance of neuromodulators has to be right, and that no doubt includes other neuromodulators as well. Attention is pleasure, one might say, as long as it's the right level of attention, and as long as the animal is in suitable condition. When it's tired or ill there's no urge to exercise either its muscles or its attentional powers. That may be where a serotonin influence comes in, augmenting or suppressing the urge to explore and investigate according to the current internal state of the organism.

The feeling of satisfaction or even joy which follows a new discovery is more difficult to explain. It's presumably triggered by the inauguration of a new schema-circuit, or the extension of an existing one. But what sort of feedback could be supplied by the potentiation of new synaptic connections to create such a circuit? Dopamine is heavily implicated in synaptic potentiation both in the ventral striatum, which might be described as recording current theories about what's going on, heavily influenced by rewarding and discouraging inputs, and in the dorsal striatum and the neocortex, where the learning is consolidated into more lasting form. It might be supposed that the activity in ventral striatum reduces fairly gradually as a situation becomes clearer, and as synaptic potentiation builds up elsewhere. But introspection indicates that sometimes the answer to a problem comes (at least to consciousness) in a flash, and animals too sometimes seem to master the idea behind an experimental test quite suddenly. Perhaps at this point there's a sudden large burst of dopamine release, which both potentiates synapses particularly firmly and produces the feeling of cognitive delight. If that is indeed what happens it seems most likely to occur when only a few last synapses have to be potentiated to create the new circuit.

The rationale behind the hypothesis lies in the fact that the 'high' experienced when many sorts of addictive drugs are first used sounds rather like the high obtained from a new idea. The former seems to depend on dopamine, so it's reasonable to suppose that the latter does too. Interestingly, it's been reported that the dopamine neurons activated by cocaine are not the same as those that respond to food rewards. That leaves open the possibility that they might be those which respond to non-material reinforcements.

I've hypothesised that dopamine's first function lay in turning an animal towards food, and then, as the locomotory pattern generator evolved, in slowing the animal down when there was a possibility of food to be investigated. Then, as distance senses developed, perhaps a new type of dopamine receptor appeared, activation of which prompted accelerated locomotion until the food cue grew stronger. It seems reasonable to suppose that the role in exploration and investigation emerged from such a beginning. Possibly the pleasure often taken in exercise - which for some people can become an addiction - is also largely attributable to dopamine.

The same neuromodulator must also have extended its role to other muscle systems. Where there are active sensory systems there must be some incentive to ensure that they are used. It seems likely to take the form of a reward derived from the result, and perhaps also from the very business of using the muscles. We humans certainly like to have something to look at, something complicated enough to require some saccades. Other primates similarly seem to be depressed by boredom. It must, indeed, be the reward to be gained from visual exploration that inspires infants to start learning how to saccade around a visual stimulus pattern, and how to control the focus of their eyes.

Dopamine is a very ancient neurotransmitter/modulator, shared with many extant multicellular species. It's found in several places in the mammalian brain, apart from the substantia nigra and the ventral tegmental area. There's a cluster of dopamine neurons in the periaqueductal grey, playing an important role in arousal and wakefulness. Another group is found in the hypothalamus, where they seem to be important in making eating an activity worth bothering about. And there are dopamine interneurons in both the retina and the olfactory bulb, influencing the operations of the larger neurons that produce the onward messages to other parts of the brain.

It's tempting to think that neurons which once ran from photoreceptors and olfactory receptors to the locomotory pattern generator, guiding an animal towards possible food, were gradually divided into separate populations, as newer neurons using glutamate, GABA and other transmitters proliferated amongst them. But that might be a speculation too far.

An unusually high degree of curiousity is certainly one of the defining characteristics of primates, and one that reaches extravagant levels in humans. We can deduce that it was language which underpinned this bit of runaway evolution, since it allows us to mine the experience of our fellows to an unprecedented extent and find out many things without the risks entailed in practical experiment. Curiousity in turn provides infants with a strong motivation for learning language, so there has been a reciprocal evolutionary influence - one that has a lot to do with the evolution of an exceptionally large brain.

I think the urge to explore and investigate, and to dream up new ideas and explanations, can be regarded as the defining feature of humankind. There's much variation between individuals, some doubtless genetic, some explained by nurture, and one person's exciting new discovery may bore someone else. But the infant's joy in studying the world around him, their pleasure in the discovery that banging a spoon on a dish makes an interesting noise, the reward we get from a decent detective story or from the next surprise in the soap opera, all these are examples of the trait, as much as the scientist's delight at their latest advance. We are all hooked to some extent on that flash of discovery – the experience the cartoonist depicts by a glowing light bulb in a balloon over a character's head.

This passion for developing all sorts of new schemata, and the large brain which goes with it, might well be seen as an exaggerated trait which parallels the peacock's tail, or the enormous antlers of the extinct Irish elk. It has led us to a situation where some of the genetic traits that promoted the survival of the fittest among our ancestors are turning into liabilities. So whether, in the long term, this defining characteristic will prove truly advantageous, or whether it will lead to our extinction, is an interesting question.

Bibliography

Abbreviations:
Ann. Rev. – Annual Review
Curr. Op. Neurobiol. – Current Opinion in Neurobiology
J. – Journal
Nat. Neurosci. - Nature Neuroscience
Nat. Rev. Neurosci. - Nature Reviews: Neuroscience
PNAS – Proceedings of the National Academy of Sciences

Chapter 1: Introduction

R. Ardrey   The Territorial Imperative   *Collins 1967*
T. Dhobzhansky   Nothing in Biology Makes Sense Except in the Light of Evolution   *American Biology Teacher Vol. 35 p.125 1973*
J. G. Frazer   The Golden Bough: a Study in Comparative Religion   *Macmillan 1890*
C. J. Jung   Man and His Symbols   *Dell Publishing 1968*
D. Morris   The Naked Ape   *Jonathan Cape 1967*
R. L. Sinsheimer   Neurophysiological Limits of Mind   *in* The Brain of Pooh: An essay on the limits of mind   *Engineering and Science Magazine Vol. 33 January 1970*
J. Z. Young   An Introduction to the Study of Man   *Clarendon Press 1971*

Chapter 2: Reason and Consciousness

F. Bartlett   Remembering: A Study in Experimental and Social Psychology   *Cambridge University Press 1932*
M. E. Bitterman   The comparative analysis of learning   *Science Vol. 188 p.699 1975*
R. C. Bolles   Learning Theory   *Holt, Rinehart & Winston 1979*
R. W. Byrne & A Whiten *eds.*   Machiavellian Intelligence   *Clarendon Press 1988*
D. Chalmers   The Conscious Mind: In Search of a Fundamental Theory   *Oxford University Press 1996*
R. E. Clark & L. R. Squire   Classical Conditioning and Brain Systems: The Role of Awareness   *Science Vol. 280 p. 77 3 April 1998*
K. J. W. Craik   The Nature of Explanation   *Cambridge University Press 1943*
A. Damasio   A Neurobiology for Consciousness   *in*   Metzinger – *see below*
D. C. Dennett   Kinds of Minds   *Weidenfeld and Nicholson 1996*
G. M. Edelman & G. Tononi   Re-entry and the Dynamic Core: Neural Correlates of Consciousness   *in*   Metzinger – *see below*
J. Goodall   In the Shadow of Man   *Houghton Mifflin 1971*
H. Harlow   The Evolution of Learning   *in* Behaviour and Evolution   *ed*. A. Roe & E. G. Sugar   *Yale University Press 1958*
H. Jerison   Evolution of the Brain and Intelligence   *Academic Press 1973*
C. McGinn   The Problem of Consciousness   *Blackwell 1991*
T. Metzinger et al.   *eds.*   Neural Correlates of Consciousness   *MIT Press 2000*

U. Neisser   Cognition and Reality: Principles and Implications of Cognitive Psychology   *W.H. Freeman 1975*
D. A. Oakley   Cerebral Cortex and Adaptive Behaviour *in* Brain Behaviour and Evolution *ed.* D. A. Oakley & H. C. Plotkin   *Methuen 1979*
J. Piaget and Bärbel Inhelder   Mémoire et Intelligence   *Presses Universitaires de France 1968*
P. Rozin   The evolution of intelligence and access to the cognitive unconscious *in* Progress in Psychobiology and Physiological Psychology   Vol. 6   *ed.* J. M. Sprague & A. N. Epstein   1976
P. Silverman   Animal Behaviour in the Laboratory   *Chapman & Hall 1978*
J. Shear *ed.*   Explaining Consciousness: The Hard Problem   *MIT Press 1977*
W. Stürzl et al.   How Wasps Acquire and Use Views for Homing   *Current Biology Vol. 26 p.470 22 February 2016*
N. Tinbergen   The Study of Instinct   *Clarendon Press 1951, 1969*
E. C. Tolman & C. H. Houzik   Introduction and removal of reward and maze performance in rats   *Psychology Vol. 4 p.257 1930*
E. D. Tolman et al.   Studies in spatial learning II: place learning versus response learning   J. Experimental Psychology   Vol. 36 (3)   p.221   June 1946

Chapter 3: Clues to Where Consciousness is at Work

P. Bateson & P. Martin   Play, Playfulness, Creativity and Innovation   *Cambridge University Press 2013*
J. G. Boal et al.   Experimental Evidence for Spatial Learning in Octopuses (*Octopus bimaculoides*)   *J. Comparative Psychology Vol. 114 No.3 p.246 September 2000*
J. Bruner, A. Jolly & K. Sylva *eds.*   Play: Its Role in Development and Evolution   *Penguin 1976*
D. Einon   Creative Play   *Duncan Petersen Publishing 1985*
R. Fagen   Animal Play Behaviour   *Oxford University Press 1981*
J. Goodall   Through a Window: Thirty Years with the Chimpanzees of Gombe   *Weidenfeld and Nicolson 1990*
K. Groos   The Play of Animals   *Appleton 1898*
H. F. Harlow & M. Harlow   Social deprivation in monkeys   *Scientific American Vol. 207 p.136 1962*
N. Humphrey   The social function of intellect   Growing Points in Ethology   1976
A. Jolly   The Evolution of Primate Behaviour   *Macmillan 1985*
A. C. Kamil & J. E. Jones   The seed-storing corvid Clark's nutcracker learns geometric relationships among landmarks   Nature Vol. 390 p.276   20 November 1997
W. Kohler   The Mentality of Apes   trans. E. Winter   *Kegan, Paul 1925*
K. Lorenz   Studies in Human and Animal Behaviour   *Macmillan 1985*
J. A. Maher   Cephalod consciousness: Behavioural Evidence   *Consciousness and Cognition Vol. 17 p.37 2008*
T. Matsuzawa *ed.*   Primate Origins of Human Cognition and Behaviour   *Springer 2001*
F. D. C. Mendes et al.   Fishing with a Bait: A Note on Behavioural Flexibility in *Cebus apella*   *Folia Primatologica Vol. 71 p.350 2000*

S. Millar The Psychology of Play *Penguin 1968*
A. R. Petersen Why Children and Young Animals Play *The Royal Danish Academy of Sciences and Letters 1988*
H. H. Peters Tool Use to Modify Calls by Wild Orang-Utans *Folia Primatologica Vol. 72 p.242 2001*
L. L. Sharpe et al. Experimental provisioning increases play in free-ranging meerkats *Animal Behaviour Vol. 64 Pt1 p.113 July 2002*
P. K. Smith *ed.* Play in Animals and Man *Blackwell 1984*
P. Smith & T. Hagan Effects of deprivation of exercise play in nursery school children *Animal Behavior Vol. 28 p.922 August 1980*
H. Spencer Principles of Psychology *1855*
B. Sutton-Smith *ed.* Play and Learning *Gardner Press 1979*
Elisabetta Visalberghi Selection of Effective Stone Tools by Wild Bearded Capuchin Monkeys *Current Biology Vol. 19 p.213 10 February 2009*
G. C. Westergaard & S. J. Suomi Modification of Clay Forms by Tufted Capuchins (*Cebus apella*) *Int. J. Primatology Vol. 18 No.3 p.455 June 1997*
R. W. Wrangham *ed.* Chimpanzee Cultures *Harvard University Press 1994*

Chapter 4: From Earliest Animals to *Homo sapiens* – A Quick Sketch

D. Arendt et al. From nerve net to nerve ring, nerve cord and brain – evolution of the nervous system *Nat. Rev. Neurosci. Vol. 17 p.61 January 2016*
M. Barluenga et al. Sympatric speciation in Nicaraguan crater lake Cichlid fish *Nature Vol. 439 p.719 9 February 2006*
S. Bearhop et al. Assortative Mating as a Mechanism for Rapid Evolution of a Migratory Divide *Science Vol. 310 p.502 21 October 2005*
D. Brawand et al. Loss of Egg Yolk Genes in Mammals and the Origin of Lactation and Placentation *PLoS Biology 6(3) e63 2008*
M. B. Brazeau & M. Friedman The origin and early phylogenetic history of jawed vertebrates *Nature Vol. 520 p.490 23 April 2015*
D. E. G. Briggs The Cambrian Explosion *Current Biology Vol. 25 R.845 5 October 2015*
A. B. Butler Evolution of Vertebrate Brains *in* Encyclopedia of Neuroscience *ed.* L. R. Squire et al. Vol. 4 p.57 *Academic Press 2009*
R. Carmody et al. Energetic consequences of thermal and nonthermal food processing *PNAS Vol. 108 p.19199 29 November 2011*
J. A. Clack Gaining Ground: The Origin and Evolution of Tetrapods *Indiana University Press 2002*
W. de Winter & C. E. Oxnard Evolutionary radiations and convergence in the structural organization of mammalian brains *Nature Vol. 409 p.710 8 February 2001*
S. Dorus et al. Accelerated evolution of nervous system genes in the origin of *Homo sapiens* *Cell Vol. 119 p.1027 29 December 2004*
R. D. Emes & S. G. N. Grant Evolution of Synaptic Complexity and Diversity *Annual Review of Neuroscience Vol. 35 p.111 2012*
E. Finney et al. Visual stimuli activate auditory cortex in the deaf *Nature Neuroscience Vol. 4 p.1171 December 2011*

S. E. Fisher & M. Ridley   Culture, Genes, and the Human Revolution   *Science Vol. 340 p.929   24 May 2013*
Y. Haile-Selassie et al.   The Pliocene hominin diversity conundrum: Do more fossils means less clarity?   *PNAS Vol. 113   p.6364   7 June 2016*
S. Harmond et al.   3.3-million-year-old tools from Lomekwi 3, West Turkana, Kenya   *Nature Vol. 521 p.310 21 May 2015*
W. Henke & I.Tattersall *eds.* Handbook of Paleoanthropology   *Springer 2007*
L. Z. Holland & N. D. Holland   Chordate origins of the vertebrate central nervous system   *Curr. Op. Neurobiol. Vol. 9 p.596   1999*
N. D. Holland et al.   Scenarios for the making of vertebrates   *Nature Vol. 520 p.450 23 April 2015*
Y. Hu et al.   Large Mesozoic mammals fed on young dinosaurs   *Nature Vol. 433 p.149 13 January 2005*
N. Jain et al.   Large-Scale Reorganization in the Somatosensory Cortex and Thalamus after Sensory Loss in Monkeys   *J. Neuroscience Vol. 28(43) p.11042 22 October 2008*
Q. Ji et al.   The earliest known eutherian mammal   *Nature Vol. 416 p.816 25 April 2002*
J. H. Kaas *ed.*   The Evolution of Nervous Systems   *Academic Press 2007*
K. V. Kardong   Vertebrates: Comparative Anatomy, Function, Evolution   *McGraw Hill 1998*
T. C. Lacalli   New perspectives on the evolution of protochordate sensory and locomotory systems, and the origin of brains and heads   *Philosophical Transactions of the Royal Society London B   vol. 356 p.1565 2001*
M. V. Leakey et al.   New fossils from Koobi Fora in northern Kenya confirm taxonomic diversity in early* Homo   Nature Vol. 488 p.201 9 August 2012*
N. Lessard et al.   Early-blind human subjects localise sound sources better than sighted subjects   *Nature Vol. 395 p.278 17 September 1998*
A. M. Lister   The role of behaviour in adaptive morphological evolution of African proboscideans   *Nature Vol. 500 p331 15 August 2012*
D. W. Loehlin & J. C. Werren   Evolution of Shape by Multiple Regulatory Changes to a Growth Gene   *Science Vol. 335 p.943 24 February 2012*
T. Martin et al.   A Cretaceous eutriconodont and integument evolution in early mammals   *Nature Vol. 526 p.380 15 October 2015*
J. A. Mohawk et al.   Central and Peripheral Circadiain Clocks in Mammals   *Ann. Rev. Neurosci. Vol. 35 p.445 2012*
A. Navarrete et al.   Energetics and the evolution of human brain size   *Nature Vol. 480 p.91 1 December 2011*
R. Nieuwenhuys et al. *eds.* The Central Nervous System of Vertebrates   *Springer 1998*
R. Glenn Northcutt   Evolving Large and Complex Brains   *Science Vol. 332 p.926 20 May 2011*
K. S. Pollard et al.   An RNA gene expressed during cortical development evolved rapidly in humans   *Nature Vol. 443 p.167 14 September 2006*
M. Ptito et al.   Alterations of the visual pathways in congenital blindness   *Experimental Brain Research   Vol. 187 p.41 2008*
P. Rakic   Evolution of the neocortex: a perspective from developmental biology   *Nat. Rev. Neurosci. Vol. 10 p.724   October 2009*

N. T. Roach et al.   Elastic energy storage in the shoulder and the evolution of high-speed throwing in *Homo*   *Nature Vol. 498 p.483 27 June 2013*
A. Roe & G. G. Simpson *eds.*   Behavior and Evolution   *Yale University Press 1958*
G. Roth & M. F. Wulliman *eds.*   Brain Evolution and Cognition   *Wiley 2001*
G. D. Ruxton & D. M. Wilkinson   Avoidance of overheating and selection for both hair loss and bipedality in hominins   *PNAS Vol. 108 p.20965 27 December 2011*
O. Seehausen et al.   Speciation through sensory drive in cichlid fish   *Nature Vol. 455 p.620 2 October 2008*
G. F. Striedter   Principles of Brain Evolution   *Sinauer Associates 2005*
H-Y. Wang et al.   Rate of Evolution in Brain-expressed Genes in Humans and Other Primates   *PLoS Biology 10.1371/journal.pbio.00500113 26 December 2006*
B. H. Weber & D. J. Depew *eds.*   The Baldwin Effect Reconsidered   *MIT Press 2003*
T. D. White et al.   Neither chimpanzee nor human: *Ardipithecus* reveals the surprising ancestry of both   *PNAS Vol. 112 p.4877 21 April 2015*
K. D. Zink & D. E. Lieberman   Impact of meat and Lower Palaeolithic food processing techniques on chewing in humans   *Nature Vol. 531 p.500 24 March 2016*

Chapter 5: An Introduction to Neurons and Nervous Systems

J. Adler   The Sensing of Chemicals by Bacteria   *Scientific American April 1976 p.40*
M. A. Becerro et al. *eds.*   Advances in Sponge Science: phylogeny, systematics, ecology   *Academic Press 2012*
J. H. Caldwell   Action Potential Initiation and Conduction in Axons   *in* Encyclopedia of Neuroscience *ed.* L.R. Squire et al. Vol. 1 *p.23 Academic Press 2009*
J. R. Cronly-Dillon & R. L. Gregory *ed.*   Vision and Visual Dysfunction   *Macmillan 1991*
M. A. England & J. Wakely   Color Atlas of the Brain and Spinal Cord   *Mosby/Elsevier 2006*
I. Fairweather *ed.*   Cell Signalling in Prokaryotes and Lower Metazoa   *Kluwer Academic Publishers 2004*
T. J. Hara *ed.*   Fish Chemoreception   *Chapman and Hall 1992*
J. H. Kaas *ed.*   Evolutionary Neuroscience   *Academic Press 2009*
T. Katsuki & R. J. Greenspan   Jellyfish nervous systems   *Current Biology Vol. 23 No.14 R592 July 2012*
P. S. Katz & R. J. Calin-Jagerman   Neuromodulation   *in* Encyclopedia of Neuroscience *ed.* L. R. Squire et al. Vol. 6 p.497 *Academic Press 2009*
N. King et al.   Evolution of Key Cell Signalling and Adhesion Protein Families Predates Animal Origins   *Science Vol. 301 p.361 18 July 2003*
J. Kremers *ed.*   The Primate Visual System: A Comparative Approach   *John Wiley 2005*
T. D. Lamb et al.   Evolution of the vertebrate eye: opsins, photoreceptors, retina and eye-cup   *Nat. Rev. Neurosci. Vol. 8 p.960 December 2007*
A. G. Liu et al.   First evidence for animal locomotion in the Ediacara Biota from the 565 Ma. Mistaken Point Formation in Newfoundland   *Geology Vol. 38 No.2 p.123*
S. A. Nichols et al.   Early evolution of animal cell signalling and cohesion genes   *PNAS Vol. 103 p.12451 15 August 2006*
A. E. Pereda   Electrical synapses and their functional interactions with chemical synapses

Nat. Rev: Neurosci. Vol. 15 p.250 April 2014
F. R. Prete ed.   Complex Worlds from Simpler Nervous Systems   *MIT Press 2004*
T. J. Ryan & S. G. N. Grant   The origin and evolution of synapses   *Nat. Rev: Neurosci. Vol. 10 p.701 October 2009*
K. Shimizu & M. Stopfer   Gap junctions   *Current Biology Vol. 23 R1026 2 December 2013*
B. Slotnick et al.   Accessory olfactory bulb function is modulated by input from the main olfactory epithelium   *European J. Neuroscience Vol. 31 p.1108 2010*
A. Welge-Lüssen et al.   Swallowing is Differentially Influenced by Retronasal Compared with Orthonasal Stimulation in Combination with Gustatory Stimuli   *Chemical Senses Vol. 34 No.6 p.499 July 2009*
H. H. Zekon   Adaptive evolution of voltage-gated sodium channels: The first 800 million years   *PNAS Vol. 109 p.10619 26 June 2012*

Chapter 6: The Basic Mechanisms of Vision

H. B. Barlow et al.   The neural mechanism of binocular depth discrimination   *J. Physiol. Vol. 193 p.327 1967*
E. D. Carterette & M. P. Friedman *eds.*   Handbook of Perception Vol. 5 Seeing   *Academic Press 1975*
C. E. Connor et al.   Shape transformation in the ventral pathway   *Curr. Op. Neurobiology Vol. 17 p.140 April 2007*
D. M. Dacey   Parallel Pathways for Spectral Coding in Primate Retina   *Ann. Rev. Neurosci. Vol. 23 p.743 2000*
J. E. Dowling   Retina: An Overview   *in* Encyclopedia of Neuroscience *ed.* L. R. Squire et al. Vol. 8 *Academic Press 2009*
T. Euler et al.   Retinal bipolar cells: elementary building blocks of vision   *Nat. Rev: Neurosci. Vol. 15 p.507 August 2014*
G. D. Field & E. J. Chichilnisky   Information Processing in the Primate Retina   *Ann. Rev. Neurosci. Vol. 30 p.1 2007*
K. R. Gegenfurter & D. C. Kiper   Color Vision   *Ann. Rev. Neurosci. Vol. 26 p.181 2003*
D. H. Hubel & T. N. Wiesel   Receptive fields, binocular interaction and functional architecture in the cat's visual cortex   *Journal of Physiology Vol. 160 p.106 1962*
D. H. Hubel   The Visual Cortex of the Brain   *Scientific American November 1963*
Bela Julesz   Foundations of Cyclopean Perception   *University of Chicago Press 1971*
J. H. Kaas & C. E. Collins *eds.*   The Primate Visual System   *CRC Press 2003*
E. H. Land   The Retinex Theory of Color Vision   *Scientific American December 1977*
R. H. Masland & T. Albright *eds.*   The Senses Vol.1 Vision   *Elsevier 2008*
J. J. Nassi & E. M. Callaway   Parallel processing strategies of the primate visual system   *Nat. Rev: Neurosci. Vol. 10 p.360 May 2009*
F. T. Qiu & R. von der Heydt   Figure and Ground in the Visual Cortex: V2 Combines Stereoscopic Cues with Gestalt Rules   *Neuron Vol. 47 p.155 7 July 2005*
G. Riddoch   Dissociation of visual perceptions due to occipital injuries, with especial reference to appreciation of movement   *Brain Vol. 40 p.15 1917*
J. Ross   The Resources of Binocular Perception   *Scientific American March 1976*

L. R Sincich & J. C. Horton   The Circuitry of V1 And V2: Integration of Color, Form and Motion   *Ann. Rev. Neurosci. Vol. 28 p.303 2005*
S. G. Solomon & P. Lennie   The machinery of colour vision   *Nat. Rev. Neurosci. Vol. 8 p.276   April 2007*
K. Tanaka   Columns for Complex Visual Object Features in the Inferotemporal Cortex: Clustering of Cells with Similar but Slightly Different Stimulus Selectivities   *Cerebral Cortex Vol. 13 p.90 January 2003*
J. S. Werner & L.M Chalupa eds.   The New Visual Neurosciences   *MIT Press 2014*
C-I Yeh et al.   Stimulus ensemble and cortical layer determine V1 spatial receptive fields   *PNAS Vol. 106 p.14652   26 August 2009*
S. Zeki   A Vision of the Brain   *Blackwell Scientific Publications   1993*
L. Zhaoping   Border ownership from intracortical interactions in visual area V2   *Neuron Vol. 47 p.147 2005*
C. M. Ziemba et al.   Selectivity and tolerance for visual texture in macaque V2   *PNAS Vol. 113 E3140 31 May 2016*

Chapter 7: Learning to See

F. Attneave   Multistability in Perception   *Scientific American   December 1971*
C. I. Baker et al.   Impact of learning on representation of parts and wholes in monkey inferotemporal cortex   *Nature Neuroscience Vol. 5 p.1210   November 2002*
T. G. R. Bower   The Visual World of Infants   *Scientific American December 1996*
T. G. R. Bower   The Object in the World of the Infant   *Scientific American October 1971*
D. C. Bradley, G. C. Chang & R. A. Andersen   Encoding of three-dimensional structure-from-motion by primate area MT neurons   *Nature Vol. 392 p.714 16 April 1998*
S. Brincat & C. E. Connor   Underlying principles of visual shape selectivity in posterior inferotemporal cortex   *Nat. Neurosci. Vol. 7 p.880 August 2004*
J. Bruner (selected and edited by J. Anglin)   Beyond the Information Given   *George Allen and Unwin 1973*
D. Chambers & D. Reisberg   Can mental images be ambiguous?   *J. Experimental Psychology: Human Perception and Performance Vol. 11 p.317 1985*
Jie Chen et al.   Rapid Integration of Tactile and Visual Information by a Newly Sighted Child   *Current Biology Vol. 26 p.1069 25 April 2016*
I. Fujita et al.   Columns for Visual Features of Objects in Inferotemporal Cortex   *Nature Vol. 360 p.343 26 November 1992*
E. J. Gibson   Principles of Perceptual Learning and Development   *Prentice-Hall 1969*
B. Gillam   Geometrical Illusions   *Scientific American January 1980*
R. L. Gregory ed.   Concepts and Mechanisms of Perception   *Duckworth 1974*
R. L. Gregory   Eye and Brain   *Weidenfeld and Nicolson 1977*
D. O. Hebb   The Organization of Behavior: A Neuropsychological Theory   *John Wiley and Sons 1949/Lawrence Erlbaum Associates 2002*
R. Held et al.   The newly sighted fail to match seen with felt   *Nat. Neurosci. Vol. 14 p.551 May 2011*
Gaetano Kanizsa   Subjective Contours   *Scientific American April 1976*
O. Kochukhova & G. Gredeback   Learning about occlusion: initial assumptions and rapid

adjustments *Cognition Vol. 105 p.26 October 2006*
Z. Kourtzi & C. E. Connor   Neural Representations for Object Perception, Structure, Category and Adaptive Coding *Ann. Rev. Neurosci. Vol. 34 p.45 2011*
E. V. Kushnerenko & M. H. Johnson   Newborn behaviour and perception *in* The Newborn Brain *ed.* H. Lagercrantz et al. Cambridge University Press 2010
D. A. Leopold & N. K. Logothetis   Activity changes in early visual cortex reflect monkey's percepts during binocular rivalry *Nature Vol. 379 p.549 8 February 1996*
A. McKyton et al. The Limits of Shape Recognition following Late Emergence from Early Blindness *Current Biology Vol. 25 p.2373 21 September 2015*
A. Messenger et al. Neuronal representations of stimulus associations develop in the temporal lobe during learning *PNAS Vol. 98 p.12239 9 October 2001*
F. Metelli Perception of Transparency *Scientific American Vol. 230 April 1974*
T. Meyer et al. Image familiarization sharpens response dynamics of neurons in inferotemporal cortex *Nat. Neurosci. Vol. 17 p.1388 October 2014*
Y. Ostrovsky et al. Visual Parsing After Recovery from Blindness *Psychological Science Vol. 20 p.1484 2009*
J. Piaget The Child's Construction of Reality *trans.* M. Cook *Routledge and Kegan Paul 1955*
M. Riesenhuber & T. Poggio Hierarchical models of object recognition in cortex *Nat. Neurosci. Vol. 2 p.1019 November 1999*
N. Sigala & N. K. Logothetis Visual categorization shapes feature selectivity in the primate temporal cortex *Nature Vol. 415 p.318 17 January 2002*
K. Tanaka Neuronal Mechanisms of Object Recognition *Science Vol. 262 p.685 29 October 1993*
K. Tsunoda et al. Complex objects are represented in macaque infero-temporal cortex by the combination of features *Nat. Neurosci. Vol. 4 p.832 August 2001*

Chapter 8: Creating the Conscious Visual Moment

P. Avanzini et al. Spatiotemporal dynamics in understanding hand-object interactions *PNAS Vol. 110 p.15878 October 2013*
I. M. Andolina et al. Corticothalamic feedback enhances stimulus response precision in the visual system *PNAS Vol. 104 p.1685 30 January 2007*
M. Bar et al. Cortical Mechanisms Specific to Explicit Visual Object Recognition *J. Cognitive Neuroscience Vol.15 no.4 p.600 2003*
C. N. Boehler et al. Rapid recurrent processing gates awareness in primary visual cortex *PNAS Vol. 105 p.8742 12 June 2008*
J. Bullier Feedback connections and conscious vision *Trends in Cognitive Sciences Vol. 5 p.369 September 2001*
H. Collewijn et al. Natural image motion: origin and change *Annals of the New York Academy of Sciences Vol. 374 p.312 1981*
A. Cowey How Does Blindsight Arise? *Current Biology Vol. 20 R702 14 September 2010*
H. Crane & T. Piantanida On Seeing Reddish Green and Yellowish Blue *Science Vol. 21 p.1078 9 September 1983*

J. Cudiero & A. Sillito   Looking back: corticothalamic feedback and early visual processing   *Trends in Neurosciences Vol. 29 p.298 June 2006*
S. Dehaene *ed.* The Cognitive Neuroscience of Consciousness   MIT Press 2001
V. de Lafuente & R. Romo   Neural correlate of subjective sensory experience gradually builds up across cortical areas   PNAS Vol. 103 p.14266 26 September 2006
A. Del Cul et al.   Brain Dynamics Underlying the Nonlinear Threshold for Access to Consciousness   PLoS Biology 5(10) e260 2007
G. Edelman & G. Tononi   A Universe of Consciousness: How Matter Becomes Imagination   Penguin 2000
C. R. Evans & D. J. Piggins   A comparison of the behaviour of geometrical shapes when viewed under conditions of steady fixation, and with apparatus for producing a stabilised retinal image   *British J. Physiological Optics   Vol. 20 p.261 1963*
C. R. Evans   Some studies of pattern perception using a stabilised retinal image   British J. Psychology Vol. 56(2-3) p.121 1965
S. R. Friedman-Hill et al.   Parietal Contributions to Visual Feature Binding: Evidence from a Patient with Bilateral Lesions   *Science Vol. 269 p.853 11 August 1995*
R. Gaillard et al.   Converging Intracranial Markers of Conscious Access   PLoS Biology *10.1371/journal.pbio.1000061 17 March 2009*
M. I. Garrido   Brain Connectivity: The Feel of Blindsight   Current Biology Vol.22 R599 7 August 2012
E. J. Gibson & A.D. Pick   An Ecological Approach to Perceptual Learning and Development   Oxford University Press 2000
R. Gregory   Seeing as thinking: An active theory of perception   Times Literary Supplement 23 June 1972
R. N. Haber & M. Hershenson   The Psychology of Visual Perception   Holt Rinehard & Winston 1980
L. Huang et al.   Characterising the Limits of Human Visual Awareness   Science Vol.317 p.823 10 August 2007
J. M. Hupé et al.   Cortical feedback improves discrimination between figure and background by V1, V2 and V3 neurons   Nature Vol.394 p.784 20 August 1998
D. Kerzel & K. R. Gegenfurter   Neuronal processing delays are compensated in the sensorimotor branch of the visual system   Current Biology Vol.13 p.1975 2003
H-K Ko et al.   Microsaccades precisely relocate gaze in a high visual acuity task   Nat. Neurosci. Vol.13 p.1549 December 2011
V. A. F. Lamme & P. R. Roelfsma   The distinct modes of vision offered by feedforward and recurrent processing in the visual cortex   Curr. Op. Neurobiol. Vol.8 p.529 1998
H. C. Lau & R. E. Passingham   Relative blindsight in normal observers and the neural correlate of visual consciousness   PNAS Vol.103 p.18763 5 December 2006
D. A. Leopold   Primary Visual Cortex: Awareness and Blindsight   Ann. Rev. Neurosci. Vol. 35 p.91 2012
B. Libet et al.   Electrical Stimulation of Cortex in Human Subjects   *in* Handbook of Sensory Physiology Vol. 2   *ed.* A. Iggo   Springer-Verlag 1973
B. Libet et al.   Control of the transition from sensory detection to sensory awareness in man by the duration of a thalamic stimulus   Brain Vol. 114 p.1731 1991

A. Maier et al.   Context-dependent perceptual modulation of single neurons in primate visual cortex   *PNAS Vol. 104 p.5620 27 March 2007*

S. Martinez-Conde et al.   The impact of microsaccades on vision: towards a unified theory of saccadic function   *Nat. Rev. Neurosci. Vol.14 p.83 February 2013*

J. N. J. McManus et al.   Adaptive shape processing in primary visual cortex   *PNAS Vol. 108 p.9739 14 June 2011*

D. Melcher & C. L. Colby   Trans-saccadic perception   *Trends in Cognitive Sciences Vol. 12 p.466 December 2008*

A. D. Milner & M. A. Goodale   The Visual Brain in Action   *Oxford University Press 1995*

P. C. Murphy et al.   Feedback connections to the Lateral Geniculate Nucleus and Cortical Response Properties   *Science Vol.286 p.1552 19 November 1999*

Ulric Neisser   Cognitive Psychology   *Appleton-Century-Crofts 1967*

D. Noton & L. Stark   Eye Movements and Visual Perception   *Scientific American Vol. 224 p.34 June 1971*

R. M. Pritchard et al.   Visual perception approached by the means of stabilised images   *Canadian J. Physiology Vol.14 p.67 1960*

V. S. Ramachandran & R. L. Gregory   Perceptual filling-in of artificically induced scotoma in periphery by normal subjects   *Nature Vol.350 p.699 1991*

M. Rucci & J. D. Victor   The unsteady eye: an information-processing stage, not a bug   *Trends in Neurosciences Vol. 38 p.195   April 2015*

C. Sergenet et al.   Timing of the brain events underlying access to consciousness during the attentional blink   *Nature Neuroscience Vol.8 p.1391 October 2005*

A. Sillito et al.   Always returning: feedback and sensory processing in visual cortex and thalamus   *Trends in Neurosciences Vol.29 p.307 June 2006*

M. Snodderley et al.   Response Variability of Neurons in Primary Visual Cortex (V1) of Alert Monkeys   *J. Neuroscience Vol.17(8) p.2914 5 April 1997*

G. Sperling   Iconic Memory   *Psychological Monographs no.4 1960*

C. L. Striemer et al.   'Real-time' obstacle avoidance in the absence of primary visual cortex   *PNAS Vol. 106 p.15996 15 September 2009*

A. Treisman & G. Gelade   A feature-integration theory of attention   *Cognitive Psychology Vol. 12 p.97 January 1980*

C. T. Trevethan et al.   Form discrimination in a case of blindsight   *Neuropsychologia Vol.45 No.9 p.2092 2007*

P. de Weerd et al.   Responses of cells in monkey visual cortex during perceptual filling-in of an artificial scotoma   *Nature Vol. 377 p.731 26 October 1995*

S. Zeki & D. H. ffytche   The Riddoch syndrome: insights into the neurobiology of conscious vision   *Brain Vol.121 p.25 1998*

S. Zhang et al.   Long-range and local circuits for top-down modulation of visual cortex processing   *Science Vol. 345 p.660 8 August 2014*

Chapter 9: The Benefits of Conscious Sensory Experience

D. Alais et al.   Visual features that vary together over time group together over space   *Nat. Neurosci. Vol. 1 p.160 June 1998*

T. B. Alder & G. J. Rose   Long-term temporal integration in the anuran auditory system   *Nat. Neurosci.   Vol.1 p.519 October 1998*
M. Boutla et al.   Short-term memory span: insights from sign language   *Nat. Neurosci. Vol.7 p.997 September 2004*
M. A. Branham & M. D. Greenfield   Flashing males win mate success   *Nature Vol. 381 p.745 27 June 1996*
F. Crick & C. Koch   Consciousness and Neuroscience   *Cerebral Cortex Vol.8 p.97 1998*
N. Kazanina et al.   The influence of meaning on the perception of speech sounds   *PNAS Vol. 103 No. 30 p.11381 25 July 2006*
G. Johansson   Visual Motion Perception   *Scientific American June 1975*
J. Y. Lettvin et al.   What the Frog's Eye Tells the Frog's Brain   *Proceedings of the Institute of Radio Engineers 1959*
G. Miller The Psychology of Communication   *Basic Books 1967, Pelican Books 1970*
K. D. Roeder   Nerve Cells and Insect Behavior   *Harvard University Press 1963/1967*
M. J. Ryan & A. S. Read   Female Responses to Ancestral Advertisement Calls in Túngara frogs   *Science Vol. 269 p.390 21 July 1995*
K Sekiyama et al.   Body image as a visuomotor transformation device revealed in adaptation to reversed vision   *Nature Vol.407 p.374 21 September 2000*
S. Sharpless & H. H. Jasper   Habituation of the arousal reaction   *Brain Vol. 79 p.655 1956*
G. M. Stratton   Some preliminary experiments on vision without inversion of the retinal image *Psychological Review Vol.4 1897*

Chapter 10: The Internal World

M. Balasko & M. Cabanac   Behaviour of juvenile lizards (*Iguana iguana*) in a Conflict Between Temperature Regulation and Palatable food   *Brain, Behavior and Evolution Vol. 52 p.257 December 1998*
A. D. Craig   How do you feel? Interoception: the sense of the physiological condition of the body   *Nat. Rev. Neurosci. Vol. 3 p.655 2002*
B. G. Galef   Social learning of food preferences in rodents: rapid appetitive learning   *Current Protocols in Neuroscience Chap. 8 p.5 2003*
H. Craig Heller et al.   The Thermostat of Vertebrate Animals   *Scientific American Vol. 239 p.88 August 1978*
G. Jones   Sensory Biology: Bats Feel the Air Flow   *Current Biology Vol. 21 R666 13 September 2011*
P. S. Kunwar et al.   Ventromedial hypothalamus neurons control a defensive emotion state   *eLife 10.7554/eLife.06633 6 March 2015*
H. Lane   The Wild Boy of Aveyron   *Allen & Unwin 1977*
R. Melzack & T. H. Scott   The effects of early experience on the response to pain   *J. Comparative Physiological Psychology Vol.50 p.155 1957*
C. B. Saper & B. B. Lowell   The hypothalamus   *Current Biology Vol.24 R1111 December 2014*

D. K. Sarko et al.   Mammalian tactile hair: divergence from a limited distribution   *Annals of the New York Academy of Science Vol.1225 p.9 April 2011*
R. Shattuck   The Forbidden Experiment: The Story of the Wild Boy of Aveyron   *Secker and Warburg 1980*
R. A. Sternbach *ed.*   The Psychology of Pain   *Academic Press 2003*
G. D. Stuber & R. A. Wise   Lateral hypothalamic circuits for feeding and reward   *Nature Neuroscience Vol. 19 p.198 February 2016*

Chapter 11: Subtler Forms of Reinforcement – Cognitive Emotions

J. S. Bruner   On Knowing: essays for the left hand   *Bellknap Harvard 1962*
S. Fraiberg   Insights from the Blind   *Basic Books USA, Souvenir Press London 1977*
E. J. Gibson   Principles of Perceptual Learning and Development   *Appleton-Century-Crofts 1969*
S. E. Glickman & R. W. Sroges   Curiosity in zoo animals   *Behavior Vol. 26 p.151 1966*
J. Kagan   The Need for Relativism   *American Psychologist Vol. 22(2) p.131 1967*
H. Keller   The Story of My Life   *Hodder & Stoughton 1958, 1970*
H. Papousek & P.Bernstein   The Functions of Conditioning Stimulation in Human Neonates and Infants *in* Stimulation in Early Infancy *ed.* A. Ambrose   *Academic Press 1969*
R. H. White   Motivation Reconsidered: The Concept of Competence   *Psychological Review no.66 p.317 1959*
R. S. Woodworth   Reinforcement of Perception   *American Journal of Psychology Vol. 60 p.119 1947*

Chapter 12: Social Emotions

P. Bateson   Models of Memory: the Case of Imprinting   *in* Brain, Perception, Memory: Advances in Cognitive Neuroscience   *ed.* J. J. Bolhuis   *Oxford University Press 2000*
P. P. G. Bateson   Imprinting and the development of preferences   *in* Stimulation in Early Infancy   *ed.* J. A. Ambrose   *Academic Press 1969*
J. J. Bolhuis & R. C. Honey   Imprinting, learning and development: from behavior to brain and back   *Trends in Neurosciences Vol. 21 p.307 1998*
J. Bowlby   Attachment and loss: Vol.1 Attachment   *Hogarth Press 1969, Pelican 1984*   Vol.2 Separation, Anxiety and Anger   *Hogarth Press 1973, Penguin 1978*   Vol.3 Sadness and Depression   *Hogarth Press 1980*
W. E. le Gros Clerk   The Antecedents of Man   *University of Edinburgh Press 1962*
A. R. Damasio   Descartes' Error: Emotion, Reason and the Human Brain   *Grosset/Putnam 1994*
H. F. Harlow   Love in Infant Monkeys   *Scientific American June 1959*
E. H. Hess   'Imprinting' in a Natural Laboratory   *Scientific American August 1972*
W. James   The Principles of Psychology   *1890*
C. McCall & T. Singer   The animal and human neuroendocrinology of social cognition, motivation and behaviour   *Nat. Neurosci. Vol. 15 p.681 May 2012*
L. A. Parr, B. M. Waller & J. Fugate   Emotional communication in primates: implications for neurobiology   *Curr. Op. Neurobiol. Vol. 15 No.6 p.716 December 2005*

P. Rauschecker & P. Marler *eds.* Imprinting and Cortical Plasticity: comparative aspects of sensitive periods   *Wiley 1987*
J. Zaki & K. Ochsner   The neuroscience of empathy: progress, pitfalls and promise   *Nat. Neurosci. Vol.15 p.675 May 2012*

Chapter 13: The Demands and Benefits of Social Living

J. H. Barkow et al.   Evolutionary Psychology and the Generation of Culture   *Oxford University Press 1992*
M. Bekoff & M. C. Wells   The Social Ecology of Coyotes   *Scientific American April 1980*
T. J. Bergman et al.   Hierarchical classification by rank and kinship in baboons   *Science Vol. 302 p.1234 2003*
C. Boesch Teaching among wild chimpanzees   *Animal Behavior Vol.41 p.530 1991*
C. Boesch et al.   Behavioural Diversity in Chimpanzees and Bonobos   *Cambridge University Press 2002*
M. Burtsev & P. Turchin   Evolution of cooperative strategies from first principles   *Nature Vol.440 p.1041 20 April 2006*
D. L. Cheney & R. M. Seyfarth   How Monkeys See the World   *Behavioral and Brain Sciences Vol. 15 p.135 1992*
D. L. Cheney & R. M. Seyfarth   Baboon Metaphysics: The Evolution of a Social Mind   *University of Chicago Press 2007*
T. H. Clutton-Brock et al.   Selfish Sentinels in Cooperative Mammals   *Science Vol. 284 p.1640 4 June 1999*
M. Derex et al.   Experimental evidence for the influence of group size on cultural complexity   *Nature Vol.503 p.389 21 November 2013*
F. de Waal   The Origins of Right and Wrong in Humans and Other Animals   *Harvard University Press 1996*
F. de Waal   Chimpanzee Politics: Power and Sex Among Apes   *Johns Hopkins University Press 2007*
R. I. M. Dunbar & S.Schultz   Evolution in the Social Brain   *Science Vol.317 p.1344 2007*
John A. Finarelli & John J. Flynn   Brain-size evolution and sociality in Carnivora   *PNAS Vol. 106 p.9345 16 June 2009*
C. Frith and D. Wolpert   The Neuroscience of Social Interaction: Decoding, Imitating and Influencing the Actions of Others   *Oxford University Press 2003*
Ashleigh S. Griffin   Naked mole-rat   *Current Biology Vol. 18 R844 23 September 2008*
W. D. Hamilton   The genetical evolution of social behaviour I & II   *Journal of Theoretical Biology Vol.7 p.1 1964*
J. K. Hamlin et al.   Social evaluatation in preverbal infants   *Nature Vol.450 p.557 22 November 2007*
A. H. Harcourt & K. J. Stuart   Gorilla Society: Conflict, Compromise and Co-operation Between the Sexes   *Nature Vol.450 p.1160 20/27 December 2007*
J. Henrich et al.   Costly Punishment Across Human Societies   *Science Vol.312 p.1767 23 June 2006*
R. A. Hernandez-Aguilar et al.   Savanna chimpanzees use tools to harvest the underground storage organs of plants   *PNAS Vol. 104 p.19210 4 December 2007*

S. Hirata & N. Morimura   Naïve Chimpanzees' (*Pan troglodytes*) Observation of Experienced Conspecifics in a Tool-Using Task   *J. Comparative Psychology Vol.114 p.291 September 2000*

A. Jolly   Lemur Social Behaviour and Primate Intelligence   *Science Vol. 153 p.501 1966*

L. V. Luncz et al.   Evidence for Cultural Differences between Neighbouring Chimpanzee Communities   *Current Biology Vol.22 p.922 22 May 2012*

W. C. McGrew   Chimpanzee Material Culture – Implications for Human Evolution   *Cambridge University Press 1992*

W. C. McGrew   Chimpanzee Technology   *Science Vol.328 p.559 30 April 2010*

M. A. Nowak et al.   The evolution of eusociality   *Nature Vol. 466 p.1057 26 August 2010*

J. S. T. Parker et al   The Mentalities of Gorillas and Orangutans   *Cambridge University Press 2004*

M. Plotnik et al.   Elephants know when they need a helping trunk in a cooperative task   *PNAS Vol. 108 p.5116 22 March 2011*

K. A. Pollard & D. T. Blumstein   Social Group Size Predicts the Evolution of Individuality   *Current Biology Vol. 27 p.413 8 March 2011*

J. D. Pruetz & P. Bertolani   Savanna Chimpanzees, *Pan Troglodytes verus*, Hunt with Tools   *Current Biology Vol. 17 p.412 6 March 2007*

I. Pyysiäinen & M. Hauser   The origins of religion: evolved adaptation or by-product?   *Trends in Cognitive Sciences Vol. 14 p.104 March 2010*

P. Richerson   Group size determines cultural complexity   *Nature Vol. 503 p.351 21 November 2013*

S. Schultz et al.   The evolution of stable sociality in primates   *Nature Vol.479 p.219 10 November 2011*

R. M. Seyfarth & D. L. Cheney   Affiliation, empathy, and the origins of the theory of mind   *PNAS Vol. 110 supplement 2 p.10349 18 June 2013*

A. Thornton & K. McAuliffe   Teaching in Wild Meerkats   *Science Vol. 313 p.227 14 July 2006*

R. L. Trivers   The evolution of reciprocal altruism   *Quarterly Review of Biology Vol.46 p.35 1971*

E. van de Waal et al.   Potent Social Learning and Conformity Shape a Wild Primate's Foraging Decisions   *Science Vol.340 p.483 26 April 2013*

J. D. van Dyken et al.   Spatial Population Expansion Promotes the Evolution of Cooperation in an Experimental Prisoner's Dilemma   *Current Biology Vol. 23 p.919 20 May 2013*

C. P. van Schaik et al.   Orangutan Cultures and the Evolution of Material Cultures   *Science Vol.299 p.102 3 January 2003*

A. Whiten et al.   Conformity to cultural norms of tool use in chimpanzees   *Nature Vol.437 p.737 29 September 2005*

A. Whiten   Incipient tradition in wild chimpanzees   *Nature Vol. 514 p.178 9 October 2014*

E. O. Wilson   The Social Conquest of Earth   *Liveright 2012*

R. W. Wrangham et al. *eds*.   Chimpanzee Cultures   *Harvard Universitiy Press 1994*

S. Yamamoto et al.   Chimpanzees' flexible targeted helping based on an understanding of conspecifics' goals   PNAS Vol.109 p.3588   28 February 2012

Chapter 14: Attention

D. Broadbent   Attention and the Perception of Speech   Scientific American Vol. 206 p.143 April 1962
J. Duncan et al.   Restricted attentional capacity within but not between sensory modalities   Nature Vol. 387 p.183 19 June 1997
W. Jones & A. Klin   Attention to eyes is present but in decline in 2-6-month-old infants later diagnosed with autism   Nature Vol.504 p.427 19/26 December 2013
J. S. Joseph et al.   Attentional requirements in a 'pre-attentive' feature search task   Nature Vol.387 p.805 19 June 1997
H. Kobayashi & S. Kohshima   Evolution of the Human Eye for Communication   in Primate Origins of Human Cognition   ed. T. Matsuzawa   Springer 2001
J. R. Lackner & M. Garrett   Resolving Ambiguity: effects of biasing context in the unattended ear   Cognition Vol. 1 no.4 p.359 1973
Z. Li   A saliency map in primary visual cortex   Trends in Cognitive Sciences Vol. 6 p.9 January 2002
G. A. Miller   Psychology: the science of mental life   Pelican 1966
N. Moray   Attention: Selective Processes in Vision and Hearing   Hutchinson Educational 1969
A. Senyu & T. Hasegawa   Direct gaze captures visuospatial attention   Visual Cognition Vol. 12 p.127 2005
F. Simion et al.   A predisposition for biological motion in the newborn baby   PNAS Vol.105 p.809 20 January 2009
Y. Sugita   Innate face processing   Curr. Op. Neurobiol. Vol. 19 p.39 2009
A. Treisman   Features and Objects in Visual Processing   Scientific American December 1986
A.Treisman & S. Gormican   Feature analysis in early vision: Evidence from search assymetries   Psychological Review Vol.95 p.15 January 1998
W. H. Zangmeister et al.   Visual Attention and Cognition   Elsevier 1996

Chapter 15: The Neurology of Movement

M. E. Anderson   Basal Ganglia: Motor Functions   in Encyclopedia of Neuroscience ed. L. R. Squire et al. Vol.2 p.105 Academic Press 2009
R. App & M. Garwicz   Anatomical and Physiological Foundations of Cerebellar Information Processing   Nat. Rev. Neurosci. Vol.6 p.297 April 2005
E. Bizzi et al.   New perspectives on spinal motor systems   Nat. Rev. Neurosci. Vol.1 p.101 November 2000
L. M. Bjursten et al.   Behavioural repertoire of cats without cerebral cortex from infancy   Exp. Brain Research Vol.25 p.115 1976
A. C. Bostan et al.   Cerebellar networks with the cerebral cortex and basal ganglia   Trends in Cognitive Sciences Vol.17 p.241 May 2013
J. Bouvier et al.   Descending Command Neurons in the Brainstem that Halt Locomotion

Cell Vol. 163 p.1191 19 November 2015

J. X. Brooks et al. Learning to expect the unexpected: rapid updating in primate cerebellum during voluntary self-motion *Nat. Neurosci. Vol. 18 p.1310 September 2013*

R. E. Burke The Central Pattern Generator for Locomotion *in Advances in Neurology Vol.87 Gait Disorders ed. E. Ruzicka et al. Lippincott Williams & Wilkins 2001*

J. L. Calton et al. Non-spatial, motor-specific activation in posterior parietal cortex *Nat. Neurosci. Vol.5 p.580 June 2001*

T. T-J. Chong et al. fMRI adaptation reveals mirror neurons in human inferior parietal cortex *Current Biology Vol.18 p.1576 2008*

M. Desmurget & A. Sirigu A parietal-premotor network for movement intentions and movement awareness *Trends in Cognitive Sciences Vol. 13 p.411 October 2009*

N. Dominici et al. Locomotor Primitives in Newborn Babies and Their Development *Science Vol. 334 p.997 18 November 2011*

R. Dubuc et al. Initiation of locomotion in lampreys *Brain Research Reviews Vol.57 p.172 January 2008*

L. Fogassi et al. Neurons responding to the sight of goal-directed hand/arm actions in the parietal area of PF(76) of the macaque monkey *Society of Neuroscience Abstracts Vol. 24 p.2575 1998*

V. Gallese et al. Action recognition in the premotor cortex *Brain Vol.119 p.593 1996*

E. Garcia-Rill et al. Activity in the mesencephalic locomotor region during locomotion *Experimental Neurosci. Vol. 82 p.609 1983*

M. Goulding Circuits controlling vertebrate locomotion: moving in new directions *Nat. Rev. Neurosci. Vol.10 p.507 July 2009*

A. M. Graybiel et al. The Basal Ganglia *Kluwer Academic/Plenum 2002*

M. Graziano The Organization of Behavioural Repertoire in Motor Cortex *Ann. Rev. Neurosci. Vol.29 p.105 2006*

M. Graziano The Intelligent Movement Machine: An Ethological Perspective on the Primate Motor System *Oxford University Press 2008*

M. Graziano et al. Complex Movements Evoked by Microstimulation of Precentral Cortex *Neuron Vol.34 p.841 30 May 2002*

D. M. Griffin et al. Corticomotoneuronal cells are 'functionally tuned' *Science Vol.350 p.667 6 November 2015*

S. Grillner et al. Mechanisms for selection of basic motor programs – roles for the striatum and pallidum *Trends in Neurosciences Vol.28 p.364 July 2005*

S. Grillner et al. Neural bases of goal-directed locomotion in vertebrates – An overview *Brain Research Reviews Vol.57 p.2 January 2008*

S. Grillner et al. The evolutionary origin of the vertebrate basal ganglia and its role in action selection *Journal of Physiology Vol. 591 p.5425 November 2013*

M. Hägglund et al. Optogenetic dissection reveals multiple rhythmogenic modules underlying locomotion *PNAS Vol. 110 p.11589 9 July 2013*

P. Han et al. Dopaminergic modulation of spinal neuronal excitability *J. Neuroscience Vol.27(48) p.13192 2007*

R. M. Harris-Warwick et al. Distributed Effects of Dopamine Modulation in the Crustacean Pyloric Network *Annals of the New York Academy of Sciences Vol.860 p.155 1998*

R. M. Harris-Warwick & A. H. Cohen   Serotonin modulates the central pattern generator for locomotion in the isolated lamprey spinal cord   *J. Experimental Biology* Vol. 116 no.1 p.27 1985

X. Jin et al.   Basasl ganglia subcircuits distinctively encode the parsing and concatenation of action sequences   *Nat. Neurosci.* Vol. 17 p.42 March 2014

L. R. Jordan et al.   Descending command systems for the initiation of locomotion in mammals   *Brain Research Reviews* Vol. 57 p.183 January 2008

E. R. Kandel et al. *eds.*   Principles of Neural Science   McGraw Hill   2013

O. Kiehn   Locomotor Circuits in the Mammalian Spinal Cord   *Ann. Rev. Neurosci.* Vol.29 p.279 2006

F. C. Lao & J. R. Fetcho   Shared versus specialized glycinergic spinal interneurons in axial motor circuits of larval zebra fish   *J. Neurosci.* Vol. 28(48) p.12982 26 November 2008

D. Le Ray et al.   Nicotinic activation of reticulospinal cells involved in the control of swimming in lampreys   *European Journal of Neuroscience* Vol. 17 p.137 2003

S. P. Liversedge et al. *eds.*   The Oxford Book of Eye Movements   Oxford University Press 2011

A. Ménard et al.   Descending GABAergic projections to the mesencephalic locomotor region in the lamprey *Petromyzon marinus*   *J. Comp. Neurol.* Vol. 501 p.260 10 March 2007

A. Ménard & S. Grillner   Diencephalic locomotor region in the lamprey – afferents and efferent control   *J. Neurophysiology* Vol.100 p.1343 September 2008

J. Mena-Segovia et al.   Pedunculopontine nucleus and basal ganglia: distant relatives or part of the same family?   *Trends in Neurosciences* Vol.27 p.585 October 2004

P. Meyrand et al.   Construction of a pattern-generating circuitry with neurons of different networks   *Nature* Vol.351 p.60   2 May 1999

A. Nambu   A new dynamic model of the cortico-basal ganglia loop   *Progress in Brain Research* Vol.143 p.461 2004

A. B. Nelson & A. C. Kreitzer   Reassessing Models of Basal Ganglia Function and Dysfunction   *Ann. Rev. Neurosci.* Vol.37 p.117   2014

F. M. Ocaña et al.   The Lamprey Pallium Provides a Blueprint of the Mammalian Motor Projections from Cortex   *Current Biology* Vol.25 p.413 16 February 2015

T. Ohyama et al.   What the cerebellum computes   *Trends in Neurosciences* Vol. 26 p.222 April 2003

N. Ramnani   The primate cortico-cerebellar system: anatomy and function   *Nat. Rev. Neurosci.* Vol. 7 p.511 July 2006

D. A. Ritter et al.   *In vivo* imaging of zebrafish reveals differences in the spinal networks for escape and swimming movements   *J. Neuroscience* Vol.21(22) p.8956 15 November 2001

G. Rizzolatti & M. Fabri-Destro   Mirror Neuron Mechanism   *in* Encyclopedia of Behavioral Neuroscience   *ed.* G. F. Koob et al.   Academic Press 2010

M. A. Rossi et al.   A GABAergic nigrotectal pathway for coordination of drinking behavior   *Nat. Neurosci.* Vol. 19 p.742 2016

D. Ryczko et al.   Forebrain dopamine neurons project down to brainstem region controlling locomotion   *PNAS* Vol. 110 E3235 20 August 2013

O. Sacks   Awakenings   Vintage Books 1973, revised edition 1990

K. Saitoh et al.   Tectal Control of Locomotion, Steering and Eye Movements in Lamprey   *J. Physiology Vol.97 p.3093 April 2007*
 H. Steiner & K. Y. Tseng   Handbook of Basal Ganglia Structure and Function   *Academic Press 2010*
M. Stephenson-Jones et al.   Evolution of the basal ganglia: dual-output pathways conserved throughout vertebrate phylogeny   *J. Comparative Neurology Vol. 520(13) p.2957 1 September 2012*
K. Takakusaki et al.   Basal ganglia efferents to the brainstem centers controlling postural muscle tone and locomotion: a new concept for understanding motor disorders in basal ganglia dysfunction   *Neuroscience Vol. 119 p.293 2003*
P. Tovote et al.   Midbrain circuits for defensive behaviour   *Nature Vol. 534 p.206 9 June 2016*

Chapter 16: The Neurology of Attention

R. A. Andersen & C. A.Bueno   Intentional Maps in Posterior Parietal Cortex   *Ann. Rev. Neurosci. Vol.25 p189 2002*
M. Arguin et al.   Divided visuo-spatial attention systems with total anterior callostomy   *Neuropsychologia Vol. 38 p.283 2000*
D. Baldauf & R. Desimone   Neural Mechanisms of Object-Based Attention   *Science Vol. 344 p.424 25 April 2014*
P. M. Bays & M. Husain   Dynamic Shifts of Limited Working Memory Resources in Human Vision   *Science Vol. 321 p.851 8 August 2008*
J. W. Bisley & M. E. Goldberg   Attention, Inattention and Priority in the Parietal Lobe   *Ann. Rev. Neurosci. Vol.33 p.1 2010*
P. Bocquillon et al.   Role of Basal Ganglia Circuits in Resisting Interference by Distractors   *PLoS ONE 10.1371/journal.pone.0034239 28 March 2012*
E. A. Buffalo et al.   A backward progression of attentional effects in the ventral stream   *PNAS Vol.107 p.361 5 January 2010*
P. J. Bushnell et al.   Effect of unilateral removal of basal forebrain cholinergic neurons on cued target detection in rats   *Behavioural Brain Research Vol.90 p.57 1998*
G. Buzsáki   Rhythms of the Brain   *Oxford University Press 2006*
R. T. Canolty et al.   Oscillatory phase coupling coordinates anatomically dispersed functional cell assemblies   *PNAS Vol. 107 p.17356 5 October 2010*
B. P. Chalfin et al.   Scaling of Neuron Number and Volume of the Pulvinar Complex in New World Primates: Comparisons with Humans, Other Primates and Mammals   *J. Comp. Neurol. Vol.504 p.265 2007*
C-Y. Chen et al.   Neuronal Response Gain Enhancement Prior to Microsaccades   *Current Biology Vol.25 p.2065 17 August 2015*
C. E. Connor et al.   Spatial Attention Effects in Macaque Area V4   *J. Neurosci. Vol.17(9) p.3201 1 May 1997*
L. Descarries et al.   Acetylcholine in the Cerebral Cortex (Progress in Brain Research 145)   *Elsevier 2004*
T. de Schotten et al.   Direct Evidence for a Parietal-Frontal Pathway Subserving Spatial Awareness in Humans   *Science Vol.309 p.2226 2005*

F. Doricchi & C. Incoccia   Seeing only the right half of the forest but cutting down all the trees?   *Nature Vol.394 p.75 2 July 1998*
T. Drew et al.   A Soft Handoff of Attention between Cerebral Hemispheres   *Current Biology Vol.24 p.1133 19 May 2014*
A. K. Engel et al.   Dynamic Predictions: Oscillations and Synchrony in Top-down Processing   *Nat. Rev. Neurosci. Vol.2 p.704 October 2001*
P. Fries et al.   Modulation of oscillatory neuronal synchronization by selective visual attention   *Science Vol. 201 p.1560 2001*
P. Fuller et al.   Reassessment of the Structural Basis of the Ascending Arousal System   *Research in Systems Neuroscience Vol.519 p.933 2011*
E. Garcia-Rill   Reticular Activating System   *in* Encyclopedia of Neuroscience   ed. L. R. Squire et al.   *Vol.8 p.137 Academic Press 2009*
G. G. Gregoriou et al.   High-Frequency, Long-Range Coupling Between Prefrontal and Visual Cortex During Attention   *Science Vol.324 p.1207 29 May 2009*
K.L.Grieve et al.   The primate pulvinar nuclei: vision and action   *Trends in Neurosciences Vol.23 p.35 January 2000*
B. J. He et al.   Breakdown of Functional Connectivity in Frontoparietal Networks Underlies Behavioural Deficits in Spatial Neglect   *Neuron Vol.53 p.905 15 March 2007*
J. E. Hopfinger et al.   The neural mechanisms of top-down attentional control   *Nat. Neurosci. Vol.3 p.284 March 2000*
B. E. Jones   Activity, modulation and role of basal forebrain cholinergic neurons innervating the cerebral cortex   *Acetylcholine in the Cerebral Cortex (Progress in Brain Research Vol.145)* ed. L. Descarries & K. Krnjevic   *p.147 2004*
S. Kinomura   Activation by Attention of the Human Reticular Formation and Thalamic Intralaminar Nuclei   *Science Vol.271 p.5112 26 January 1996*
R. J. Krauzlis et al.   Superior Colliculus and Visual Spatial Attention   *Ann. Rev. Neurosci. Vol.36 p.165 2013*
P. Lakatos et al.   Entrainment of Neuronal Oscillations as a Mechanism of Attentional Selection   *Science Vol.320 p.110 4 April 2008*
B. Machner et al.   Impact of dynamic bottom-up features and top-down control on the visual exploration of moving real-world scenes in hemispheric neglect   *Neuropsychologia Vol.50 p.2415 2012*
J. H. R. Maunsell & S. Treue   Feature-based attention in visual cortex   *Trends in Neurosciences Vol.29 p.317 June 2006*
R. M. McPeek & E. L. Keller   Deficits in saccade target selection after inactivation of superior colliculus   *Nat. Neurosci. Vol.7 p.757 July 2004*
T. Moore et al.   Visuomotor Origins of Covert Spatial Attention   *Neuron Vol.40 p.671 13 November 2003*
T. Moore & M. Fallay   Microstimulation of the frontal eye field and its effects on covert spatial attention   *J. Neurophysiology Vol.91 p.152 2004*
B. Noudoost et al.   Topdown control of visual attention   *Curr. Op. Neurobiol. Vol.20 p.183 April 2010*
G. H. Patel et al.   Functional evolution of new and expanded attentional networks in humans   *PNAS Vol.112 p.9454 28 July 2015*
L. Pinto et al.   Fast modulation of visual perception by basal forebrain cholinergic neu

rons *Nat. Neurosci. Vol.16 p.1857 December 2013*
A. T. Popescu et al.   Phasic dopamine release in the medial prefrontal cortex enhances stimulus discrimination   *PNAS Vol. 113 E3169 31 May 2016*
G. Purushothaman et al.   Gating and control of primary visual cortex by pulvinar   *Nat. Neurosci. Vol.15 p.905 June 2012*
I. H. Robertson et al.   Phasic alerting of neglect patients overcomes their spatial deficit in visual awareness   *Nature Vol.395 p.169   10 September 1998*
E. Rodriguez et al.   Perception's shadow: long-distance synchronization of human brain activity   *Nature Vol. 3997 p.430   4 February 1999*
P. Schatpour et al.   A human intracranial study of long-range oscillatory coherence across a frontal-occipital-hippocampal brain net work during visual object processing   *PNAS Vol.105 p.4399   18 March 2008*
A. B. Scheibel   Reticular Formation and the Brain Stem *in* Encyclopedia of Neuroscience *ed.* L. R. Squire et al. Vol. 8 p.145 *Academic Press 2009*
S. Shipp   The brain circuitry of attention   *Trends in Cognitive Sciences Vol.8 p.223 May 2004*
L. A. Schwartz & L. Luo   Organization of the Locus Coeruleus-Norepinephrine System   *Current Biology Vol.25 R1050 2 November 2015*
J. H. Siegle et al.   Gamma-range synchronization of fast-spiking interneurons can enhance detection of tactile stimuli   *Nat. Neurosci. Vol.17 p.1371 October 2014*
A. M. Sillito & H. E. Jones   The role of the thalamic reticular nucleus in visual processing   *Thalamus and Related Systems Vol.4(1) p.1 2008*
J. Silvanto et al.   The Perceptual and Functional Consequences of Parietal Top-Down Modulation on the Visual Cortex   *Cerebral Cortex Vol. 19 p.327 February 2009*
J. Skelhorn et al.   Eyespots   *Current Biology Vol. 26 R52 25 January 2016*
J. C. Snow et al.   Impaired attentional selection following lesions to human pulvinar: Evidence for homology between human and monkey   *PNAS Vol.106 p.4054 10 March 2009*
M. Stopfer et al.   Impaired odour discrimination on desynchronisation of odour-encoding neural assemblies   *Nature Vol.390 p.70   6 November 1997*
A. Treisman   Feature binding, attention and object perception   *Philosophical Transactions of the Royal Society B Vol.353 p.1295   1998*
S. Treue & J.H.R. Maunsell   Attentional modulation of visual motion processing in cortical areas MT and MST   *Nature Vol.382 p.539 8 August 1996*
M. Usher & N. Donnelly   Visual synchrony affects binding and segmentation in perception   *Nature Vol. 394 p.179   9 July 1998*
K. G. Usunoff et al.   Brainstem afferent connections of the amygdala in the rat with special reference to a projection from the parabigeminal nucleus: a fluorescent retrograde tracing study   *Anatomy and Embryology Vol. 212 p.475 9 June 2006*
P. Vuilleumier   Faces call for attention: evidence from patients with visual exctinction   *Neuropsychologia Vol.38 no.5 p.693   May 2000*
R. D. Wimmer et al.   Thalamic control of sensory selection in divided attention   *Nature Vol.526 p.705 29 October 2015*
T. Womelsdorf & P. Fries   The role of neuronal synchronization in selective attention   *Curr. Op. Neurobiol. Vol. 17 p.154 April 2007*

G. B. Young   Coma   *Annals of the New York Academy of Sciences Vol.1157 p.32 March 2009*
X-J. Yu et al.   Change detection by thalamic reticular neurons   *Nat. Neurosci. Vol.12 p.1165 September 2009*

Chapter 17: Neurology of Reward

R. Adolphs et al.   Amygdala damage impairs emotional memory for gist but not details of complex stimuli   *Nat. Neurosci. Vol.8 p.512   April 2005*
A. K. Anderson & E. A. Phelps   Lesions of the human amygdala impair enhanced perception of emotionally salient events   *Nature Vol.411 p.395   17 May 2001*
M. F. Barbano & M. Cador   Opioids for hedonic experience and dopamine to get ready for it   *Psychopharmacology Vol.191 p.497   2007*
J. A. Bartz et al.   Social effects of oxytocin in humans: context and person matter   *Trends in Cognitive Sciences Vol.15 p.301 July 2011*
A. Bechara et al.   Emotion, Decision Making and the Orbitofrontal Cortex   *Cerebral Cortex Vol.10. p295   March 2000*
B. C. Bernhardt & T. Singer   The Neural Basis of Empathy   *Ann. Rev. Neurosci. Vol.35 p.1   2012*
E. S. Bromberg-Martin & O. Hikosaka   Midbrain Dopamine Neurons Signal Preference for Advance Information about Upcoming Rewards   *Neuron Vol.63 p.119   16 July 2009*
H. J. Campbell   The Pleasure Areas   *Eyre Methuen   1973*
L. Cantin et al.   Cocaine is Low on the Value Ladder for Rats: Possible Evidence for Resilience to Addiction   *PLoS one 5(7) e11592   2010*
R. M. Carelli et al.   Evidence that Separate Neural Circuits in the Nucleus Accumbens Encode Cocaine Versus 'Natural' (Water and Food) Reward   *J. Neurosci. Vol.20 no.11 p.4255   2000*
E. Choleris et al.   Microparticle-based delivery of oxytocin receptor antisense DNA in the medial amygdala blocks social recognition in female mice   *PNAS Vol.104 p.4670 13 March 2000*
H. D. Critchley & S. N. Garfinkel   Interactions between visceral afferent signalling and stimulus processing   *Frontiers in Neuroscience Vol. 9 p.286   2015*
J. W. Dalley et al.   Nucleus Accumbens D2/3 Receptors Predict Trait Impulsivity and Cocaine Reinforcement   *Science Vol.315 p.1267   2 March 2007*
A. D. Damasio   Towards a neuropathology of emotion and mood   *Nature Vol.386 p.769   24 April 1997*
P. Dayan & Q. J. M. Huys   Serotonin in Affective Control   *Ann. Rev. Neurosci. Vol.34 p.95   2009*
M. De Biasi & J. A. Dani   Reward, Addiction, Withdrawal to Nicotine   *Ann. Rev. Neurosci. Vol.34 p.105 2011*
J. Debiec & R. M. Sullivan   Intergenerational transmission of emotional trauma through amygdala-dependent mother-to-infant transfer of specific fear   *PNAS Vol.111 p.12222 19 August 2014*
R. C. deCharms et al.   Control over brain activation and pain learned by using real-time functional MRI   *PNAS Vol.102 p.18626 20 December 2005*

C. K. W. De Dreu et al.   The Neuropeptide Oxytocin Regulates Parochial Altruism in Intergroup Conflict Among Humans   Science Vol.328 p.1408   11 June 2010

P. Durieux et al.   D$_2$R striatopallidal neurons inhibit both locomotor and drug reward processes   Nat. Neurosci. Vol. 12 p.393 April 2009

R. P. Epstein et al.   Dopamine D4 receptor (D4DR) exon III polymorphism associated with the human personality trait of novelty seeking   Nature Genetics Vol. 12 p.78   12 January 1996

D. Edwards et al.   Social Status Sculpts Activity of Crayfish Neurons   Science Vol.271 p.366   19 January 1996

S. Edwards et al.   Addiction-Related Alterations in D$_1$ and D$_2$ Dopamine Receptors Behavioral Responses Following Chronic Cocaine Self-Adminstration   Neuropsychopharmacology Vol.32 p.354 2007

B. J. Everitt & T. W. Robbins   Neural systems of reinforcement for drug addiction: from actions to habits to compulsion   Nat. Neurosci. Vol.8 p.1481   November 2005

C. D. Fiorillo et al.   Discrete Coding of Reward Probability and Uncertainty by Dopamine Neurons   Science Vol.299 p.1898   21 March 2003

M. J. Frank et al.   Prefrontal and striatal dopaminergic genes predict individual differences in exploration and exploitation   Nat. Neurosci. Vol.12 p.1062 August 2009

S. Frühholz et al.   Asymmetrical effects of unilateral right or left damage on auditory cortical processing of vocal emotions   PNAS Vol.112 p.1583   3 February 2015

J. L. Garrison et al.   Oxytocin/Vasopressin-Related Peptides Have an Ancient Role in Reproductive Behavior   Science Vol.338 p.540   26 October 2012

N. Gaznick et al.   Basal Ganglia Plus Insula Damage Yields Stronger Disruption of Smoking Than Basal Ganglia Damage Alone   Nicotine and Tobacco Research doi:10.1093/ntr/ntt172 29 October 2013

A. M. Graybiel   The basal ganglia: learning new tricks and loving it   Curr. Op. Neurobiol. Vol.15 no.6 p.638 December 2005

S. B. Hamman et al.   Amygdala activity related to enhanced memory for pleasant and aversive stimuli   Nat. Neurosci. Vol.2 p.289 March 1999

M. M. Heinricher et al.   Descending control of nociception: Specificity, Recruitment and Plasticity   Brain Research Reviews Vol.60 p.214 2009

K. Hikosaka & M. Watanabe   Delay Activity of Orbital and Lateral Prefrontal Neurons of the Monkey Varying with Different Rewards   Cerebral Cortex Vol.10 p.263 March 2000

O. Hikosaka et al.   Basal Ganglia Circuits for Reward and Value-Guided Behavior   Ann. Rev. Neurosci. Vol.37 p.289   2014

B. Holmes   A gene for boozy mice   New Scientist Issue 2047 14 September 1996

T. Kahnt et al.   The neural code of reward anticipation in human orbitofrontal cortex   PNAS Vol.107 p.6010 March 2010

I. Katona & T. F. Freund   Multiple Functions of Endocannabinoid Signalling in the Brain   Ann. Rev. Neurosci. Vol. 35 p.529   2012

R. Kawashima et al.   The human amygdala plays an important role in gaze monitoring: A PET study   Brain Vol. 122 p.779   1999

M. Kosfeld et al.   Oxytocin increases trust in humans   Nature Vol. 435 p.673 2 June 2005

A. Kravitz et al.   Distinct roles for direct and indirect pathway striatal neurons in

reinforcement *Nat. Neurosci. Vol.15 p.816 June 2012*
M-J. Kreek et al. Genetic influences on impulsivity, risk taking, stress responsivity and vulnerability to drug abuse and addiction *Nat. Neurosci. Vol.8 p.1451 November 2005*
E. Kross et al. Social rejection shares somatosensory representations with physical pain *PNAS Vol.108 p.6270 12 April 2011*
J. LeDoux The Emotional Brain *Weidenfeld and Nicolson 1998*
J. K. Lerch-Haner et al. Serotonergic transcriptional programming determines maternal behaviour and offspring survival *Nat. Neurosci. Vol.11 p.1001 September 2008*
M. M. Lim et al. Enhanced partner preference in promiscuous species by manipulating the expression of a single gene *Nature Vol.429 p.754 17 June 2004*
D. Lin et al. Functional identification of an aggression locus in the mouse hypothalamus *Nature Vol.470 p.221 10 February 2011*
H-S. Liu et al. Dorsolateral caudate nucleus differentiates cocaine from natural reward-associated contextual cues *PNAS Vol.110 p.4093 12 March 2013*
R. Maldonado et al. Absence of opiate rewarding effects in mice lacking D2 receptors *Nature Vol.388 p.586 7 August 1997*
R. Maldonado et al. Involvement of the Endocannabinoid System in Drug Addiction *Trends in Neurosciences Vol.29 p.225 April 2006*
M. Matsumoto & O. Hikosaka Lateral habenula as a source of negative reward signals in dopamine neurons *Nature Vol. 447 p.1111 28 June 2007*
M. Matsumoto & O. Hikosaka Two types of dopamine neuron distinctly convey positive and negative emotional signals *Nature Vol.459 p.837 11 June 2009*
C. McCall & T. Singer The animal and human neuroendocrinology of social cognition, motivation and behaviour *Nat. Neurosci. Vol.15 p.681 May 2012*
L. A. McGraw & L. R. Young The prairie vole: an emerging model organism for understanding the social brain *Trends in Neurosciences Vol. 33 p.103 February 2010*
K. W. Miyazaki et al. Optogenetic Activation of Dorsal Raphe Serotonin Neurons Enhances Patience for Future Rewards *Current Biology Vol.24 p.2033 8 September 2014*
D. Mobbs et al. Humour Modulates the Mesolimbic Reward Centres *Neuron Vol. 40 p.1041 4 December 2003*
A. Moles et al. Deficit in Attachment Behaviour in Mice Lacking the µ-Opioid Receptor Gene *Science Vol.304 p.1983 25 June 2004*
J. S. Morris et al. A subcortical pathway to the right amygdala mediating 'unseen' fear *PNAS Vol.96 p.1680 February 1999*
N. H. Naqvi et al. Damage to the Insula Disrupts Addiction to Cigarette Smoking *Science Vol.315 p.531 26 January 2007*
J. Olds Pleasure Centres in the Brain *Scientific American Vol. 195 p.105 October 1956*
J. Olds and P. Milner Positive reinforcement produced by electrical stimulation of the septal area and other regions of rat brain *J. Comp. and Physiol. Psychol. Vol.47 No.6 p.419 1954*
S. Pecina et al. Hyperdopaminergic mutant mice have higher 'wanting' but not 'liking' for sweet rewards *J. Neurosci. Vol.23 p.9395 2003*
E. A. Phelps Emotion and Cognition: Insights from Studies of the Human Amygdala *Ann. Rev. of Psychology Vol. 57 p.27 2006*

W. Ren et al.   The indirect pathway of the nucleus accumbens shell amplifies neuropathic pain   *Nat. Neurosci. Vol. 19 p.220 February 2016*
J. Roeper   Dissecting the diversity of midbrain dopamine neurons   *Trends in Neurosciences Vol.36 p.336 June 2013*
C. M. Root et al.   The participation of cortical amygdala in innate, odour-driven behaviour   *Nature Vol.515 p.269 13 November 2014*
M. Roy et al.   Cerebral and spinal modulation of pain by emotions   *PNAS Vol.106 p.20900 8 December 2009*
W. Schultz   Multiple Reward Signals in the Brain   *Nat Rev. Neurosci. Vol.1 p.199 December 2000*
D. W. Self et al.   Opposite Modulation of Cocaine-Seeking by D_1-Like and D_2-Like Dopamine Receptor Agonists   *Science Vol.27 p.1586 15 March 1996*
M. Tamietto et al.   Unseen facial and bodily expression trigger fast emotional reactions   *PNAS Vol. 106 p.17661 20 October 2009*
L. Tellez et al.   Separate circuitries encode the hedonic and nutritional values of sugar   *Nature Neuroscience Vol. 19 p.465 March 2016*
J. Tihonen et al.   Altered striatal dopamine re-uptake site densities in habitually violent and non-violent alcoholics   *Nature Medicine Vol. 1 p.654 July 1995*
P. N. Tobler et al.   Adaptive Coding of Reward Value by Dopamine Neurons   *Science Vol. 307 p1642 11 March 2005*
L. Tremblay & W. Schultz   Relative reward preference in primate orbitofrontal cortex   *Nature Vol. 398 p.704 22 April 1999*
N. D. Volkow & T-K. Li   Drug addiction: the neurobiology of behaviour gone awry   *Nat. Rev. Neurosci. Vol. 5 p.963 December 2004*
P. J. Whalen et al.   Human Amygdala Responsivity to Masked Fearful Eye Whites   *Science Vol. 306 p.2061 17 December 2004*
N. M. White & N. Hiroi   Preferential localization of self-stimulation sites in striosomes/patches in the rat striatum   *PNAS Vol. 95 p.6486 26 May 1998*
B. J. Williams et al.   Comparison of two positive reinforcing stimuli: Pups and cocaine throughout the postpartum period   *Behavioural Neuroscience Vol. 115 p.683 2001*
R. A. Wise   Role of brain dopamine in food reward and reinforcement   *Philosophical Transactions of the Royal Society B Vol. 361 p.1149 2006*
R. A. Wise   Roles for nigrostriatal – not just mesocorticolimbic – dopamine in reward and addiction   *Trends in Neurosciences Vol. 32 p.517 October 2009*
M. E. Wolf   Synaptic mechanisms underlying persistent cocaine craving   *Nat. Rev. Neurosci. Vol. 17 p.351 June 2016*
Z. Xu et al.   Effects of enriched environment on morphine-induced reward in mice   *Experimental Neurology Vol. 204 p.714 27 January 2007*
T. Yamamoto   Brain Regions Responsible for the Expression of Conditioned Taste Aversion in Rats   *Chemical Senses Vol.32 p.105 January 2007*
L. J. Young & Z. Wang   The neurobiology of pair bonding   *Nat. Neurosci. Vol. 7 p.1049 October 2004*
J. Zaki & K. Ochsner   The neuroscience of empathy: progress, pitfalls and promise   *Nat. Neurosci. Vol. 15 p.675 May 2012*
K. A. Zalocusky et al.   Nucleus accumbens D2R cells signal prior outcomes and control

risky decision-making   *Nature* Vol. 531 p.642 31 March 2016
Q. Y. Zhou & R. D. Palmiter   Dopamine-deficient mice are severely hypoactive, adipsic and aphagic   *Cell* Vol.83 p.1197 1995

Chapter 18: Learning

R. Araya et al.   Activity-dependent dendritic spine neck changes are correlated with synaptic strength   *PNAS* Vol.111 E2895 29 July 2014
J. S. Bakin & N. M. Weinberger   Induction of a physiological memory in the cerebral cortex by stimulation of the nucleus basalis   *PNAS* Vol. 93 p.11219 October 1996
T. D. Barnes et al.   Activity of striatal neurons reflects dynamic encoding and recoding of procedural memories   *Nature* Vol.437 p.1158   20 October 2005
J. D. Berke & S. E. Hyman   Addiction, Dopamine and the Molecular Mechanisms of Memory   *Neuron* Vol.25 p.515 March 2000
T.V.P. Bliss and T. Lømo   Long-lasting potentiation of synaptic transmission in the dentate area of the anaesthetised rabbit following stimulation of the perforant path   *J. Physiology* Vol. 232 p.331 1973
T. Bliss et al. eds.   LTP: Long Term Potentiation: enhancing neuroscience for 30 years   Oxford University Press 2004
B. L. Bloodgood & B. L. Sabatini Ca^{2+} signalling in dendritic spines   *Curr. Op. Neurobiology* Vol.17 p.345 June 2007
C. Bocklisch et al.   Cocaine Disinhibits Dopamine Neurons by Potentiation of GABA Transmission in the Vental Tegmental Area   *Science* Vol. 341 p.1521 27 September 2013
C. R. Bramham & D. G. Wells   Dendritic mRNA: transport, translocation and function   *Nat. Rev. Neurosci.* Vol. 8 p.776 October 2007
B. Brembs et al.   Operant Reward Learning in *Aplysia*: Neural Correlates and Mechanisms   *Science* Vol.296 p.1706 31 May 2002
P. Calabresi et al.   Hyperkinetic disorders and loss of synaptic downscaling   *Nat. Neurosci.* Vol. 19 p.868 July 2016
P. Caroni et al.   Structural plasticitiy upon learning: regulation and functions   *Nat. Rev. Neurosci.* Vol. 13 p.478 July 2012
R. C. Carroll et al.   Role of AMPA receptor endocytosis in synaptic plasticity   *Nat. Rev. Neurosci.* Vol.2 p.315 May 2001
M. A. Castro-Alamancos et al.   Different Forms of Synaptic Plasticity in Somatosensory and Motor Areas of the Neocortex   *J. Neurosci.* Vol.15 p.5324 July 1995
J-P. Changeux   Nicotine addiction and nicotinic receptors: lessons from genetically modified mice   *Nat. Rev. Neurosci.* Vol. 11 p.389 June 2010
B. T. Chen et al.   Cocaine but not natural reward self-administration nor passive cocaine infusion produces persistent LTP in the VTA   *Neuron* Vol. 59(2) p.288 31 July 2008
J. L. Chen et al.   Structural basis for the role of inhibition in facilitating adult brain plasticity   *Nat. Neurosci.* Vol. 14 p.587 May 2011
V. Chevaleyre et al.   Endocannabinoid-Mediated Synaptic Plasticity in the CNS   *Ann. Rev. Neurosci.* Vol.29 p.37 2006

S. Ciocchi et al.   Encoding of conditioned fear in central amygdala inhibitory circuits   *Nature Vol. 468 p277 11 November 2011*

G. L. Collingridge et al.   Long-term depression in the CNS   *Nat. Rev. Neurosci. Vol.11 p.459 July 2010*

Y. Dudai   The Restless Engram: Consolidation Never Ends   *Ann. Rev. Neurosci. Vol. 35 p.227 2012*

S. W. Flavell & M. D. Greenberg   Signaling Mechanisms Linking Neuronal Activity to Gene Expression and Plasticity of the Nervous System   *Ann. Rev. Neurosci. Vol.31 p.563 2008*

M. Fu et al.   Repetitive motor learning induces co-ordinated formation of clustered dendritic spines *in vivo*   *Nature Vol. 483 p.92 19 February 2012*

A. M. Graybiel   Habits, Rituals and the Evaluative Brain   *Ann. Rev. Neurosci. Vol.31 p.359 2008*

M. S. Grubb & J. Burrone   Activity-dependent relocation of the axon initial segment fine-tunes neuronal excitability   *Nature Vol. 465 p.1070   24 June 2010*

Y. Hayashi & A. K. Majewska   Dendritic Spine Geometry: Functional Implication and Regulation   *Neuron Vol.46 p.529 19 May 2005*

M. Heine et al.   Surface Mobility of Postsynapatic AMPARS Tunes Synaptic Transmission   *Science Vol. 320 p.201 11 April 2008*

A. Holtmaat et al.   Imaging of experience-dependent structural plasticity in the mouse neocortex in vivo   *Behavioural Brain Research Vol. 192(1) p.20 1 September 2008*

H. Hu et al.   Emotion Enhances Learning via Norepinephrine Regulation of AMPA-Receptors Trafficking   *Cell Vol.131 p.160 5 October 2011*

Y-Y. Huang & E. R. Kandel   The Influence of Prior Synaptic Activity on the Induction of Long-Term Potentiation   *Science Vol.255 p.730 7 February 1992*

S. E. Hyman et al.   Neural Mechanisms of Addiction: The Role of Reward-Related Learning and Memory   *Ann. Rev. Neurosci. Vol.29 p.565 2006*

J. P. Johansen et al.   Hebbian and neuromodulatory mechanisms interact to trigger associative memory formation   *PNAS Vol.111 E5584 23 December 2014*

F. Kazanetz et al.   Transition to Addiction Is Associated with a Persistent Impairment in Plasticity   *Science Vol.328 p.1709 25 June 2010*

J. T. Kittler & S. J. Moss *eds.*   The Dynamic Synapse: Molecular Methods in Ionotropic Receptor Biology   *CRC Press 2006*

B. Kolb et al.   Amphetamine or cocaine limits the ability of later experience to promote structural plasticity in the neocortex and nucleus accumbens   *PNAS Vol.100 p.10523 2003*

A. J. Koleske   Molecular mechanisms of dendrite stability   *Nat. Rev. Neurosci. Vol. 14 p.536 August 2013*

E. A. Kramár et al.   Synaptic evidence for the efficacy of spaced learning   *PNAS Vol. 109 p.5121 27 March 2012*

H. Kuba et al.   Presynaptic activity regulates Na$^+$ channel distribution at the axon initial segment   *Nature Vol. 465 p.1075 24 June 2010*

H-K. Lee et al.   Phosphorylation of the AMPA Receptor GluR1 Subunit Is Required for Synaptic Plasticity and Retention of Spatial Memory   *Cell Vol. 112 p.631 7 March 2003*

J. J. Letzkuks et al.   A disinhibitory microcircuit for associative fear learning in the auditory cortex   *Nature Vol. 480 p.331   15 December 2011*

Q-S. Liu et al.   Repeated cocaine exposure *in vivo* facilitates LTP induction in midbrain dopamine neurons   *Nature Vol.437 p.1027 13 October 2005*

C. Lohmann et al.   Transmitter-evoked local calcium release stabilizes developing dendrites   *Nature Vol. 418 p.177 11 July 2002*

H. D. Mansvelder & D. S. McGehee   Long-Term Potentiation of Excitatory Inputs to Brain Reward Areas by Nicotine   *Neuron Vol.27 p.349 August 2000*

S. Maren   Long-term potentiation in the amygdala: a mechanism for emotional learning and memory   *Trends in Neurosciences Vol.22 p.561 December 1999*

H. Markram et al.   Regulation of Synaptic Efficacy by Coincidence of Postsynaptic APs and EPSPs   *Science Vol.275 p.213   10 January 1997*

A. R. O. Martins & R. C. Froemke   Coordinated forms of noradrenergic plasticity in the locus coeruleus and primary auditory cortex   *Nat. Neurosci. Vol.18 p.1483 October 2015*

P. Massey & Z. I. Bashir   Long-term depression: multiple forms and implications for brain function   *Trends in Neurosciences Vol. 30 p.176   April 2007*

M. Matsuzaki et al.   Structural basis of long-term potentiation in single dendritic spines   *Nature Vol.429 p.761 17 June 2004*

F. J. Munoz-Cuevas et al.   Cocaine-induced structural plasticity in frontal cortex correlates with conditioned place preference   *Nat. Neurosci. Vol.16 p.1367 October 2013*

T. Nakamura et al.   Synergistic Release of Ca^{2+} from IP_3-sensitive Stores Evoked by Synaptic Activation of MGluRs Paired with Backward-Propagating Potentials   *Neuron Vol. 24 p727 November 1999*

J. Noguchi et al.   Spine-Neck Geometry Determines NMDA Receptor-Dependent CA^{2+} Signaling in Dendrites   *Neuron Vol.46 p.609 19 May 2005*

F. S. Nugent et al.   Opioids block long-term potentiation of inhibitory synapses   *Nature Vol.446 p.1086 26 April 2007*

A. T. Popescu et al.   NMDA-dependent facilitation of corticostriatal plasticity by the amygdala   *PNAS Vol.104 p.341 2 January 2007*

C. R. Raymond   LTP forms 1, 2, and 3: different mechanisms for the 'long' in long-term potentiation   *Trends in Neurosciences Vol.30 p.167   April 2007*

T. F. Roberts et al.   Rapid spine stabilization and synaptic enhancement at the onset of behavioural learning   *Nature Vol.463 p.948 18 February 2010*

J. A. Rossato et al.   Dopamine Controls Persistence of Long-Term Memory Storage *Science Vol.325 p.1017 21 August 2009*

S. J. Russo et al.   The addicted synapse: mechanisms of synaptic and structural plasticity in nucleus accumbens   *Trends in Neurosciences Vol. 33 p.267 June 2010*

B. Scelfo et al.   Learning-related long-term potentiation of inhibitory synapses in the cerebellar cortex   *PNAS Vol.105 p.769 15 January 2008*

S-H Shi et al.   Rapid Spine Delivery and Redistribution of AMPA Receptors After Synaptic NMDA Receptor Activation   *Science Vol. 284 p.1811 11 June 1999*

L. R. Squire & E. R. Kandel   Memory: From Mind to Molecules   *Scientific American Library 1999*

D. D. Stettler et al.   Axons and Synaptic Boutons Are Highly Dynamic in Adult Visual Cortex   *Neuron Vol. 49 p.877 16 March 2006*

O. Steward & E. M. Schuman   Protein Synthesis at Synaptic Sites on Dendrites   *Ann. Rev. Neurosci. Vol. 24 p.299 2001*
M. V. Storozhuk & V.F.Castellucci   Aplysia – Gill and Siphon Withdrawal Reflex Network   *Neuroscience Vol.1(9) p.291 1999*
G. Stuart & B. Sakmann   Active propagation of somatic action potentials into neocortical pyramidal cell dendrites   *Nature Vol. 367 p.69 6 January 1994*
N. Toni et al.   LTP promotes formation of multiple spine synapses between a single axon terminal and a dendrite   *Nature Vol.402 p.421 25 November 1999*
G. G. Turigiano   Homeostatic plasticity in neuronal networks: the more things change, the more they stay the same   *Trends in Neurosciences Vol. 22 p.221 May 1999*
A. R. Webb et al.   Mother's voice and heartbeat sound elicit auditory plasticity in the human brain before full gestation   *PNAS Vol. 112 p.3152 10 March 2015*
J. S. Wiegert & T. G. Oertner   Long-term depression triggers the selective elimination of weakly integrated synapses   *PNAS Vol.110 E4510   19 November 2013*
I. Willuhn et al.   Hierarchical recruitment of phasic dopamine signaling in the striatum during the progression of cocaine use   *PNAS Vol. 109 p.20703   12 December 2012*
S. Yagishita et al.   A critical time window for dopamine actions on the structural plasticity of dendritic spines   *Science Vol. 345 p.1616 26 September 2014*
Y. Yazaki-Sugiyama et al.   Bidirectional plasticity in fast-spiking GABA circuits by visual experience   *Nature Vol. 462 p.218   12 November 2009*
P. Zheng et al.   Opposite modulation of cortical N-methyl-D-aspartate receptor-mediated responses by low and high concentrations of dopamine   *Neuroscience Vol. 91 p.527 1999*

Chapter 19: Contextual and Episodic Memory

A-L. R. Adlam et al.   Dissociation between recognition and recall in developmental amnesia   *Neuropsychologia Vol.47 p.2207 2009*
J. P. Aggleton & M. W. Brown   Interleaving brain systems for episodic and recognition memory   *Trends in Cognitive Sciences Vol.10 p.455 October 2006*
J. P. Aggleton   Understanding retrosplenial amnesia: Insights from animal studies   *Neuropyschologia Vol.48 p.2328 2010*
J. A. Ainge et al.   Hippocampal place cells encode intended destination and not a discriminative stimulus in a conditional T-maze task   *Hippocampus Vol.22 p.534 March 2012*
P. Anderson et al. eds.   The Hippocampus Book   *Oxford University Press 2007*
I. E. T. de Araujo et al.   A View Model Which Accounts for the Spatial Fields of Hippocampal Primate Spatial View Cells and Rat Place Cells   *Hippocampus Vol.11 p.699 2001*
A. Attardo et al.   Impermanence of dendritic spines in live adult CA1 hippocampus   *Nature Vol.523 p.592   30 July 2015*
J. C. Augustinack et al.   H.M.'s contributions to neuroscience: a review and autopsy studies   *Hippocampus Vol.24 p.11267 November 2014*
M. D. Barense et al.   Medial Temporal Lobe Activity During Complex Discrimination of Faces, Objects, and Scenes: Effects of Viewpoint   *Hippocampus Vol.20 p.839 2010*

J. Barrash et al. The neuroanatomical correlates of route learning impairment *Neuropsychologia Vol.38 p.820 2000*

T. Bartsch et al. CA1 neurons in the human hippocampus are critical for autobiographical memory, mental time travel, and autonoetic consciousness *PNAS Vol. 108 p.17562 18 October 2011*

T. V. P. Bliss & G. L. Collingridge A synaptic model of memory: long-term potentiation in the hippocampus *Nature Vol.361 p.31 7 April 1993*

C. N. Boccara et al. Grid cells in pre- and parasubiculum *Nat. Neurosci. Vol.13 p.987 August 2010*

T. Bonnevie et al. Grid cells require excitatory drive from the hippocampus *Nat. Neurosci. Vol.16 p.309 March 2012*

M. Braun et al. Lesions Affecting the Right Hippocampus Differentially Impair Short-Term Memory of Spatial and Non-Spatial Associations *Hippocampus Vol.21 p.309 2011*

G. Buzsáki Our skewed sense of space *Science Vol.347 p.612 6 February 2015*

K. A. Cameron et al. Human Hippocampal Neurons Predict How Well Word Pairs Will be Remembered *Neuron Vol.30 p.289 April 2001*

M. J. Chadwick et al. A Goal Direction Signal in the Human Entorhinal/Subicular Region *Current Biology Vol.25 p.87 5 January 2015*

L. Cipolotti et al. Recollection and familiarity in dense hippocampal amnesia: A case study *Neuropsychologia Vol.44 p489 2006*

R. E. Clark et al. Rats with Lesions of the Hippocampus are Impaired on the Delayed Nonmatching-to-Sample Task *Hippocampus Vol. 11 p.176 2001*

S. Corkin Permanent Present Tense: The Unforgettable Life of the Amnesic Patient, H. M. *Basic Books 2013*

R. Czajkowski et al. Encoding and storage of spatial information in the retrosplenial cortex *PNAS Vol. 111 p.8661 3 June 2014*

L. Davichi & S. Du Brow How the hippocampus preserves order: the role of prediction and context *Trends in Cognitive Sciences Vol. 19 p.92 February 2015*

C. F. Doeller et al. Parallel striatal and hippocampal systems for landmarks and boundaries in spatial memory *PNAS Vol.105 p.5915 15 April 2008*

G. Dragoi & G. Buzsáki Temporal Encoding of Place Sequences by Hoppocampal Cell Assemblies *Neuron Vol.50 p.150 6 April 2006*

G. Dragoi & S. Tonegawa Preplay of future place cell sequences by hippocampal cellular assemblies *Nature Vol.469 p.397 20 January 2011*

D. Dupret et al. The reorganization and reactivation of hippocampal maps predict spatial memory performance *Nat. Neurosci. Vol. 13 p.995 August 2010*

H. Eichenbaum Time cells in the hippocampus: a new dimension for mapping memories *Nat. Rev. Neurosci. Vol.15 p.732 November 2014*

R. A. Epstein Parahippocampal and retrosplenial contributions to spatial navigation *Trends in Cognitive Sciences Vol.12 p.388 October 2008*

A. Finkelstein et al. Three-dimensional head-direction coding in the bat brain *Nature Vol. 517 p.159 8 January 2015*

L. M. Frank et al. Hippocampal Plasticity across Multiple Days of Exposure to Novel Environments *J. Neuroscience Vol.24 p.7681 1 September 2004*

P. W. Frankland & B. Bontempi   The Organisation of Recent and Remote Memories   *Nat. Rev. Neurosci. Vol.6 p.119   February 2005*
Y. Furuya et al.   Place-Related Neuronal Activity in the Monkey Parahippocampal Gyrus and Hippocampal Formation During Virtual Navigation   *Hippocampus Vol.24 p.113   January 2014*
M. Fyhn et al.   Spatial Representation in the Entorhinal Cortex   *Science Vol. 305 p.1259   27 August 2004*
M. Fyhn et al.   Hippocampal remapping and grid realignment in entorhinal cortex   *Nature Vol. 446 p.190   8 March 2007*
S. Ge et al.   A Critical Period for Enhanced Synaptic Plasticity in Newly Generated Neurons of the Adult Brain   *Neuron Vol. 54 p.55   24 May 2007*
Y. Ge et al.   Hippocampal long-term depression is required for the consolidation of spatial memory   *PNAS Vol. 107 p.16697   21 August 2010*
A. S. Gupta et al.   Segmentation of spatial experience by hippocampal theta sequences   *Nat. Neurosci. Vol. 15 p.1032   July 2012*
R. E. Hampson et al.   Distribution of spatial and nonspatial information in dorsal hippocampus   *Nature Vol. 402 p.610   9 December 1999*
T. Hartley et al.   The Hippocampus is Required for Short-term Topographical Memory in Humans   *Hippocampus Vol.17 p.34   January 2007*
H. T. Ito et al.   A prefrontal-thalamo-hippocampal circuit for goal-directed spatial navigation   *Nature Vol.522 p.50   4 June 2015*
S. Jadhav et al.   Awake Hippocampal Sharp-Wave Ripples Support Spatial Memory   *Science Vol. 336 p.1454   15 June 2012*
G. Janzen & M. van Turennout   Selective neural representation of objects relevant for navigation   *Nat. Neurosci. Vol.7 p.673   June 2004*
K. J. Jeffery ed.   The Neurobiology of Spatial Behaviour   *Oxford University Press   2003*
A. Johnson & A. D. Relish   Neural Ensembles in CA2 Transiently Encode Paths Forward of the Animal at a Decision Point   *J. Neuroscience Vol. 27(45) p.12176   2007*
M. W. Jones & M. A. Wilson   Theta Rhythms Coordinate Hippocampal-Prefrontal Interactions in a Spatial Memory Task   *PLoS biol. 3(12) e402   2005*
J. B. Julian et al.   The Occipital Place Area is Causally Involved in Representing Environmental Boundaries during Navigation   *Current Biology Vol. 26 p.1104   25 April 2015*
N. Kee et al.   Preferential incorporation of adult-generated granule cells into spatial memory networks in the dentate gyrus   *Nat. Neurosci. Vol. 10 p.355   March 2007*
N. J. Killian et al.   Saccade direction encoding in the primate entorhinal cortex during visual exploration   *PNAS Vol. 112 p.15743   22 December 2015*
E. Kropff et al.   Speed cells in the medial entorhinal cortex   *Nature Vol.523 p.419   23 July 2015*
J. Krupic et al.   Grid cell symmetry is shaped by environmental geometry   *Nature Vol.518 p.232   12 February 2015*
B.. Laeng et al.   The Eyes Remember It: Oculography and Pupillometry during Recollection in Three Amnesic Patients   *J. Cognitive Neuroscience Vol.19:11 p.1888   2007*
A. S. Lee et al.   A double dissociation revealing bidirectional competition between striatum and hippocampus during learning   *PNAS Vol.. 105 p.17163   4 November 2008*
P-P. Lenck-Santini et al.   Evidence for a Relationship Between Place-cell Spatial Firing and

Spatial Memory Performance   *Hippocampus Vol. 11 p.377   2001*
S. Leutgeb et al.   Independent Codes for Spatial and Episodic Memory in Hippocampal Neuronal Ensembles   *Science Vol. 309 p.619   22 July 2005*
E. A. Maguire et al.   London Taxi Drivers and Bus Drivers: A Structural MRI and Neuropsychological Analysis   *Hippocampus Vol. 16 p.1091   2006*
B. L. McNaughton et al.   Path integration and the neural basis of the 'cognitive map'   *Nat. Rev. Neurosci. Vol.7 p.663   August 2006*
B. Milovojevic et al.   Insight Reconfigures Hippocampal-Prefrontal Memories   *Current Biology Vol. 25 p.821   30 March 2015*
F. Mingaud et al.   The Hippocampus Plays a Critical Role at Encoding Discontinuous Events for Subsequent Declarative Memory in Mice   *Hippocampus Vol. 17 p.264 April 2007*
J. D. Monaco et al.   Attentive scanning behaviour drives one-trial potentiation of hippocampal place fields   *Nat. Neurosci. Vol. 17 p.725 May 2014*
E. I Moser et al.   Place Cells, Grid Cells and the Brain's Spatial Representation System   *Ann. Rev. Neurosci. Vol. 31 p.69 2008*
L. Nadel & W. J. Jacobs   The Role of the Hippocampus in PTSD, Panic and Phobia *in* The Hippocampus: Functions and Clinical Relevance *ed.* N. Kato   *Elsevier Science BV   1996*
Z. Nedelska et al.   Spatial navigation impairment is proportional to right hippocampal volume   *PNAS Vol. 109 p.2590   14 February 2012*
D. M. Nielson et al.   Human hippocampus represents space and time during retrieval of real-world memories   *PNAS Vol. 112 p.11078 1 September 2015*
J. O'Keefe & J. Dostrovsky   The hippocampus as a spatial map. Preliminary evidence from unit activity in the freely moving rat   *Brain Research Vol. 34 p.171 1971*
J. O'Keefe & L. Nadel   The Hippocampus as a Cognitive Map   *Oxford University Press 1979*
J. O'Keefe and M. L. Reece   Phase relationship between hippocampal place units and the EEG theta rhythm   *Hippocampus Vol.3 p.317   1993*
J. O'Keefe and N. Burgess   Geometric determinants of the place fields of hippocampal neurons   *Nature Vol. 381 p.425 30 May 1996*
S. M. O'Mara & E. T. Rolls   View-Responsive Neurons in the Primate Hippocampus   *Abstracts, Society for Neuroscience Vol. 19 Pt. 1 p.800   1993*
R. Paz et al.   A neural substrate in the human hippocampus for linking successive events   *PNAS Vol. 107 p.6046 30 March 2010*
J. M. Pearce et al.   Hippocampal lesions disrupt navigation based on cognitive maps but not heading vectors   *Nature Vol. 396 p.75 5 November 1998*
B. E. Pfeiffer & D. J. Foster   Hippocampal place-cell sequences depict future paths to remembered goals   *Nature Vol. 497 p.74 2 March 2013*
P. Y. Risold & L. W. Swanson   Structural Evidence for Functional Domains in the Rat Hippocampus   *Science Vol. 272 p.1484*
S. Robinson et al.   Involvement of retrosplenial cortex in forming functional associations between multiple sensory stimuli   *Behavioral Neuroscience Vol. 125 p.578 2011*
C. Rochfort et al.   Cerebellum Shapes Hippocampal Spatial Code   *Science Vol. 334 p.385 21 October 2011*
E.T. Rolls et al.   Spatial View Cells and the Representation of Space in the Primate Hip

pocampus *Hippocampus Vol. 9 no.4 p.467 1999*
C. J. Sandoval et al. When Are New Hippocampal Neurons, Born in the Adult Brain, Integrated into the Network That Processes Spatial Information *PLoS one 6(3) e17689 2011*
F. Sargolini Conjunctive Representation of Position, Direction and Velocity in Entorhinal Cortex *Science Vol. 312 p. 759 5 March 2006*
C. Schmidt-Hieber Enhanced synaptic plasticity in newly generated granule cells of the adult hippocampus *Nature Vol. 429 p.184 13 May 2004*
W. B. Scoville & B. Milner Loss of recent memory after bilateral hippocampal lesions *J. Neurology, Neurosurgery and Psychiatry Vol. 20 p.11 1957*
T. Solstad et al. Representation of Geometric Borders in the Entorhinal Cortex *Science Vol. 322 p.1865 19 December 2008*
E. Y. Song Role of Active Movement in Place-Specific Firing of Hippocampal Neurons *Hippocampus Vol. 15 p.8 2005*
L. R. Squire & J. T. Wixted The Cognitive Neuroscience of Human Memory Since H. M. *Ann. Rev. Neurosci. Vol. 34 p.259 2011*
B. P. Staresina & L. Davichi Mind the Gap: Binding Experiences Across Space and Time in the Human Hippocampus *Neuron Vol. 63 p.267 30 July 2009*
B. A. Strange et al. Functional organization of the hippocampal longitudinal axis *Nat. Rev. Neurosci. Vol. 15 p.655 October 2014*
N. Takahashi et al. Pure topographic disorientation due to right retrosplenial lesion *Neurology Vol. 49 no.2 p.464 August 1997*
S. Tronel et al. Spatial learning sculpts the dendritic arbor of adult-born hippocampal neurons *PNAS Vol. 107 p.7963 27 April 2010*
S. D. Vann et al. What does the retrosplenial cortex do? *Nat. Rev. Neurosci. Vol. 10 p.792 November 2009*
F. Vargha-Khadem et al. Differential Effects of Early Hippocampal Pathology on Episodic and Semantic Memory *Science Vol. 277 p.376 18 July 1997*
I. V. Viskontas et al. The Neural Correlates of Recollection: Hippocampal Activation Declines as Episodic Memory Fades *Hippocampus Vol. 19 p.265 2000*
G. Winocur et al. An Investigation of the Effects of Hippocampal Lesions in Rats on Pre- and Postoperatively Acquired Spatial Memory in a Complex Environment *Hippocampus Vol. 20 p.1350 2010*
Y. Zhang et al. Bilateral vestibular deafferentation impairs performance in a spatial forced alternation task in rats *Hippocampus Vol. 17 p.253 April 2007*
S. M. Zola-Morgan & L. R. Squire The primate hippocampal formation: evidence for a time-limited role in memory storage *Science Vol. 250 p.288 12 October 1999*

Chapter 20: Sleep, Dreams and Memory

S. J. Aton et al. Visual experience and subsequent sleep induce sequential plastic changes in putative inhibitory and excitatory cortical neurons *PNAS Vol. 110 p.3101 19 February 2013*
D.C. Barnes et al. Odor Fear Conditioning Modifies Piriform Cortex Local Field Potentials Both during Conditioning and during Post-Conditioning *Sleep PLoS ONE*

10.1371/journal.pone.0018130   23 March 2011
T. Brashers-Krug et al.   Consolidation in human motor memory   Nature   Vol. 382   p.252   1996
F. Crick & G. Mitchison   The Function of Dream Sleep   Nature   Vol. 304   p.111   14 July 1983
W.C. Dement   Some Must Watch While Some Must Sleep   W. H. Freeman   1972
V. Ego-Stengel & M. A. Wilson   Disruption of Ripple-Associated Hippocampal Activity During Rest Impairs Spatial Learning in the Rat   Hippocampus   Vol.20   p.1   2010
D. R. Euston et al.   Fast-Forward Playback of Recent Memory Sequences in Prefrontal Cortex During Sleep   Science   Vol. 218   p.1147   16 November 2007
C. Evans and E. A. Newman   Landscapes of the Night   Gollancz   1983
C. A. Everson   Functional consequences of sustained sleep deprivation in the rat   Behavioural Brain Research   Vol. 69   p.43   1995
K. M. Fenn et al.   Consolidation during sleep of perceptual learning of spoken language   Nature   Vol. 2245   p.614   9 October 2003
D. J. Foster and M. A. Wilson   Reverse replay of behavioural sequences in hippocampal cells during the awake state   Nature   Vol. 440   p.680   30 March 2006
M. G. Frank et al.   Sleep enhances plasticity in the developing visual cortex   Neuron   Vol. 30   p.275   April 2001
N. P. Franks   General anaesthesia: from molecular targets to neuronal pathways of sleep and arousal   Nat. Rev. Neurosci.   Vol. 9   p.370   May 2008
S. Freud   The Interpretation of Dreams   1900;   trans. J. Strachey   Hogarth Press 1953; Pelican Books 1976
L. Fuentemilla et al.   Hippocampus-Dependent Strengthening of Targeted Memories via Reactivation during Sleep in Humans   Current Biology   Vol. 23   p.1769   23 September 2013
F. Haist et al.   Consolidation of human memory revealed over decades by functional magnetic resonance imaging   Nature Neuroscience   Vol. 4   p.1139   November 2001
Y. Han et al.   Selective Activation of Cholinergic Basal Forebrain Neurons Induces Immediate Sleep-wake Transitions   Current Biology   Vol. 24   p.693   17 March 2014
O. K. Hassani et al.   Melanin-concentrating hormone neurons discharge in a reciprocal manner to orexin neurons across the sleep-wake cycle   PNAS   Vol. 106   p.2418   14 March 2009
E. Hennevin et al.   Processing of learned information in paradoxical sleep: relevance for memory   Behavioural Brain Research   Vol. 69   p.125   1995
C. G. Herrera et al.   Hypothalamic feedforward inhibition of thalamocortical network controls arousal and consciousness   Nature Neuroscience   Vol. 19   p.290   February 2016
J. Allan Hobson   Dreaming: An Introduction to the Science of Sleep   Oxford University Press   2002
J. Allan Hobson & E. F. Pace-Schott   The cognitive neuroscience of sleep: neuronal systems of consciousness and learning   Nat. Rev. Neurosci.   Vol. 3   p.679   September 2002
K. L. Hoffman & B. L. McNaughton   Sleep on it: cortical reorganisation after-the-fact   Trends in Neurosciences   Vol. 25   p.1   January 2002
N. Humphrey   Nature's Psychologists   in   Consciousness and the Physical World   ed. B. D. Josephson & V.S. Ramachandran   Duckworth   1979

L. Imeri & M. R. Opp   How (and why) the immune system makes us sleep   *Nat. Rev. Neurosci.* Vol. 10 p.199   March 2009

M. Korman et al.   Daytime sleep condenses the time course of motor memory consolidation   *Nat. Neurosci.* Vol. 9 p.910   September 2008

J. Lara-Carrasco et al.   Impact of REM sleep deprivation and dreaming on emotional adaptation to negative stimuli   *Sleep* Vol.30 Abstracts Supplement A375 2007

J. Lapierre et al.   Cortical acetylcholine release is lateralised during asymmetrical slow wave sleep in northern fur seals   *Sleep* Vol. 30 Abstracts Suppement A28 2007

J. A. Lesku & N. C. Rattenborg   Avian Sleep   *Current Biology* Vol. 24 R12   6 January 2014

C. M. Lewis et al.   Learning sculpts the spontaneous activitiy of the resting human brain   *PNAS* Vol. 106 p.17558   13 October 2009

P. A. Lewis & S. J. Durrant   Overlapping memory replay during sleep builds cognitive schemata   *Trends in Cognitive Sciences* Vol. 15 p.343 August 2011

P. A. Lewis et al.   The impact of overnight consolidation upon memory for emotional and neutral encoding contexts   *Neuropsychologia* Vol. 49 p.2619 August 2011

A. Lim et al.   Deep brain recording and stimulation of the human pedunculopontine nucleus – identification of PGO waves and modulation of REM sleep   *Sleep* Vol. 30 Abstracts Supplment A2 2007

K. Louie & M. A. Wilson   Temporallly Structured Replay of Awake Hippocampal Ensemble Activity during Rapid Eye Movement Sleep   *Neuron* Vol. 29 p.145   January 2001

P. Macquet   The Role of Sleep in Learning and Memory   *Science* Vol. 294 p.1048 2 November 2001

G. A. Marks et al.   A functional role for REM sleep in brain maturation   *Behavioural Brain Research* Vol. 69 9.1 1995

D. Marr   Simple memory: a theory for archicortex   *Philosophical Transactions of the Royal Society of London B* Vol. 262 p.123 1971

S. Mednick et al.   Sleep-dependent learning: a nap is as good as a night   *Nature Neurosci.* Vol. 6 p697 July 2003

D. Miyamoto et al.   Top-down cortical input during NREM sleep consolidates perceptual memory   *Science* Vol. 352 p.1315 10 June 2016

M. Morrissey et al.   Consequences of neonatal REM sleep deprivation   *Sleep* Vol. 30 A27 2007

T. A. Nielsen et al.   Immediate and delayed incorporation of events into dreams: further replication and implications for dream function   *J. Sleep Research* Vol. 13 p.327 2004

E. F. Pace-Schott & J. A. Hobson   The Neurobiologoy of Sleep: Genetics, Cellular Physiology and Subcortical Networks   *Nat. Rev. Neurosci.* Vol. 3 p.591 August 2002

C. Pavlides & J.Winson   Influence of hippocampal place cell firing in the awake state on the activity of these cells during subsequent sleep episodes   *J. Neuroscience* Vol. 9 p.2907 1989

H. Poincaré   The Foundations of Science   trans. G. B. Halstead   Science Press 1924

W. Ramadan et al.   Hippocampal Sharp/Wave Ripples during Sleep for Consolidation of Associative Memory   PLoS ONE 10.1371/jounal.pone.0006697 20 August 2009

B. H. Rasch et al.   Combined Blockade of Cholinergic Receptors Shifts the Brain from Stimulus Encoding to Memory Consolidation   *J. Cognitive Neurosci.* Vol. 18:5 p.793

2006

B. H. Rasch et al.   Odor Cues During Slow-Wave Sleep Prompt Declarative Memory Consolidation   *Science Vol. 315 p.1426 9 March 2007*

G. F. Reed   Sensory Deprivation *in* Aspects of Consciousness *ed.* G. Underwood & R. Stevens   *Academic Press 1982*

S. Ribeiro et al.   Long-Lasting Novelty-Induced Neuronal Reverberation during Slow-Wave Sleep in Multiple Forebrain Areas   *PLoS Biology 10.1371/journal.pbio.0020024 20 January 2004*

W. Robert   Der Traum als Naturnotwendigkeit erklärt   *Hamburg 1886*

H. P. Roffwarg et al.   Ontogenesis of Sleep Cycles *in* Early Human Development *ed.* S. J. Hutt & C. Hutt   *Oxford University Press 1973*

N. F. Ruby et al.   Hippocampal-dependent learning requires a functional circadian system   *PNAS Vol. 105 p.15593 8 October 2008*

C. B. Saper et al.   The sleep switch: hypothalamic control of sleep and wakefulness   *Trends in Neurosciences Vol. 24 p.726 December 2001*

M. Schatzman   Sleeping on problems really can solve them   *New Scientist p.416 11 August 1983*

J. M. Siegel   Clues to the functions of mammalian sleep   *Nature Vol. 437 p.1264 27 October 2005*

J. M. Siegel   Do all animals sleep?   *Trends in Neurosciences Vol. 31 p.208 April 2008*

W. E. Skaggs & B. L. McNaughton   Replay of Neuronal Firing Sequences in Rat Hippocampus During Sleep Following Spatial Experience   *Science Vol. 271 p.1870 29 March 1996*

C. Smith and S. Butler   Paradoxical sleep at selective times following training is necessary for learning   *Physiology and Behaviour Vol. 29 p.469 1982*

B. P. Staresina et al.   Awake reactivation predicts memory in humans   *PNAS Vol. 110 p.21159 24 December 2013*

R. Stickgold et al.   Sleep-Induced Changes in Associative Memory   *J. Cognitive Neurosci. Vol. 11 p.182 March 1999*

R. Stickgold et al.   Replaying the Game: Hypnagogic Images in Normals and Amnesics   *Science Vol. 290 p.350 13 October 2000*

R. Stickgold et al.   Sleep, Learning and Dreams: Off-line Memory Reprocessing   *Science Vol. 294 p.1052 2 November 2001*

R. Stickgold & M. P. Walker *eds.* The Neuroscience of Sleep   *Academic Press 2009*

R. Stickgold & M. P. Walker   Sleep-dependent memory triage: evolving generalization through selective processing   *Nat. Neurosci. Vol. 16 p.139 February 2013*

G. R. Taylor   The Natural History of the Mind   *Secker and Warburg 1979*

V. V. Vyazovskiy et al.   Molecular and electrophysical evidence for net synaptic potentiation in wake and depression in sleep   *Nat. Neurosci. Vol. 11 p.200 February 2008*

U. Wagner et al.   Sleep inspires insight   *Nature Vol. 427 p.352 22 January 2004*

E. J. Walmsley et al.   Dreaming of a Learning Task is Associated with Enhanced Sleep-Dependent Memory Consolidation   *Current Biology Vol. 20 p.850 11 May 2010*

M. Xu et al.   Basal forebrain circuit for sleep-wake control   *Nat. Neurosci. Vol. 18 p.1641 November 2015*

G. Yang et al.   Sleep promotes branch-specific formation of spines after learning

Science Vol. 344 p.1173 6 June 2014

Chapter 21: The Long-Term Memory Store

T. J. Andrews et al.   Activity in the Fusiform Gyrus Predicts Conscious Perception of Rubin's Vase-Face Illusion   *Neuroimage Vol. 17 p.890 2002*
G. Avidan et al.   Impaired holistic processing in congenital prosopagnosia   *Neuropsychologia Vol. 49 p.2541 2011*
E. H. Aylward et al.   Brain Activation during Face Perception: Evidence of a Developmental Change   *J. Cognitive Neuroscience Vol. 17 No.2 p.308 2005*
P. Bartolomeo et al.   Visually- and motor-based knowledge of letters: evidence from a pure alexic patient   *Neuropsychologia Vol. 40 no.8 p.1363 2002*
P. Belin et al.   Voice-selective areas in human auditory cortex   *Nature Vol. 403 p.309 20 January 2000*
U. Bellugi & E. S.Klima   Language, spatial cognition and the brain   in Cognition, Computation and Consciousness   ed. M. Ito et al.   *Oxford University Press 1997*
M. Bilalic et al.   It Takes Two – Skilled Recognition of Objects Engages Lateral Areas in Both Hemispheres   *PLoS ONE 10.1371/journal.pone.0016202 24 January 2011*
J. Billino et al.   Cortical networks for motion processing: Effects of focal brain lesions on perception of different motion types   *Neuropsychologia Vol. 47 p.2133 2009*
S. C. Blank et al.   Speech Production: Wernicke, Broca and beyond   *Brain Vol. 125 p.1829 2001*
A. Blaisi et al.   Early Specialisation for Voice and Emotion Processing in the Infant Brain   *Current Biology Vol. 21 p.1220 26 July 2011*
S. Bookheimer   Functional MRI of Language: New Approaches to Understanding the Cortical Organisation of Semantic Processing   *Ann. Rev. Neurosci. Vol. 25 p.151 2002*
K. H. Britten   Mechanisms of Self-Motion Perception   *Ann. Rev. Neurosci. Vol. 31 p.389 2008*
T. Busigny et al.   Holistic perception of the individual face is specific and necessary: Evidence from an extensive case study of acquired prosopagnosia   *Neuropsychologia Vol. 48 p.4057 2010*
Q. Cai et al.   Complementary hemispheric specialization for language production and visuospatial attention   *PNAS Vol. 110 E322 22 January 2013*
G. A. Calvert et al.   Activation of Auditory Cortex During Silent Lipreading   *Science Vol. 276 p.593 25 April 1997*
B. Calvo-Merino et al.   Action observation and acquired motor skills: an fMRI study with expert dancers   *Cerebral Cortex Vol. 15 p.1243 2005*
F. Campanella et al.   Naming manipulable objects: Anatomy of a category specific effect in left temporal tumours   *Neuropsychologia Vol. 48 p.1597 2010*
M. Carreiras et al.   Neural processing of a whistled language   *Nature Vol. 433 p.31 6 January 2005*
O. Collignon et al.   Functional specialization for auditory-spatial processing in the occipital cortex of congenitally blind humans   *PNAS Vol. 108 p.4435 15 March 2011*
O. Collignon et al.   Long-Lasting Crossmodal Cortical Reorganisation Triggered by Brief Postnatal Visual Deprivation   *Current Biology Vol. 25 p.2379 21 September 2015*

J. A. Collins & I. R. Olson   Beyond the FFA: The role of the anterior temporal lobes in face processing   *Neuropsychologia Vol. 61 p.65 2014*
A. C. Connolly et al.   The representation of biological classes in the human brain   *J. Neuroscience Vol. 32 p.2608 2012*
S. Curtiss   Genie: A Psycholinguistic Study of a Modern-Day "Wild Child"   Academic Press 1977
A. D'Anselmo et al.   Hemispheric asymmetries in bilinguals: Tongue similarity affects lateralization of second language   *Neuropsychologia Vol. 51 No.7 p.1187 2013*
S. Dehaene & L. Cohen   The unique role of the visual word form area in reading   *Trends in Cognitive Sciences Vol. 15 p.254 June 2011*
P. E. Downing et al.   A Cortical Area Selective for Visual Processing of the Human Body   *Science Vol. 293 p.2470 28 September 2001*
A. D. Engell & J. V. Haxby   Facial expression and gaze-direction in human superior temporal sulcus   *Neuropsychologia Vol. 45 p.3234 2007*
E. Fedorenko et al.   Language and Domain-General Regions Lie Side by Side within Broca's Area   *Current Biology Vol. 22 p.2059 6 November 2012*
L. Franco & R. W. Sperry   Hemisphere Lateralisation for Cognitive Processing of Geometry   *Neuropsychologia Vol. 15 p.107 1977*
A. D. Friederici   Pathways to language: fiber tracts in the human brain   *Trends in Cognitive Sciences Vol. 13 p.175 2009*
V. A. Fromkin   Slips of the Tongue   *Scientific American Vol. 229 p.110 December 1973*
S. Freud   The Psychopathology of Everyday Life   1901
M. G. Funnell   The calculating hemispheres: Studies of a split-brain patient   *Neuropsychologia Vol. 45 No.1 p.2378 2007*
M. Gamer et al.   The human amygdala drives reflexive orienting towards facial features   *Current Biology Vol. 23 R.917 21 October 2013*
L. Garrido et al.   Developmental phonagnosia: A selective deficit of vocal identity recognition   *Neuropsychologia Vol. 47 p.123 2009*
J. Ge et al.   Cross-language differences in the brain network subserving intelligible speech   *PNAS Vol. 112 p.2972 10 March 2015*
L. Georgy et al.   The superior colliculus is sensitive to gestalt-like stimulus configurations in hemispherectomy patients   *Cortex Vol. 81 p.151 August 2016*
C. L. Grady et al.   Early visual deprivation from congenital cataracts disrupts activity and functional connectivity in the face network   *Neuropsychologia Vol. 57 p.122 2014*
J. R. Hanley & S. Peters   A dissociation between the ability to print and the ability to write discursively in lower case letters   *Cortex Vol.32 p.737 December 1996*
J. Hart Jr. et al.   Category-specific naming deficit following cerebral infarction   *Nature Vol. 316 p.439 1 August 1985*
B. M. Harvey et al.   Topographic Representation of Numerosity in the Human Parietal Cortex   *Science Vol. 341 p.1123 6 September 2012*
J. Haxby et al.   The distributed human neural system for face perception   *Trends in Cognitive Sciences Vol. 4 p.223 June 2000*
W. Hirstein & V. S. Ramachandran   Capgras syndrome: a novel probe for understanding the neural representation of the identity and familiarity of persons   *Proceedings of the Royal Society B Vol. 264 p.437 22 March 1997*

X. Huang et al.   Progressive maturation of silent synapses governs the duration of a critical period   *PNAS Vol. 112 E3131   16 June 2015*

M. H. Johnson   Subcortical Face Processing   *Nat. Rev. Neurosci. Vol. 6 p.766   October 2005*

D. Kemmerer   The spatial and temporal meanings of English prepositions can be independently impaired   *Neuropsychologia Vol. 43 no.5 p.797 2005*

Z. Kourtzi & N. Kanwisher   Activation in Human MT/MST by Static Images with Implied Motion   *J. Cognitive Neuroscience Vol. p.48   January 2000*

R. Le Grand et al.   Expert face processing requires visual input to the right hemisphere during infancy   *Nat. Neurosci. Vol. 6 p.1108   October 2003*

D. A. Leopold et al.   Norm-based face encoding by single neurons in the monkey inferotemporal cortex   *Nature Vol. 442 p.572   3 August 2006*

F. Liégeois et al.   Language after hemispherectomy in childhood: Contributions from memory and intelligence   *Neuropsychologia Vol. 46 p.3101   2008*

G. Loffler et al.   fMRI evidence for the neural representation of faces   *Nat. Neurosci. Vol. 8 p.1386   October 2005*

M. Makinodan et al.   A Critical Period for Social-Experience Dependent Oligodendrocyte Maturation and Myelination   *Science Vol. 337 p.1357   14 September 2012*

J. S. Morris et al.   Occipitotemporal activation evoked by the perception of human bodies is modulated by the presence or absence of the face   *Neuropsychologia Vol. 44 no. 10 p.1919   2006*

J. S. Morris et al.   Differential extrageniculate and amygdal responses to presentation of emotional faces in a cortically blind field   *Brain Vol. 1234 p.1241   June 2001*

M. Moscovitch et al.   What is so special about face recognition? Nineteen experiments on a person with visual object agnosia and dyslexia but normal face recognition   *J. Cognitive Neurosci. Vol. 9 p.555   1997*

H. E. Moss & L. K. Tyler   A progressive category-specific semantic deficit for non-living things   *Neuropsychologia Vol. 38 p.60   January 2000*

T. Nakano   Facilitation of face recognition through the retino-tectal pathway   *Neuropyschologia Vol. 51 no.10 p.2043   August 2013*

K. Nelissen et al.   Observing Others: Multiple Action Representation in the Frontal Lobe   *Science Vol. 310 p.332   14 October 2005*

E. E. Nelson et al.   Growing pains and pleasures: how emotional learning guides development   *Trends in Cognitive Sciences Vol. 18 p.99   February 2014*

H. J. Neville et al.   Cerebral organization for language in deaf and hearing subjects: Biological constraints and effects of experience   *PNAS Vol. 107 p.7539   20 April 2010*

A. Nieder   The neuronal code for number   *Nat. Rev. Neurosci. Vol. 17 p.366   June 2016*

M. F. Paredes et al.   Extensive migration of young neurons into the infant human frontal lobe   *Science Vol. 354   7 October 2016*

K. Patterson et al.   Where do you know what you know? The representation of semantic knowledge in the human brain   *Nat. Rev. Neurosci. Vol. 8 p.976   December 2007*

M. V. Peelen & P. E. Downing   The neural basis of visual body perception   *Nat. Rev. Neurosci. Vol. 8 p.636   August 2007*

J. W. Pierce et al.   Configurational coding, familiarity and the right hemisphere advantage for face recognition in sheep   *Neuropsychologia Vol. 38 p.475   2000*

D. Pitcher at el.   Combined TMS and fMRI Reveal Dissociable Cortical Pathways for Dynamic and Static Face Perception   *Current Biology Vol. 24 p.2066   8 September 2014*
R. Rajimehr et al.   An anterior face patch in human cortex, predicted by macaque maps   *PNAS Vol. 106 p.1995   12 January 2009*
T. C. Rickard et al.   The Calculating Brain: an fMRI study   *Neuropsychologia Vol 38 p325   2000*
D. E. Rumelhart   Introduction to Human Information Processing   *John Wiley and Sons 1977*
O. W. Sacks   The Man Who Mistook His Wife for a Hat   *Summit Books 1985*
K. S. Scherg et al.   Visual category-selectivity for faces, places and objects emerges along different developmental trajectories   *Developmental Science Vol. 10 F15   2007*
A. Schlack & T. D. Albright   Remembering Visual Motion: Neural Correlates of Associative Plasticity and Motion Recall in Cortical Area MT   *Neuron Vol. 53 No.6 p.881   15 March 2007*
J. S. Siegel et al.   Disruptions of network connectivity predict impairment in multiple behavioral domains after *stroke*   *PNAS Vol.113 E4367   26 July 2016*
N. Sigala & N. K. Logothetis   Visual categorization shapes feature selectivity in the primate temporal cortex   *Nature Vol. 415 p.318   17 January 2002*
W. K. Simmons et al.   A common neural substate for perceiving and knowing about colour   *Neuropsychologia Vol. 45 No.12 p.2802   2007*
M. Siniscalchi et al.   Hemispheric specialisation in dogs for processing different acoustic stimuli   *PLoS one 3(10) e3349   2008*
E. Striem-Amit et al.   Reading with Sounds: Sensory Substitution Selectively Activates the Visual Word Form Area in the Blind   *Neuron Vol. 76 p.640   8 November 2012*
L. Thaler et al.   Neural correlates of Natural Human Echolocation in Early and Late Blind Echolocation Experts   *PLoS one 6(5) e20162   2011*
D. Tsao   A Dedicated System for Processing Faces   *Science Vol. 314 p.72   6 October 2006*
G. Van Belle et al.   Whole not hole: Expert face recognition requires holistic perception   *Neuropsychologia Vol. 48 p.2620   2010*
Y. Wang et al.   Dynamic neural architecture for social knowledge retrieval   *PNAS Vol. 114 E3305 18 April 2017*
R. Zahn et al.   Social concepts are represented in the superior anterior temporal cortex   *PNAS Vol. 104 p.6430   2007*

Chapter 22: Where the Thinking's Done

K. Amemori & A. M. Graybiel   Localized microstimulation of pregenual cingulate cortex induces negative decision-making   *Nat. Neurosci. Vol. 15 p.776   May 2012*
C. Amiez & M. Petrides   Selective involvement of the mid-dorsolateral prefrontal cortex in the coding of the serial order of visual stimuli in working memory   *PNAS Vol. 104 p.13786 2007*
D. M. Amodio & C. D. Frith   Meeting of minds: the medial frontal cortex and social cognition   *Nat. Rev. Neurosci. Vol. 7 p.268 April 2006*
D. Badre et al.   Hierarchical cognitive control deficits following damage to the human

frontal lobe   *Nat. Neurosci. Vol. 12 p.515   April 2009*
H. Barbas & B. Zikopoulos   The Prefrontal Cortex and Flexible Behaviour   *The Neuroscientist Vol. 13 No. 5 p.532   2007*
A. K. Barbey et al.   An evolutionarily adaptive neural architecture for social reasoning   *Trends in Neurosciences Vol. 32 p.603   December 2009*
U. Basten et al.   How the brain integrates costs and benefits during decision making   *PNAS Vol. 107 p.21767   14 December 2010*
A. Bechara   The neurology of social cognition   *Brain Vol. 125 p.1673 August 2003*
E. A. Boschin et al.   Essential functions of the primate frontopolar cortex in cognition   *PNAS Vol. 112 E1020   3 March 2015*
M. Botvinick et al.   Conflict monitoring and anterior cingulate cortex: an update   *Trends in Cognitive Sciences Vol. 8 p.539   December 2004*
M. Brass et al.   Imitative Response Tendencies in Patients with Frontal Lesions   *Neuropsychology Vol. 17 p.265 2003*
M. Brass and P. Haggard   To Do or Not to Do: The Neural Signature of Self-Control   *J. Neuroscience Vol.. 27(34) p.9141   22 August 2007*
U. Braun et al.   Dynamic reconfiguration of frontal brain networks during executive cognition in humans   *PNAS Vol. 112 p.1168   15 September 2015*
T. J. Bronowski et al.   Cognitive Deficit Caused by Regional Depletion of Dopamine in Prefrontal Cortex of Rhesus Monkey   *Science Vol. 205 p.929   1979*
J. W. Brown & T. S. Braver   Learned Predictions of Error Likelihood in the Anterior Cingulate Cortex   *Science Vol. 307 p.118   18 February 2005*
K. A. Burke et al.   The role of the orbitofrontal cortex in the pursuit of happiness and more specific rewards   *Nature Vol. 454 p.340   17 July 2008*
T. J. Buschman and E. K. Miller   Top-Down Versus Bottom-Up Control of Attention in the Prefrontal and Posterior Parietal Cortices   *Science Vol. 315 p.1860   30 March 2007*
G. Chahine et al.   On the role of the anterior prefrontal cortex in cognitive 'branching': An fMRI study   *Neuropsychologia Vol.77 p.421   2015*
L. L. Chao & R. T. Knight   Contribution of Human Prefrontal Cortex to Delay Performance   *J. Cognitive Neuroscience Vol. 10 no.2 p.167   March 1998*
S. Charron & E. Keochlin   Divided Representation of Concurrent Goals in the Human Frontal Lobes   *Science Vol. 328 p.360 16 April 2010*
K. Christoff & J. D. E. Gabriele   The frontopolar cortex and human cognition: evidence for a rostrocaudal hierarchical organisation within the human prefrontal cortex   *Psychobiology Vol. 28 p.168   2000*
K. Christoff et al.   Neural Basis of Spontaneous Thought Processes   *Cortex Vol. 40 p.6623   2004*
A. K. Churchland et al.   Decision-making with multiple alternatives   *Nat. Neurosci. Vol. 11 p.693 June 2008*
M. W. Cole et al.   Global Connectivity of Prefrontal Cortex Predicts Cognitive Control and Intelligence   *J. Neuroscience Vol. 32(26) p.8988   27 June 2012*
S. M. Courtney et al.   An Area Specialized for Spatial Working Memory in Human Frontal Cortex   *Science Vol. 279 p.1347   27 February 1998*
P. L. Croxson et al.   Cholinergic modulation of a specific memory function of prefrontal cortex   *Nat. Neurosci. Vol. 12 p.1510   December 2011*

K. R. Daffner & M. M. Mesulam   The central role of the prefrontal cortex in directing attention to novel events   Brain Vol. 123(5) p.927   May 2000

H. Damasio et al.   The Return of Phineas Gage: Clues About the Brain from the Skull of a Famous Patient   Science Vol. 264 p.1102   20 May 1994

V. de Lafuente & R. Romo   Neural correlates of subjective sensory experience   Nat. Neurosci. Vol. 8 p.1698   December 2005

E. K. Diekhof et al.   The orbitofrontal cortex and its role in the assignment of behavioural significance   Neuropsychologia Vol. 49 p.984   2011

W. C. Drevets et al.   Subgenual prefrontal cortex abnormalities in mood disorders   Nature Vol. 386 p.824   24 April 1997

N. Durstewitz et al.   Dopamine-Mediated Stabilization of Delay-Period Activity in a Network Model of Prefrontal Cortex   J. Neurophysiology Vol. 83 No.3 p.1733   2000

N. Eisenberger et al.   Does Rejection Hurt? An fMRI Study of Social Exclusion   Science Vol. 302 p.237   10 October 2003

G. N. Elston   Pyramidal cells of the frontal lobe: all the more spinous to think with   J. Neuroscience Vol. 20(18) RC95   2000

A. Etkin et al.   Emotional Processing in anterior cingulate and medial prefrontal cortex   Trends in Cognitive Sciences Vol. 15 p.85   February 2011

B. Figner et al.   Lateral prefrontal cortex and self-control in intertemporal choice   Nat. Neurosci. Vol. 13 p.538   May 2010

D. J. Freedman et al.   Categorical Representation of Visual Stimuli in the Primate Prefrontal Cortex   Science Vol. 291 p.312   12 January 2001

N. Fujii & A. M. Graybiel   Representation of Action and Sequence Boundaries by Macaque Prefrontal Cortical Neurons   Science Vol. 301 p.1246   29 August 2003

S. Funahishi et al.   Prefrontal neuronal activity in rhesus monkeys performing a delayed anti-saccade task   Nature Vol. 365 p.753   21 October 1993

J. Fuster   Prefrontal neurons in networks of executive memory   Brain Research Bulletin Vol. 52 no.5 p331   2000

P. S. Goldman-Rakic   Dynamics of cortical memory mechanisms in Integrative and Molecular Approach to Brain Function   ed. M. Ito & Y. Miyashita   Elsevier 1996

R. Z. Goldstein & N. D. Volkow   Dysfunction of the prefrontal cortex in addiction: neuroimaging findings and clinical implications   Nat. Rev. Neurosci. Vol. 12 p.652   November 2011

F. Grabenhorst & E. T. Rolls   Value, pleasure and choice in the ventral prefrontal cortex   Trends in Cognitive Sciences Vol. 15 p.56   February 2011

J. Gray et aL.   Neural mechanisms of general fluid intelligence   Nat. Neurosci. Vol. 6 p.316   March 2003

J. R. Gray et al.   Integration of emotion and cognition in the lateral prefrontal cortex   PNAS Vol. 99 p.4115   19 March 2002

J. Grèzes et al.   Brain mechanisms for inferring deceit in the actions of others   J. Neuroscience Vol. 24 p.5500   2004

M. Hampson et al.   Brain Connectivity Related to Working Memory Performance   J. Neuroscience Vol. 26(51) p.13338   20 December 2006

B. Y. Hayden et al.   Neuronal basis of sequential foraging decisions in a patchy environment   Nat. Neurosci. Vol. 14 p.933   July 2011

G. Herbet et al.   Disrupting posterior cingulate connectivity disconnects consciousness from the external environment   *Neuropsychologia Vol. 56 p.2239   2014*

K. Hikosaka & M. Watanabe   Delay Activity of Orbital and Lateral Prefrontal Neurons of the Monkey Varying with Different Rewards   *Cerebral Cortex Vol. 10 p.263   March 2000*

C. B. Holroyd et al.   Dorsal anterior cingulate cortex shows fMRI response to internal and external error signals   *Nat. Neurosci. Vol. 7 p.497 May 2004*

J. D. Howard et al.   Identity-specific coding of future rewards in the human orbitofrontal cortex   *PNAS Vol. 112 p.5195 21 April 2015*

M. Hsu et al.   Neural Systems Responding to Degrees of Uncertainty in Human Decision-making   *Science Vol. 310 p.1680 9 December 2005*

S. A. Huettel et al.   Perceiving patterns in random series: dynamic processing of sequence in prefrontal cortex   *Nat. Neurosci. Vol. 5 p.485 May 2002*

S. W. Kennerley et al.   Optimal decision making and the anterior cingulate cortex   *Nat. Neurosci. Vol. 9 p.940 July 2006*

T. Kahn et al.   The neural code of reward anticipation in human orbitofrontal cortex   *PNAS Vol. 107 p.6010 30 March 2010*

Y. Kikuchi-Yorioka & T. Sawaguchi   Parallel visuospatial and audiospatial working memory processes in the monkey dorsolateral prefrontal cortex   *Nat. Neurosci. Vol. 3 p.1075 November 2000*

E. Koechlin et al.   The role of the anterior prefrontal cortex in human cognition   *Nature Vol. 339 p.148   13 May 1999*

C. C. Lapish et al.   Successful choice behavior is associated with distinct and coherent network states in anterior cingulate cortex   *PNAS Vol. 105 p.1196 3 26 August 2008*

H. C. Lau & R. E. Passingham   Relative blindsight in normal observers and the neural correlate of visual consciousness   *PNAS Vol. 103 p.18763 5 December 2006*

R. Levy & P. S. Goldman-Rakic   Segregation of working memory functions within the dorsolateral prefrontal cortex   *Exp. Brain Research Vol. 133 p.23 2000*

F. A. Mansouri et al.   Conflict-induced behavioural adjustment: a clue to the executive function of the prefrontal cortex   *Nat. Rev. Neurosci. Vol. 10 p.141 February 2009*

D. A. Markovitz et al.   Multiple component networks support working memory in prefrontal cortex   *PNAS Vol. 112 p.11084 1 September 2015*

C. O. Martin et al.   The effects of vagus nerve stimulation on decision making   *Cortex Vol. 40 p.605 2004*

K. Matsumoto et al.   Neuronal Correlates of Goal-Based Motor Selection in the Prefrontal Cortex   *Science Vol. 301 p.229 11 July 2003*

H. C. Meyer & D. J. Bucci   Imbalanced activity in the orbitofrontal cortex and nucleus accumbens impairs behavioral inhibition   *Current Biology Vol. 26 p.2834*

G. A. Miller   The magical number seven, plus or minus two: some limits for processing information   *Psychological Review Vol. 63 p.81 1956*

J. Moll et al.   Human fronto-mesolimbic networks guide decisions about charitable donation   *PNAS Vol. 103 p.15623 17 October 2006*

K. Nakahara et al.   Functional MRI of Macaque Monkeys Performing a Cognitive Set-Shifting Task   *Science Vol. 295 p.1532 22 February 2002*

N. A. Nasrallah et al.   Long-term risk preference and suboptimal decision making

following adolescent alcohol use   PNAS Vol. 106 p.17600   13 October 2009
J. O'Doherty et al.   Abstract reward and punishment representations in the human orbitofrontal cortex   Nat. Neurosci. Vol. 4 p.95 January 2001
M. O'Neill & W. Schultz   Coding of Reward Risk by Orbitofrontal Neurons Is Mostly Distinct from Coding of Reward Value   Neuron Vol.68 p.789   18 November 2010
S. P. óScalaidhe & P.S. Goldman-Rakic   Memory fields in prefrontal cortex of the macaque   Abstracts, Society for Neuroscience   1993
S. Otani ed.   Prefrontal Cortex: From Synaptic Plasticity to Cognition   Kluwer 2004
C. Padoa-Schioppa & J. A. Assad   Neurons in the orbitofrontal cortex encode economic value   Nature Vol. 441 p.223 11 May 2006
J. Panksepp   Feeling the Pain of Social Loss   Science Vol. 302 p.237 10 October 2003
R. E. Passingham & S. P. Wise   The Neurobiology of Prefrontal Cortex: Anatomy, Evolution and the Origin of Insight   Oxford University Press 2012
G. Rainer et al.   Selective representation of relevant information by neurons in the primate prefrontal cortex   Nature Vol. 393 p.577   11 June 1998
N. Ramnani & A. M. Owen   Anterior Prefrontal Cortex: Insights into Function from Anatomy and Neuroimaging   Nat. Rev. Neurosci. Vol. 5 p.184   March 2004
S. C. Rao et al.   Integration of What and Where in the Primate Prefrontal Cortex   Science Vol. 276 p.821   2 May 1997
D. S. Rizzuto et al.   Spatial selectivity in human ventrolateral prefrontal cortex   Nature Neurosci. Vol. 8 p.415   April 2005
M. Roca et al.   The role of Area 10 (BA 10) in human multitasking and in social cognition: A lesion study   Neuropsychologia Vol. 49 p.3225   2011
M. R. Roesch & C. R. Olsen   Neuronal Activity Related to Reward Value and Motivation in Primate Frontal Cortex   Science Vol. 30 p.307   9 April 2004
S. Rossi et al.   Prefrontal cortex in long-term memory: an 'interference' approach using magnetic stimulation   Nat. Neurosci. Vol. 4 p.948 September 2001
D. S. Ruchkin et al.   Working memory retention systems: a state of activated long-term memory   Behavioral and Brain Sciences Vol. 26 p.709   2003
P. H. Rudebeck et al.   A Role for the Macaque Anterior Cingulate Gyrus in Social Valuation   Science Vol. 313 p.1310 1 September 2006
P. H. Rudebeck et al.   Separate neural pathways process different decision costs   Nat. Neurosci. Vol. 9 p.1161 September 2006
M. D. Rugg et al.   Differential activation of the prefrontal cortex in successful and unsuccessful memory retrieval   Brain Vol. 119 p.2073   1996
M.F.S. Rushworth et al.   Contrasting roles for cingulate and orbitofrontal cortex in decisions and social behaviour   Trends in Cognitive Sciences Vol. 11 p.168   April 2007
K. Sakai et al.   Active maintenance in prefrontal area 46 creates distractor-resistant memory   Nat. Neurosci. Vol. 5 p.479   May 2002
T. Sawaguchi et al.   Effects of dopamine antagonists on neuronal activity related to a delayed response task in monkey prefrontal cortex   J. Neurophysiology Vol. 63 p.1401 1990
A. Shenhav et al.   Anterior cingulate engagement in a foraging task reflects choice difficulty, not foraging value   Nat. Neurosci. Vol. 17 p.1249   September 2014
G. Schoenbaum et al.   A new perspective on the role of the orbitofrontal cortex in

adaptive behaviour   *Nat. Rev. Neurosci. Vol. 10 p.885 December 2009*
P. T. Schoenemann   Evolution of the Size and Functional Areas of the Brain   *Ann. Rev. Anthropology Vol. 35 p.379 2006*
W. Schultz et al.   Reward Processing in Primate Orbitofrontal Cortex and Basal Ganglia *Cerebral Cortex Vol. 10 p.272 March 2000*
A. Schüz & R. Miller *eds*. Cortical Areas: Unity and Diversity *Taylor & Francis 2002*
K. Semendeferi et al.   Prefrontal Cortex in Humans and Apes: A Comparative Study of Area 10   *American J. Physical Anthropology Vol. 114 p.224   2001*
K. Semendeferi et al.   Spatial Organization of Neurons in the Frontal Pole Sets Humans Apart from Great Apes   *Cerebral Cortex Vol. 21 p.1485   7 June 2011*
A. J. Shackman et al.   The integration of negative affect, pain and cognitive control in the cingulate cortex   *Nat. Rev. Neurosci. Vol. 12 p.154 March 2011*
T. Sharot et al.   Neural mechanisms mediating optimism bias   *Nature Vol. 450 p.102 1 November 2007*
K. Shima et al.   Categorisation of behavioural sequences in the prefrontal cortex *Nature Vol. 445 p.315 18 January 2007*
N. Sigala et al.   Hierarchical coding for sequential task events in the monkey prefrontal cortex   *PNAS Vol. 105 p.11969 26 August 2008*
J. S. Simms et al.   Anterior prefrontal cortex and the recollection of contextual information   *Neuropsychologia Vol. 43 no. 12 p.1774   2005*
B. M. Slotnick   Disturbances of maternal behavior in the rat following lesions of the cingulate cortex   *Behaviour Vol. 29 no.2/4 p.204   1967*
J. B. Smaers et al.   Frontal White Matter Volume is Associated with Brain Enlargement and Higher Structural Connectivity in Anthropoid Primates   *PLoS ONE 10.1371/journal.pone.0009123   9 February 2010*
M-H. Sohn et al.   The role of prefrontal cortex and posterior parietal cortex in task switching   *PNAS Vol. 97 p.13443 2000*
D. T. Stuss & R. T Knight *eds*.   Principles of Frontal Lobe Function   *Oxford University Press 2002*
S. M. Tom et al.   The Neural Basis of Loss Aversion in Decision-Making Under Risk *Science Vol. 315 p.515 26 January 2007*
H. Tomita et al.   Top-down signal from prefrontal cortex in executive control of memory retrieval   *Nature Vol. 401 p.699   14 October 1999*
S. Tsujimoto et al.   Frontal pole cortex: encoding ends at the end of the endbrain *Trends in Cognitive Sciences Vol. 15 p.169   April 2011*
S. Vijayraghavan et al.   Inverted-U dopamine D1 receptor actions on prefrontal neurons engaged in working memory   *Nat. Neurosci. Vol. 10 p.376 March 2007*
E. K. Vogel et al.   Neural measures reveal individual differences in controlling access to working memory   *Nature Vol. 438 p.500 24 November 2005*
B. Voytek & R. T. Knight   Prefrontal and basal ganglia contributions to visual working memory   *PNAS Vol. 107 p.18167 19 October 2020*
J. D. Wallis et al.   Single neurons in prefrontal cortex encode abstract rules   *Nature Vol. 411 p.953   21 June 2001*
M. E. Walton et al.   Calculating the Cost of Acting in Frontal Cortex   *Annals of the New York Academy of Sciences Vol. 1104 p.340   2007*

M. Wang et al.   Selective D$_2$ receptor actions on the functional circuits of working memory   Science Vol. 303 p.853   6 February 2004

G. V. Williams & P. S. Goldman-Rakic   Modulation of memory fields by dopamine D$_1$ receptors in prefrontal cortex   Nature Vol. 376 p.572   17 August 1995

K. Wunderlich et al.   Neural computations underlying action-based decision making in the human brain   PNAS Vol. 106 p.17199   6 October 2009

J. Xie & C. Padoa-Schioppa   Neuronal remapping and circuit persistence in economic decisions   Nat. Neurosci. Vol. 19 p.855 2016

T-X. Xu & W-D. Yao   D$_1$ and D$_2$ dopamine receptors in separate circuits cooperate to drive asssociative long-term potentiation in the prefrontal cortex   PNAS Vol. 107 p.16366 14 September 2010

D. H. Zald & S. L. Rauch eds.   The Orbitofrontal Cortex   Oxford University Press 2006

T. P. Zanto et al.   Causal role of the prefrontal cortex in top-down modulation of visual processing and working memory   Nat. Neurosci. Vol. 14 p.656   May 2011

Chapter 23: More About the Basal Ganglia

C. T. Anderson et al.   Sublayer-specific microcircuits of corrticospinal and corticostriatal neurons in motor cortex   Nat. Neurosci. Vol. 13 p.737   June 2010

M. E. Avale et al.   Interplay of β$_2$ nicotinic receptors and dopamine pathways in the control of spontaneous locomotion   PNAS Vol. 105 p.15991   14 October 2008

H. S. Bateup et al.   Distinct subclasses of medium spiny neurons differentially regulate striatal motor behaviors   PNAS Vol. 107 p.14845   17 August 2010

J. Ben Arous et al.   Molecular and Sensory Basis of a Food Related Two-State Behavior in C. Elegans   PLoS ONE 10.1371/journal.pone.0007584 23 October 2009

R. Bock et al.   Strengthening the accumbal indirect pathway promotes resilience to compulsive cocaine use   Nat. Neurosci. Vol. 16 p.632   May 2013

J. J. Bolhuis et al.   Twitter evolution: converging mechanisms in birdsong and human speech   Nat. Rev. Neurosci. Vol. 11 p.747   November 2010

S. M. Boye et al.   Disruption of dopaminergic transmission in nucleus accumbens core inhibits the locomotor stimulant effects of nicotine and D-amphetamine in rats   Neuropharmacology Vol. 40 No.6 p.792 May 2001

P.Calabresi et al.   Direct and indirect pathways of basal ganglia: a critical reappraisal   Nat. Neurosci. Vol. 17 p.1022 August 2014

E. C .Clark et al.   Mammal-Like Striatal Functions in Anolis   Brain Behavior and Evolution Vol. 56 p.249   2000

R. Cools   Role of Dopamine in the Motivational and Cognitive Control of Behavior   The Neuroscientist Vol. 14 No.1 p.73   2014

G. Cui et al.   Concurrent activation of striatal direct and indirect pathways during action initiation   Nature Vol. 494 p.238   14 February 2013

J. J. Day & R. M. Carelli   The Nucleus Accumbens and Pavlovian Reward Learning   The Neuroscientist Vol. 13 No.2 p.148   2007

M. de Bono & A. V. Maricq   Neuronal Substrates of Complex Behaviors in C. elegans   Ann. Rev. Neurosci. Vol. 28 p.451   2005

J. DeGutis & M. D'Esposito   Network changes in the transition from initial learning to

well-practised visual categorization   *Frontiers in Human Neuroscience Vol. 3(4)   2009*
S. de Wit et al.   Corticostriatal Connectivity Underlies Individual Differences in the Balance between Habitual and Goal-Directed Action   *J. Neuroscience Vol. 32(35) p.12066   29 August 2012*
E. Düzel et al.   Brain Oscillations and Memory   *Curr. Op. Neurobiol. Vol. 20 p.143   2010*
A. K. Engel et al.   Where's the Action? The pragmatic turn in cognitive science   *Trends in Cognitive Sciences Vol. 17 p.202   May 2013*
S. J. Fallon et al.   Differential optimal dopamine levels for set-shifting and working memory in Parkinson's disease   *Neuropsychologia Vol. 77 p.42   2015*
E. H. Feinberg & M. Meister   Orientation columns in the mouse superior colliculus   *Nature Vol. 519 p.229   12 March 2015*
M. J. Frank et al.   Hold Your Horses: Impulsivity, Deep Brain Stimulation and Medication in Parkinsonism   *Science Vol. 318 p.1309   23 November 2007*
R. Gangler & A. L. Bilgrani eds.   Nematode Behaviour   *CABI Publishing   2004*
A. H. Gittis & A.C. Kreitzer   Striatal microcircuitry in movement disorders   *Trends in Neurosciences Vol. 35 p.557   September 2012*
A. González et al.   Evidence for Shared Features in the Organization of the Basal Ganglia in Tetrapods: Studies in Amphibians   *European J. Morphology Vol. 37 p.151   1999*
J. M. Gray et al.   A circuit for navigation in *Caenorhabditis elegans*   *PNAS Vol. 102 p.3184   1 March 2005*
S. Grillner & B. Robertson   The Basal Ganglia Over 500 Million Years   *Current Biology Vol. 26 R1088   24 October 2016*
O. Hikosaka & M. Isoda   Switching from automatic to controlled behavior: cortico-basal ganglia mechanisms   *Trends in Cognitive Sciences Vol. 14 p.155 April 2010*
M. Howe et al.   Habit learning is associated with major shifts in frequency of oscillatory activity and synchronized spike firing in striatum   *PNAS Vol. 108 p.16801   4 October 2011*
M. W. Howe & D. A. Dombeck   Rapid signalling in distinct dopaminergic axons during locomotion and reward   *Nature Vol. 535 p.505   28 July 2016*
X. Jin & R. M. Costa   Start/stop signals emerge in nigrostriatal circuits during sequence learning   Nature Vol. 466 p.457   *22 July 2010*
E. A. Kabotyanski et al.   Modulation of Fictive Feeding by Dopamine and Serotonin in *Aplysia*   *J. Neurophysiology Vol. 83 p.374   January 2000*
K. Kondabolu et al.   Striatal cholinergic interneurons generate beta and gamma oscillations in the corticostriatal circuit and produce motor deficits   *PNAS Vol. 113 E3159   31 May 2016*
A. K. Koslov et al.   Gating of steering signals through phasic modulation of reticulospinal neurons during locomotion   *PNAS Vol. 111 p.3591   4 March 2014*
A. V. Kravitz, L. D. Tye & A. C. Kreitzer   Distinct roles for direct and indirect pathway striatal neurons in reinforcement   *Nat. Neurosci. Vol. 15 p.816   June 2012*
G. J. Kress et al.   Convergent cortical innervation of striatal projection neurons   *Nat. Neurosci. Vol. 16 p665 June 2013*
J. G. McHaffie et al.   Subcortical loops through the basal ganglia   *Trends in Neurosciences Vol 28 p.401 August 2005*

L. Medina   Basal Ganglia: Evolution   *in* Encyclopedia of Neuroscience   *ed*. L. R. Squire et al. Vol. 2 p.67   *Academic Press 2009*

R. Miller & J. R. Wickens *eds*.   Brain Dynamics and the Striatal Complex   *CRC Press 2000*

A. B. Nelson & A. C. Kreitzer   Reassessing Models of Basal Ganglia Function and Dysfunction   *Ann. Rev. Neurosci. Vol. 37 p.117 2014*

J. O'Doherty et al.   Dissociable Roles of Ventral and Dorsal Striatum in Instrumental Conditioning   *Science Vol. 304 p. 452 16 April 2004*

T. Okubo et al.   Growth and splitting of neural sequences in songbird vocal development   *Nature Vol. 528 p.352   17 December 2015*

M. Parent & A. Parent   Axonal collateralization in primate basal ganglia and related thalamic nuclei   *Thalamus and Related Systems Vol. 2 p.71   2002*

J. A. Parkinson et al.   Nucleus accumbens dopamine depletion impairs both acquisition and performance of appetitive Pavlovian approach behaviour: implications for mesoaccumbens dopamine function   *Behavioural Brain Research Vol. 137 p.149 2002*

A. Pasupathy & E. K. Miller   Different time courses of learning-related activity in the prefrontal cortex and striatum   *Nature Vol. 443 p.873   24 February 2005*

C. S. Prat et al.   Basal ganglia impairments in autism spectrum disorder are related to abnormal signal gating to prefrontal cortex   *Neuropsychologia Vol. 91 p.268-281   2016*

K. A. Quinlan et al.   Cholinergic Modulation of the Locomotor Network in the Lamprey Spinal Cord   *J. Neurophysiology Vol. 92 p.1536   19 May 2004*

P. Redgrave et al.   Goal-directed and habitual control in the basal ganglia: implications for Parkinson's disease   *Nat. Rev. Neurosci. Vol. 11 p.760   November 2010*

T. Riemensperger et al.   Behavioral consequences of dopamine deficiency in the *Drosophila* central nervous system   *PNAS Vol. 108 p.834   11 January 2011*

A. Roberts et al.   Origin of excitatory drive to a spinal locomotor network   *Brain Research Reviews Vol. 57 p.22   2008*

A. F. Sadikot & V. V. Rymar   The primate centromedian-parafascicular complex: Anatomical organization with a note on neuromodulation   *Brain Research Bulletin Vol. 78 p.122   2009*

A. Saunders et al.   A direct GABAergic output from the basal ganglia to prefrontal cortex   *Nature Vol. 521 p.85   7 May 2015*

E. R. Sawin et al.   *C. elegans'* Locomotory Rate Is Modulated by the Environment through a Dopaminergic Pathway and by Experience Through a Serotonergic Pathway   *Neuron Vol. 26 p.619 June 2000*

R. Schmidt et al.   Canceling actions involves a race between basal ganglia pathways   *Nat. Neurosci. Vol. 16 p.1178 August 2013*

G. E. Schneider   Contrasting visual functions of tectum and cortex in the golden hamster   *Psychologische Forschung 1967*

G. E. Schneider   Brain Structure and Its Origins in Development and in Evolution of Behavior and the Mind   *MIT Press 2014*

C. R. Slater   Vertebrate Patterns of Neuromuscular Connections   *in* Encyclopedia of Neuroscience *ed.* L. R. Squire et al. Vol. 6   *Academic Press   2009*

Y. Smith et al.   The thalamostriatal system: a highly specific network of the basal ganglia circuitry   *Trends in Neurosciences Vol. 27 p.520   September 2004*

M. Stephenson-Jones et al.   Evolutionary Conservation of the Basal Ganglia as a Common Vertebrate Mechanism for Action Selection   *Current Biology Vol. 21 p.1081 1 July 2011*

N. J. Strausfeld & F. Hirth   Deep Homology of Arthropod Central Complex and Vertebrate Basal Ganglia   *Science Vol. 340 p.157   12 April 2013*

J. M. Tepper & J. P. Bolam   Functional diversity and specificity of neostriatal neurons   *Curr. Op. Neurobiol. Vol. 14 p.685   6 December 2004*

J. M. Tepper et al.   GABAergic microcircuits in the neostriatum   *Trends in Neurosciences Vol. 27 p.662   November 2004*

S. Threlfall et al.   Striatal Dopamine Release is Triggered by Synchronised Activity in Cholinergic Interneurons   *Neuron Vol. 75 p.58   12 July 2012*

A. B. L. Tort et al.   Dynamic cross-frequency coupling of local field potential oscillations in rat striatum and hippocampus during performance of a T-maze task   *PNAS Vol. 105 p.20517   11 December 2012*

I. Ubeda-Bañon et al.   Vomeronasal inputs to the rodent ventral striatum   *Brain Research Bulletin Vol. 75 p.467   2008*

M. R. van Schouwenburg et al.   Cognitive flexibility depends on white matter microstructure of the basal ganglia   *Neuropsychologia Vol. 53 No.1 p171   2014*

A. Vidal-Gadea et al.   *Caenorhabditis elegans* selects distinct crawling and swimming gaits via dopamine and serotonin   *PNAS Vol. 108 p.17504   18 October 2011*

L. Wilkinson et al.   The role of the basal ganglia and its cortical connections in sequence learning: Evidence from implicit and explicit sequence learning in Parkinson's disease   *Neuropsychologia Vol. 47 p.2564   2009*

D. Xiao & H. Barbas   Circuits through prefrontal cortex, basal ganglia, and ventral anterior nucleus map pathways beyond motor control   *Thalamus and Related Systems Vol. 2 p.325   2004*

Chapter 25: The Thalamus

S. D. Fisher & John N. J. Reynolds   The intralaminar thalamus – an expressway linking visual stimuli to circuits determining agency and action selection   *Frontiers in Behavioral Neuroscience Vol. 8 p.115   2014*

J. H. Goldberg et al.   Basal ganglia output to the thalamus: still a paradox   *Trends in Neurosciences Vol. 36 p.695   December 2013*

N. Gonzalo et al.   The parafascicular thalamic complex and basal ganglia circuitry   *Thalamus and Related Systems Vol. 1 p.341   2002*

S. N. Haber & R. Calzavara   The cortico-basal ganglia integrative network: The role of the thalamus   *Brain Research Bulletin Vol. 78 p.69   2009*

A. S. Mitchell et al.   Neurotoxic lesions of the medial mediodorsal nucleus of the thalamus disrupt reinforcer devaluation effects in rhesus monkeys   *J. Neuroscience Vol. 27(42) p.11289   17 October 2007*

A. S. Mitchell & J. Dalrymple-Alford   Lateral and anterior thalamic lesions impair independent memory systems   *Learning and Memory Vol. 13(3) p.388   May 2006*

G. L. Poirier et al.   Anterior thalamic lesions produce chronic and profuse transcriptional deregulation in retrosplenial cortex: a model of retrosplenial hypoactivity and covert pa

thology   *Thalamus and Related Systems*   Vol. 4(1) p.59   2008
M. B. Pritz   The Thalamus of Reptiles and Mammals: Similarities and Differences   *Brain Behavior and Evolution*   Vol. 46 p.197   1995
K. Reinhold et al.   Distinct recurrent versus afferent dynamics in cortical visual processing   *Nat. Neurosci.*   Vol. 18 p.1789   December 2015
Y. B. Saalmann et al.   The Pulvinar Regulates Information Transmission Between Cortical Areas Based on Attention Demands   *Science*   Vol. 337 p.753   10 August 2012
Y. B. Saalmann   Intralaminar and medial thalamic influences on cortical synchrony, information transmission and cognition   *Frontiers in Systems Neuroscience*   Vol. 8 p.83   2014
S. M. Sherman   Thalamus plays a central role in ongoing cortical functioning   *Nat. Neurosci.*   Vol. 19 p.533   April 2016
S. M. Sherman & R. W. Guillery   A New View from the Thalamus   MIT Press   2013
L. Sun et al.   Human anterior thalamic nuclei are involved in emotion-attention interaction   *Neuropsychologia*   Vol. 78 p.88   2015
W. Sun et al.   Thalamus provides layer 4 of primary visual cortex with orientation and direction-tuned inputs   *Nat. Neurosci.*   Vol. 19 p.308   February 2016
V. Sziklas & M. Petrides   Contribution of the Anterior Thalamic Nuclei to Conditional Learning in Rats   *Hippocampus*   Vol. 17 p.456   2007
Y. D. Van Der Werf et al.   The intralaminar and midline nuclei of the thalamus: anatomical evidence for participation in processes of arousal and awareness   *Brain Research Reviews*   Vol. 39 p.107   2002
T. R. Vidyasagar et al.   Subcortical orientation biases explain orientation selectivity of visual cortical cells   *Physiological Reports*   Vol. 3(4) e12374   April 2015

Chapter 24: Overview and Some More Speculations

R. A. Barton & C. Venditti   Rapid Evolution of the Cerebellum in Humans and Other Great Apes   *Current Biology*   Vol. 24 p.2440   20 October 2014
K. C. Bickhart et al.   The amygdala as a hub in brain networks that support social life   *Neuropsychologia*   Vol. 63 p.235   2014
H. A. Burgess et al.   Distinct Retinal Pathways Drive Spatial Orientation Behaviors in Zebrafish Navigation   *Current Biology*   Vol. 20 p.381   23 February 2010
B. Cardinaud et al.   Early Emergence of Three Dopamine D_1 Receptor Subtypes in Vertebrates   *J. Biol. Chem.*   Vol. 272 No. 5 p.2778   31 January 1997
S. H. A. Chen & J. E. Desmond   Temporal dynamics of cerebro-cerebellar network recruitment during a cognitive task   *Neuropsychologia*   Vol. 43 No. 9 p.1227   2005
F. E. Cooper et al.   Distinct critical cerebellar subregions for components of verbal working memory   *Neuropsychologia*   Vol. 50 p.189   2012
M. del C. de Arriba & M. A. Pombal   Afferent connections of the optic tectum in lampreys: an experimental study   *Brain, Behaviour and Evolution*   Vol. 69 p.37   January 2007
W. E. DeCoteau et al.   Learning-related coordination of striatal and hippocampal theta rhythms during acquisition of a procedural maze task   *PNAS*   Vol. 104 p.5644   2007
M. T. Eckart et al.   Dorsal hippocampal lesions boost performance in the rat sequential

reaction time task   *Hippocampus Vol. 22 p.1202 May 2012*
F. Fabbro et al.   Long-term neuropsychological deficits after cerebellar infarctions in two young adult twins   *Neuropsychologia Vol. 42 No. 4 p.536 2004*
T. G. Fujii et al.   Auditory Responses to Vocal Sounds in the Songbird Nucleus Taeniae of the Amygdala and the Adjacent Arcopallium   *Brain, Behavior and Evolution Vol. 87 p.275 September 2016*
A. Johnson et al.   Integrating hippocampus and striatum in decision-making   *Curr. Op. Neurobiol. Vol. 17 p.697 2007*
O. Kiehn ed.   Neuronal Mechanisms of Locomotor Activity   *Annals of the New York Academy of Sciences Vol. 860 1998*
J. Kremkow et al.   Neuronal nonlinearity explains greater visual spatial resolution for darks than lights   *PNAS Vol. 111 p.3170 25 February 2014*
E. Lanuza et al.   Sexual pheromones and the evolution of the reward system of the brain: The chemosensory function of the amygdala   *Brain Research Bulletin Vol. 75 p.460 2008*
J. E. LeDoux   The slippery slope of fear   *Trends in Cognitive Sciences Vol. 17 p.155 April 2013*
F. A. Middleton & P. L. Strick   Cerebellar output: motor and cognitive channels   *Trends in Cognitive Sciences Vol. 2 p.348 September 1998*
D. E. Nee & J. Jonides   Neural correlates of access to short-term memory   *PNAS Vol. 105 p.14228 2 October 2008*
R. D. Proville et al.   Cerebellar involvement in cortical sensorimotor circuits for the control of voluntary movements   *Nat. Neurosci. Vol. 17 p.1233 September 2014*
J. D. Schmahmann   From Movement to Thought: Anatomic Substrates of the Cerebellar Contribution to Cognitive Processing   *Human Brain Mapping Vol. 4 p.174 1996*
J. D. Schmahmann & J. C. Sherman   The cerebellar cognitive affective syndrome   *Brain Vol. 121 p.561 1998*
J. B. Smaers et al.   Modeling the evolution of cortico-cerebellar systems in primates   *Annals of the New York Academy of Sciences Vol. 1225 p.176 April 2011*
K. S. Sugamori et al.   Invertebrate Dopamine Receptors   *in* Dopamine Receptors and Transporters: pharmacology, structure and function   *ed*. H. B. Niznik   *Marcel Dekker 1994*
N. Virji-Babul et al.   Neural correlates of action understanding in infants: influence of motor experience   *Brain and Behavior Vol. 2 p.237 May 2012*
G. Von de Ernde et al. *eds*.   The Senses of Fish   *Kluwer Academic/Narosa 2004*
S. Wang et al.   Neurons in the human amygdala selective for perceived emotion   *PNAS Vol. 111 p.3110 29 July 2014*

Chapter 26: Some Pivotal Innovations

R. J. Addante et al.   Neurophysiological evidence for a recollection impairment in amnesia patients that leaves familiarity intact   *Neuropsychologia Vol. 50 p.3004 2012*
L. M. Aplin et al.   Experimentally induced innovations lead to persistent culture via conformity in wild birds   *Nature Vol. 518 p.538 26 February 2015*
S. Ballesta & J-R. Duhamel   Rudimentary empathy in macaques' social decision-making

PNAS Vol. 112 p.15516 22 December 2015
G. S. Berns et al. Brain Regions Responsive to Novelty in the Absence of Awareness *Science* Vol. 276 p.1272 23 May 1997
L. Bonini & P. F. Ferrari Evolution of mirror systems: a simple mechanism for complex cognitive functions *Annals of the New York Academy of Sciences* Vol. 1225 p.166 April 2011
B. Bowles et al. Impaired familiarity with preserved recollection after anterior lobe resection that spares the hippocampus *PNAS* Vol. 104 p.16382 9 October 2007
B. Bowles et al. Double dissociation of selective recollection and familiarity impairments following two different surgical treatments for temporal-lobe epilepsy *Neuropsychologia* Vol. 48 p.2640 2010
B. Bowles et al. Impaired assessment of cumulative lifetime familiarity for object concepts after left anterior temporal-lobe resection that includes perirhinal cortex but spares the hippocampus *Neuropsychologia* Vol. 90 p.170 2016
C. Busettini et al. Vergence Eye Movements *in* Encyclopedia of Neuroscience *ed.* L. R. Squire et al. Vol. 10 p.75 Academic Press 2009
V. Caggiano et al. View-Based Encoding of Actions in Mirror Neurons of Area F5 in Macaque Premotor Cortex *Current Biology* Vol. 21 p.144 25 January 2011
M. de Haan et al. Human memory development and its dysfunction after early hippocampal injury *Trends in Neurosciences* Vol. 29 p.374 July 2006
F. B. M. de Waal Animal Conformists *Science* Vol. 340 p.437 26 April 2013
R. A. Diana et al. Imaging recollection and familiarity in the medial temporal lobe: a three-component model *Trends in Cognitive Sciences* Vol. 11 p.379 September 2007
U. Dimberg et al. Unconscious facial reactions to emotional facial expressions *Psychological Science* Vol. 11 p.86 January 2000
R. Egger et al. Robustness of sensory-evoked excitation is increased by inhibitory inputs to distal apical tuft dendrites *PNAS* Vol. 112 p.14072 10 November 2015
L. M. Herman Vocal, Social and Self-Imitation by Bottlenosed Dolphins *in* Imitation in Animals and Artefacts *ed.* K. Dautenhahn & C. L. Nehaniv MIT Press 2002
M. Iacoboni & M. Dapretto The mirror neuron system and the consequences of its dysfunction *Nat. Rev. Neurosci.* Vol. 7 p.942 December 2006
A. Jack et al. Subcortical contributions to effective connectivity in brain networks supporting imitation *Neuropsychologia* Vol. 49 p.3689 2011
A. Kafkas & D. Montaldi Two separate but interacting neural systems for familiarity and novelty detection: A dual-route system *Hippocampus* Vol. 24 p.516 May 2014
J. M. Kilner & R. N. Lemon What We Know Currently About Mirror Neurons *Current Biology* Vol. 23 R1057 2 December 2013
C. B. Kirwan et al. Medial temporal lobe activity can distinguish between old and new stimuli independently of overt behavioral choice *PNAS* Vol. 106 p.14517 25 August 2009
A. G. Leventhal et al. GABA and Its Agonists Improved Cortical Visual Function in Senescent Monkeys *Science* Vol. 300 p.812 2 May 2003
A. N. Meltzoff & M. K. Moore Imitation of facial and manual gestures by human neonates *Science* Vol. 198 p.75 1977
A. N. Meltzoff & W. Prinz *eds.* The Imitative Mind: Development, Evolution and Brain

Bases   *Cambridge Universitiy Press*   2002

M. Myowa-Yamakoshi et al.   Imitation in neonatal chimpanzees (*Pan troglodytes*)   *Developmental Science* Vol. 7 p437   September 2004

A. Norenzayan   Big Gods: How Religion Transformed Cooperation and Conflict   *Princeton University Press* 2015

J. M. Plotnik et al.   Self-recognition in an Asian elephant   *PNAS* Vol. 103 p.17053   7 November 2006

H. Prior et al.   Mirror-Induced Behavior in the Magpie (*Pica pica*): Evidence of Self-Recognition   *PLoS Biol.* 10.1371/journal.pbio.006202 19 August 2008

B. G. Purzycki et al.   Moralistic gods, supernatural punishment and the expansion of human sociality   *Nature* Vol. 530 p.327   18 February 2016

K. B. Quast et al.   Developmental broadening of inhibitory sensory maps   *Nature Neuroscience* Vol. 20 p.189   February 2017

A. Z. Rajala et al.   Rhesus Monkeys (*Macaca mulatta*) Do Recognise Themselves in the Mirror: Implications for the Evolution of Self-Recognition   *PLoS ONE* 10.1371/journal.pone.0012865 29 September 2010

G. Rizzolatti & C. Sinigaglia   The functional role of the parieto-frontal mirror circuit: interpretations and misinterpretations   *Nat. Rev. Neurosci.* Vol. 11 p.264   April 2010

T. A. Schoenfeld & T. A. Cleland   Anatomical Contributions to Odorant Sampling and Representation in Rodents: Zoning in on Sniffing Behavior   *Chemical Senses* Vol. 31 p.131   February 2006

C. E. Schroeder et al.   Dynamics of Active Sensing and Perceptual Selection   *Curr. Op. Neurobiol.* Vol. 20 p.172   2010

K. Schulze et al.   Test of a motor theory of long-term auditory memory   *PNAS* Vol. 109 p.1721   17 April 2012

M. W. Self et al.   Distinct Roles of the Cortical Layers of Area V1 in Figure-Ground Separation   *Current Biologoy* Vol. 23 p.2121   4 November 2013

D. M. Small et al.   Sensory Neuroscience: Taste Responses in Primary Olfactory Cortex   *Current Biology* Vol. 23 R158   18 February 2013

M. Smear et al.   Perception of sniff phase in mouse olfaction   *Nature* Vol. 479 p.397   17 November 2011

N. J. Sofroniew & K. Svoboda   Whisking   *Current Biology* Vol. 25 R137   16 February 2015

E. R. Soucy et al.   Precision and diversity in an odor map on the olfactory bulb   *Nature Neuroscience* Vol. 12 p.210   February 2009

A. H. Taylor et al.   New Caledonian Crows Learn the Functional Properties of Novel Tool Types   *PLoS one* 6(12) e26887   2011

A. Tiriac et al.   Rapid Whisker Movements in Sleeping Newborn Rats   *Current Biology* Vol. 22 p.2075   6 November 2012

R. P. G. van Gompel et al. *eds.*   Eye-Movements: A window on mind and brain   *Elsevier* 2007

T. van Kerkoerle et al.   Alpha and gamma oscillations characterize feedback and feed-forward processing in monkey visual cortex   *PNAS* Vol. 111 p.14332   7 October 2014

Y. Vázquez et al.   Transformation of the neural code for tactile detection from thalamus to cortex   *PNAS* Vol. 110 E2635   9 July 2013

A. J. Watrous et al.   Frequency-specific network connectivity increases underlie accurate spatiotemporal memory retrieval   *Nat. Neurosci. Vol. 16 p.349 March 2013*
R. H. Wurtz et al.   Thalamic pathways for active vision   *Trends in Cognitive Sciences Vol. 15 p.177 April 2011*
N-L. Xu et al.   Nonlinear dendritic integration of sensory and motor input during an active sensing task   *Nature Vol. 492 p.247 13 December 2012*
R. B. Yaffe et al.   Reinstatement of distributed cortical oscillations occurs with precise spatiotemporal dynamics during successful memory retrieval   *PNAS Vol. 111 p.12727 30 December 2014*
R. Yuste et al.   The cortex as a central pattern generator   *Nat. Rev. Neurosci. Vol. 6 p.477 June 2005*
A. Zangaladze et al.   Involvement of visual cortex in tactile discrimination of orientation   *Nature Vol. 401 p.587 7 October 1999*

Chapter 27: A Tentative History of Consciousness

G. Bargary & K. J. Mitchell   Synaesthesia & cortical connectivity   *Trends in Neurosciences Vol. 31 p.335 July 2008*
J. D. Howard et al.   Odor quality coding and categorization in the posterior piriform cortex   *Nat. Neurosci. Vol. 12 p.932 July 2009*
J. S. Isaacson   Odor representations in mammalian cortex   *Curr. Op. Neurobiol. Vol. 20 p328 2010*
C. Koch et al.   Neural correlates of consciousness: progress and problems   *Nat. Rev. Neurosci. Vol. 17 p.307 May 2016*
A. T. Schaefer & T. W. Margrie   Spatiotemporal representations in the olfactory system   *Trends in Neurosciences Vol. 30 p.92 March 2007*
A. Schurger et al.   Cortical activity is more stable when sensory stimuli are consciously perceived   *PNAS Vol. 112 E2093 21 April 2015*
J. Simner & J. Ward   The taste of words on the tip of the tongue   *Nature Vol. 444 p.483 23 November 2006*
C. Sinke et al.   Inside a synesthete's head: A functional connectivity analysis with graphene-color synesthetes   *Neuropyschologia Vol. 50 p.3363 2012*
N. Tsuchiya & R. Adophus   Emotion and Consciousness   *Trends in Cognitive Sciences Vol. 11 p.158 April 2007*
T. M. van Leeuwen et al.   Synaesthetic Colour in the Brain: Beyond Colour Areas. A Functional Magnetic Resonance Imaging Study of Synaesthetes and Matched Controls   *PLoS ONE 10.1371/journal.pone.0012074 10 August 2010*
J. Ward & P. Meijer   Visual experiences in the blind induced by an auditory sensory substitution device   *Consciousness and Cognition Vol. 19 p.292 2010*

Chapter 28: Language

D. Bickerton   Language and Human Behaviour   *UCL Press 1976*
D. Bickerton   Roots of Language   *Karoma 1981*
D. Bickerton   Language and Species   *University of Chicago Press 1990*

F. Boas   Handbook of American Indian Languages   *Smithsonian Institution   Bureau of American Ethology   1911*
M. D. Bodamer & R. A. Gardner   How Cross-Fostered Chimpanzees (*Pan troglodytes*) Initiate and Maintain Conversations   *J. Comp. Psychol. Vol. 116 No. 1 p.151   2002*
W. H. Calvin & D. Bickerton   Reconciling Darwin and Chomsky with the Human Brain   *MIT Press   2000*
C. Cantalupo   Asymmetric Broca's area in great apes   *Nature Vol. 414 p.505   29 November 2001*
E. A. Cartmill et al.   Quality of early parent input predicts child vocabulary 3 years later   *PNAS Vol. 110 p.11278   9 July 2013*
Ernst Cassirer   An Essay on Man   *Yale University Press   1944*
N. Chomsky   Cartesian Linguistics   *Harper & Row   1996*
N. Chomsky   Syntactic Structures   *Mouton & Co.   1997*
N. Chomsky   The Architecture of Language   *Oxford University Press   2000*
S. Curtiss   Genie: A Psycholinguistic Study of a Modern-Day 'Wild Child'   *Academic Press   1977*
C. Darwin   A Biographical Sketch of an Infant   Classics in the History of Psychology - an internet resource developed by C. D. Green
C. Darwin   The Descent of Man   *1871*
T. W. Deacon   The Symbolic Species: The Co-Evolution of Language and the Brain   *W. W. Norton   1997*
M. Fedzechkina et al.   Language learners restructure their input to facilitate efficient communication   *PNAS Vol. 109 p.17897   30 October 2012*
A. Fernald   The Perceptual and Affective Salience of Mothers' Speech to Infants   *in* The Origins and Growth of Communication   ed. L. Feagan et al.   *Ablex   1984*
T. Fitch   The Evolution of Language   *Cambridge Unversity Press   2010*
R. Fouts with S. Mills   Next of Kin   *Michael Joseph   1997*
A. D. Friederici   The cortical language circuit: from auditory perception to sentence comprehension   Trends in Cognitive Sciences Vol. 16 p.262   *May 2012*
R. A. & B. T. Gardner   Teaching Sign-Language to a Chimpanzee   *Science Vol. 165 p.664   1969*
R. Gil-da-Costa et al.   Species-specific calls activate homologs of Broca's and Wernicke's areas in the macaque   *Nat. Neurosci. Vol. 9 p.1064   August 2006*
R. Jackendoff   Foundations of Language: Brain, Meaning, Grammar, Evolution   *Oxford University Press   2002*
V. M. Janik   Cognitive skills in bottlenose dolphin communication   *Trends in Cognitive Sciences Vol. 17 p.157   April 2013*
M. L. A Jensvold & R. A. Gardner   Interactive Use of Sign Language by Cross-Fostered Chimpanzees (*Pan troglodytes*)   *J. Cogn. Psychol. Vol. 114 No.4 p335   2000*
U. Jürgens   The Neural Control of Vocalization in Mammals: A Review   *Journal of Voice Vol. 23 No.1 p.1   2009*
N. Kawai & T. Matsuzawa   Numerical memory span in a chimpanzee   *Nature Vol. 403 p.39   6 January 2000*
E. S. Klima & U. Bellugi   The Signs of Language   *Harvard Univesity Press   1980*
D. Lipkind et al.   Stepwise acquisition of vocal combinatorial capacity in songbirds and

human infants   *Nature* Vol. 498 p.104 6 June 2013
G. A. Miller   The Science of Words   *Scientific American Library*   1991
T. Nazzi & A. Gopnik   Linguistic and cognitive abilities in infancy: when does language become a tool for categorisation?   *Cognition* Vol. 80 p.B11 2001
K. Ouattara et al.   Campbell's monkeys concatenate vocalizations in context-specific call sequences   *PNAS* Vol. 106 p.22026 December 2009
F. Patterson & E. Linden   The Education of Koko   *Andre Deutsch*   1982
I. M. Pepperburg   The Alex Studies: Cognitive and Communicative Abilities of Grey Parrots   *Harvard University Press*   2000
A. J. Premack & D. Premack   Teaching Language to an Ape   *Scientific American* October 1972
E. Sapir & D. Goodman *eds*.   Selected writings of Edward Sapir in Language, Culture and Personality   *University of California Press*   1985
S. Savage-Rumbaugh & R. Lewin   Kanzi – The Ape at the Brink of the Human Mind   *Doubleday*   1994
S. Savage-Rumbaugh et al.   Apes, Language and the Human Mind   *Oxford University Press*   1998
S. Savage-Rumbaugh et al.   Language, Speech, Tools and Writing: A Cultural Imperative   *J. Consciousness Studies* Vol. 8 p.273 2001
E. S. Savage-Rumbaugh & L. A. Baker   Vocal Production by a Language-Competent *Pan paniscus*   *International J. Primatology* Vol. 24 p.1 2003
P. Segerdal et al.   Kanzi's Primal Language   *Palgrave Macmillan* 2005
A. Senghas et al.   Children Creating Core Properties of Language: Evidence from an Emerging Sign Language in Nicaragua   *Science* Vol. 305 p.1779 17 September 2004
M. F. Small   Ay up, a chimp wi' an accent   *New Scientist*   4 June 1994
C. Snow & C. Ferguson *eds*.   Talking to Children: Language Input and Acquisition   *Cambridge University Press*   1977
S. K. Watson et al.   Vocal Learning in the Functionally Referential Food Grunts of Chimpanzees   *Current Biology* Vol. 25 0.495 16 February 2015
B. Whorf   The Relation of Habitual Thought and Behaviour to Language   *in* Language, Culture and Personality: Essays in Memory of Edward Sapir   *University of Utah Press* 1941
F. Xu   The role of language in acquiring object kind concepts in infancy   *Cognition* Vol. 85 p.223 2002
K. Zuberbühler et al.   Conceptual Semantics in a Nonhuman Primate   *J. Comp. Psychol.* Vol. 113 p.33 1999

Chapter 29: The Evolution of Language Ability

S. H. Ambrose   Paleolithic Technology and Human Evolution   *Science* Vol. 291 p.1749 2 March 2001
M. Arbib   From Mirror Neurons to Complex Imitation in the Evolution of Language and Tool Use   *Ann. Rev. Anthropology* Vol. 40 p.257 2011
K. Arnold & K. Zuberbühler   Meaningful call combinations in a non-human primate   *Current Biology* Vol. 18 No.5 R202 2008

J. H. Barkow et al. *eds.*   The Adapted Mind   *Oxford University Press 1992*
R. Dunbar   Grooming, Gossip and the Evolution of Language   *Faber 1996*
R. Dunbar et al. *eds.*   The Evolution of Culture   *Edinburgh University Press 1999*
W. T. Fitch   The evolution of speech: a comparative review   *Trends in Cognitive Science Vol.4 p.258   July 2000*
A. A. Ghazanfar et al.   Cineradiography of Monkey Lip-Smacking Reveals Putative Precursors of Speech Dynamics   *Current Biology Vol. 22 p.1176   10 July 2012*
S. Goldin-Meadow & C. Mylander   Spontaneous sign systems created by deaf children in two cultures   *Nature Vol. 391 p.278   15 January 1998*
M. Gluckman   Gossip and Scandal   *Current Anthropology Vol. 4 p.307   1963*
B. J. King *ed.*   The Origins of Language: What Nonhuman Primates Can Tell Us   *School of American Research Press 1999*
C. Knight et al. *eds.*   The Evolutionary Emergence of Language: Social Function and the Origins of Linguistic Form   *Cambridge University Press 2000*
P. Lieberman   Toward an Evolutionary Biology of Language   *Bellknap Press of Harvard University Press 2006*
M. A. Nowak et al.   Computational and evolutionary aspects of language   *Nature Vol. 417 p.611   2 June 2002*
R. R. Provine   Curious Behavior: Yawning, Laughing, Hiccuping and Beyond   *Bellknap Press 2012*
E. Sapir   Culture, Language and Personality   *University of California Press 1970*
Y. Shtyrov et al.   Automatic ultrarapid activation and inhibition of cortical motor systems in spoken word comprehension   *PNAS Vol. 111 E1918   6 May 2014*
D. Stout & T. Chaminade   Stone tools, language and the brain in human evolution   *Philosophical Transactions of the Royal Society B Vol. 367 p.75   21 November 2011*
J. Winawer et al.   Russian blues reveal effects of language on color discrimination   *PNAS Vol. 104 p.7780   8 May 2007*
J. Xu et al.   Symbolic gestures and spoken language are processed by a common neural system   *PNAS Vol. 106 p.20664   8 December 2009*

Chapter 30: Endword

R. Adolphs   The unsolved problems of neuroscience   *Trends in Cognitive Science Vol. 19 p.173   April 2015*
C. M. Bergey et al.   Dopamine pathway is highly diverged in primate species that differ markedly in social behavior   *PNAS Vol. 113 p.6178   31 May 2016*
C. A. Botero et al.   The ecology of religious beliefs   *PNAS Vol. 111 p.16784   25 November 2014*
A. Dance   Connectomes make the map   *Nature Vol. 526 p.147   1 October 2015*
J. Freund et al.   Emergence of Individuality in Genetically Identical Mice   *Science Vol. 340 p.756   10 May 2013*
M. F. Glasser et al.   A multi-modal parcellation of human cerebral cortex   *Nature Vol. 536 p.171 11 August 2016*
M. Z. Koubeiss et al.   Electrical stimulation of a small brain area reversibly disrupts consciousness   *Epilepsy and Behavior Vol. 37 p.32 August 2014*

E. L. MacLean   Unraveling the evolution of uniquely human cognition   *PNAS Vol. 113 p.6348   7 June 2016*

D. Milardi et al.   Cortical and Subcortical Connections of the Human Claustrum Revealed In Vivo by Constrained Spherical Deconvolution Tractography   Cerebral Cortex Vol. 25 p.406 September 2013

E. Pennisi   The power of personality   Science Vol. 352 p.644   6 May 2016

R. A. Poldrack & M. J. Farah   Progress and challenges in probing the human brain   *Nature Vol. 526 p.371 15 October 2015*

V. Riedl et al.   Metabolic connectivity mapping reveals effective connectivity in the resting human brain   *PNAS Vol. 113 p.428   12 January 2016*

E. Underwood   The Brain's Identity Crisis (new tools for classifying neurons)   *Science Vol. 349 p.575   7 August 2015*

An extended bibliography will be found at www.ancestryofreason.com - for readers who want to pursue any of the subjects covered further, or to find alternative interpretations of some of the evidence.

Glossary

acetylcholine Neurotransmitter associated with attention, deployed from pedunculopontine nucleus, basal forebrain, and within striatum. Also the transmitter used at the neuromuscular junction.
action potential A wave of depolarisation which reaches the axon terminals and has the potential to cause excitation in the neurons to which they connect.
agonist Substance which can be used to activate a neuronal receptor artificially.
altricial Used of young birds which are born incapable of looking after themselves and must be fed by their parents.
amacrine cells Retinal interneurons using various transmitters which form a network at the level of the connections between bipolar cells and ganglion cells.
amniotes Reptiles, mammals and birds - in which the embryo is protected by a waterproof, fluid-containing sac, the amnion.
AMPA receptors Receptors which respond to the neurotransmitter glutamate, opening to allow sodium or both sodium and calcium ions into a neuron. Named for the selective agonist α-amino-3-hydroxy-5-methyl-4-isoxazolepropionic acid.
Amphioxus One of the few surviving non-vertebrate chordates.
amygdala Brain structure implicated in reinforcement, emotion and in identifying conspecifics.
anamniotes Amphibians and fish, whose embryos lack a waterproof wrapping.
antagonist A substance used to block a neuronal receptor artifically.
apical Pyramidal neurons receive one type of information through their apical dendrites, inputs from other sources through their basal dendrites.
Australopithecines A name covering several species of extinct hominins.
autonomic Refers to the part of the nervous system that controls heartbeats, digestion and all such automatic muscle action.
axon Projection from the neuronal body which carries message to other neurons.
axon initial segment Sector richly supplied with ionic gates; when a wave of depolarisation reaches this far it's likely to continue to the end of the axon.
BA10 Short for Brodmann's area 10. The bit of neocortex at the very front, particularly large in primates. Also called the frontal pole.
basal forebrain Group of nuclei in the lower part of the telencephalon.
basal ganglia Group of structures closely linked with the thalamus and the pedunculopontine nucleus, and in mammals with the neocortex.
basal nucleus of Meynert Also known as nucleus basalis or just basal nucleus. Part of basal forebrain, it sends GABA and acetylcholine axons to neocortex.
basilar membrane Stiff membrane which separates the two liquid-filled tubes of the ear and transmits vibrations from one to the other.
bipolar cell Neuron with dendrites clustered on the end of a single branch,

which leaves the soma or body of the cell on the opposite side from the axon.
brainstem The relatively narrow part of the mammalian brain below telencephalon and diencephalon - includes midbrain and hindbrain.
CA1, CA2, CA3 Three components of the hippocampus, subdivisions of what an early neuroanatomist designated *cornu Ammonis*. (Ammon was a name given to Jupiter when he appeared in the form of a ram.)
caecilians A group of amphibian species, wormlike in form.
cannabinoids Natural neurotransmitters. Their receptors can also be activated by substances such as cannabis and marijuana.
cartilage Tough elastic substance, largely converted into bone as the vertbrate embryo develops, though it remains as connective tissue in the joints.
cartilaginous fish Class of fish in which cartilage is not replaced by bone as the embryo develops. Includes sharks.
caudate nucleus One of the structures that make up the ventral striatum.
cephalopods A class of animals belonging to the mollusc phylum. Includes octopus, squid, cuttlefish and nautilus.
cetaceans Order of aquatic mammals: includes dolphins, porpoises and whales.
cerebellum The conspicuous structure at the back of the vertebrate hindbrain. (Literally, little cerebrum, because of its appearance.)
cerebrum The mammalian forebrain.
chemoreceptor A sensory receptor activated by a particular chemical structure.
chordates The phylum to which vertebrates belong.
cilia Delicate projections from the outside of a cell which can propel a single-celled organism through water, or cause fluid to flow in a multicellular organism.
cingulate cortex The central part of the neocortex, curled around the corpus callosum, underneath the neocortical lobes.
cognition From the Latin *cognoscere*, to know. Refers to conscious mental processes of perception, knowledge and reasoning. Recognition is re-cognition.
cognitive map A mental model of the environment supporting navigation.
colliculus One of two pairs of 'little hills' at the back of the midbrain.
conditioned stimulus A stimulus-pattern which has come to signal the likelihood of something rewarding or noxious, and prompts anticipatory action.
corpus callosum Thick rope of nerve-fibres, or axons, which connects the two cerebral hemispheres.
cortex Outer layer of the telencephalon, or of the cerebellum. Composed of cell bodies, dendrites and local axons.
corticospinal tract Pathway composed of axons from motor areas of neocortex which terminate on neurons in the spinal cord.
dendrite Protrusion from a neuron on which input synapses are situated.
dentate gyrus Part of the hippocampal formation.
depotentiation Process of weakening a potentiated synapse, so that messages

are passed less easily.
depression As applied to a synapse, a weakening of effectiveness. Involves slightly different mechanisms from depotentiation.
dichotic Simultaneous application of different patterns of sound to each ear.
dichoptic The simultaneous exposure of different images to each eye.
diencephalon The part of the brain that develops from the second of five bumps in the neural cord of the embryo. Includes thalamus, hypothalamus, habenula.
dopamine An important and ancient neuromodulator/transmitter.
dorsal Towards the back part of an animal, i.e. in most animals the upper part.
dorsolateral prefrontal cortex The part of prefrontal cortex at the top and side. Important in attention.
efference copy When an axon ordering action has a collateral branch which goes to another area, the message carried by the latter is termed an efference copy.
electroencephalogram The record produced by an electroencephalograph.
electroencephalograph An instrument which records the pattern of electrical activity detectable by electrodes placed on the scalp.
entorhinal cortex Ancient part of cortex through which hippocampus and neo-cortex communicate about spatial relationships.
enzymes Proteins that function as catalysts in biochemical processes, reducing the amount of energy needed to achieve them.
epithelium A layer or layers of closely packed cells covering an exterior or interior area of the body.
eon Geological term. An eon is divided into eras, an era into periods, a period into epochs, an epoch into ages.
eukaryotic Used where cells have the genetic material arranged as chromosomes and confined within a membrane-bounded nucleus. All multicellular organisms are eukaryotic.
eutherian Mammals which have a placenta and give birth to well-developed young - the great majority around today.
extant In existence at the current time.
excitatory post-synaptic potential A post-synaptic excitatory current which may join with others induced elsewhere. Together they may develop sufficient power to reach the axon initial segment and trigger an action potential.
extrastriate cortex Primary visual cortex, V1, is stripey, thanks to dark-coloured cells in some layers. Extrastriate cortex refers to the other, adjacent visual areas.
filter-feeder Feeds by filtering tiny particles of nourishment out of the water.
fMRI   functional Magnetic Resonance Imaging - a scanning method that tracks the intensity of blood flow across the brain, showing where activity is greatest.
forebrain Telencephalon and diencephalon - which develop from the front two

of the five bumps that form in the neural cord of the vertebrate embryo.
fovea Small area at the centre of the retina in which (in species with highly developed visual systems) photoreceptors are very densely packed
frontal lobe Frontmost of the four lobes into which deep fissures divide each side of the neocortex.
fusiform gyrus Area of temporal lobe at the end of the visual pathway, dealing with stimulus-patterns requiring fine distinctions.
GABA An inhibitory transmitter. The initials stand for gamma-aminobutyric acid.
ganglion cell Retinal cell which collects the data transmitted from photoreceptors via bipolar cells.
genome Complete DNA specification of a species, or the complete set of genes in one organism.
glia Cells that surround long axons with myelin, providing insulation and support.
globus pallidus, internal and external Components of the basal ganglia.
glomerulus Olfactory receptors communicate to glomeruli in the olfactory bulb.
glutamate Amino acid which functions as the major excitatory transmitter in vertebrates. In protein-eating species it also excites a gustatory receptor.
glycine Inhibitory neurotransmitter.
gustatory To do with taste.
gyrus One of the outward curves of the neocortex, divided by sulci. (q.v.)
habituation The gradual reduction found in the response of many neurons as stimulation is repeated.
Hertz (or Hz.) Denotes cycles or units per second.
hindbrain The part of the brain that grows from the fourth and fifth bumps at the top of the embryonic nerve-cord.
hippocampus A structure in the older part of the telencephalon, important for recording larger patterns, registered over a period of time.
hippocampal formation The hippocampus plus structures adjacent to and closely connected with it - subiculum, presubiculum, parasubiculum.
histamine A neurotransmitter/modulator associated with arousal. Also found in the body, associated with the immune system.
homeostasis Metabolic equilibrium
hominids Includes great apes and all the species which have evolved since our ancestral split from the apes.
hominins All hominids apart from the apes.
homologous Used of animal parts which have a common evolutionary origin though their functions may have come to differ.
horizontal cells Inhibitory interneurons of the retina at the level of the receptor connections to bipolar cells.

hormones Substances carried in the bloodstream to convey messages around the body. Some also function in the brain as neuromodulators.
hypothalamus Structure in the diencephalon which controls various aspects of the internal economy and the release of hormones.
inferior colliculus Bump at the back of the midbrain. Deals with auditory and somatic input, before passing on the results to superior colliculus and neocortex.
inferotemporal Lower part of the temporal lobe.
inhibition Process which makes a neuron less likely to fire.
inhibitory postsynaptic potential Immediate result of inhibitory input to a dendrite. Makes it less likely that excitatory inputs will lead to an action potential in the axon.
intralaminar nuclei of the thalamus A number of relatively small nuclei disposed along the central line of the thalamus.
invertebrate Any species without a bony or cartilaginous support along its back.
in vitro Literally 'in glass'. Refers to experiments carried out on brain slices maintained in a lifelike condition.
ion An atom which carries a positive charge as a result of losing one or two electrons, or a negative charge as a result of gaining one or two.
ionotropic receptors Neuronal receptors which admit a flow of ions to the cell.
kainate receptor Type of glutamate receptor which can be fooled by kainate.
koniocellular Composed of very tiny cells.
lateral geniculate nucleus of the thalamus Small nucleus on the side of the thalamus. Processes input from retina and passes it on to visual cortex.
locus coeruleus Small blueish nucleus in the brainstem. Sends norepinephrine-deploying axons through most of the rest of the brain.
MRI Magnetic Resonance Imaging.
MT Middle Temporal cortex. It specialises in deciphering visual movement.
magnocellular Composed of large cells. As in the magnocellular layers of the lateral geniculate nucleus, which receive from retinal ganglion cells that respond to moving stimuli, and give their name to the pathway carrying this data.
marsupials Mammals whose young are born in a very immature state, and must be protected in a marsupium, or pouch.
mechanoreceptor A sensory receptor activated by mechanical impact.
medial Refers to the inner part of the brain, along the front-to-back axis.
medial prefrontal cortex The inner portion of prefrontal cortex.
medium spiny neurons The major neurons of the striatum, the input part of the basal ganglia. They send an inhibitory message to the output stations.
metabotropic receptors Receptors which activate a cascade of reactions inside a neuron, modifying the way it responds to ionotropic inputs.
microsaccades Very small eye movements.
midbrain The part of the brain that grows from the third (middle) bump at the

front of the nerve-cord in the developing embryo.
molecule The smallest superatomic unit: a structure of two or more atoms held together by interlocking among their outer electrons.
monotremes Mammals which lay eggs, such as the duck-billed platypus.
myelin The white stuff which forms an insulating sheath around axons.
NMDA receptors Receptors activated when the arrival of glutamate coincides with a depolarisation of the adjoining membrane caused by input at nearby AMPA receptors. This shifts the magnesium ion that usually blocks the channel through which sodium and/or calcium ions must flow. Named for the selective artificial agonist, N-methyl-D-aspartate, which can substitute for glutamate.
neocortex Six-layered sheet of cells forming outer part of the mammalian telencephalon.
neural Alternative form of neuronal.
neurogenesis Production of new neurons.
neuromodulator Engages with a metabotropic receptor to modify the situation inside a neuron, and thus the response to a neurotransmitter.
neurons Cells that transmit messages rapidly to activate muscles or process sensory data. Nerves are formed of extended neuronal axons, bunched together.
neuropeptides Small molecules which contribute to signalling between neurons, often released in conjunction with other transmitters.
neurotransmitter Substance that engages with a neuronal receptor to open a channel and let in a flow of ions.
nociceptors Receptors which respond to sensory inputs representing harmful or potentially harmful events.
non-REM sleep The four stages of quiet sleep, which contrast with Rapid Eye Movement sleep.
norepinephrine Neuromodulator deployed from the locus coeruleus and the subcoeruleus. Important in alertness and attention.
nucleus Compact structure in the brain. Also the central, membrane-bound compartment of the eukaryotic cell which contains the chromosomes.
nucleus accumbens One of the input structures of the basal ganglia, it receives information from all the areas concerned with desirable or rewarding input.
nucleus gigantocellularis A nucleus in the reticular formation which receives inputs of the sort that prompt alertness and attention.
occipital lobe The rearmost of the four lobes of the neocortical hemisphere.
olfactory bulb The structure to which the axons of the olfactory receptors run.
opioids Neuromodulators which operate the receptors that are artifically stimulated by opium and similar drugs.
optic tectum Non-mammals' equivalent of the superior colliculus.
orbitofrontal cortex The part of prefrontal cortex behind the eyes, particularly

concerned with weighing up values and with emotions.
orthonasal Describes the current of air on its way in through the nose.
oxytocin A neurohormone important to maternal and mating behaviour.
parabigeminal nucleus Nucleus close to the superior colliculus which specialises in locating singularities in the visual input.
parahippocampal place area Part of the parahippocampal cortex, specialises in registering and remembering stimulus-patterns which can function as landmarks.
parietal lobe The one behind the frontal lobe, above the temporal and in front of the occipital. Somatosensory cortex is at the front. The posterior part relates egocentric sensory input to the objective outside world, and is much involved in the direction of attention, and in the calculations necessary to direct action.
Parkinson's disease A progressive disorder characterised by inadequate production of dopamine, especially in the substantia nigra pars compacta.
parvocellular Composed of small neurons. Used of the four layers of the lateral geniculate nucleus of the thalamus which receive input from the retinal ganglion cells that fire continuously in response to a stationary stimulus, and of the pathways to visual cortex that run through these layers.
pedunculopontine nucleus Nucleus in the lower brain whence axons descend to the spinal cord, influencing posture and muscle tone, and to higher regions of the brain, influencing alertness.
periacqueductal grey An area located around the cerebral aqueduct in the lower midbrain. Essential to consciousness. Transmits pain signals.
perirhinal cortex The region through which neocortex communicates to the hippocampus about sensory stimulus-patterns.
PET scan Positron Emission Tomography. X-rays of the brain are made after a rapidly decaying radioactive substance has been consumed.
pharynx The space at the back of the throat through which air is transported to the larynx and food to the oesophagus.
pheromone Odour which evokes a genetically determined response to a potential sexual partner, food or a predator.
photon A quantum or minimal unit of light.
photoreceptor A sensory receptor activated by light.
phylum A major taxonomic grouping in the animal kingdom, such as chordates, arthropods, or molluscs.
piriform cortex The largest part of olfactory cortex.
plasticity Used of synapses to indicate a capacity for modification by learning, through potentiation or depression.
pons A part of the hindbrain, situated between midbrain and medulla.
postsynaptic On the dendrite (receptor) side of a synapse.
postsynaptic excitatory potential An excitation which travels up a dendrite and may combine with those initiated at other synapses to reach cell body and axon.

postsynaptic inhibitory potential The effect of an inhibitory input to a neuron. Reduces the chances of excitatory inputs reaching the axon.
potentiation The process by which synapses are rendered larger, stronger and more effective as a result of use.
pre- Neuroscientists use this prefix in both temporal and spatial senses.
precocial Used of birds which can move around and feed themselves almost as soon as they are hatched.
prefrontal cortex The part of the frontal lobe in front of the motor areas
premotor cortex Just in front of primary motor cortex. Particularly involved when new movements are being mastered, and with visually guided movements.
presynaptic On the axon (transmitter-releasing) side of a synapse.
pretectum The area in front of the tectum. Contains nuclei responsible for several of the non-visual responses to light, such as adjusting pupil size.
primate The mammalian order currently comprising prosimians, monkeys, apes and *Homo*.
project to Of a neuron, to send its axon to. An axon may be called a projection.
proprioceptive The sense which reports on how the body is currently disposed.
prosopagnosia An inability to distinguish among individual faces.
pulvinar A large thalamic nucleus which plays an important part in attention.
putamen One of the structures which form the ventral striatum.
pyramidal neuron The major type of excitatory neuron in the neocortex. Has a bunch of dendrites at its apex, while others spread from the base, the two sets bringing information from different sources.
raphe nuclei A set of nuclei in the lower brainstem using serotonin as transmitter.
receptor The structure which opens a gate to allow a flow of ions into a cell (an ionotropic receptor) or sets off a train of reactions affecting the cell's metabolism and responsiveness to ionotropic input (a metabotropic receptor).
red nucleus A nucleus which receives input from motor cortex and cerebellum and despatches messages to the spinal cord via the rubrospinal pathway.
reinforcement A sensory input which encourages repetition of a preceding action.
REM sleep Rapid Eye Movement sleep, also known as paradoxical sleep. Dreams are most likely to occur during REM sleep.
retina The complex sheet of neurons at the back of the vertebrate eye and the photoreceptors behind them.
reticular alerting system General alerting system which includes an assortment of nuclei in the lower part of the brain. Activated by attention-demanding events reported via the spinal cord, or by messages from prefrontal cortex.
reticular nucleus of the thalamus A blanket-like nucleus wrapped around much of the thalamus which filters the information passed on to the neocortex.

reticulospinal tract The pathway which carries impulses from the reticular formation at the rear of the brain to the motor neurons of the spinal cord.
retronasal the current of air which is breathed out.
retrosplenial cortex The part of the neocortex tucked behind and under the bend (the splenium) at the back of the corpus callosum.
ribosome The cell structure on which proteins are assembled.
rubrospinal tract The pathway from the red nucleus to the spinal cord.
saccades The visible movements of the eye from one fixation to another.
schema (*pl.* schemata) A mental model or category. A concept.
sclera The white of the eye
semantic Relating to meaning.
semiotic Non-verbal communicative signals such as gesture and facial expression.
sensitisation A process by which a neuron becomes temporarily more responsive to input after two or more excitations following in quick succession.
serotonin A neuromodulator which is deployed by the raphe nuclei in the lower part of the brain. Of very ancient origin, its effects are varied and widespread, through a particularly large assortment of receptor types.
skeletal muscle Muscle attached to bone, as opposed to visceral muscle.
somatosensory The mechanical senses of the body, such as touch, and joint and limb position.
short wave ripple event A brief, speeded-up replay of a pattern of neuronal activation which occurred during earlier experience. Happens during sleep or quiet rest, and helps to transfer new learning to longer-term memory.
slow wave sleep Deep sleep, shown by large waves on the electroencephalogram.
spinal fluid Flows around the interior of the spinal cord and the brain's ventricles.
spines Tiny protrusions on the dendrites of several types of neuron, on which the synapses are located. (Actually more rounded than spiny.)
stellate cells Star-shaped cells. One of two visually distinct types of excitatory cells in the neocortex, the other being the more numerous pyramidal cells.
stereoscope Shows pictures taken from slightly different viewpoints to left and right eye.
striatum The input sectors of the basal ganglia. Ventral striatum connects to sensory areas of cortex, dorsal striatum to motor and prefrontal cortex.
striate cortex Primary visual cortex, V1, which looks stripey under the micoscope.
stimulus That which prompts a response in a sensory or neuronal receptor.
striosomes Small compartments within the striatum revealed by staining.
substantia innominata A structure which forms part of the basal forebrain.

substantia nigra pars compacta A component of the basal ganglia which deploys dopamine as its main output transmitter.
substantial nigra pars reticulata An output station of the basal ganglia which uses the inhibitory transmitter GABA for its output.
subthalamic nucleus Part of the hyperdirect route through the basal ganglia.
sulcus A groove such as those which divide the surface of the neocortex into gyri.
superior colliculus A bump on the back of the brainstem which does a large proportion of the visual processing in many mammals (as well as other vertebrates). Contributes to the control of eye movements.
synapse Structure by which the axon of one neuron communicates to a dendrite of the next, in most cases by a chemical transmitter, allowing many messages to be combined. Neurons may also be connected by electrical synapses, or gap junctions, where the current flows straight through to the next neuron.
tachistoscope Instrument which displays an image on a screen for a very brief and very accurately measured moment.
tectum The dorsal part of the midbrain (from the Latin for roof), containing the superior and inferior colliculi.
telencephalon The bit of the brain that grows from the frontmost bump in the embryonic nerve-cord. In mammals it includes neocortex, hippocampus, basal forebrain, and parts of the amygdala and of the basal ganglia.
temporal lobe The cortical lobe at the side of the brain, devoted mainly to the pathway which extracts information about visual objects from the data provided by the occiptial lobe, and to similar processing of auditory input.
tetrapods Literally, vertebrates with four feet, but includes those which have transformed feet into hands, wings or flippers.
thalamus Major structure of the diencephalon, quite complex in mammals. All sensory data other than olfactory input is channelled through it to cortex. Also provides pathways from basal ganglia and other subcortical structures to cortex.
tonic Continuous.
Tourette syndrome Disorder of motor control involving too much dopamine.
transcranial magnetic stimulation A method of temporarily disrupting, from the outside, the normal patterns of activation in a small area of brain.
unconditioned stimulus A stimulus to which an animal is genetically programmed to react.
V1, V2, V3 and V4 The first stations of the cortical visual pathway.
vasopressin Neuropeptide which influences social behaviour and pair bonding.
ventral Towards the lower or stomach side of an animal.
ventral tegmental area Area in midbrain from which a network of long axons distributes dopamine all over the forebrain, especially to prefrontal cortex.
ventricles Chambers in the brain through which cerebrospinal fluid circulates.

vertebrates Animals with a backbone, usually of bone, in some cases of cartilage. Subsection of the phylum known as chordates.
vesicle Small membranous sac in which substances are transported around a cell. In neurons neurotransmitter is delivered to the synapse in vesicles.
vestibular organ Organ just behind the ear which registers changes in the pull of gravity, and movements of the head.
vestibulo-spinal tract Pathway which conveys vestibular information to neurons of the spinal cord, allowing for rapid adjustments of stance.
voltage-gated channel A channel in the neuronal membrane which opens when a current of ions flows past and briefly alters the electrical charge on the membrane, after ionotropic receptors have been activated at synapses.
vomeronasal organ Found in many mammals and reptiles. Its very sensitive receptors respond to some of the odours that elicit hardwired responses.